Optimal Structural Analysis

Optimal Structural Analysis

Second edition

A. Kaveh
Iran University of Science and Technology, Iran

John Wiley & Sons, Ltd   **Research Studies Press Limited**

Copyright © 2006   Research Studies Press Limited, 16 Coach House Cloisters, 10 Hitchin Street, Baldock, Hertfordshire, SG7 6AE

Published by   John Wiley & Sons, Ltd., The Atrium, Southern Gate, Chichester, West Sussex PO19 8SQ, England
                        Telephone   (+44) 1243 779777

Email (for orders and customer service enquiries): cs-books@wiley.co.uk
Visit our Home Page on www.wiley.com

This work is a co-publication between Research Studies Press Limited and John Wiley & Sons, Ltd.

All Rights Reserved. No part of this publication may be reproduced, stored in a retrieval system or transmitted in any form or by any means, electronic, mechanical, photocopying, recording, scanning or otherwise, except under the terms of the Copyright, Designs and Patents Act 1988 or under the terms of a licence issued by the Copyright Licensing Agency Ltd, 90 Tottenham Court Road, London W1T 4LP, UK, without the permission in writing of the Publisher. Requests to the Publisher should be addressed to the Permissions Department, John Wiley & Sons Ltd, The Atrium, Southern Gate, Chichester, West Sussex PO19 8SQ, England, or emailed to permreq@wiley.co.uk, or faxed to (+44) 1243 770620.

Designations used by companies to distinguish their products are often claimed as trademarks. All brand names and product names used in this book are trade names, service marks, trademarks or registered trademarks of their respective owners. The Publisher is not associated with any product or vendor mentioned in this book.

This publication is designed to provide accurate and authoritative information in regard to the subject matter covered. It is sold on the understanding that the Publisher is not engaged in rendering professional services. If professional advice or other expert assistance is required, the services of a competent professional should be sought.

Other Wiley Editorial Offices

John Wiley & Sons Inc., 111 River Street, Hoboken, NJ 07030, USA

Jossey-Bass, 989 Market Street, San Francisco, CA 94103-1741, USA

Wiley-VCH Verlag GmbH, Boschstr. 12, D-69469 Weinheim, Germany

John Wiley & Sons Australia Ltd, 42 McDougall Street, Milton, Queensland 4064, Australia

John Wiley & Sons (Asia) Pte Ltd, 2 Clementi Loop #02-01, Jin Xing Distripark, Singapore 129809

John Wiley & Sons Canada Ltd, 22 Worcester Road, Etobicoke, Ontario, Canada M9W 1L1

Wiley also publishes its books in a variety of electronic formats. Some content that appears in print may not be available in electronic books.

Library of Congress Cataloging-in-Publication Data

Kaveh, A. (Ali), 1948-

　Optimal structural analysis / A. Kaveh. -- 2nd ed.

　　p. cm. -- (RSP bird)

　Includes bibliographical references and index.

　ISBN-13: 978-0-470-03015-8 (cloth : alk. paper)

　ISBN-10: 0-470-03015-1 (cloth : alk. paper)

　1. Structural analysis (Engineering) 2. Structural optimization. I. Title. II. Series.

TA645.K38 2006

624.1'71--dc22
　　　　　　　　　　　2006014662

British Library Cataloguing in Publication Data

A catalogue record for this book is available from the British Library

ISBN-13 978-0-470-03015-8 (HB)
ISBN-10 0-470-03015-1 (HB)

Typeset in 10/12pt Times New Roman by Laserwords Private Limited, Chennai, India
Printed and bound in Great Britain by TJ International, Padstow, Cornwall
This book is printed on acid-free paper responsibly manufactured from sustainable forestry
in which at least two trees are planted for each one used for paper production.

CONTENTS

Foreword of the first edition			xvi
Preface			xvii
List of Abbreviations			xix
1.	**Basic Concepts and Theorems of Structural Analysis**		**1**
1.1	Introduction		1
	1.1.1	Definitions	1
	1.1.2	Structural Analysis and Design	4
1.2	General Concepts of Structural Analysis		4
	1.2.1	Main Steps of Structural Analysis	4
	1.2.2	Member Force and Displacements	6
	1.2.3	Member Flexibility and Stiffness Matrices	8
1.3	Important Structural Theorems		11
	1.3.1	Work and Energy	11
	1.3.2	Castigliano's Theorem	14
	1.3.3	Principle of Virtual Work	15
	1.3.4	Contragradient Principle	18
	1.3.5	Reciprocal Work Theorem	19
	Exercises		20
2.	**Static Indeterminacy and Rigidity of Skeletal Structures**		**23**
2.1	Introduction		23
2.2	Mathematical Model of a Skeletal Structure		25
2.3	Expansion Process for Determining the Degree of Statical Indeterminacy		27
	2.3.1	Classical Formulae	27
	2.3.2	A Unifying Function	28

vi   CONTENTS

		2.3.3	An Expansion Process	28
		2.3.4	An Intersection Theorem	29
		2.3.5	A Method for Determining the DSI of Structures	30
2.4	The DSI of Structures: Special Methods			33
2.5	Space Structures and their Planar Drawings			35
		2.5.1	Admissible Drawing of a Space Structure	35
		2.5.2	The DSI of Frames	37
		2.5.3	The DSI of Space Trusses	38
		2.5.4	A Mixed Planar drawing - Expansion Method	39
2.6	Rigidity of Structures			41
2.7	Rigidity of Planar Trusses			45
		2.7.1	Complete Matching Method	45
		2.7.2	Decomposition Method	47
		2.7.3	Grid-form Trusses with Bracings	48
2.8	Connectivity and Rigidity			50
	Exercises			50
3.	**Optimal Force Method of Structural Analysis**			**53**
3.1	Introduction			53
3.2	Formulation of the Force Method			54
		3.2.1	Equilibrium Equations	54
		3.2.2	Member Flexibility Matrices	57
		3.2.3	Explicit Method for Imposing Compatibility	60
		3.2.4	Implicit Approach for Imposing Compatibility	62
		3.2.5	Structural Flexibility Matrices	64
		3.2.6	Computational Procedure	64
		3.2.7	Optimal Force Method	69
3.3	Force Method for the Analysis of Frame Structures			70
		3.3.1	Minimal and Optimal Cycle Bases	71
		3.3.2	Selection of Minimal and Subminimal Cycle Bases	72

	3.3.3	Examples	79
	3.3.4	Optimal and Suboptimal Cycle Bases	81
	3.3.5	Examples	84
	3.3.6	An Improved Turn-Back Method for the Formation of Cycle Bases	87
	3.3.7	Examples	88
	3.3.8	An Algebraic Graph-Theoretical Method for Cycle Basis Selection	91
	3.3.9	Examples	93
3.4		Conditioning of the Flexibility Matrices	97
	3.4.1	Condition Number	98
	3.4.2	Weighted Graph and an Admissible Member	101
	3.4.3	Optimally Conditioned Cycle Bases	101
	3.4.4	Formulation of the Conditioning Problem	103
	3.4.5	Suboptimally Conditioned Cycle Bases	104
	3.4.6	Examples	107
	3.4.7	Formation of $\mathbf{B}_0$ and $\mathbf{B}_1$ matrices	109
3.5		Generalised Cycle Bases of a Graph	115
	3.5.1	Definitions	115
	3.5.2	Minimal and Optimal Generalized Cycle Bases	118
3.6		Force Method for the Analysis of Pin-jointed Planar Trusses	119
	3.6.1	Associate Graphs for Selection of a Suboptimal GCB	119
	3.6.2	Minimal GCB of a Graph	122
	3.6.3	Selection of a Subminimal GCB: Practical Methods	123
3.7		Force Method of Analysis for General Structures	125
	3.7.1	Flexibility Matrices of Finite Elements	125
	3.7.2	Algebraic Methods	131
	Exercises		139

4. Optimal Displacement Method of Structural Analysis — 141

- 4.1 Introduction — 141
- 4.2 Formulation — 142
  - 4.2.1 Coordinate Systems Transformation — 142
  - 4.2.2 Element Stiffness Matrix using Unit Displacement Method — 146
  - 4.2.3 Element Stiffness Matrix using Castigliano's Theorem — 150
  - 4.2.4 Stiffness Matrix of a Structure — 153
  - 4.2.5 Stiffness Matrix of a Structure: An Algorithmic Approach — 158
- 4.3 Transformation of Stiffness Matrices — 160
  - 4.3.1 Stiffness Matrix of a Bar Element — 161
  - 4.3.2 Stiffness Matrix of a Beam Element — 163
- 4.4 Displacement Method of Analysis — 166
  - 4.4.1 Boundary Conditions — 168
  - 4.4.2 General Loading — 169
- 4.5 Stiffness Matrix of a Finite Element — 173
  - 4.5.1 Stiffness Matrix of a Triangular Element — 173
- 4.6 Computational Aspects of the Matrix Displacement Method — 176
  - 4.6.1 Algorithm — 176
  - 4.6.2 Example — 178
- 4.7 Optimally Conditioned Cutset Bases — 180
  - 4.7.1 Mathematical Formulation of the Problem — 181
  - 4.7.2 Suboptimally Conditioned Cutset Bases — 182
  - 4.7.3 Algorithms — 183
  - 4.7.4 Example — 184
- Exercises — 186

5. Ordering for Optimal Patterns of Structural Matrices: Graph Theory Methods — 191

- 5.1 Introduction — 191
- 5.2 Bandwidth Optimisation — 192

5.3	Preliminaries		194
5.4	A Shortest Route Tree and its Properties		196
5.5	Nodal Ordering for Bandwidth Reduction		197
	5.5.1	A Good Starting Node	198
	5.5.2	Primary Nodal Decomposition	201
	5.5.3	Transversal P of an SRT	201
	5.5.4	Nodal Ordering	202
	5.5.5	Example	202
5.6	Finite Element Nodal Ordering for Bandwidth Optimisation		203
	5.6.1	Element Clique Graph Method (ECGM)	204
	5.6.2	Skeleton Graph Method (SGM)	205
	5.6.3	Element Star Graph Method (ESGM)	208
	5.6.4	Element Wheel Graph Method (EWGM)	209
	5.6.5	Partially Triangulated Graph Method (PTGM)	211
	5.6.6	Triangulated Graph Method (TGM)	212
	5.6.7	Natural Associate Graph Method (NAGM)	214
	5.6.8	Incidence Graph Method (IGM)	217
	5.6.9	Representative Graph Method (RGM)	218
	5.6.10	Discussion of the Analysis of Algorithms	220
	5.6.11	Computational Results	221
	5.6.12	Discussions	223
5.7	Finite Element Nodal Ordering for Profile Optimisation		224
	5.7.1	Introduction	224
	5.7.2	Graph Nodal Numbering for Profile Reduction	226
	5.7.3	Nodal Ordering with Element Clique Graph (NOECG)	230
	5.7.4	Nodal Ordering with Skeleton Graph (NOSG)	230
	5.7.5	Nodal Ordering with Element Star Graph (NOESG)	232
	5.7.6	Nodal Ordering with Element Wheel Graph (NOEWG)	232
	5.7.7	Nodal Ordering with Partially Triangulated Graph (NOPTG)	232

	5.7.8	Nodal Ordering with Triangulated Graph (NOTG)	233
	5.7.9	Nodal Ordering with Natural Associate Graph (NONAG)	233
	5.7.10	Nodal Ordering with Incidence Graph (NOIG)	234
	5.7.11	Nodal Ordering with Representative Graph (NORG)	234
	5.7.12	Nodal Ordering with Element Clique Representative Graph (NOECRG)	236
	5.7.13	Computational Results	236
	5.7.14	Discussions	240
5.8	Element Ordering for Frontwidth Reduction		241
	5.8.1	Definitions	242
	5.8.2	Different Strategies for Frontwidth Reduction	244
	5.8.3	Efficient Root Selection	246
	5.8.4	Algorithm for Frontwidth Reduction	249
	5.8.5	Complexity of the Algorithm	252
	5.8.6	Computational Results	253
	5.8.7	Discussions	256
5.9	Element Ordering for Bandwidth Optimisation of Flexibility Matrices		256
	5.9.1	An Associate Graph	257
	5.9.2	Distance Number of an Element	257
	5.9.3	Element Ordering Algorithms	258
5.10	Bandwidth Reduction for Rectangular Matrices		260
	5.10.1	Definitions	260
	5.10.2	Algorithms	262
	5.10.3	Examples	262
	5.10.4	Bandwidth Reduction of Finite Element Models	264
5.11	Graph-Theoretical interpretation of Gaussian Elimination		266
	Exercises		269

6. Ordering for Optimal Patterns of Structural Matrices: Algebraic Graph Theory Methods — 273

- 6.1 Introduction — 273
- 6.2 Adjacency Matrix of a Graph for Nodal Ordering — 273
    - 6.2.1 Basic Concepts and Definition — 273
    - 6.2.2 A Good Starting Node — 277
    - 6.2.3 Primary Nodal Decomposition — 277
    - 6.2.4 Transversal P of an SRT — 277
    - 6.2.5 Nodal Ordering — 278
    - 6.2.6 Example — 278
- 6.3 Laplacian Matrix of a Graph for Nodal Ordering — 279
    - 6.3.1 Basic Concepts and Definitions — 279
    - 6.3.2 Nodal Numbering Algorithm — 282
    - 6.3.3 Example — 283
- 6.4 A Hybrid Method for Ordering — 284
    - 6.4.1 Development of the Method — 284
    - 6.4.2 Numerical Results — 285
    - 6.4.3 Discussions — 290
- Exercises — 291

7. Decomposition for Parallel Computing: Graph Theory Methods — 293

- 7.1 Introduction — 293
- 7.2 Earlier Works on Partitioning — 294
    - 7.2.1 Nested Dissection — 294
    - 7.2.2 A modified Level-Tree Separator Algorithm — 294
- 7.3 Substructuring for Parallel Analysis of Skeletal Structures — 295
    - 7.3.1 Introduction — 295
    - 7.3.2 Substructuring Displacement Method — 296
    - 7.3.3 Methods of Substructuring — 298
    - 7.3.4 Main Algorithm for Substructuring — 300

xii   CONTENTS

	7.3.5	Examples	301
	7.3.6	Simplified Algorithm for Substructuring	304
	7.3.7	Greedy Type Algorithm	305
7.4	Domain Decomposition for Finite Element Analysis		305
	7.4.1	Introduction	306
	7.4.2	A Graph-Based Method for Subdomaining	307
	7.4.3	Renumbering of Decomposed Finite Element Models	309
	7.4.4	Complexity Analysis of the Graph-Based Method	310
	7.4.5	Computational Results of the Graph-Based Method	312
	7.4.6	Discussions on the Graph-Based Method	315
	7.4.7	Engineering-Based Method for Subdomaining	316
	7.4.8	Genre Structure Algorithm	317
	7.4.9	Example	320
	7.4.10	Complexity Analysis of the Engineering-Based Method	323
	7.4.11	Computational Results of the Engineering-Based Method	325
	7.4.12	Discussions	328
7.5	Substructuring: Force Method		330
	7.5.1	Algorithm for the Force Method Substructuring	330
	7.5.2	Examples	333
7.6	Substructuring for Dynamic Analysis		336
	7.6.1	Modal Analysis of a Substructure	336
	7.6.2	Partitioning of the Transfer Matrix $\mathbf{H}(w)$	338
	7.6.3	Dynamic Equation of the Entire Structure	338
	7.6.4	Examples	342
	Exercises		346

8.	**Decomposition for Parallel Computing: Algebraic Graph Theory Methods**		**349**
8.1	Introduction		349
8.2	Algebraic Graph Theory for Subdomaining		350
	8.2.1	Basic Definitions and Concepts	350
	8.2.2	Lanczos Method	354
	8.2.3	Recursive Spectral Bisection Partitioning Algorithm	359
	8.2.4	Recursive Spectral Sequential-Cut Partitioning Algorithm	362
	8.2.5	Recursive Spectral Two-way Partitioning Algorithm	362
8.3	Mixed Method for Subdomaining		363
	8.3.1	Introduction	363
	8.3.2	Mixed Method for Graph Bisection	364
	8.3.3	Examples	369
	8.3.4	Discussions	371
8.4	Spectral Bisection for Adaptive FEM; Weighted Graphs		371
	8.4.1	Basic Concepts	372
	8.4.2	Partitioning of Adaptive FE Meshes	374
	8.4.3	Computational Results	376
8.5	Spectral Trisection of Finite Element Models		378
	8.5.1	Criteria for Partitioning	378
	8.5.2	Weighted Incidence Graphs for Finite Element Models	380
	8.5.3	Graph Trisection Algorithm	381
	8.5.4	Numerical Results	387
	8.5.5	Discussions	389
8.6	Bisection of Finite Element Meshes using Ritz and Fiedler Vectors		389
	8.6.1	Definitions and Algorithms	390
	8.6.2	Graph Partitioning	390
	8.6.3	Determination of Pseudo-Peripheral Nodes	391
	8.6.4	Formation of an Approximate Fiedler Vector	391
	8.6.5	Graph Coarsening	392

		8.6.6	Domain Decomposition using Ritz and Fiedler Vectors	393
		8.6.7	Illustrative Example	393
		8.6.8	Numerical Results	397
		8.6.9	Discussions	401
	Exercises			401
9.	**Decomposition and Nodal Ordering of Regular Structures**			**403**
9.1	Introduction			403
9.2	Definitions of Different Graph Products			404
		9.2.1	Boolean Operations on Graphs	404
		9.2.2	Cartesian Product of Two Graphs	404
		9.2.3	Strong Cartesian Product of Two Graphs	407
		9.2.4	Direct Product of Two Graphs	409
9.3	Eigenvalues of Graphs Matrices for Different Products			410
		9.3.1	Kronecker Product	410
		9.3.2	Cartesian Product	411
		9.3.3	Strong Cartesian Product	414
		9.3.4	Direct Product	417
		9.3.5	Second Eigenvalues for Different Graph Products	419
9.4	Eigenvalues of **A** and **L** Matrices for Cycles and Paths			421
		9.4.1	Computing λ_2 for Laplacian of Regular Models	424
		9.4.2	Algorithm	425
9.5	Numerical Examples			426
		9.5.1	Examples for Cartesian Product	426
		9.5.2	Examples for Strong Cartesian Product	430
		9.5.3	Examples for Direct Product	431
9.6	Spectral Method for Profile Reduction			433
		9.6.1	Algorithm	433
		9.6.2	Examples	433

9.7	Non-Compact Extended p-Sum	435
	Exercises	436

Appendix A   Basic Concepts and Definitions of Graph Theory   437

A.1	Introduction	437
A.2	Basic Definitions	437
A.3	Vector Spaces Associated with a Graph	445
A.4	Matrices Associated with a Graph	448
A.5	Directed Graphs and their Matrices	456
A.6	Graphs Associated with Matrices	458
A.7	Planar Graphs: Euler's Polyhedron Formula	459
A.8	Maximal Matching in Bipartite Graphs	462

Appendix B   Greedy Algorithm and its Applications   465

B.1	Axiom System for a Matroid	465
B.2	Matroids Applied to Structural Mechanics	467
B.3	Cocycle Matroid of a Graph	470
B.4	Matroid for Null Basis of a Matrix	471
B.5	Combinatorial Optimisation: the Greedy Algorithm	472
B.6	Application of the Greedy Algorithm	473
B.7	Formation of Sparse Null Bases	474

References   477

Index   495

Index of Symbols   505

Foreword of the first edition

This book will be welcome as an application of discrete mathematics rather than the more usual calculus-based methods of analysis. The subject of graph theory has become important in science and engineering through its strong links with matrix analysis and computer science. At first glance it seems extraordinary that such abstract material should have quite practical applications. However, as the author makes clear, the early relationship between graph theory and skeletal structures is now obvious: the structure of the mathematics is well suited to the structure of the physical problem. In fact could there be any other way of dealing with this structural problem? The engineer studying these applications of structural analysis has either to apply the computer programs as a black box, or to become involved in graph theory, matrices and matroids. This book is addressed to those scientists and engineers, and their students, who wish to understand the theory.

The book is written in an attractive dynamic style that immediately goes to the heart of each subtopic. The many worked examples and exercises will help the reader to appreciate the theory. The book is likely to be of interest to pure and applied mathematicians who use and teach graph theory, as well as to those students and researchers in structural engineering science who will find it to be necessary professional reading.

P.C. Kendall
University of Sheffield
United Kingdom

Preface

Recent advances in structural technology require greater accuracy, efficiency and speed in the analysis of structural systems, referred to as *"Optimal Structural Analysis"* in this book. It is therefore not surprising that new methods have been developed for the analysis of the structures with complex configurations.

The requirement of accuracy in analysis has been brought about by need for demonstrating structural safety. Consequently, accurate methods of analysis had to be developed since conventional methods, although perfectly satisfactory, when used on simple structures, have been found inadequate when applied to complex and large-scale structures. Another reason why greater accuracy is required results from the need to achieve efficient and optimal use of the material, i.e. optimal design.

The methods of analysis that meet the requirements mentioned above, employ matrix algebra and graph theory, which are ideally suited for modern computational mechanics. Although this text deals primarily with analysis of structural engineering systems, it should be recognized that these methods are also applicable to other types of structures. The concepts presented in this book are not only applicable to skeletal structures, but can equally be used for the analysis of other systems such as hydraulic and electrical networks. These concepts are also extended to finite element methods.

The author has been involved in various developments and applications of graph theory and matroids in the last three decades. The present book contains part of this research suitable for various aspects of matrix structural analysis. Other well-known methods in the fields relevant to the subject matter of this book are also presented.

In Chapter 1 the most important concepts and theorems are presented. Chapter 2 contains novel approaches for determining the degree of static indeterminacy of structures and provides systematic methods for studying the connectivity properties of structures. Rigidity of planar trusses is also briefly studied. In Chapter 3 a through study of the force method is presented. Methods are developed for the formation of highly sparse and well-conditioned flexibility matrices. Chapter 4 provides simple and efficient methods for construction of stiffness matrices. The formation of well-conditioned stiffness matrices are also dealt with briefly. In Chapters 5 and 6 banded, variable banded and frontal methods are investigated. Efficient methods are presented for both node and element ordering. Many new graphs are introduced for transforming the connectivity properties of finite element models onto graph models. Chapters 7 and 8 include powerful graph theory and algebraic graph theory methods for decomposition of structures and finite element meshes ideal for parallel processing, and parallel dynamic analysis of structures is also briefly studied. Chapter 9 is devoted to the most recent results concerning efficient calculation of eigenvalues and eigenvectors of regular structures using different graph products. These methods have applications in decomposing and nodal ordering of space structures and finite element models. In all the chapters many examples are designed to make the text easier to be understood.

Appendix A contains basic graph theory definitions and concepts, and Appendix B introduces matroids with special emphasis on the Greedy algorithm and its applications in structural mechanics.

I would like to take this opportunity to acknowledge a deep sense of gratitude to a number colleagues and friends who in different ways helped in the preparation of this book. Mr. J.C. de C. Henderson formerly of Imperial College of Science and Technology first introduced me to the subject with most stimulating discussions on various aspects of topology and combinatorial mathematics. Dr. A.C. Cassell discussed many concepts on structures and helped writing the early papers on applications of graph theory in structural analysis. Professor F. Ziegler encouraged me to write this book and enabled me to complete it in the happy and stimulating atmosphere of his institute at the Technical University of Vienna. My gratitude is extended to Professor P.C. Kendall, formerly of Sheffield University and Mrs V.A. Wallace for constructive comments on the first edition of this book. My special thanks are due to Mrs C. Holmes, publishing director of Research Studies Press Limited, for editing the second edition and her unfailing kindness in the course of the preparation of this book. I would like to express my sincere thanks to Mrs Debbie Cox, the assistant editor of John Wiley & Sons Ltd for her continuous assistance in the final stage of preparing this book, and Laserwords Private Limited for their contribution to the typesetting and project management of this book.

I would like to thank my former Ph.D. and M.Sc. students, Dr. H.A. Rahimi Bondarabady, Dr. H. Rahami, Dr. A. Mokhtarzadeh, Dr. A. Davaran, G.R. Roosta, I. Gaderi, for their help in various aspects of writing this book.

My warmest gratitude is due to my wife Mrs Leopoldine Kaveh for proof reading the first edition and for her continued support in the course of preparing both editions.

Every effort has been made to render the book error free. However, the author would appreciate any remaining errors being brought to his attention.

A. Kaveh,
Tehran

List of Abbreviations

CLN	Cycle Length Number	82
DKI	Degree of Kinematic Indeterminacy	24
DOF	Degrees of Freedom	142
DSI	Degree of Static Indeterminacy	23
ECGM	Element Clique Graph Method	204
ESGM	Element Star Graph Method	208
EWGM	Element Wheel Graph Method	209
FEA	Fixed End Actions	169
FEM	Finite Element Mesh (Model)	203
GCB	Generalized Cycle Basis	117
$HSRT_c^{n_0}$	Heeled SRT	242
IGM	Incidence Graph Method	217
IN	Incidence Number	82
NAGM	Natural Associate Graph Method	214
NOECG	Nodal Ordering with Element Clique Graph	230
NOECRG	Nodal Ordering with Element Clique Representative Graph	236
NOESG	Nodal Ordering with Element Star Graph	232
NOEWG	Nodal Ordering with Element Wheel Graph	232
NOIG	Nodal Ordering with Incidence Graph	234
NONAG	Nodal Ordering with Natural Associate Graph	233

LIST OF ABBREVIATIONS

NOPTG	Nodal Ordering with Partially Triangulated Graph	232
NORG	Nodal Ordering with Representative Graph	234
NOSG	Nodal Ordering with Skeleton Graph	230
NOTG	Nodal Ordering with Triangulated Graph	233
PC	Personal Computer	107
PTGM	Partially Triangulated Graph Method	211
RGM	Representative Graph Method	218
SES	Self-Equilibrating Stress	55
SGM	Skeleton Graph Method	206
SRT	Shortest Route Tree	460
TGM	Triangulated Graph Method	212

CHAPTER 1

Basic Concepts and Theorems of Structural Analysis

1.1 INTRODUCTION

In this chapter, basic definitions, concepts and theorems of structural analysis are presented. These theorems are employed in the following chapters and are very important for their understanding.

For an analytical determination of the distribution of internal forces and displacements, under prescribed external loading, a solution to the basic equations of the theory of structures should be obtained, satisfying the boundary conditions. In the matrix methods of structural analysis, one must also use these basic equations.

In order to provide a ready reference for the development of the general theory of matrix structural analysis, the most important basic theorems are introduced in this chapter, and illustrated through simple examples.

1.1.1 DEFINITIONS

A *structure* can be defined as a body that resists external effects (loads, temperature changes and support settlements) without undue deformation. Building frames, industrial building, halls, towers, bridges, dams, reservoirs, tanks, channels and pavements are typical structures that are of interest to civil engineers.

Optimal Structural Analysis   A. Kaveh
© 2006 Research Studies Press Limited

The underlying principles for the analysis of other structures are more or less the same. Airplane, missile and satellite structures are of interest to the aviation engineer. The analysis and design of a ship is interesting for a naval architect. A machine engineer should be able to design machine parts. However, in this book only structures that are of interest to structural engineers will be studied.

A structure can be considered to be an assemblage of members and nodes. Structures with clearly defined members are known as *skeletal structures*. Planar and space trusses, planar and space frames, single- and double-layer grids are examples of skeletal structures; see Figure 1.1.

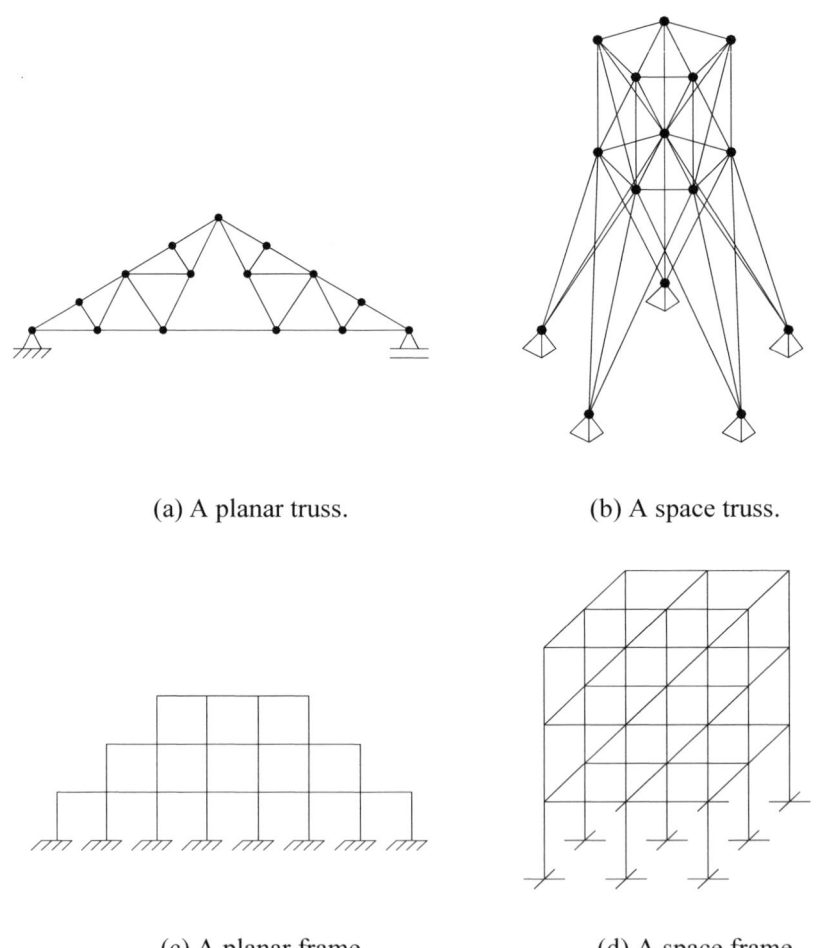

(a) A planar truss.   (b) A space truss.

(c) A planar frame.   (d) A space frame.

BASIC CONCEPTS AND THEOREMS OF STRUCTURAL ANALYSIS   3

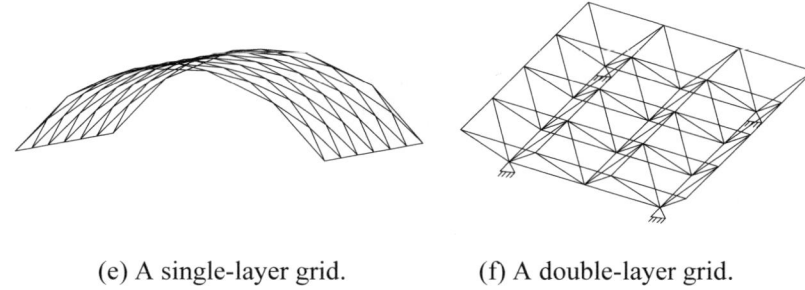

(e) A single-layer grid.   (f) A double-layer grid.

Fig. 1.1 Examples of skeletal structures.

Structures that must be artificially divided into members (elements) are called *continua*. Concrete domes, dams, plates and pavements are examples of continua; see Figure 1.2.

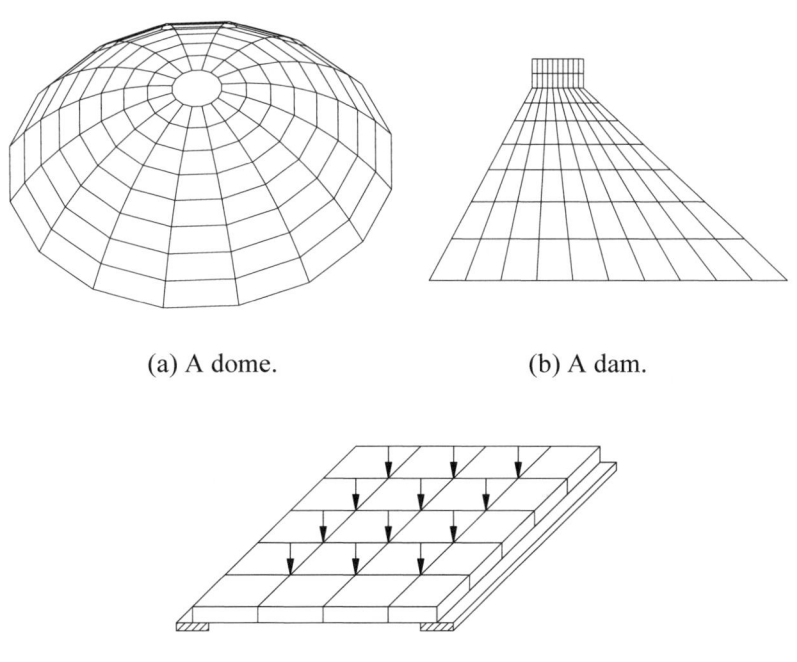

(a) A dome.   (b) A dam.

(c) A plate.

Fig. 1.2 Examples of continua.

4 OPTIMAL STRUCTURAL ANALYSIS

1.1.2 STRUCTURAL ANALYSIS AND DESIGN

Structural analysis is the determination of the response of a structure to external effects such as loading, temperature changes and support settlements. *Structural design* is the selection of a suitable arrangement of members, and a selection of materials and member sections, to withstand the stress resultants (internal forces) of a specified set of loads, and satisfy the specified displacement constraints. Figure 1.3 is a simple illustration of the cycle of structural analysis and design.

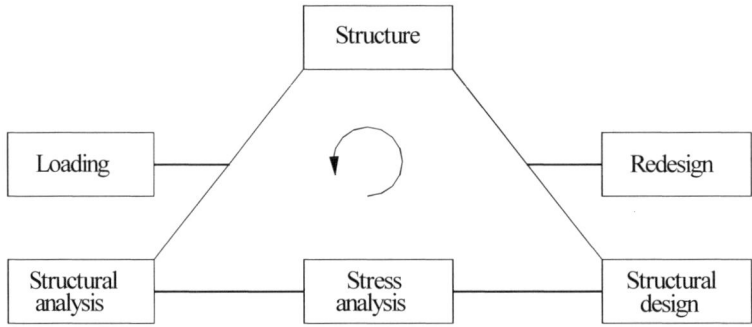

Fig. 1.3 The cycle of analysis and design.

Structural theories may be classified from different points of view as follows:

Static versus dynamic
Planar versus space
Linear versus non-linear
Statically determinate versus statically indeterminate.

In this book, static analyses of linear structures are mainly discussed for the statically determinate and indeterminate cases; however, a brief section is also included on dynamic analysis of structures.

1.2 GENERAL CONCEPTS OF STRUCTURAL ANALYSIS

1.2.1 MAIN STEPS OF STRUCTURAL ANALYSIS

A correct solution of a structure should satisfy the following requirements:

1. Equilibrium: The external forces applied to a structure and the internal forces induced in its members are in equilibrium at each node.

BASIC CONCEPTS AND THEOREMS OF STRUCTURAL ANALYSIS

2. Compatibility: The members should deform so that they fit together.

3. Force–displacement relationship: The internal forces and deformations satisfy the stress–strain relationship of the member.

For structural analysis, two basic methods are in use:

Force method: In this method, some of the internal forces and/or reactions are taken as primary unknowns, called *redundants*. Then the stress–strain relationship is used to express the deformations of the members in terms of external and redundant forces. Finally, by applying the compatibility condition that the deformed members must fit together, a set of linear equations yields the values of the redundant forces. The stresses in the members are then calculated and the displacements at the nodes in the direction of external forces are found. This method is also known as the *flexibility method* and *compatibility approach*.

Displacement method: In this method, the displacements of the nodes necessary to describe the deformed state of the structure are taken as unknowns. The deformations of the members are then calculated in terms of these displacements, and by using the stress–strain relationship, the internal forces are related to them. Finally, by applying the equilibrium equations at each node, a set of linear equations is obtained, the solution of which results in the unknown nodal displacements. This method is also known as the *stiffness method* and *equilibrium approach*.

For choosing the most suitable method for a particular structure, the number of unknowns is one of the main criteria. A comparison of the force and displacement methods can be made by calculating the degree of static and kinematic indeterminacies, respectively. As an example, for the truss structure shown in Figure 1.4(a), the number of redundants is 2 in the force method, while the number of unknown displacements is 9 for the displacement approach. For the 3×3 planar frame shown in Figure 1.4(b), the static indeterminacy and the kinematic indeterminacy are 27 and 36, respectively. For the simple six-bar truss of Figure 1.4(c), the number of unknowns for the force and displacement methods is 4 and 2, respectively. Methods for calculating the indeterminacies are discussed in Chapter 2. The number of unknowns is not the only consideration: another criterion for choosing the most suitable method is the conditioning of the flexibility and stiffness matrices, which will be dealt with in Chapters 3 and 4.

6   OPTIMAL STRUCTURAL ANALYSIS

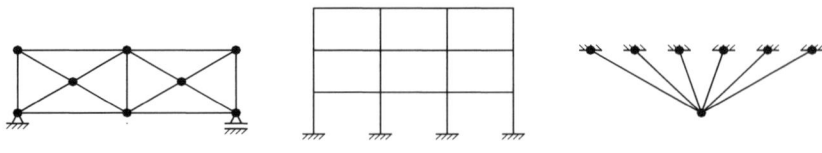

(a) A planar truss.   (b) A planar frame.   (c) A simple truss.

Fig. 1.4 Some simple structures.

1.2.2 MEMBER FORCES AND DISPLACEMENTS

A structure can be considered as an assembly of its members, subject to external effects. These effects will be considered as external loads applied at nodes, since any other effect can be reduced to such equivalent nodal loads. The state of stress in a member (internal forces) is defined by a vector,

$$\mathbf{r}_m = \{r_1^k \ r_2^k \ r_3^k \ ... \ r_n^k \}^t, \tag{1-1}$$

and the associated member deformation (distortion) is designated by a vector,

$$\mathbf{u}_m = \{u_1^k \ u_2^k \ u_3^k \ ... \ u_n^k \}^t, \tag{1-2}$$

where n is the number of force or displacement components of the *k*th member (element), and t shows the transposition. Some simple examples of typical elements, common in structural mechanics, are shown in Figure 1.5.

(a) Bar element.   (b) Beam element.

BASIC CONCEPTS AND THEOREMS OF STRUCTURAL ANALYSIS   7

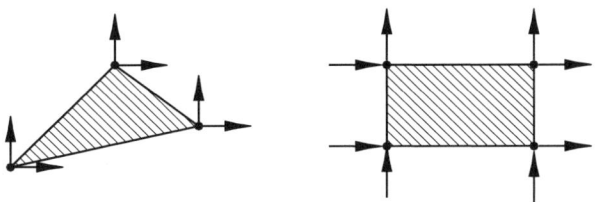

(c) Triangular plane stress element.   (d) Rectangular plane stress element.

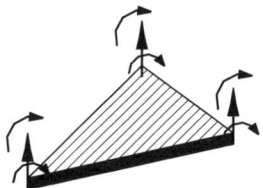 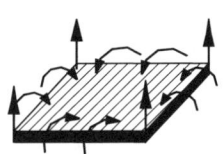

(e) Triangular plate bending element.   (f) Rectangular plate bending element.

Fig. 1.5 Some simple elements.

The relation between member forces and displacements can be written as

$$\mathbf{r}_m = \mathbf{k}_m \mathbf{u}_m \text{ or } \mathbf{u}_m = \mathbf{f}_m \mathbf{r}_m, \qquad (1\text{-}3)$$

where $\mathbf{k}_m$ and $\mathbf{f}_m$ are called *member stiffness* and *member flexibility matrices*, respectively. Obviously, $\mathbf{k}_m$ and $\mathbf{f}_m$ are related as follows:

$$\mathbf{k}_m \mathbf{f}_m = \mathbf{I}. \qquad (1\text{-}4)$$

Flexibility matrices can be written only for members supported in a stable manner, because rigid body motion of the undefined amplitude would otherwise result from application of applied loads. These matrices can be written in as many ways as there are stable and statically determinate support conditions.

The stiffness and flexibility matrices can be derived using different approaches. For simple members like bar elements and beam elements, methods based on the principles of strength of materials or classical theory of structures will be sufficient. However, for more complicated elements the principle of virtual work or alternatively variational methods can be employed. In this section, only simple members are studied, and further considerations will be presented in Chapters 3 and 4.

8 OPTIMAL STRUCTURAL ANALYSIS

1.2.3 MEMBER FLEXIBILITY AND STIFFNESS MATRICES

Consider a bar element as shown in Figure 1.6 that carries only axial forces and has two components of member forces. From the equilibrium,

$$N_m^L + N_m^R = 0, \tag{1-5}$$

and only one end force needs to be specified in order to determine the state of stress throughout the member. The corresponding deformation of the member is simply the elongation. Hence,

$$r_m^1 = N_m^R, \text{ and } u_m^1 = \delta_m^R. \tag{1-6}$$

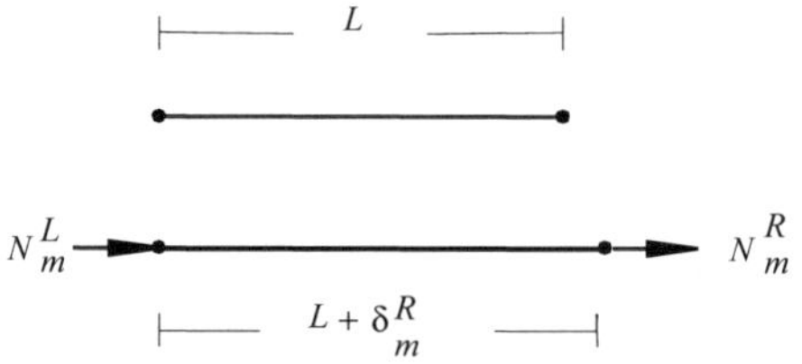

Fig. 1.6 Internal forces and deformation of a bar element.

From Hooke's law, $N_m^R = \dfrac{EA}{L}\delta_m^R$, and therefore

$$\mathbf{f}_m = \frac{L}{EA} \text{ and } \mathbf{k}_m = \frac{EA}{L}. \tag{1-7}$$

Now consider a prismatic beam of a planar frame with length L and bending stiffness *EI*. The internal forces are shown in Figure 1.7.

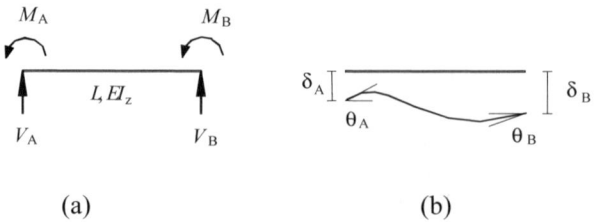

(a)   (b)

Fig. 1.7 End forces and deflected position of a beam element.

BASIC CONCEPTS AND THEOREMS OF STRUCTURAL ANALYSIS

This element is assumed to be subjected to four end forces, as shown in Figure 1.7(a), and the deflected shape and position is illustrated in Figure 1.7(b). Four end forces are related by the following two equilibrium equations:

$$V_A + V_B = 0, \tag{1-8a}$$

$$M_A + M_B + V_B L = 0 \tag{1-8b}$$

Therefore, only two end-force components should be specified as internal forces. Some possible choices for $\mathbf{r}_m$ are $\{M_A, M_B\}$, $\{V_B, M_B\}$ and $\{V_A, M_A\}$.

Using classical formulae, such as those of the strength of materials or slope-deflection equations of the theory of structures, the force–displacement relationships can be established. As an example, the flexibility matrix for a prismatic beam supported as a cantilever is obtained using the differential equation of the elastic deformation curve as follows:

$$\frac{d^2 v}{dx^2} = \frac{M_z}{EI_z} = \frac{1}{EI_z}[V_B(L-x) + M_B].$$

The integration of the above equation leads to

$$\frac{dv}{dx} = \frac{1}{EI_z}[V_B(Lx - \tfrac{1}{2}x^2) + M_B x] + C_1,$$

and integrating again we have

$$v = \frac{1}{EI_z}[V_B(\tfrac{1}{2}Lx^2 - \tfrac{1}{6}x^3) + \tfrac{1}{2}M_B x^2] + C_1 x + C_2.$$

Using the boundary conditions at A as

$$\left[\frac{dv}{dx}\right]_{x=0} = 0 \text{ and } [v]_{x=0} = 0,$$

we have $C_1 = 0$ and $C_2 = 0$.

Substituting these constants we have

$$v = \frac{1}{EI_z}[V_B(\tfrac{1}{2}Lx^2 - \tfrac{1}{6}x^3) + \tfrac{1}{2}M_B x^2],$$

$$\frac{dv}{dx} = \frac{1}{EI_z}[V_B(Lx - \tfrac{1}{2}x^2) + M_B x].$$

For $x = L$, the displacement and rotation of end B are obtained as

$$\delta_B = \frac{V_B L^3}{3EI_z} + \frac{M_B L^2}{2EI_z} \quad \text{and} \quad \theta_B = \frac{V_B L^2}{2EI_z} + \frac{M_B L}{EI_z}.$$

Using $I_z = I$, the above relationships in matrix form become

$$\begin{bmatrix} \delta_B \\ \theta_B \end{bmatrix} = \begin{bmatrix} u_m^1 \\ u_m^2 \end{bmatrix} = \begin{bmatrix} \dfrac{L^3}{3EI} & \dfrac{L^2}{2EI} \\ \dfrac{L^2}{2EI} & \dfrac{L}{EI} \end{bmatrix} \begin{bmatrix} V_B \\ M_B \end{bmatrix},$$

or

$$\mathbf{f}_m = \frac{L^2}{6EI}\begin{bmatrix} 2L & 3 \\ 3 & 6/L \end{bmatrix}. \tag{1-9}$$

Using a similar method, for a simply supported beam with two moments acting at the two ends, we have

$$\mathbf{f}_m = \begin{bmatrix} \dfrac{L}{3EI} & -\dfrac{L}{6EI} \\ -\dfrac{L}{6EI} & \dfrac{L}{3EI} \end{bmatrix} = \frac{L}{6EI}\begin{bmatrix} 2 & -1 \\ -1 & 2 \end{bmatrix}. \tag{1-10}$$

If the axial forces are also included as member forces, then $\mathbf{r}_m^t = [N_B \ V_B \ M_B]$ and $\mathbf{r}_m^t = [N_B \ M_A \ M_B]$, as shown in Figure 1.8. The above matrices become

BASIC CONCEPTS AND THEOREMS OF STRUCTURAL ANALYSIS 11

$$\mathbf{f}_m = \begin{bmatrix} \dfrac{L}{EA} & 0 & 0 \\ 0 & \dfrac{L^3}{3EI} & \dfrac{L^2}{2EI} \\ 0 & \dfrac{L^2}{2EI} & \dfrac{L}{EI} \end{bmatrix} \text{ and } \mathbf{f}_m = \begin{bmatrix} \dfrac{L}{EA} & 0 & 0 \\ 0 & \dfrac{L}{3EI} & -\dfrac{L}{6EI} \\ 0 & -\dfrac{L}{6EI} & \dfrac{L}{3EI} \end{bmatrix} \quad (1\text{-}11)$$

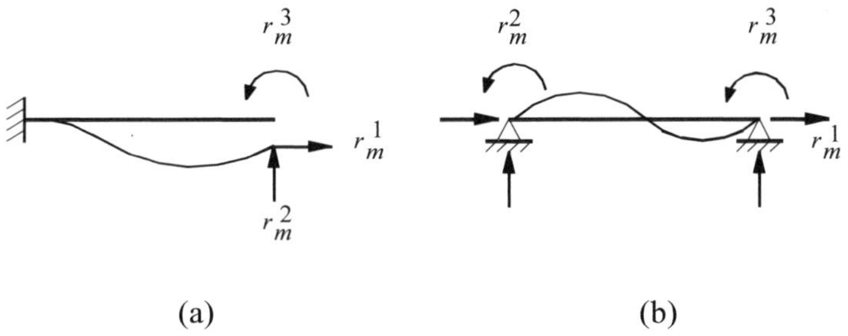

(a)   (b)

Fig. 1.8 Two sets of end forces and displacements for a beam element.

The corresponding stiffness matrices are

$$\mathbf{k}_m = \begin{bmatrix} \dfrac{EA}{L} & 0 & 0 \\ 0 & \dfrac{12EI}{L^3} & -\dfrac{6EI}{L^2} \\ 0 & -\dfrac{6EI}{L^2} & \dfrac{4EI}{L} \end{bmatrix} \text{ and } \mathbf{k}_m = \begin{bmatrix} \dfrac{EA}{L} & 0 & 0 \\ 0 & \dfrac{4EI}{L} & \dfrac{2EI}{L} \\ 0 & \dfrac{2EI}{L} & \dfrac{4EI}{L} \end{bmatrix} \quad (1\text{-}12)$$

It should be mentioned that both flexibility and stiffness matrices are symmetric, on account of the Maxwell–Betti reciprocal work theorem proven in the next section. More general methods for the derivation of member flexibility and stiffness matrices will be studied in Chapters 3 and 4.

1.3 IMPORTANT STRUCTURAL THEOREMS

1.3.1 WORK AND ENERGY

The work, δW, of a force $\mathbf{r}$ acting through a change in displacement $\mathbf{du}$ in the direction of that force is the product $\mathbf{rdu}$.

12   OPTIMAL STRUCTURAL ANALYSIS

Consider a general load–displacement relationship as shown in Figure 1.9(a). The area under this curve represents the work done, denoted by W. The area above this curve is the complementary work designated by W^*.

For a total displacement of u_1, the total work is given by

$$W = \int_0^{u_1} \mathbf{r}\,d\mathbf{u}, \tag{1-13}$$

and the complementary work is

$$W^* = \int_0^{r_1} \mathbf{u}\,d\mathbf{r}. \tag{1-14}$$

For a linear case, as shown in Figure 1.9(b), we have

$$W = W^*. \tag{1-15}$$

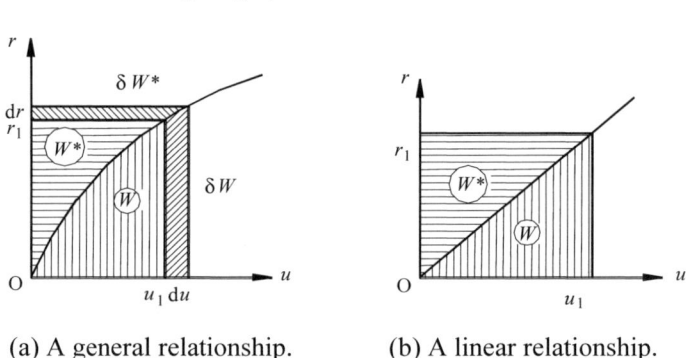

(a) A general relationship.   (b) A linear relationship.

Fig. 1.9 Force–displacement relationships.

In this book, it is assumed that the loads are applied to a structure in a gradual manner, and attention is limited to linear behaviour. Therefore, the load–displacement relationship is as shown in Figure 1.9(b), and the relation can be expressed as

$$\mathbf{r} = \mathbf{k}\mathbf{u}, \tag{1-16}$$

where $\mathbf{k}$ is a constant. The work in Figure 1.9(b) can be written as

$$W = \frac{1}{2}r_1 u_1. \tag{1-17}$$

BASIC CONCEPTS AND THEOREMS OF STRUCTURAL ANALYSIS   13

Forces and displacements at a point are both represented by vectors, and their work is represented as a dot product. In matrix notations, however, the work can be written as

$$W = \frac{1}{2}\mathbf{r}^t\mathbf{u}. \tag{1-18}$$

Using Eq. (1-3),

$$W = \frac{1}{2}\mathbf{u}^t\mathbf{k}^t\mathbf{u} = \frac{1}{2}\mathbf{u}^t\mathbf{k}\mathbf{u}. \tag{1-19}$$

Similarly, W^* can be calculated as

$$W^* = \frac{1}{2}\mathbf{r}^t\mathbf{fr}. \tag{1-20}$$

Consider the stress–strain relationship as illustrated in Figure 1.10(a). The area under this curve represents the density of the strain energy, and when integrated over the volume of the member (or structure) results in the strain energy U. The area to the left of the stress–strain curve is the density of the complementary strain energy, and by integrating over the member (or structure) the complementary energy U^* is obtained. For the linear stress–strain relationship as shown in Figure 1.10(b), $U = U^*$.

Since the work done by external actions on an elastic system is equal to the strain energy stored internally in the system (work-energy law),

$$W = U \text{ and } W^* = U^*. \tag{1-21}$$

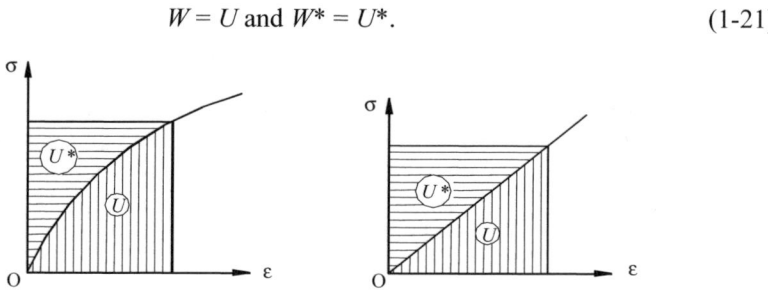

(a) A general stress–strain relationship.   (b) Linear stress–strain relationship.

Fig. 1.10 Stress–strain relationships.

1.3.2 CASTIGLIANO'S THEOREMS

Consider the force–displacement curve in Figure 1.9(a), and suppose an imaginary displacement δu_i is imposed on the system. The work done, δW, under the action of r_i in moving through δu_i is equal to

$$\delta W = r_i \delta u_i. \tag{1-22}$$

Using Eq. (1-21), and taking limit, we get the first theorem of Castigliano as

$$\frac{\partial U}{\partial q_i} = r_i, \tag{1-23}$$

which can be stated as follows [24]:

> The partial derivative of the strain energy with respect to a displacement is equal to the force applied at the point and along the considered displacement.

Similarly, if the system is subjected to an imaginary force δr_i along the displacement u_i, then the complementary work done δW^* is equal to

$$\delta W^* = u_i \delta r_i = \delta U^*, \tag{1-24}$$

and in the limit, the second theorem of Castigliano is obtained as

$$\frac{\partial U^*}{\partial r_i} = u_i. \tag{1-25}$$

> The partial derivative of the complementary strain energy with respect to a force is equal to the displacement at the point where the force is applied and directed along the action of the force.

For the linear case, $U^* = U$ and therefore Eq. (1-25) becomes

$$\frac{\partial U}{\partial r_i} = u_i. \tag{1-26}$$

BASIC CONCEPTS AND THEOREMS OF STRUCTURAL ANALYSIS   15

1.3.3 PRINCIPLE OF VIRTUAL WORK

The *principle of virtual work* is a very powerful means for deducing the conditions of compatibility and equilibrium [5], and it can be stated as follows:

The work done by a set of external forces **P** acting on a structure in moving through the associated displacements **v**, is equal to the work done by some other set of forces **R**, that is statically equivalent to **P**, moving through associated displacements **u**, that are compatible with **v**. Associated forces and displacements have the same lines of actions.

Using a statically admissible set of forces and the work equation, the compatibility relations between the deformations and displacements can be derived. Alternatively, employing a compatible set of displacements and the work equation, one obtains the equations of equilibrium between the forces. These approaches are elegant and practical.

Dummy-Load Theorem: This theorem can be used to determine the conditions of compatibility. Suppose that the deformed shape of each member of a structure is known. Then it is possible to find the deflection of the structure at any point by using the principle of virtual work. For this purpose, a dummy load (usually unit load) is applied at the point and in the direction of required displacement, which is why it is also known as the *unit load method*. The dummy-load theorem can be stated as follows:

$$\begin{Bmatrix} \text{applied} \\ \text{dummy} \\ \text{load} \end{Bmatrix} \times \begin{Bmatrix} \text{actual displacement} \\ \text{of structure where external} \\ \text{dummy load is applied} \end{Bmatrix} = \begin{Bmatrix} \text{internal forces} \\ \text{statically equivalent to} \\ \text{the applied dummy load} \end{Bmatrix} \times \begin{Bmatrix} \text{actual} \\ \text{deformation} \\ \text{of elements} \end{Bmatrix}$$

It should be noted that the dummy-load theorem is a condition on the geometry of the structure. In fact, once the deformations of elements are known, one can draw the deflected shape of the structure, and the results obtained for the deflections will agree with those of the dummy-load theorem.

Example 1: Consider a truss as shown in Figure 1.11. It is desired to measure the vertical deflection at node C when the structure is subjected to a certain loading.

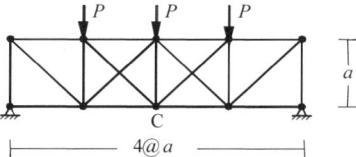

Fig. 1.11 A planar truss.

16 OPTIMAL STRUCTURAL ANALYSIS

A unit load is applied at C, and a set of internal forces statically equivalent to the unit load is chosen. However, for such equivalent internal forces, there exists a wide choice of systems, since there are several numbers of structural possibilities that can sustain the load at C. Three examples of such a system are shown in Figure 1.12(a–c).

Obviously, system (a) will need a lot of calculation because it is statically indeterminate.

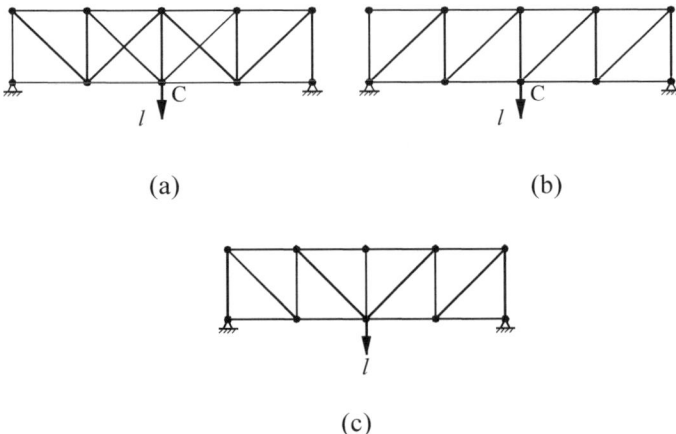

Fig. 1.12 Three different systems capable of supporting the dummy load.

System (c) is used here, since it has a smaller number of members than (b), and symmetry is also preserved. Internal forces of the members in this system shown in Figure 1.13 are

$\mathbf{r} = \{-1/2, \sqrt{2}/2, -1/2, \sqrt{2}/2, \sqrt{2}/2, -1/2, \sqrt{2}/2, -1/2, 1/2, 1/2, -1/2, -1, -1, -1/2\}^t$.

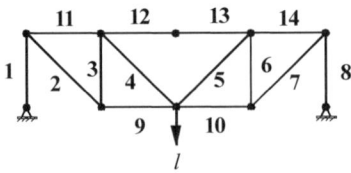

Fig. 1.13 Internal forces equivalent to unit dummy load.

BASIC CONCEPTS AND THEOREMS OF STRUCTURAL ANALYSIS   17

Measuring the elongation in members of this system containing 14 bars, and using the dummy-load theorem, we have

$$(\tfrac{1}{2})(0)+(1)(v_c)+(\tfrac{1}{2})(0) = v_c = -\tfrac{1}{2}e_1 + \tfrac{\sqrt{2}}{2}e_2 + -\tfrac{1}{2}e_3 + \tfrac{\sqrt{2}}{2}e_4 + \tfrac{\sqrt{2}}{2}e_5 - \tfrac{1}{2}e_6$$
$$+ \tfrac{\sqrt{2}}{2}e_7 - \tfrac{1}{2}e_8 + \tfrac{1}{2}e_9 + \tfrac{1}{2}e_{10} - \tfrac{1}{2}e_{11} - e_{12} - e_{13} - \tfrac{1}{2}e_{14}.$$

Dummy-Displacement Theorem: This method is usually used to find the applied external forces when the internal forces are known. In order to obtain the external force at a particular point, one subjects the structure to a unit displacement at that point in the direction of the force and chooses any set of deformations compatible with the unit displacement. Then, from the principle of work, the dummy-displacement theorem can be stated as follows:

$$\begin{Bmatrix}\text{dummy displacement applied}\\ \text{in the direction of unknowns}\\ \text{actual external forces}\end{Bmatrix} \times \begin{Bmatrix}\text{actual}\\ \text{external}\\ \text{forces}\end{Bmatrix} = \begin{Bmatrix}\text{deformation of elements}\\ \text{compatible with}\\ \text{dummy displacement}\end{Bmatrix} \times \begin{Bmatrix}\text{actual}\\ \text{internal}\\ \text{forces}\end{Bmatrix}$$

This method is also known as the *unit displacement method*.

Example 2: For the truss studied in Example 1, it is required to find the magnitude of P by measuring the internal forces in the members of the truss.

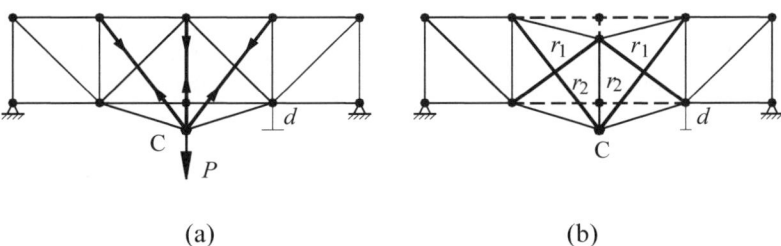

(a)   (b)

Fig. 1.14 Element deformations equivalent to unit dummy displacement.

Again, many systems can be chosen; two of these are illustrated in Figure 1.14(a) and (b). In these systems, the internal forces to be measured are shown in bold lines. Owing to the symmetry, in both cases only two measurements are needed. Applying the dummy-displacement theorem to system (a), we get

$$Pd = r_1 d\tfrac{\sqrt{2}}{2} + r_2 d + r_1 d\tfrac{\sqrt{2}}{2} = d(\sqrt{2}r_1 + r_2).$$

1.3.4 CONTRAGRADIENT PRINCIPLE

Consider two statically equivalent force systems **R** and **P** related by a linear transformation as

$$\mathbf{R} = \mathbf{BP}, \tag{1-27}$$

R is considered to have more entries than **P**, that is, there are solutions to **R** for which **P** is zero. Associated with **R** and **P** let there be two sets of displacements **v** and **u**, respectively. These are compatible displacements and therefore the work done in each system is the same, that is,

$$\mathbf{P}^t\mathbf{u} = \mathbf{R}^t\mathbf{v}. \tag{1-28}$$

From Eq. (1-27),

$$\mathbf{R}^t = \mathbf{P}^t\mathbf{B}^t. \tag{1-29}$$

Therefore,

$$\mathbf{P}^t\mathbf{u} = \mathbf{P}^t\mathbf{B}^t\mathbf{v}. \tag{1-30}$$

Since **P** is arbitrary, we have

$$\mathbf{u} = \mathbf{B}^t\mathbf{v}. \tag{1-31}$$

Equations (1-27) and (1-31) will be used in the formulation of the force method.

In a general structure, if member forces **R** are related to external nodal loads **P**, similar to Eq. (1-27), then according to the contragradient principle [5], the member distortions **v** and nodal displacement **u** will be related by an equation similar to Eq. (1-31).

If two displacement systems **u** and **v** are related by a linear transformation as

$$\mathbf{v} = \mathbf{Cu}, \tag{1-32}$$

and **R** and **P** are statically equivalent forces, then equating the work done for compatible displacements we have

$$\mathbf{P}^t\mathbf{u} = \mathbf{R}^t\mathbf{v} = \mathbf{R}^t\mathbf{Cu}. \tag{1-33}$$

Again **u** is arbitrary and

$$\mathbf{P} = \mathbf{C}^t\mathbf{R}. \tag{1-34}$$

BASIC CONCEPTS AND THEOREMS OF STRUCTURAL ANALYSIS 19

Equations (1-32) and (1-34) are employed in the formulation of the displacement method.

For a statically determinate structure,

$$\mathbf{P} = \mathbf{B}^{-1}\mathbf{R}, \tag{1-35}$$

and therefore

$$\mathbf{C}^{\mathrm{t}} = \mathbf{B}^{-1}. \tag{1-36}$$

1.3.5 RECIPROCAL WORK THEOREM

Consider a structure as shown in Figure 1.15(a) subjected to a set of loads, $\{P_1, P_2, \ldots, P_m\}$. The same structure is considered under the action of a second set of loads $\{Q_1, Q_2, \ldots, Q_n\}$, as shown in Figure 1.15(b). The reciprocal work theorem can be stated as follows:

> The work done by $\{P_1, P_2, \ldots, P_m\}$ through displacements $\{\delta_1, \delta_2, \ldots, \delta_m\}$ produced by $\{Q_1, Q_2, \ldots, Q_n\}$ is the same as the work done by $\{Q_1, Q_2, \ldots, Q_n\}$ through displacements $\{\Delta_1, \Delta_2, \ldots, \Delta_n\}$ produced by $\{P_1, P_2, \ldots, P_m\}$; that is,

$$\sum_{i=1}^{m} P_i \delta_i = \sum_{j=1}^{n} Q_j \Delta_j. \tag{1-37}$$

When single loads P and Q are considered, Eq. (1-37) reduces to

$$P\delta_i = Q\Delta_j, \tag{1-38}$$

and for the case where $P = Q$, one obtains

$$\delta_i = \Delta_j. \tag{1-39}$$

Equation (1-39) is known as *Betti's law*, and can be stated as follows:

> The deflection at point i due to a load at point j is the same as deflection at j when the same load is applied at i.

20   OPTIMAL STRUCTURAL ANALYSIS

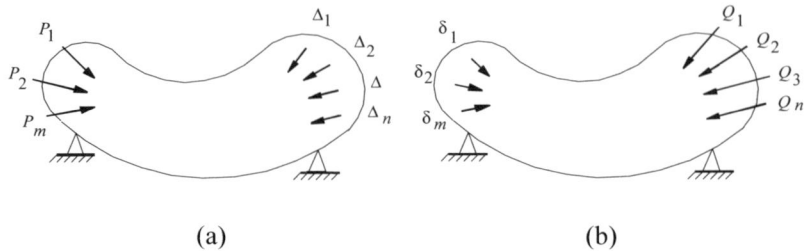

(a)   (b)

Fig. 1.15 A structure subjected to two sets of loads.

The proof of the reciprocal work theorem is constructed by equating the strain energy of the structure in two different loading sequences [217]. In the first sequence, both sets of loads are applied simultaneously, while in the second sequence, loads $\{P_1, P_2, \ldots, P_m\}$ are applied first, followed by the application of the second set of loads $\{Q_1, Q_2, \ldots, Q_n\}$.

EXERCISES

1.1   Using slope-deflection equations, write the stiffness matrix for a prismatic beam.

1.2   Develop the flexibility matrix for a beam element, using simple supports for the element.

1.3   Show that the alternative forms of the beam flexibility matrices yield the same complementary energy.

1.4   Use Castigliano's theorem to find the vertical deflection at the tip of the following craned cantilever. The members are steel with elastic modulus 200.0 GN/m², and the section properties are $A = 4.0 \times 10^3$ mm² and $I = 30.0 \times 10^6$ mm².

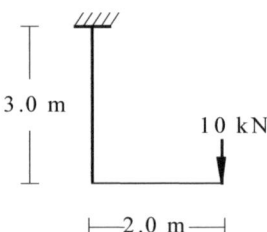

1.5 Use the dummy-load method to calculate the horizontal and vertical displacements at node C of the following truss. The members are of steel with elastic modulus 200.0 GN/m². Their cross-sections are of two types: members 2, 3, 5 have cross-sectional areas of 200.0 mm², and those for members 1, 4, 6 are 750.0 mm².

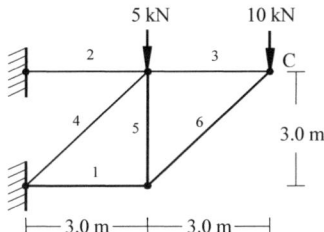

CHAPTER 2
Static Indeterminacy and Rigidity of Skeletal Structures

2.1 INTRODUCTION

Skeletal structures are the most common type of structures encountered in civil engineering practice. These structures sustain the applied loads mainly by virtue of their topology, that is, the way the members are connected to each other (connectivity). Therefore, topology plays a vital role in their design. The first step in the design of such structures is to provide sufficient rigidity and to make it reliable, but this partly depends on the degrees of static indeterminacy (DSI) of the structures. One method of calculating the DSI is to use classical formulae such as those given in Timoshenko and Young [217]; however, the application of these formulae usually provides only a small part of the necessary topological properties. The methods presented in this chapter provide us with a powerful means to understand the distribution of indeterminacy within a structure, and make the study of the load distribution feasible. The concepts presented are efficient both in terms of the optimal force method of structural analysis, as discussed in Chapter 3, and the methods of optimal structural design.

In this chapter, simple and general methods are presented for calculating the DSI of different types of skeletal structures, such as rigid-jointed planar and space frames, pin-jointed planar trusses and ball-jointed space trusses. Euler's polyhedra formula is then used to develop very efficient special methods for determining the DSI of different types of structures. These methods provide an insight into the topological properties of the structures.

Optimal Structural Analysis   A. Kaveh
© 2006 Research Studies Press Limited

In the analysis of skeletal structures, three different properties are encountered, which are classified as topological, geometrical and material. A separate study of these properties results in considerable simplification of the analysis and leads to a clear understanding of the structural behaviour. This chapter is confined to the study of those topological properties of skeletal structures that are needed to study displacement and force methods. The number of equations to be solved in the two methods may differ widely for the same structure. This number depends on the size of the flexibility and the stiffness matrices. These matrix orders are the same as the DSI and the degree of kinematic indeterminacy (DKI) of a structure, respectively. Obviously, the method that leads to the required results with the least amount of computational time and storage should be used for the analysis of a given structure. Thus, the comparison of the DSI and the DKI may be the main criterion for selecting the method of analysis.

The DKI of a structure, also known as its total number of degrees of freedom, can easily be obtained by summing up the degrees of freedom of its nodes. A node of planar and space trusses has 2 and 3 degrees of freedom, respectively. For planar and space frames, these numbers are 3 and 6, respectively. Single-layer grids also have 3 degrees of freedom for each node.

For determining the DSI of structures, numerous formulae depending on the kinds of members or types of joints have been given, for example [217]. The use of these classical formulae, in general, requires counting the number of members and joints, which becomes a tedious process for multi-member and/or complex-pattern structures; moreover, the count provides no additional information about connectivity.

Henderson and Bickley [77] related the DSI of a rigid-jointed frame to the first Betti number (cyclomatic number) of its graph model S. Generalising the Betti's number to a linear function and using an expansion process, Kaveh [94] developed a general method for determining the DSI and the DKI of different types of skeletal structures. Special methods have also been developed for transforming the topological properties of space structures to those of their planar drawings, in order to simplify the calculation of their degrees of static indeterminacy [106,107].

A DSI equal to or greater than zero is a necessary condition for rigidity; however, it is by no means sufficient. Therefore, rigidity requires a separate careful study. This property was studied by pioneering structural engineers such as Henneberg [80] and Müller–Breslau [167]. The methods that they developed for examining the rigidity of skeletal structures are useful for the study of structures either with a small number of joints and members, or possessing special connectivity properties. Rigid-jointed structures (frames), when supported in an appropriate form and containing no release, are always rigid. Therefore, only truss structures need to be studied for rigidity.

Various types of methods have been employed for the study of rigidity; however, the main approaches are either algebraic or combinatorial. A comprehensive discussion of algebraic methods may be found in the work of Pellegrino and Calladine [175]. The first combinatorial approach to the study of rigidity is due to Laman [144], who found the necessary and sufficient conditions for a graph to be rigid, when its members and nodes correspond to rigid rods (bars) and rotatable pin-joints of a planar truss. Certain types of planar trusses have been studied for rigidity by Bolker and Crapo [14], Roth and Whiteley [190] and Crapo [32].

Although Laman theoretically solved the problem of rigidity for planar trusses, no algorithm was given to check whether a given graph was rigid. Two combinatorial algorithms developed by Lovász and Yemini [156], and Sugihara [214] and Tay [216] showed that they are interrelated. Some studies have recently been made with a view to extending the concepts developed for planar trusses to those of space trusses. However, the results obtained are incomplete and apply only to special classes of space trusses. Therefore, only planar trusses are studied in this chapter.

It should be noted that various methods for determining the DSI of structures are a by-product of the general methods developed here. The method of expansion and its control at each step, using the intersection theorem of this chapter, provides a powerful tool for further studies in the field of structural analysis.

2.2 MATHEMATICAL MODEL OF A SKELETAL STRUCTURE

The mathematical model of a structure is considered to be a finite, connected graph S. There is a one-to-one correspondence between the elements of the structure and the members (edges) of S. There is also a one-to-one correspondence between the joints of the structure and the nodes of S, except for the support joints of some models.

For frame structures, shown in Figures 2.1 (a1) and (a2), two graph models are considered. For the first model, all the support joints are identified as a single node called a *ground node*, as shown in Figures 2.1 (b1) and (b2). For the second model, all the joints are connected by an artificial arbitrary spanning tree, termed *ground tree*, as shown in Figures 2.1 (c1) and (c2).

(a1) A plane frame.   (b1) First model with a ground node.   (c1) Second model with a ground tree.

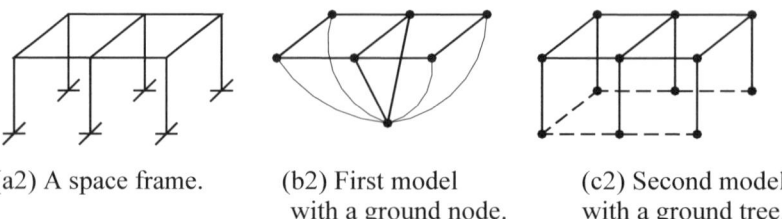

(a2) A space frame. (b2) First model with a ground node. (c2) Second model with a ground tree.

Fig. 2.1 Frame structures and their mathematical models.

Truss structures shown in Figures 2.2 (a1) and (a2) are assumed to be supported in a statically determinate fashion (Figures 2.2 (b1) and (b2)), and the effect of additional supports can easily be included in calculating the DSI of the corresponding structures. Alternatively, artificial members can be added as shown in Figures 2.2 (c1) and (c2) to model the components of the corresponding supports. For a fixed support, two members and three members are considered for planar and space trusses, respectively, and one member is used for representing a roller.

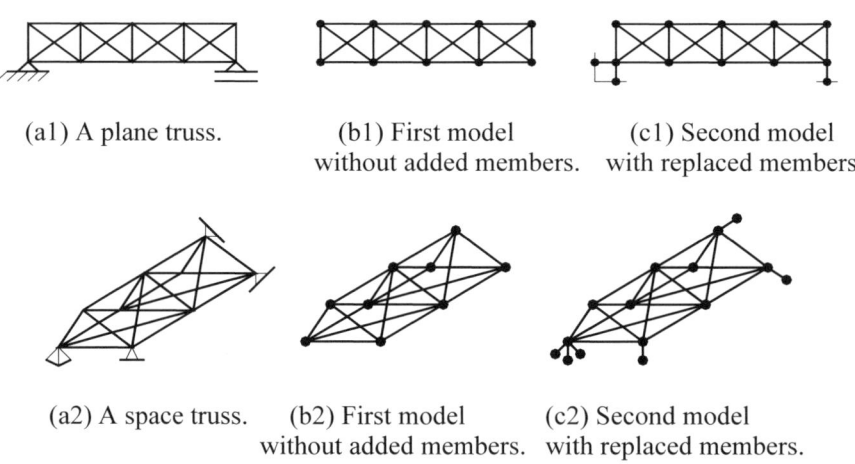

(a1) A plane truss. (b1) First model without added members. (c1) Second model with replaced members.

(a2) A space truss. (b2) First model without added members. (c2) Second model with replaced members.

Fig. 2.2 Trusses and their graph models.

The skeletal structures are considered to be in perfect condition; that is, planar and space trusses have pin and ball joints only. Obviously, the effect of extra constraints or releases can be taken into account in determining their DSI and also in their analysis, as shown in the work by Mauch and Fenves [161].

2.3 EXPANSION PROCESS FOR DETERMINING THE DEGREE OF STATIC INDETERMINACY

The *DKI* of a structure is the number of independent displacement components (translations and rotations) required for describing a general state of deformation of the structure. The *DKI* is also referred to as the *total degrees of freedom* of the structure. On the other hand, the DSI (redundancy) of a structure is the number of independent force components (forces and moments) required for describing a general equilibrium state of the structure. The DSI of a structure can be obtained by subtracting the number of independent equilibrium equations from the number of its unknown forces.

2.3.1 CLASSICAL FORMULAE

Formulae for calculating the DSI of various skeletal structures can be found in textbooks on structural mechanics, for example, the DSI of a planar truss, denoted by $\gamma(S)$, can be calculated from,

$$\gamma(S) = M(S) - 2N(S) + 3, \qquad (2\text{-}1)$$

where S is supported in a statically determinate fashion (internal indeterminacy). For extra supports (external indeterminacy), $\gamma(S)$ should be further increased by the number of additional unknown reactions.

A similar formula holds for space trusses:

$$\gamma(S) = M(S) - 3N(S) + 6. \qquad (2\text{-}2)$$

For planar and space frames, the classical formulae is given as,

$$\gamma(S) = \alpha\,[M(S) - N(S) + 1], \qquad (2\text{-}3)$$

where all supports are modelled as a datum (ground) node, and $\alpha = 3$ or 6 for planar and space frames, respectively.

All these formulae require counting a great number of members and nodes, which makes their application impractical for multi-member and complex-pattern structures. These numbers provide only a limited amount of information about the connectivity properties of structures. In order to obtain additional information, the methods developed in the following sections will be utilised.

2.3.2 A UNIFYING FUNCTION

All the existing formulae for determining DSI have a common property, namely their linearity with respect to $M(S)$ and $N(S)$. Therefore, a general unifying function can be defined as,

$$\gamma(S) = aM(S) + bN(S) + c\gamma_0(S), \tag{2-4}$$

where $M(S)$, $N(S)$ and $\gamma_0(S)$ are the numbers of members, nodes and components of S, respectively. The coefficients a, b and c are integer numbers depending on both the type of the corresponding structure and the property, which the function is expected to represent. For example, $\gamma(S)$ with appropriate values for a, b and c may describe the DSI of certain types of skeletal structures, as shown in Table 2.1. For $a = 1$, $b = -1$ and $c = 1$, $\gamma(S)$ becomes the first Betti number $b_1(S)$ of S, as described in Appendix A.

Table 2.1 Coefficients of $\gamma(S)$ for different types of structures.

Type of structure	a	b	c
Plane truss	+1	−2	+3
Space truss	+1	−3	+6
Plane frame	+3	−3	+3
Space frame	+6	−6	+6

2.3.3 AN EXPANSION PROCESS

An expansion process, in its simplest form, has been used by Müller–Breslau [167] for re-forming structural models, such as simple planar and space trusses. In this expansion process, the properties of typical subgraphs, selected in each step to be joined to the previously expanded subgraph, guarantee the determinacy of the simple truss. These subgraphs consist of two and three concurrent bars for planar and space trusses, respectively.

The idea can be extended to other types of structures, and more general subgraphs can be considered for addition at each step of the expansion process. A cycle, a planar subgraph and a subgraph with prescribed connectivity properties are examples of these, which are employed in this book. For example, the planar truss of Figure 2.3(a) can be formed in four steps, joining a substructure S_i with $\gamma(S_i) = 1$ as shown in Figure 2.3(b), sequentially, as illustrated in Figure 2.3(c).

STATIC INDETERMINACY 29

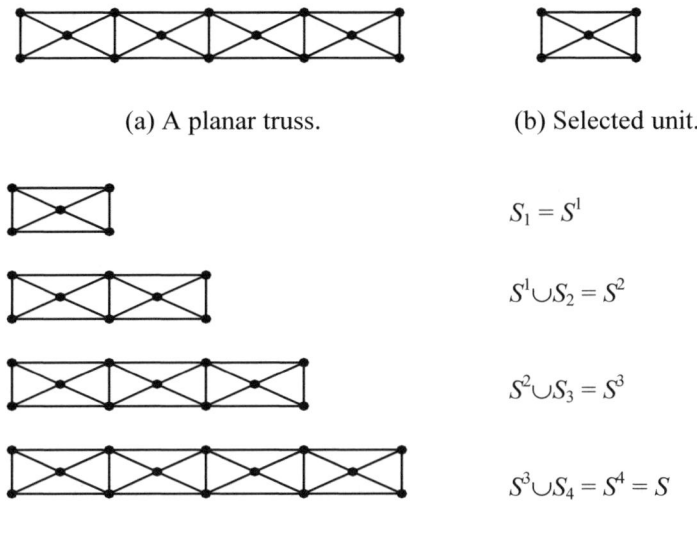

(a) A planar truss. (b) Selected unit.

$S_1 = S^1$

$S^1 \cup S_2 = S^2$

$S^2 \cup S_3 = S^3$

$S^3 \cup S_4 = S^4 = S$

(c) The process of expansion as $S_1 = S^1 \to S^2 \to S^3 \to S^4 = S$.

Fig. 2.3 Process for the formation of a planar truss.

2.3.4 AN INTERSECTION THEOREM

In a general expansion process, a subgraph S_i may be joined to another subgraph S_j in an arbitrary manner. For example, $\gamma(S_i)$ or $\gamma(S_j)$ may have any arbitrary value and the union $S_i \cup S_j$ may be a connected or a disjoint subgraph. The intersection $S_i \cap S_j$ may also be connected or disjoint. It is important to find the properties of $S_1 \cup S_2$ having the properties of S_1, S_2 and $S_1 \cap S_2$. The following theorem provides a correct calculation of the properties of $S_i \cup S_j$. In order to have the formula in its general form, q subgraphs are considered in place of two subgraphs.

Theorem (Kaveh [94]): Let S be the union of q subgraphs $S_1, S_2, S_3, ..., S_q$ with the following functions being defined:

$$\gamma(S) = aM(S) + bN(S) + c\gamma_0(S),$$

$$\gamma(S_i) = aM(S_i) + bN(S_i) + c\gamma_0(S_i) \qquad i = 1,2,...,q,$$

$$\gamma(A_i) = aM(A_i) + bN(A_i) + c\gamma_0(A_i) \qquad i = 2,3,...,q,$$

where $A_i = S^{i-1} \cap S_i$ and $S^i = S_1 \cup S_2 \cup ... \cup S_i$. Then

$$[\gamma(S) - c\gamma_0(S)] = \sum_{i=1}^{q}[\gamma(S_i) - c\gamma_0(S_i)] - \sum_{i=2}^{q}[\gamma(A_i) - c\gamma_0(A_i)] \qquad (2\text{-}5)$$

For proof, the interested reader may refer to Kaveh [113].

Special Case: If S and each of its subgraphs considered for expansion (S_i for $i = 1$, ..., q) are non-disjoint (connected), then Eq. (2-5) can be simplified as

$$\gamma(S) = \sum_{i=1}^{q}\gamma(S_i) - \sum_{i=2}^{q}\bar{\gamma}(A_i), \qquad (2\text{-}6)$$

where $\bar{\gamma}(A_i) = aM(A_i) + bN(A_i) + c$.

For calculating the DSI of a multi-member structure, one normally selects a repeated unit of the structure and joins these units sequentially in a connected form. Therefore, Eq. (2-6) can be applied in place of Eq. (2-5) to obtain the overall property of the structure.

2.3.5 A METHOD FOR DETERMINING THE DSI OF STRUCTURES

Let S be the union of its repeated and/or simple pattern subgraphs S_i ($i = 1, ..., q$). Calculate the DSI of each subgraph, using the appropriate coefficients from Table 2.1. Now perform the union-intersection method with the following steps:

Step 1: Join S_1 to S_2 to form $S^2 = S_1 \cup S_2$, and calculate the DSI of their intersection $A_2 = S_1 \cap S_2$. The value of $\gamma(S^2)$ can be found using Eq. (2-5) or Eq. (2-6), as appropriate.

Step 2: Join S_3 to S^2 to obtain $S^3 = S^2 \cup S_3$, and determine the DKI or DSI of $A_3 = S^2 \cap S_3$. As in Step 1, calculate $\gamma(S^3)$.

Step k: Subsequently join S_{k+1} to S^k, calculating the DSI of $A_{k+1} = S^k \cap S_{k+1}$ and evaluating the magnitude of $\gamma(S^{k+1})$.

Repeat Step k until the entire structural model $S = \bigcup_{i=1}^{q} S_i$ has been re-formed and its DSI determined.

In the above expansion process, the value of q depends on the properties of the substructures (subgraphs) that are considered for re-forming S. These subgraphs have either simple patterns for which $\gamma(S_i)$ can easily be calculated, or the DSIs of which are already known.

In the process of expansion, if an intersection A_i itself has a complex pattern, further refinement is also possible; that is, the intersection can be considered as the union of simpler subgraphs.

Example: Let S be the graph model of a space frame. This graph can be considered as 27 subgraphs S_i as shown in Figure 2.4(a), connected to each other to form a graph $S = \bigcup_{i=1}^{27} S_i$. The interfaces of S_i (i = 1, ..., 27) are shown in Figure 2.4(b), in which some of the members are omitted for the sake of clarity.

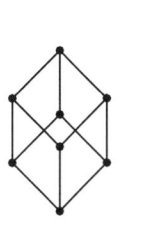 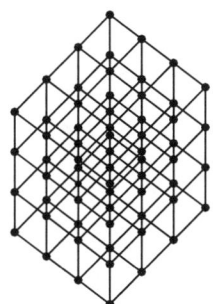

(a) A subgraph S_i of S.   (b) $S = \bigcup_{i=1}^{27} S_i$ without some of its members.

Fig. 2.4 A space structure S.

The expansion process consists of joining 27 subgraphs S_i one at a time. In this process, the selected subgraphs can have three different types of intersection, which are shown in Figure 2.5(a). In order to simplify the counting and the recognition of the types of interfaces, S is re-formed storey by storey. For the first storey, a 3 × 3 table is used to show the types of intersections occurring in the process of expansion. The numbers on each box designate the type of intersection, as shown in Figure 2.5(b). Similar tables are used for the second storey and the third storey of S, as shown in Figure 2.5(b).

Thus, there exist 6 intersections of type A_i^1, 12 intersections of type A_i^2 and 8 intersections of type A_i^3.

32   OPTIMAL STRUCTURAL ANALYSIS

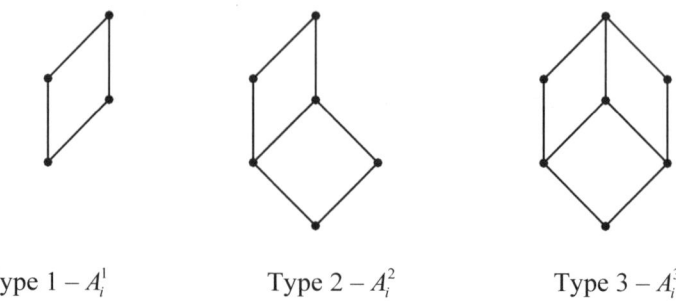

Type 1 – A_i^1 Type 2 – A_i^2 Type 3 – A_i^3

(a) Three different types of intersections.

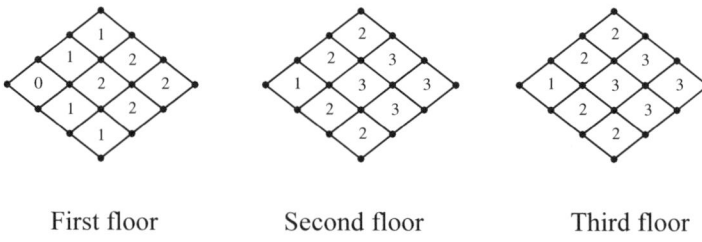

First floor   Second floor   Third floor

(b) Types of intersections after the completion of each storey.

Fig. 2.5 Intersections and their types.

Since each S_i is a connected subgraph, and is in the process of expansion, S_i is kept connected, and a simplified Eq. (2-5) can be employed:

$$\gamma(S) = \sum_{i=1}^{27} \gamma(S_i) - \sum_{i=2}^{27} \bar{\gamma}(A_i).$$

As previously shown

$$\sum_{i=2}^{27} \bar{\gamma}(A_i) = \sum_{i=2}^{7} \bar{\gamma}(A_i^1) + \sum_{i=8}^{19} \bar{\gamma}(A_i^2) + \sum_{i=20}^{27} \bar{\gamma}(A_i^3).$$

The intersections A_i^2 and A_i^3 can be decomposed as

$$A_i^2 = A_i^1 \cup A_i^1 \text{ and } A_i^3 = A_i^2 \cup A_i^1.$$

The DSI of S can now be calculated as follows:

$$\gamma(S_i) = 6(12 - 8 + 1) = 6 \times 5 = 30.$$

Using Eq. (2-3),

$$\overline{\gamma}(A_i^1) = 6(4 - 4 + 1) = 6,$$

$$\overline{\gamma}(A_i^2) = 6 \times 1 + 6 \times 1 - 6 \times 0 = 12,$$

$$\overline{\gamma}(A_i^3) = 6 \times 2 + 6 \times 1 - 6 \times 0 = 18,$$

hence: $\gamma(S) = 27(30) - [6(6) + 12(12) + 8(18)] = 486.$

The expansion process becomes very efficient for structures with repeated patterns. Counting is reduced considerably by this method. As an example, the use of the classical formula for finding the DSI of S in the above example requires counting 124 members and 64 nodes, which is a task involving possible errors.

2.4 THE DSI OF STRUCTURES: SPECIAL METHODS

In this section, using Euler's polyhedron formula (Appendix A), some useful theorems are stated, which provide a simple means for calculating the DSI of various types of skeletal structures. For proofs for these theorems, interested readers may refer to Kaveh [113].

Theorem 1: For a fully triangulated planar truss (except the exterior boundary), the internal DSI is the same as the number of its internal nodes:

$$\gamma(S) = N_i(S). \qquad (2\text{-}7)$$

For trusses with non-triangulated internal regions (Figure 2.6(a)), let $M_c(S)$ be the number of members required for the completion of the triangulation of the internal regions of S, then:

$$\gamma(S) = N_i(S) - M_c(S). \qquad (2\text{-}8)$$

The number of members required for triangulation of a polygon is constant and independent of the way it is triangulated. This is why Eq. (2-8) can easily be established.

34 OPTIMAL STRUCTURAL ANALYSIS

For planar trusses, the crossing point of two bars can be identified as an additional dummy node. This is illustrated in Figure 2.6(a), where the truss is drawn as in Figure 2.6(b). The application of Eq. (2-7) leads to $\gamma(S) = N_i(S) = 1$. The addition of a dummy node at crossing point as in Figure 2.6(c) has the same effect and results in $\gamma(S) = N_i(S) = 1$.

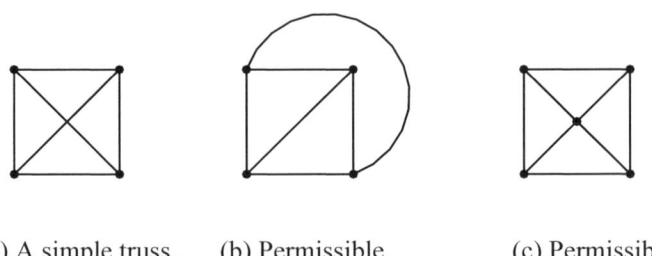

(a) A simple truss.   (b) Permissible.   (c) Permissible.

Fig. 2.6 A planar truss and two admissible models.

However, when three bars cross at a point, as illustrated in Figure 2.7(a), the addition of a dummy node is not permissible, since it leads to $\gamma(S) = N_i(S) = 1$, while a correct drawing of S results in three internal nodes with three non-triangulated regions. The addition of three bars completes the triangulation, and therefore $\gamma(S) = N_i(S) - M_c(S) = 3 - 3 = 0$, which is the correct DSI for the given truss.

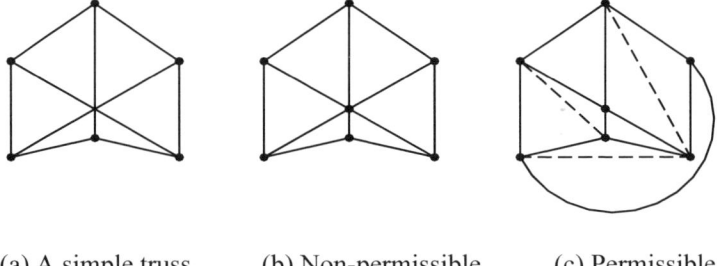

(a) A simple truss.   (b) Non-permissible.   (c) Permissible.

Fig. 2.7 A planar truss and its non-admissible and admissible models.

Once the internal DSI of a structure has been found, the external DSI resulting from additional supports can be easily added, to obtain the total DSI.

Example 2: Let S be a planar truss as shown in Figure 2.8. Triangulation of the internal region in an arbitrary manner requires nine members, shown as dashed lines.

STATIC INDETERMINACY 35

Therefore:

$$\gamma(S) = N_i(S) - M_c(S) = 72 - 9 = 63.$$

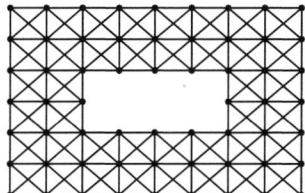
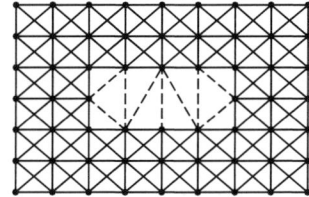

(a) A planar truss S. (b) Triangulated S.

Fig. 2.8 A general planar truss and its triangulation.

Theorem 2: The DSI of a planar rigid-jointed frame S is equal to three times the number of its internal regions, that is,

$$\gamma(S) = 3R_i(S). \tag{2-9}$$

Theorem 3: A ball-jointed space truss drawn (embedded) on a sphere is internally statically determinate, if all the created regions are triangles.

As an example, a ball-jointed truss with S of Figure 2.2 (b2) as its graph model, is statically determinate.

2.5 SPACE STRUCTURES AND THEIR PLANAR DRAWINGS

The topological properties of space structures can be transformed into those of their planar drawings, thus simplifying the counting process for the calculation of the DSI for space structures.

2.5.1 ADMISSIBLE DRAWING OF A SPACE STRUCTURE

A *drawing* S^p of a graph S in the plane is a mapping of the nodes of S to distinct points of S^p, and the members of S to open arcs of S^p, such that

(i) the image of no member contains that of any node;

(ii) the image of a member (n_i, n_j) joins the points corresponding to n_i and n_j.

A drawing is called *admissible (good)* if the members are such that

(iii) no two arcs with a common end point meet;

36 OPTIMAL STRUCTURAL ANALYSIS

(iv) no two arcs meet in more than one point;

(v) no three arcs meet at a common point.

The configurations prohibited by these three conditions are shown in Figure 2.9.

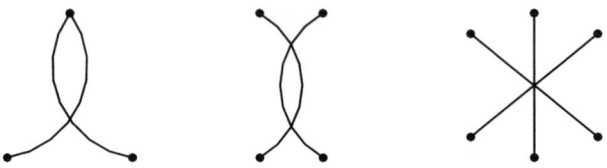

Fig. 2.9 The prohibited configurations.

A point of intersection of two members in a drawing is called a *crossing*, and the *crossing number* $c(S^P)$ of a graph S is the number of crossings in any admissible drawing of S in the plane. An *optimal* drawing in a given surface is one that exhibits the least possible crossings. In this book, we use only admissible drawings in the plane, but not necessarily optimal.

As an example, different admissible drawings of a space model, shown in Figure 2.10(a), are illustrated in Figures 2.10 (b–d). The crossing points are marked by ×. The number of crossing points for each case is also provided.

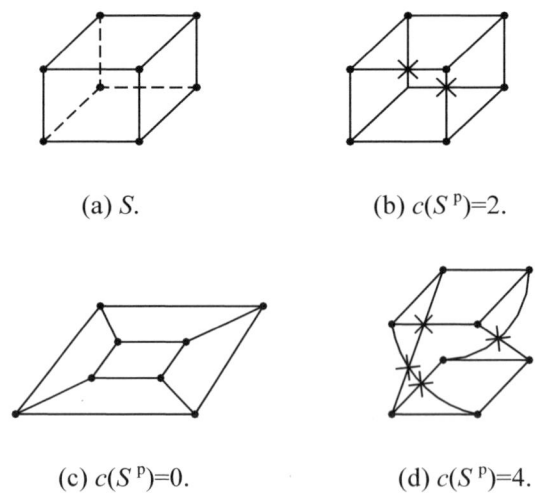

(a) S. (b) $c(S^P)=2$.

(c) $c(S^P)=0$. (d) $c(S^P)=4$.

Fig. 2.10 Admissible drawings of S.

STATIC INDETERMINACY   37

In this example, an optimal drawing corresponds to $c(S^P) = 0$, which is shown in Figure 2.10(c).

2.5.2 THE DSI OF FRAMES

For a rigid-jointed space frame, the DSI can be determined by using

$$\gamma(S) = 6[M(S) - N(S) + 1]. \tag{2-10}$$

Counting the nodes in a drawing of S on the plane produces no problem; however, recognising and counting the members can be very cumbersome. The following theorem transforms this procedure to counting the crossing nodes and regions of S^p, in place of members and nodes of S.

Theorem: For a space rigid-jointed frame, the DSI is given by:

$$\gamma(S) = 6[R_i(S^P) - c(S^P)]. \tag{2-11}$$

Example 1: A simple space frame is considered, as shown in Figure 2.11(a). Both models introduced in Section 2.2 are employed in Figures 2.11(a) and (b). For the first model with ground node $\gamma(S) = 6(5 - 1) = 24$, and for the second model with ground tree the same result as $\gamma(S) = 6(6 - 2) = 24$ is obtained.

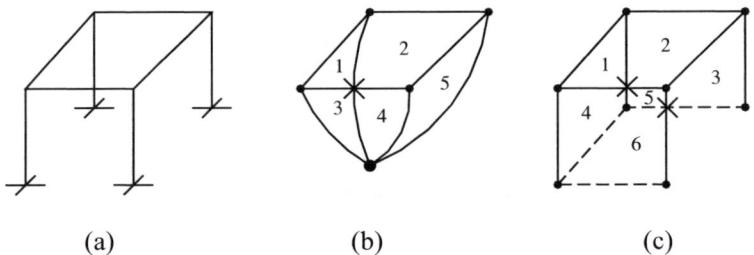

(a)   (b)   (c)

Fig. 2.11 A simple frame and its drawing with different models.

Example 2: Let S be the graph model of a space frame with a ground tree, as shown in Figure 2.12(a). A drawing of S is shown in Figure 2.12(b), for which $c(S^P) = 12$ and $R_i(S^P) = 33$, resulting in $\gamma(S) = 6(33 - 12) = 126$.

38 OPTIMAL STRUCTURAL ANALYSIS

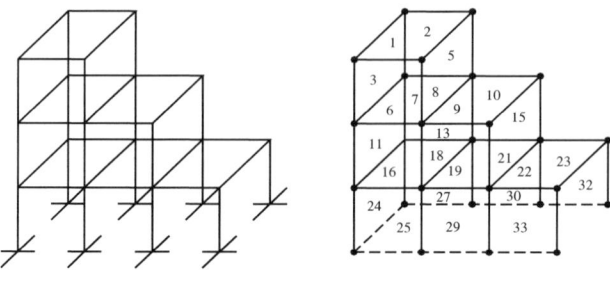

(a) A space frame S.   (b) A drawing of S.

Fig. 2.12 A space frame and its drawing.

2.5.3 THE DSI OF SPACE TRUSSES

Ball-jointed space trusses are often multi-member structures in the form of double- and triple-layer grids. The following theorem simplifies the calculation of the DSI for these structures.

Theorem: For a ball-jointed space truss supported in a statically determinate fashion, the DSI is given by,

$$\gamma(S) = c(S^p) - M_c(S^p), \qquad (2\text{-}12)$$

where $M_c(S^p)$ is the number of members required for the full triangulation of S^p.

Example 1: A space structure in the form of the skeleton of a cube is considered as shown in Figure 2.13(a). An optimal drawing of S is illustrated in Figure 2.13(b). Since a fully triangulated truss can be drawn on the plane with no additional member, $\gamma(S) = 0$ and Theorem 3 of Section 2.4 therefore follows.

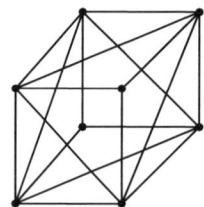 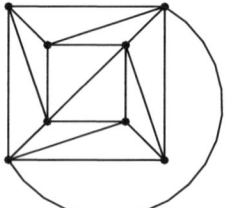

(a) A fully triangulated space truss.   (b) Its optimal drawing.

Fig. 2.13 A fully triangulated space truss and its optimal drawing.

STATIC INDETERMINACY   39

Example 2: Consider a space ball-jointed double-layer grid S as shown in Figure 2.14(a), an admissible drawing of which is depicted in Figure 2.14(b).

This drawing contains 12 crossing points, and for a full triangulation, $M_c(S^P) = 9$ members are added, as shown by dashed lines.

From Eq. (2-12) we have

$$\gamma(S) = c(S^P) - M_c(S^P) = 12 - 9 = 3.$$

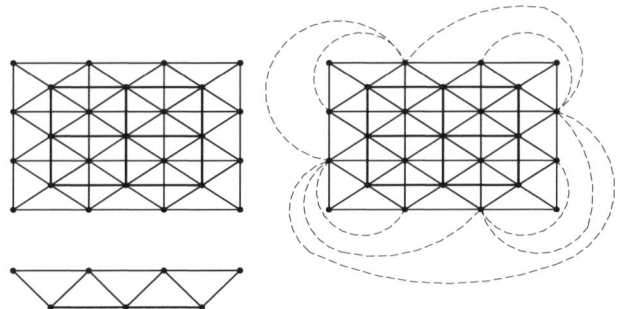

(a) A double-layer grid S.   (b) An arbitrary drawing of S.

Fig. 2.14 A space truss and its planar drawings.

It should be noted that the addition of dashed lines to complete the triangulation of the exterior region (unbounded cycle) is not necessary, since an m-polygon can be triangulated by $m-3$ members. Therefore, one can use,

$$\gamma(S) = c(S^P) - \bar{M}_c(S^P) - m + 3, \qquad (2\text{-}13)$$

where $\bar{M}_c(S^P)$ is the number of members required for a full triangulation of bounded regions.

2.5.4 A MIXED PLANAR DRAWING-EXPANSION METHOD

A mixed method can now be designed for determining the DSI of complex-pattern and/or large-scale structures. Repeated units of the structure are considered, and the theorems of Sections 2.2 and 2.5 are applied to find their indeterminacies. An expansion process is then used to determine the DSI of the entire structure. This method may be illustrated by means of the simple space frame shown in Figure 2.15(a).

40   OPTIMAL STRUCTURAL ANALYSIS

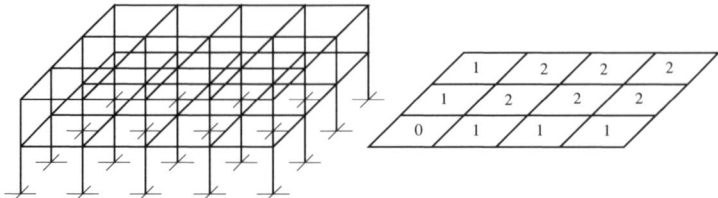

(a) A space structure S.

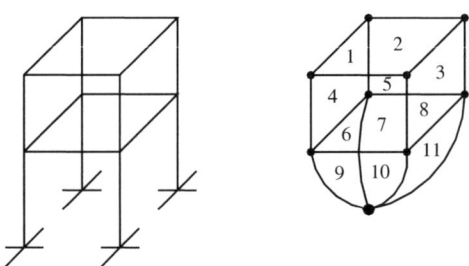

(b) A subgraph S_i of S and its graph model.

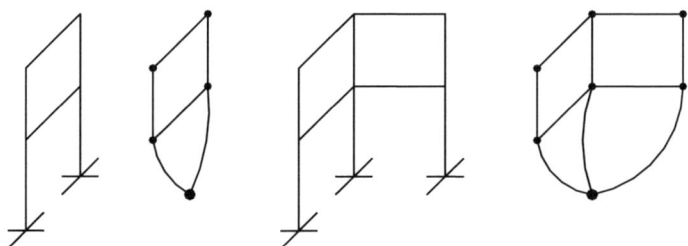

(c) Intersection Type 1 – A_i^1.   (d) Intersection Type 2 – A_i^2.

Fig. 2.15 A space frame S, its unit, and intersections.

A simple unit is considered in Figure 2.15(b), for which models with both ground node and tree node are illustrated. Eq. (2-11) results in:

$$\gamma(S_i) = 6(11-3) = 6 \times 8 = 48.$$

STATIC INDETERMINACY   41

Two types of intersections are illustrated in Figure 2.15(c) and (d), and their DSIs are calculated as

$$\overline{\gamma}(A_i^1) = 6[2] = 12 \text{ and } \overline{\gamma}(A_i^2) = 6[4] = 24.$$

The DSI of the entire structure is now determined as

$$\gamma(S) = 12 \times 48 - [5 \times 12 + 6 \times 24] = 12 [48 - 17] = 12 \times 31 = 6 \times 62 = 372.$$

For support conditions, one could consider the structure comprising complete cubes, and after calculating the DSI as $\gamma(S) = 6 \times 74$, 12 cycles corresponding to the ground cycles are contracted to obtain $\gamma(S) = 6 [74 - 12] = 6 \times 62 = 372$.

The DSI provides useful information about the connectivity properties of structures; however, it does not guarantee the rigidity of the structures. Therefore, additional studies are required, which are partially considered in the remainder of this chapter.

2.6 RIGIDITY OF STRUCTURES

The rigidity of trusses can be studied at different levels. The first level is combinatorial – is the graph of joints and members (bars) correct? The second level is geometrical – are the placements of joints appropriate? The third level is mechanical – are the selected materials and methods of construction suitable? Attention will be devoted to a first-level rigidity analysis of planar trusses. For this purpose, simplifying assumptions and definitions are made as follows.

Consider a planar truss composed of rigid members and pinned joints. Each joint connects the end nodes of two or more members in such a way that the mutual angles of the members change freely if the other ends are left free. Such an assumption is adequate for the first-level analysis of the rigidity. Let $M(S)$ and $N(S)$ denote the set of members and nodes of the graph model S of a truss. Denote the Cartesian coordinates of a node $n_i \in N(S)$ by (x_i, y_i). The number of members and nodes of S, as before, is also denoted by $M(S)$ and $N(S)$, respectively.

A member connecting n_i to n_j constrains the movement of S in such a way that the distance between these two nodes remains constant, that is,

$$(x_i - x_j)^2 + (y_i - y_j)^2 = const. \tag{2-14}$$

Differentiating this equation with respect to time t, we get,

$$(x_i - x_j)(\dot{x}_i - \dot{x}_j) + (y_i - y_j)(\dot{y}_i - \dot{y}_j) = 0, \tag{2-15}$$

where the dot above x and y denotes the differentiation with respect to t. Equation (2-15) implies that the relative velocity should be perpendicular to the member, that is, no member is stretched or compressed. Writing all such equations for the members of S, the following system of linear equations is obtained,

$$\mathbf{Hw} = \mathbf{0}, \tag{2-16}$$

where $\mathbf{H}$ is a $M(S) \times 2N(S)$ constant matrix and $\mathbf{w}$ is a column vector of unknown variables $\mathbf{w} = \{\dot{x}_1 \ \dot{y}_1 \ ... \ \dot{x}_{N(S)} \dot{y}_{N(S)}\}^t$, t denoting the transpose. A vector $\mathbf{w}$, which satisfies Eq. (2-16), is called an *infinitesimal displacement* of S. The infinitesimal displacements of S with respect to point-wise addition and multiplication by scalars, form a linear vector space $R^{2N(S)}$. The rigid body motion in a plane is a three-dimensional subspace of this linear space. The co-dimension of this subspace of rigid motions in the space of all infinitesimal motions is called the *degrees of freedom* of S, denoted by $f(S)$. The structure S is *rigid*, if $f(S) = 0$.

As an example, consider a truss as shown in Figure 2.16. For this truss, matrix $\mathbf{H}$ and vector $\mathbf{w}$ can be written as:

$$\mathbf{H} = \begin{bmatrix} x_i - x_j & x_j - x_i & 0 & y_i - y_j & y_j - y_i & 0 \\ x_i - x_k & 0 & x_k - x_i & y_i - y_k & 0 & y_k - y_i \\ 0 & x_j - x_k & x_k - x_j & 0 & y_j - y_k & y_k - y_j \end{bmatrix}. \tag{2-17}$$

and

$$\mathbf{w} = \{\dot{x}_i \ \dot{x}_j \ \dot{x}_k \ \dot{y}_i \ \dot{y}_j \ \dot{y}_k\}^t.$$

The entries of $\mathbf{H}$ are real and linear functions of the nodal coordinates of the corresponding graph.

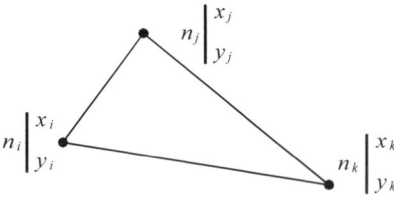

Fig. 2.16 A triangular planar truss.

It should not be thought that the rigidity of S requires Eq. (2-16) to have only the trivial solution $\mathbf{w} = \mathbf{0}$. Any rigid body motion with non-trivial $\mathbf{w}$ also satisfies this equation. As an example, consider a translation of the entire S specified by a vector $\{a, b\}^t$, that is, $\dot{x}_i = \dot{x}_j = \dot{x}_k = a$ and $\dot{y}_i = \dot{y}_j = \dot{y}_k = b$. Obviously $\mathbf{Hw} = \mathbf{0}$ still holds, since the sum of the first (or second) three columns of **H** is zero. Therefore, rank (**H**) < 2$N(S)$. The rigid body motion subspace in the plane has dimension 3, and for any truss we have rank (**H**) ≤ 2$N(S)$ − 3. However, if rank (**H**) = 2$N(S)$ − 3, then S is called *rigid*, and for rank (**H**) < 2$N(S)$ − 3, it is *non-rigid*. In the above example, rank (**H**) = 2 × 3 − 3 = 3 holds, and therefore a triangular planar truss is rigid.

Now consider other examples, as shown in Figure 2.17. The truss shown in Figure 2.17(a) is rigid, while the one in Figure 2.17(b) is not rigid, although their underlying graphs are the same. The assignment of velocities, indicated by arrows, forms an infinitesimal displacement because it does not violate Eq. (2-16). The nodes without arrows are assumed to have zero velocities. Similarly, though Figure 2.17(c) and Figure 2.17(d) have the same graph models, (c) is rigid but (d) is not rigid. It should be noted that an infinitesimal displacement does not always correspond to an actual movement of a mechanism. The truss (b) deforms mechanically, while truss (d) violates only Eq. (2-16).

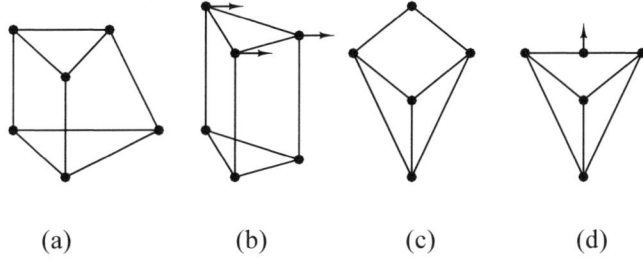

(a)   (b)   (c)   (d)

Fig. 2.17 Rigid and non-rigid planar trusses.

The nodes of a structure S are in general position if $x_1, y_1, x_2, y_2 ..., x_{N(S)}, y_{N(S)}$ are algebraically independent over the rational field. When the nodes are in general position, the definition of algebraic dependence shows that a sub-determinant of matrix **H** is 0, if and only if it is identically **0**, when $x_1, y_1, x_2, y_2 ..., x_{N(S)}, y_{N(S)}$ are considered as variables. Therefore if the nodes of S are in general position, the linear independence of Eq. (2-16) depends only on the underlying graph, and consequently, the rigidity also depends only on the graph model of the structure. From now on, it is assumed that the nodes of S are in general position.

For a ball-jointed space truss, Eq. (2-16) can be written in a general form to include a third dimension z. For such a case, a rigid body motion in space is a six-dimensional subspace of $R^{3N(S)}$. Therefore, a space truss will be rigid, if rank (**H**) = 3$N(S)$ − 6, and non-rigid, if rank (**H**) < 3$N(S)$ − 6.

Suppose S is the graph model of a planar truss whose joints are in general position. A graph S is called *stiff*, if the corresponding truss is rigid. For any $X \subseteq M$, let

$\rho_S(X)$ be the rank of the submatrix of **H** consisting of the rows associated with the members of X. X is called *generically independent*, if $\rho_S(X) = |X|$, and *generically dependent*, if $\rho_S(X) < |X|$, where $|X|$ denotes the cardinality of X.

For any subset X of M(S), define,

$$\mu_S(X) = -M(X) + 2N(X) - 3, \qquad (2\text{-}18)$$

where $|M(X)| = |X|$. Then the following basic theorem on rigidity can be stated:

Theorem 1 (Laman [144]): The graph S is generically independent if, and only if, $\mu_S(X) \geq 0$ for any non-empty subset X of M(S).

Corollary 1: S is stiff if, and only if, there exists $M' \subseteq M(S)$ such that $|M'| = 2N(S) - 3$ and $\mu_S(X) \geq 0$ for every non-empty subset X of M'.

Corollary 2: S is stiff and generically independent if and only if:

(a) $\mu_S(X) = 0$, and

(b) $\mu_S(X) \geq 0$ for every non-empty subset X of M(S).

Using $\gamma(S) = M(S) - 2N(S) + 3 = -\mu_S(S)$, Theorem 1 can be restated as follows:

The graph S is generically independent if, and only if, $\gamma(S_i) \leq 0$ for every subgraph S_i of S, as shown in Figure 2.18(a). The graph S is stiff if, and only if there is a covering subgraph $\overline{S}$ of S such that $\gamma(\overline{S}) = 0$ and $\gamma(S_i) \leq 0$ for every non-empty subgraph S_i of $\overline{S}$, as shown in Figure 2.18(b), which is a statically indeterminate structure.

Finally, the graph S is stiff and generically independent if and only if:

(a) $\gamma(S) = 0$, and

(b) $\gamma(S_i) \leq 0$ for every subgraph S_i of S, as shown in Figure 2.18(c), which is a statically determinate truss.

Unfortunately, the application of Theorem 1 requires $2^{M(S)}$ steps to determine whether a graph is generically independent. In the following sections, two methods are described for an efficient recognition of generic independence, which were developed by Sugihara [214], and Lovász and Yemini [156].

STATIC INDETERMINACY   45

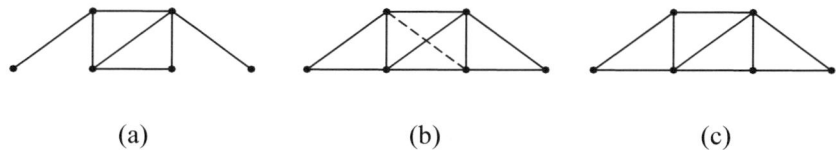

(a)   (b)   (c)

Fig. 2.18 (a) Generically independent, (b) stiff and (c) generically independent, and stiff graphs.

2.7 RIGIDITY OF PLANAR TRUSSES

2.7.1 COMPLETE MATCHING METHOD

Planar trusses are frequently used in structural engineering, and therefore a method suitable for both determinate and indeterminate trusses is presented in the following text, for checking the rigidity of these structures. The algorithm is polynomial bounded and uses complete matching of a specially constructed bipartite graph for the recognition of generic independence.

Definitions: Let $B(S) = (A, E, B)$ be a bipartite graph with node sets A, B and member set E. A subset E' of E is called a *complete matching* with respect to A, if the end nodes of members in E' are distinct and if every node in A is an end node of some members in E'.

For $X \subseteq A$, let $\Gamma(X)$ be the set of all those nodes in B that are connected to nodes in X by some members of E. As described in Appendix A, a graph has complete matching if, and only if, $|X| \leq |\Gamma(X)|$ for every $X \subseteq A$. Examples of matching and complete matching are depicted in Figures 2.19(a) and (b), respectively.

Using these definitions, the method is described as follows:

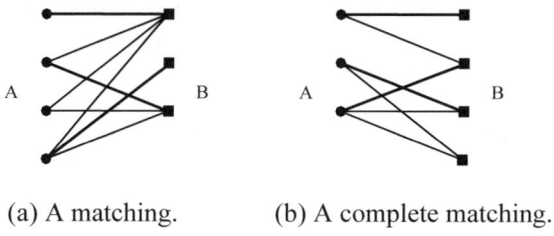

(a) A matching.   (b) A complete matching.

Fig. 2.19 Examples of matching.

Construct a bipartite graph of S. For this purpose, let S be a graph with $N(S)$ nodes and $M(S)$ members. The corresponding node set and member set are shown with the same symbols. For each node n_i of S, let p_i and q_i be two distinct symbols.

46   OPTIMAL STRUCTURAL ANALYSIS

Then let $B(S) = (A^*, E^*, B^*)$ be the bipartite graph whose node sets A^* and B^* and member set E^* are defined as

$A^* = M(S)$,

$B^* = \{p_1, q_1, p_2, q_2, \ldots, p_{N(S)}, q_{N(S)}\}$,

$E^* = \{(m, p_i), (m, q_i), (m, p_j), (m, q_j) \mid m = \{n_i, n_j\} \in M(S)\}$.

This bipartite graph is now augmented as follows:

Let t_1, t_2 and t_3 be three distinct symbols. Then, for any $1 \leq i < j \leq N(S)$, let $B_{ij}(S) = (A^*, E_{ij}, B^*)$ be the new bipartite graph constructed from $B(S)$ by the addition of three nodes and three members in the following manner:

$A_c^* = A^* \cup \{t_1, t_2, t_3\}$,

$E_{ij} = E^* \cup \{(t_1, p_i), (t_2, q_i), (t_3, p_j)\}$.

For any $Z \subseteq A_c^*$, denote by $\Gamma_{ij}(Z)$ the set of nodes of B^*, which are connected to elements of Z by members in E_{ij}. For any $X \subseteq A_c^*$, note that $2N(X) = |\Gamma_{ij}(X)|$. Then the following theorem can be proved.

Theorem 2 (Sugihara [214]): The graph model S is generically independent if, and only if, for any i and j ($1 \leq i < j \leq N(S)$), $B_{ij}(S) = (A_c^*, E_{ij}, B^*)$ has a complete matching with respect to A_c^*.

The proof can be found in the original paper of Sugihara [214] or Kaveh [113].

A complete matching of $B_{ij}(S) = (A_c^*, E_{ij}, B^*)$ can be found by Hopcroft and Karp's algorithm [82]. The number of $B_{ij}(S)$ is proportional to $M(S) \times M(S)$. As an example, consider the graph S as shown in Figure 2.20(a).

The bipartite graph of S is depicted in Figure 2.20(b), and a typical complete matching $B_{23}(S)$ is illustrated in bold lines. The examination of all $B_{ij}(S)$ for $1 \leq i < j \leq 4$ shows that complete matching exists and S is a generically independent graph.

STATIC INDETERMINACY 47

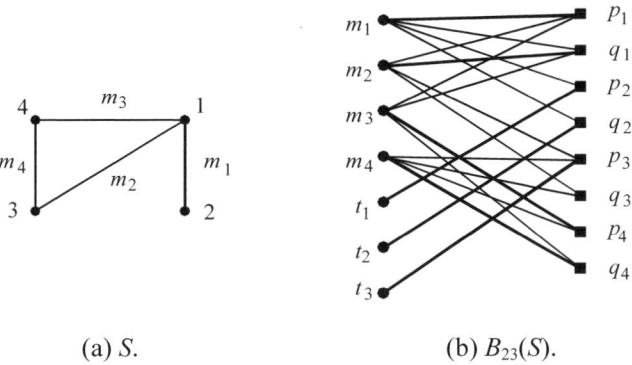

(a) S.   (b) $B_{23}(S)$.

Fig. 2.20 A complete matching $B_{23}(S)$ of a graph S.

The above method is quite general, and it is applicable to statically determinate and indeterminate structures. Another approach for the recognition of determinate trusses is due to Lovász and Yemini [156], which is described in the following section.

2.7.2 DECOMPOSITION METHOD

A graph S is generically independent, if doubling any member of S results in a new graph, which is the union of two spanning forests. A spanning forest is the union of k trees, containing all the nodes of S. This is the result of a special case of the following theorem of Nash–Williams [168].

Theorem 3: A graph S has a k member-disjoint spanning forest if, and only if, $M(S_i) \leq k[N(S_i) - 1]$ for every partition of $N(S)$.

Consider $k=2$, then $M(S_i) \leq 2N(S_i) - 2$ for every $S_i \subseteq S$. If a member is added to S_i without increasing its nodes $S_i' = S_i \cup m$, then $M(S_i') - 2N(S_i') + 3 \leq 0$, that is, $\mu(S_i') \geq 0$ for every $S_i \subseteq S$. This verifies the above method for checking the generic independence of S.

As an example, the above method is applied to check the graph shown in Figure 2.21(a) for generic independence. It can be seen that doubling any member of S leads to a graph that can be decomposed into two forests. The members for one of these forests, which have become spanning trees, are shown in bold lines in Figure 2.21(b)

48   OPTIMAL STRUCTURAL ANALYSIS

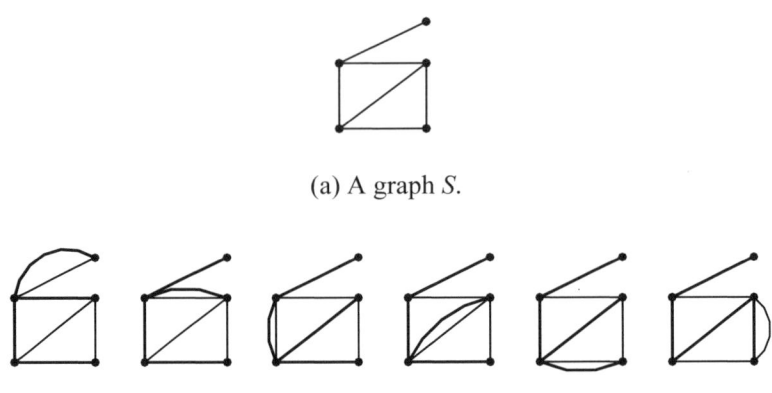

(a) A graph S.

(b) Decomposition of $S \cup m_i$.

Fig. 2.21 The generic independence check of S.

The above two seemingly different methods, are mathematically interrelated. A proof of this fact can be found in [216].

2.7.3 GRID-FORM TRUSSES WITH BRACINGS

There is a special kind of planar truss known as a *grid-form truss* whose rigidity is easier to control. Consider a planar truss consisting of square panels with or without diagonal members, an example of which is shown in Figure 2.22(a). The *bipartite graph B(S)* of S is constructed as follows:

Associate one vertex with each row of squares and use the notation $r_1, r_2, ..., r_m$ for them. With each column of squares, associate one vertex denoted by $c_1, c_2, ..., c_n$ as depicted in Figure 2.22(b). Connect r_i to c_j by an edge if the corresponding squares in S have a diagonal member; then the graph obtained in this manner is called the *bipartite graph B(S)* of S.

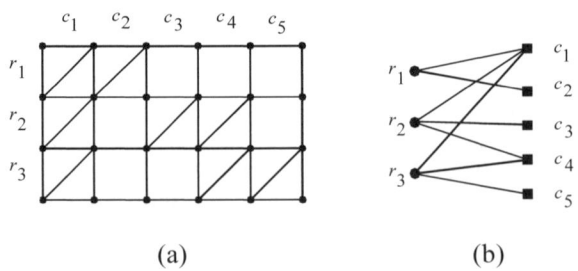

Fig. 2.22 A planar truss and its bipartite graph B(S).

STATIC INDETERMINACY   49

It is easy to prove that S is rigid if, and only if, the corresponding bipartite graph $B(S)$ is a connected graph. For this purpose, consider a square grid-form truss with two rows and two columns as depicted in Figure 2.23(a). A series of deformations can now be performed as shown in Figure 2.23(b).

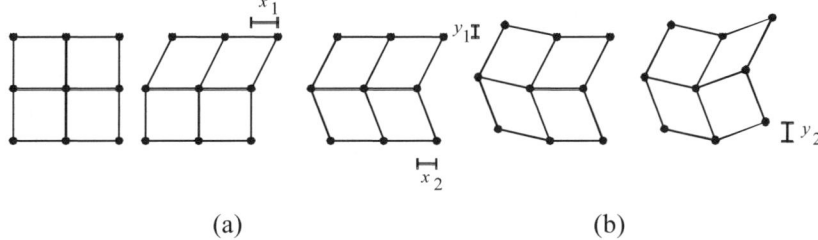

(a)   (b)

Fig. 2.23 A square grid-form truss and its deformation components.

Obviously, an arbitrary deformation of the truss can be considered as a combination of these deformation components. Now if a diagonal member is added to one of the squares, say the square corresponding to r_1 and c_1, then the deformation still takes place; however, a constraint in the form of $x_1 = y_1$ is imposed. If sufficient diagonal members are added, then $x_1 = x_2 = y_1 = y_2$, and no square deforms relative to the squares, that is, the entire truss will be rigid (if it is properly supported). This argument holds for any square grid-form truss with m rows and n columns. Since the nodes of $B(S)$ correspond to $x_1, x_2, \ldots, x_m, y_1, y_2, \ldots, y_n$ and the adjacency of r_i and c_j in $B(S)$ corresponds to the equality of $x_i = x_j$, the result follows.

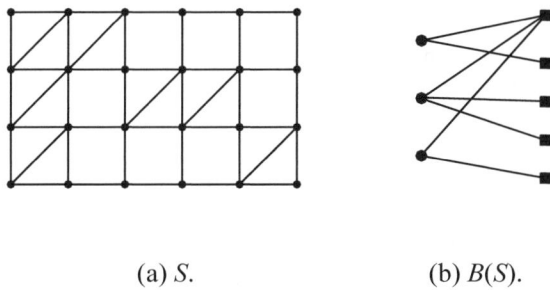

(a) S.   (b) $B(S)$.

Fig. 2.24 A γ-tree and its tree bipartite graph.

If $B(S)$ is a spanning tree, then the corresponding S is generically independent and stiff. It can also be proved that the static indeterminacy of S is the same as the first Betti number of $B(S)$, as in Kaveh [94]. Figure 2.24(a) shows a rigid γ-tree for a planar truss, the corresponding bipartite graph of which is illustrated in Figure 2.24(b).

2.8 CONNECTIVITY AND RIGIDITY

The *member-connectivity* $e(S)$ of a connected graph S is the smallest number of members, the removal of which disconnects S. When $e(S) \geq k$, then the graph S is called *k-member-connected*. A similar definition can be obtained for *node-connectivity* $v(S)$ by replacing "members" with "nodes".

Attempts have been made to relate the connectivity of a graph to its rigidity. Some partial results have been obtained; however, no general approach is found for such an interrelation. Some of the results obtained by Lovász and Yemini [156] are outlined in the following:

Thorem 5: Every 6-connected graph is stiff.

The proof can be found in the original paper of Lovász and Yemini [156] or Kaveh [113].

Finally, it should be noted that many attempts have recently been made to extend these ideas presented for planar trusses to space trusses; however, no concrete result applicable to general space graphs has so far been obtained. Many open problems remain for further research, if pure graph-theoretical methods are to be developed for the recognition of the rigidity of space trusses. The theory of matroids, which is briefly introduced in Appendix B and in [230], seems to be a promising tool for the future study of rigidity.

EXERCISES

2.1   Use an expansion process to find the DSI of a 3 × 4 planar truss S as shown. The unit to be considered for expansion is also given.

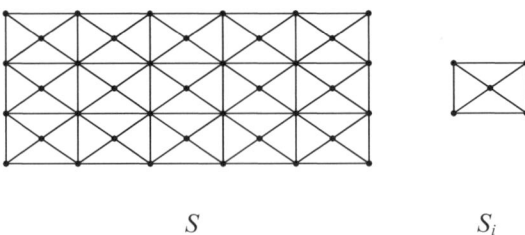

S S_i

2.2   If the truss in the previous example is an m × n planar truss, determine the corresponding DSI.

2.3   Derive Eq. (2-6) from Eq. (2-5).

STATIC INDETERMINACY   51

2.4   Prove that for determining the DSI of a planar truss, the crossing point of any two members can be regarded as an extra node. If the crossing members are more than two, why does such an operation become incorrect?

2.5   Find the DSI of the following planar truss using three different methods: classical, modification and triangulation:

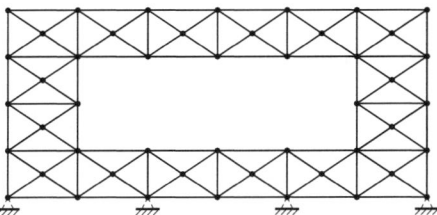

2.6   Determine the DSI of S_i in Figure 2.4 using its planar drawing. Consider S_i first as a space truss and secondly as a space frame.

2.7   Use Sugihara's matching method to verify the generic independence of the following truss models:

(a)   (b)

2.8   Determine the DSI of the following double-layer grid. Suppose S is supported in a statically determinate fashion:

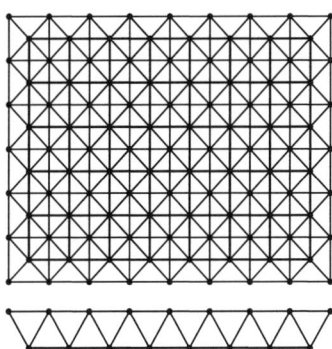

2.9 Employ Lovász and Yemini's decomposition method to verify the generic independence of the following truss:

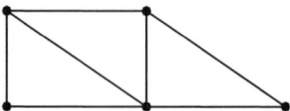

2.10 Examine the rigidity of the following trusses:

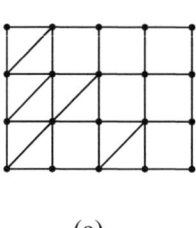   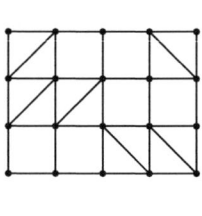   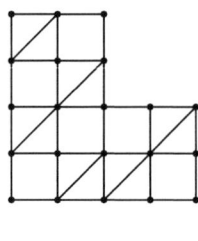

(a)   (b)   (c)

CHAPTER 3
Optimal Force Method of Structural Analysis

3.1 INTRODUCTION

This chapter is devoted to the progress made in the force method of structural analysis in recent years and summarises the state of the art. Efficient methods are developed leading to highly sparse flexibility matrices. The methods are mainly developed for frame structures; however, extensions are made to general skeletal structures and finite element analysis of continuum.

The force method of structural analysis, in which the member forces are used as unknowns, is appealing to engineers, since the properties of members of a structure most often depend on the member forces rather than joint displacements. This method was used extensively until 1960. After this, the advent of the digital computer and the amenability of the displacement method for computation attracted most researchers. As a result, the force method and some of the advantages it offers in optimisation and non-linear analysis, have been neglected.

Five different approaches are adopted for the force method of structural analysis, which will be classified as follows:

1. Topological force methods

2. Combinatorial force methods

Optimal Structural Analysis   A. Kaveh
© 2006 Research Studies Press Limited

3. Algebraic force methods

4. Mixed algebraic–combinatorial force methods

5. Integrated force method.

Topological methods have been developed by Henderson [76] and Maunder [162] for rigid-jointed skeletal structures using manual selection of the cycle bases of their graph models. Methods suitable for computer programming are due to Kaveh [94,100,103–105,111]. These topological methods are generalised to cover all types of skeletal structures, such as rigid-jointed frames, pin-jointed planar trusses and ball-jointed space trusses [98,113]. Algebraic topology is employed extensively in the work of Langefors [145,146].

Algebraic methods have been developed by Denke [36], Robinson [186], Topçu [219], Kaneko et al. [92], Soyer and Topçu [211], and mixed algebraic–topological methods have been used by Gilbert and Heath [60], Coleman and Pothen [28,29] and Pothen [181].

The integrated force method has been developed by Patnaik [172,173], in which the equilibrium equations and the compatibility conditions are satisfied simultaneously in terms of the force variables.

3.2 FORMULATION OF THE FORCE METHOD

In this section, a matrix formulation using the basic tools of structural analysis – equilibrium, compatibility and load–displacement relationships – is described. The notations are chosen from those most commonly utilised in structural mechanics.

3.2.1 EQUILIBRIUM EQUATIONS

Consider a structure S with M members and N nodes that is $\gamma(S)$ times statically indeterminate. Select $\gamma(S)$ independent unknown forces as redundants. These unknown forces can be selected from external reactions and/or internal forces of the structure. Denote these redundants by

$$\mathbf{q} = \{q_1, q_2, ..., q_{\gamma(S)}\}^t. \quad (3-1)$$

Remove the constraints corresponding to redundants, in order to obtain the corresponding statically determinate structure, known as the *basic* (*released* or *primary*) *structure* of S. Obviously, a basic structure should be rigid. Consider the joint loads as

OPTIMAL FORCE METHOD OF STRUCTURAL ANALYSIS 55

$$\mathbf{p} = \{p_1, p_2, ..., p_n\}^t, \qquad (3\text{-}2)$$

where n is the number of components for applied nodal loads.

Now the stress resultant distribution $\mathbf{r}$, due to the given load $\mathbf{p}$, for a linear analysis by the force method can be written as

$$\mathbf{r} = \mathbf{B}_0 \mathbf{p} + \mathbf{B}_1 \mathbf{q}, \qquad (3\text{-}3)$$

where $\mathbf{B}_0$ and $\mathbf{B}_1$ are rectangular matrices each having m rows, and n and $\gamma(S)$ columns, respectively, m being the number of independent components for member forces. $\mathbf{B}_0\mathbf{p}$ is known as a *particular solution*, which satisfies equilibrium with the imposed load, and $\mathbf{B}_1\mathbf{q}$ is a *complementary solution*, formed from a maximal set of independent self-equilibrating stress systems (SESs), known as a *statical basis*.

Example 1: Consider a planar truss, as shown in Figure 3.1(a), which is two times statically indeterminate. EA is taken to be the same for all the members.

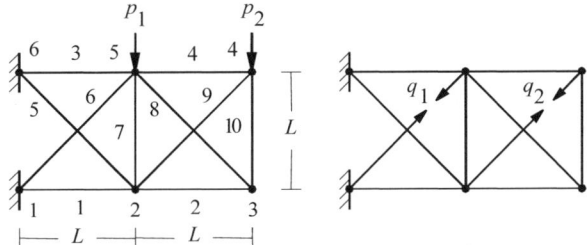

(a) A planar truss.   (b) The selected unknown forces.

Fig. 3.1 A statically indeterminate planar truss.

One member force and one component of a reaction may be taken as redundants. Alternatively, two member forces can also be selected as unknowns, as shown in Figure 3.1(b). With the latter selection, the corresponding $\mathbf{B}_0$ and $\mathbf{B}_1$ matrices can now be obtained by applying unit values of p_i ($i = 1, 2$) and q_j ($j = 1, 2$), respectively:

$$\mathbf{B}_0^t = \begin{bmatrix} -1 & 0 & 0 & 0 & \sqrt{2} & 0 & -1 & 0 & 0 & 0 \\ -2 & -1 & +1 & 0 & \sqrt{2} & 0 & -1 & \sqrt{2} & 0 & -1 \end{bmatrix},$$

and

$$\mathbf{B}_1^t = \begin{bmatrix} -1/\sqrt{2} & 0 & -1/\sqrt{2} & 0 & +1 & +1 & -1/\sqrt{2} & 0 & 0 & 0 \\ 0 & -1/\sqrt{2} & 0 & -1/\sqrt{2} & 0 & 0 & -1/\sqrt{2} & +1 & +1 & -1/\sqrt{2} \end{bmatrix}.$$

The columns of $\mathbf{B}_1$ form a statical basis of S. The underlying subgraph of a typical SES (for $q_2 = 1$) is shown in bold lines in Figure 3.1(b).

Example 2: Consider a portal frame shown in Figure 3.2(a), which is three times statically indeterminate.

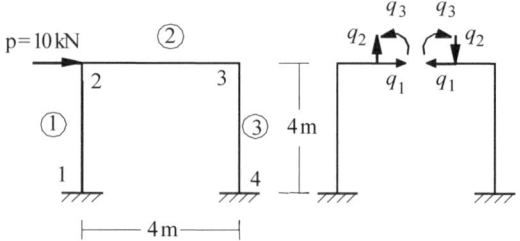

(a) A portal frame S.   (b) The basic structure of S.

Fig. 3.2 A statically indeterminate frame.

This structure is made statically determinate by an imaginary cut at the middle of its beam. The unit value of external load p_1 and each of the bi-actions q_i ($i = 1, 2, 3$) lead to the formation of $\mathbf{B}_0$ and $\mathbf{B}_1$ matrices, in which the two end bending moments (M_i, M_j) of a member are taken as its member forces. Using the sign convention introduced in Chapter 1, $\mathbf{B}_0$ and $\mathbf{B}_1$ matrices are formed as

$$\mathbf{B}_0^t = \begin{bmatrix} +4 & 0 & 0 & 0 & 0 & 0 \end{bmatrix},$$

and

$$\mathbf{B}_1^t = \begin{bmatrix} +4 & 0 & 0 & 0 & 0 & -4 \\ -2 & +2 & -2 & -2 & +2 & -2 \\ -1 & +1 & -1 & +1 & -1 & +1 \end{bmatrix}.$$

The columns of $\mathbf{B}_1$ form a statical basis of S, and the underlying subgraph of each SES is a cycle, as illustrated in bold lines in Figure 3.2(b). Notice that three SESs can be formed on each cycle of a planar frame.

In both of the above examples, particular and complementary solutions are obtained from the same basic structure. However, this is not a necessary requirement, as imagined by some authors. In fact, a particular solution is any solution satisfying equilibrium with the applied loads, and a complementary solution comprises of any maximal set of independent self-equilibrating systems. The latter is a basis of a vector space over the field of real numbers, known as a *complementary solution space*; see Henderson and Maunder [78].

Using the same basic structure is equivalent to searching for a cycle basis of a graph, but restricting the search to fundamental cycles only, which is convenient but not efficient when the structure is complex or when cycle bases with specific properties are needed.

As an example, consider a three-storey frame as shown in Figure 3.3(a). A cut system as shown in Figure 3.3(b) corresponds to a statical basis, containing three SESs formed on each element of the cycle basis shown in Figure 3.3(b). However, the same particular solution can be employed with a statical basis formed on the cycles of the basis shown in Figure 3.3(c).

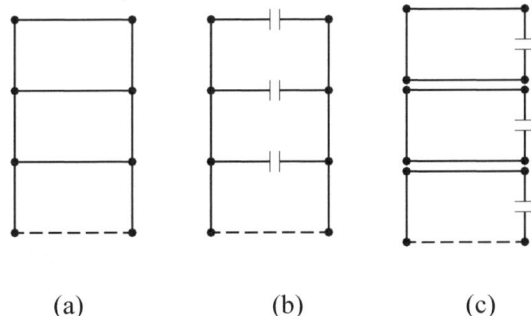

(a)   (b)   (c)

Fig. 3.3 A three-storey frame with different cut systems.

A basic structure need not be selected as a determinate one. For a redundant basic structure, one may obtain the necessary data either by analysing it first for the loads **p** and bi-actions $q_i = 1$ ($i = 1, 2, ..., \gamma(S)$) or by using existing information.

3.2.2 MEMBER FLEXIBILITY MATRICES

In the force method of analysis, the determination of the member flexibility matrix is an important step. A number of alternative methods are available for the formation of displacement–force relationships describing the flexibility properties of the members. Four such approaches are as follows:

58   OPTIMAL STRUCTURAL ANALYSIS

1. Inversion of the force–displacement relationship
2. Unit load method
3. Castigliano's theorem
4. Solution of differential equations for member displacements.

In the following, the unit load method is briefly described for the formation of the flexibility matrices:

Consider a general element with n member forces,

$$\mathbf{r}_m^t = \{r_1, r_2, ..., r_n\}, \qquad (3-4)$$

and member displacements,

$$\mathbf{u}_m^t = \{u_1, u_2, ..., u_n\}. \qquad (3-5)$$

A typical component of the displacement u_i can be found using the unit load method as follows:

$$u_i = \iiint_V \bar{\sigma}_i^t \varepsilon \, dV, \qquad (3-6)$$

where $\bar{\sigma}_i$ represents the matrix of statically equivalent stresses due to a unit load in the direction of r_i, and ε is the exact strain matrix due to all applied forces $\mathbf{r}_m$. The unit loads can be used in turn for all the points where member force are applied, and therefore,

$$u_m = \iiint_V \bar{\sigma}^t \varepsilon \, dV, \qquad (3-7)$$

where,

$$\bar{\sigma} = \{\bar{\sigma}_1 \ \bar{\sigma}_2 \ ... \ \bar{\sigma}_n\}^t. \qquad (3-8)$$

For a linear system,

$$\sigma = \mathbf{c}\mathbf{r}_m, \qquad (3-9)$$

where $\mathbf{c}$ is the stress distribution due to unit forces $\mathbf{r}_m$.

The stress–strain relationship can be written as follows:

$$\varepsilon = \phi\sigma = \phi\mathbf{c}\mathbf{r}_m. \tag{3-10}$$

Substituting in Eq. (3-7), we have,

$$\mathbf{u}_m = \iiint_V \bar{\sigma}^t \phi \mathbf{c}\, dV\, \mathbf{r}_m \tag{3-11}$$

or,

$$\mathbf{u}_m = \mathbf{f}_m \mathbf{r}_m, \tag{3-12}$$

where

$$\mathbf{f}_m = \iiint_V \bar{\sigma}^t \phi \mathbf{c}\, dV, \tag{3-13}$$

represents the element flexibility matrix.

The evaluation of $\bar{\sigma}$ representing the exact stress distribution due to the forces $\mathbf{r}_m$, may not be possible, and hence an approximate relationship should be used. Usually the matrix $\mathbf{c}$ is selected such that it will satisfy at least the equations of equilibrium. Denoting this approximate matrix by $\bar{\mathbf{c}}$ and using $\bar{\sigma} = \bar{\mathbf{c}}$, we have

$$\mathbf{f}_m = \iiint_V \bar{\mathbf{c}}^t \phi \bar{\mathbf{c}}\, dV. \tag{3-14}$$

This equation will be used for the derivation of the flexibility matrices of some finite elements in the following sections.

For a bar element of a space truss, however, the flexibility matrix can be easily obtained using Hooke's law as already discussed in Chapter 1. For a beam element ij of a space frame, y and z axes are taken as the principal axes of the beam's cross section; see Figure 3.4. The forces of end j are selected as a set of independent member forces, and the element is considered to be supported at point i. The axial, torsional, and flexural behaviours in respective planes are uncoupled, and therefore, one needs to only consider the flexibility relationships for four separate members:

1. An axial force member (along x-axis)

2. A pure torsional member (about x-axis)

3. A beam bent about y-axis

4. A beam bent about z-axis.

60 OPTIMAL STRUCTURAL ANALYSIS

Direct adaptation of the flexibility relationships derived in Chapter 1 gives the following 6 × 6 flexibility matrix:

$$\mathbf{f}_m = \begin{bmatrix} L/EA & 0 & 0 & 0 & 0 & 0 \\ 0 & L^3/3EI_z & 0 & 0 & 0 & L^2/2EI_z \\ 0 & 0 & L^3/3EI_y & 0 & -L^2/2EI_y & 0 \\ 0 & 0 & 0 & L/GJ & 0 & 0 \\ 0 & 0 & -L^2/2EI_y & 0 & L/EI_y & 0 \\ 0 & L^2/2EI_z & 0 & 0 & 0 & L/EI_z \end{bmatrix}, \quad (3\text{-}15)$$

where G is the shear modulus, I_y and I_z are the moments of inertia with respect to y and z axes, respectively. J is the Saint-Venant torsion constant of the cross section.

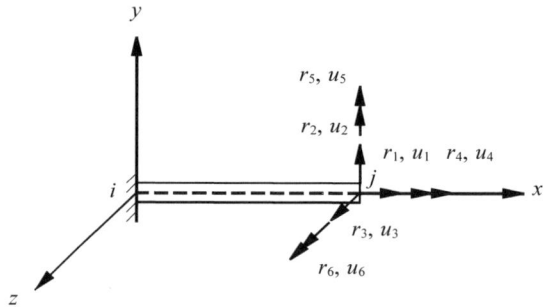

Fig. 3.4 A beam element and selected independent member forces.

3.2.3 EXPLICIT METHOD FOR IMPOSING COMPATIBILITY

The compatibility equations in the actual structure will now be derived. Using the displacement–load relationship for each member, and collecting them in the diagonal of the unassembled flexibility matrix $\mathbf{F}_m$, one can write member distortions as follows:

$$\mathbf{u} = \mathbf{F}_m \mathbf{r} = \mathbf{F}_m \mathbf{B}_0 \mathbf{p} + \mathbf{F}_m \mathbf{B}_1 \mathbf{q}. \quad (3\text{-}16)$$

In matrix form,

$$[\mathbf{u}] = [\mathbf{F}_m][\mathbf{B}_0 \ \mathbf{B}_1]\begin{bmatrix} \mathbf{p} \\ \mathbf{q} \end{bmatrix}. \quad (3\text{-}17)$$

From the contragradient principle of Chapter 1,

$$[\mathbf{v}] = \begin{bmatrix} \mathbf{B}_0^t \\ \mathbf{B}_1^t \end{bmatrix} [\mathbf{u}]. \tag{3-18}$$

Combining Eq. (3-17) and Eq. (3-18), we have

$$\begin{bmatrix} \mathbf{v}_0 \\ \mathbf{v}_c \end{bmatrix} = \begin{bmatrix} \mathbf{B}_0^t \\ \mathbf{B}_1^t \end{bmatrix} [\mathbf{F}_m][\mathbf{B}_0 \ \mathbf{B}_1] \begin{bmatrix} \mathbf{p} \\ \mathbf{q} \end{bmatrix}, \tag{3-19}$$

in which $\mathbf{v}_0$ contains the displacements corresponding to the force components of $\mathbf{p}$, and $\mathbf{v}_c$ denotes the relative displacements of the cuts for the basic structure. Performing the multiplication,

$$\begin{bmatrix} \mathbf{v}_0 \\ \mathbf{v}_c \end{bmatrix} = \begin{bmatrix} \mathbf{B}_0^t \mathbf{F}_m \mathbf{B}_0 & \mathbf{B}_0^t \mathbf{F}_m \mathbf{B}_1 \\ \mathbf{B}_1^t \mathbf{F}_m \mathbf{B}_0 & \mathbf{B}_1^t \mathbf{F}_m \mathbf{B}_1 \end{bmatrix} \begin{bmatrix} \mathbf{p} \\ \mathbf{q} \end{bmatrix}. \tag{3-20}$$

Defining

$$\begin{aligned} \mathbf{D}_{00} &= \mathbf{B}_0^t \mathbf{F}_m \mathbf{B}_0, & \mathbf{D}_{10} &= \mathbf{B}_0^t \mathbf{F}_m \mathbf{B}_1, \\ \mathbf{D}_{01} &= \mathbf{B}_1^t \mathbf{F}_m \mathbf{B}_0, & \mathbf{D}_{11} &= \mathbf{B}_1^t \mathbf{F}_m \mathbf{B}_1, \end{aligned} \tag{3-21}$$

the expansion of Eq. (3-8) leads to

$$\mathbf{v}_0 = \mathbf{D}_{00}\mathbf{p} + \mathbf{D}_{01}\mathbf{q}, \tag{3-22}$$

and

$$\mathbf{v}_c = \mathbf{D}_{10}\mathbf{p} + \mathbf{D}_{11}\mathbf{q}. \tag{3-23}$$

Consider now the compatibility conditions as

$$\mathbf{v}_c = 0. \tag{3-24}$$

Equation (3-24) together with Eq. (3-23) leads to

$$\mathbf{q} = -\mathbf{D}_{11}^{-1}\mathbf{D}_{10}\mathbf{p}. \tag{3-25}$$

Substituting in Eq. (3-22), we have

$$\mathbf{v}_0 = [\mathbf{D}_{00} - \mathbf{D}_{01}\mathbf{D}_{11}^{-1}\mathbf{D}_{10}]\mathbf{p} = \mathbf{F}\mathbf{p}, \quad (3\text{-}26)$$

and the stress resultant in a structure can be obtained as

$$\mathbf{r} = [\mathbf{B}_0 - \mathbf{B}_1\mathbf{D}_{11}^{-1}\mathbf{D}_{10}]\mathbf{p}. \quad (3\text{-}27)$$

3.2.4 IMPLICIT APPROACH FOR IMPOSING COMPATIBILITY

A direct application of the work principle of Chapter 1 can also be used to impose the compatibility conditions in an implicit form as follows:

Since the structure is considered to be linearly elastic, a linear relation exists between the unknown forces $\mathbf{q}$ and the applied forces $\mathbf{p}$, that is,

$$\mathbf{q} = \mathbf{Q}\mathbf{p}, \quad (3\text{-}28)$$

where $\mathbf{Q}$ is a transformation matrix, which is still unknown.

Equation (3-3) can now be written as

$$\mathbf{r} = \mathbf{B}_0\mathbf{p} + \mathbf{B}_1\mathbf{Q}\mathbf{p} = (\mathbf{B}_0 + \mathbf{B}_1\mathbf{Q})\mathbf{p} = \mathbf{B}\mathbf{p}. \quad (3\text{-}29)$$

Using the work theorem,

$$\mathbf{P}^t\mathbf{v} = \mathbf{r}^t\mathbf{u} = \mathbf{p}^t\mathbf{B}^t\mathbf{u}. \quad (3\text{-}30)$$

Now a set of suitable internal forces, $\mathbf{r}^*$, is considered that is statically equivalent to the external loads. From the work principle,

$$\mathbf{p}^t\mathbf{v} = \mathbf{r}^{*t}\mathbf{u}, \quad (3\text{-}31)$$

or

$$\mathbf{p}^t\mathbf{v} = \mathbf{p}^t\mathbf{B}_0^t\mathbf{u}. \quad (3\text{-}32)$$

A comparison of the above two equations leads to

$$\mathbf{p}^t\mathbf{B}^t\mathbf{u} = \mathbf{p}^t\mathbf{B}_0^t\mathbf{u}. \quad (3\text{-}33)$$

Substituting $\mathbf{u} = \mathbf{F}_m\mathbf{BP}$ in the above equation, we have

$$\mathbf{p}^t\mathbf{B}^t\mathbf{F}_m\mathbf{Bp} = \mathbf{p}^t\mathbf{B}_0^t\mathbf{F}_m\mathbf{Bp}. \qquad (3\text{-}34)$$

This holds for any $\mathbf{p}$, and therefore

$$\mathbf{B}^t\mathbf{F}_m\mathbf{B} = \mathbf{B}_0^t\mathbf{F}_m\mathbf{B}. \qquad (3\text{-}35)$$

From Eq. (3-29) by transposition,

$$\mathbf{B}^t = \mathbf{B}_0^t + \mathbf{Q}^t\mathbf{B}_1^t, \qquad (3\text{-}36)$$

and therefore,

$$(\mathbf{B}_0^t + \mathbf{Q}^t\mathbf{B}_0^t)\mathbf{F}_m\mathbf{B} = \mathbf{B}_0^t\mathbf{F}_m\mathbf{B}, \qquad (3\text{-}37)$$

or

$$\mathbf{Q}^t\mathbf{B}_0^t\mathbf{F}_m(\mathbf{B}_0 + \mathbf{B}_1\mathbf{Q}) = \mathbf{0}, \qquad (3\text{-}38)$$

or

$$\mathbf{Q}^t(\mathbf{B}_1^t\mathbf{F}_m\mathbf{B}_0 + \mathbf{B}_1^t\mathbf{F}_m\mathbf{B}_1\mathbf{Q}) = \mathbf{0}. \qquad (3\text{-}39)$$

Using the notation introduced in Eq. (3-9), we have

$$\mathbf{Q}^t(\mathbf{D}_{10} + \mathbf{D}_{11}\mathbf{Q}) = \mathbf{0}, \qquad (3\text{-}40)$$

or

$$\mathbf{D}_{10} + \mathbf{D}_{11}\mathbf{Q} = \mathbf{0}. \qquad (3\text{-}41)$$

Therefore,

$$\mathbf{Q} = -\mathbf{D}_{11}^{-1}\mathbf{D}_{10}, \qquad (3\text{-}42)$$

and

$$\mathbf{q} = -\mathbf{D}_{11}^{-1}\mathbf{D}_{10}\mathbf{p}, \qquad (3\text{-}43)$$

and Eq. (3-14) is obtained as in the previous approach.

64 OPTIMAL STRUCTURAL ANALYSIS

3.2.5. STRUCTURAL FLEXIBILITY MATRICES

The overall flexibility matrix of a structure can be expressed as

$$\mathbf{v} = \mathbf{F}\mathbf{p}. \tag{3-44}$$

Pre-multiplying the above equation by $\mathbf{p}^t$, we have

$$\mathbf{p}^t \mathbf{F} \mathbf{p} = \mathbf{p}^t \mathbf{B}_0^t \mathbf{F}_m \mathbf{B} \mathbf{p}. \tag{3-45}$$

Since $\mathbf{p}$ is arbitrary,

$$\mathbf{F} = \mathbf{B}_0^t \mathbf{F}_m \mathbf{B}, \tag{3-46}$$

or
$$\mathbf{F} = \mathbf{B}_0^t \mathbf{F}_m (\mathbf{B}_0 + \mathbf{B}_1 \mathbf{Q}), \tag{3-47}$$

or
$$\mathbf{F} = \mathbf{B}_0^t \mathbf{F}_m \mathbf{B}_0 - \mathbf{B}_0^t \mathbf{F}_m \mathbf{B}_1 \mathbf{D}_{11}^{-1} \mathbf{D}_{10}. \tag{3-48}$$

Since $\mathbf{F}_m$ is symmetric, it follows that

$$\mathbf{B}_0^t \mathbf{F}_m \mathbf{B}_1 = \mathbf{B}_1^t \mathbf{F}_m^t \mathbf{B}_0 = \mathbf{D}_{10}^t. \tag{3-49}$$

Therefore, the *overall flexibility matrix* (known also as the influence matrix) of the structure is obtained as

$$\mathbf{F} = \mathbf{D}_{00} - \mathbf{D}_{10}^t \mathbf{D}_{11}^{-1} \mathbf{D}_{10}, \tag{3-50}$$

and $\mathbf{D}_{11} = \mathbf{B}_1^t \mathbf{F}_m \mathbf{B}_1 = \mathbf{G}$ is also referred to as the *flexibility matrix* of the structure. In this book, the properties of $\mathbf{G}$ will be studied, since its pattern is the most important factor in optimal analysis of the structure by the force method.

Equation (3-34) can now be used to calculate the nodal displacements.

3.2.6 COMPUTATIONAL PROCEDURE

The sequence of computational steps for the force method can be summarised as follows:

1. Construct $\mathbf{B}_0$ and obtain $\mathbf{B}_0^t$.

2. Construct $\mathbf{B}_1$ and obtain $\mathbf{B}_1^t$.

3. Form unassembled flexibility matrix $\mathbf{F}_m$.

4. Form $\mathbf{F}_m \mathbf{B}_0$ followed by $\mathbf{F}_m \mathbf{B}_1$.

OPTIMAL FORCE METHOD OF STRUCTURAL ANALYSIS 65

5. Calculate $\mathbf{D}_{00}$, $\mathbf{D}_{10}^t$, $\mathbf{D}_{10}$ and $\mathbf{D}_{11}$, sequentially.

6. Compute $-\mathbf{D}_{11}^{-1}$.

7. Calculate $\mathbf{Q} = -\mathbf{D}_{11}^{-1}\mathbf{D}_{10}$.

8. Form $\mathbf{B}_1\mathbf{Q}$ and find $\mathbf{B} = \mathbf{B}_0 + \mathbf{B}_1\mathbf{Q}$.

9. Form $\mathbf{D}_{10}^t\mathbf{Q}$ and find $\mathbf{D}_{00} + \mathbf{D}_{10}^t\mathbf{Q}$.

10. Compute the internal forces as $\mathbf{r} = \mathbf{Bp}$.

11. Compute nodal displacements as $\mathbf{v}_0 = \mathbf{Fp}$.

Example 3: In this example, the complete analysis of the truss of Example 1 will be given.

$\mathbf{B}_0$ and $\mathbf{B}_1$ matrices are already formed in Example 1 of Section 3.2.1. The unassembled flexibility matrix can be constructed as follows:

$$\mathbf{F}_m = \frac{L}{EA}\begin{bmatrix} 1 & & & & & & & & \\ & 1 & & & & & & & \\ & & 1 & & & & 0 & & \\ & & & 1 & & & & & \\ & & & & \sqrt{2} & & & & \\ & & & & & \sqrt{2} & & & \\ & & & & & & 1 & & \\ & 0 & & & & & & \sqrt{2} & \\ & & & & & & & & \sqrt{2} \\ & & & & & & & & & 1 \end{bmatrix}.$$

Using the above matrix and the matrices from Example 1, we have

$$\mathbf{D}_{11} = \frac{L}{EA}\begin{bmatrix} 2\sqrt{2}+3/2 & 1/2 \\ 1/2 & 2\sqrt{2}+2 \end{bmatrix},$$

and

$$\mathbf{D}_{10} = \frac{L}{EA}\begin{bmatrix} 2+2/\sqrt{2} & 2+2/\sqrt{2} \\ 1/\sqrt{2} & 2+3/\sqrt{2} \end{bmatrix}.$$

Substituting in Eq. (3-25), we have

$$\begin{bmatrix} q_1 \\ q_2 \end{bmatrix} = -\begin{bmatrix} 2\sqrt{2}+3/2 & 1/2 \\ 1/2 & 2\sqrt{2}+2 \end{bmatrix}^{-1} \begin{bmatrix} 2+2/\sqrt{2} & 2+2/\sqrt{2} \\ 1/\sqrt{2} & 2+3/\sqrt{2} \end{bmatrix} \begin{bmatrix} p_1 \\ p_2 \end{bmatrix}.$$

Taking $p_1 = p_2 = P$ for simplicity and solving the above equations, we have

$$q_1 = -1.43P \text{ and } q_2 = -1.17P.$$

Equation (3.3) is then used to calculate the member forces as

$$\mathbf{r} = \{r_1 \ r_2 \ r_3 \ r_4 \ r_5 \ r_6 \ r_7 \ r_8 \ r_9 \ r_{10}\}^t =$$

$$\{-1.95P \ -0.17P \ 2.05P \ 0.83P \ 1.36P \ -1.44P \ -0.12P \ 0.24P \ -1.17P \ -0.17P\}^t.$$

Nodal displacements can be found using Eq. (3-25).

Example 4: In this example, the complete analysis of the frame in Example 2 is given.

$\mathbf{B}_0$ and $\mathbf{B}_1$ matrices are already formed in Example 2 of Section 3.2.1. The unassembled flexibility matrix of the structure, using the sign convention introduced in Chapter 1, is formed as

$$\mathbf{F}_m = \frac{L}{6EI} \begin{bmatrix} 2 & -1 & & & & & \\ -1 & 2 & & & & & \\ & & 2 & -1 & & & \\ & & -1 & 2 & & & \\ & & & & 2 & -1 \\ & & & & -1 & 2 \end{bmatrix}.$$

Substituting in Eq. (3-21), we have

$$\mathbf{D}_{11} = \frac{L}{6EI} \begin{bmatrix} 64 & 0 & -24 \\ 0 & 56 & 0 \\ -24 & 0 & 18 \end{bmatrix},$$

and

$$\mathbf{D}_{10} = \frac{L}{6EI}\begin{bmatrix} 32 \\ -24 \\ -12 \end{bmatrix}.$$

The inverse of $\mathbf{D}_{11}$ is computed as

$$\mathbf{D}_{11}^{-1} = \frac{6EI}{L}\begin{bmatrix} 18/576 & 0 & 3/72 \\ 0 & 576 & 0 \\ 3/72 & 0 & 1/9 \end{bmatrix},$$

and $\mathbf{Q}$ can be obtained as

$$\mathbf{Q} = -\mathbf{D}_{11}^{-1}\mathbf{D}_{10} = \begin{bmatrix} -1/2 \\ +3/7 \\ 0 \end{bmatrix}.$$

Matrix $\mathbf{B}$ is now computed as

$$\mathbf{B} = \begin{bmatrix} 4 \\ 0 \\ 0 \\ 0 \\ 0 \\ 0 \end{bmatrix} + \begin{bmatrix} +4 & -2 & -1 \\ 0 & +2 & +1 \\ 0 & -2 & -1 \\ 0 & -2 & +1 \\ 0 & +2 & -1 \\ -4 & -2 & +1 \end{bmatrix}\begin{bmatrix} -1/2 \\ +3/7 \\ 0 \end{bmatrix},$$

and finally by using Eq. (3-17) the member forces are obtained as

$$\mathbf{r} = \{+11.43 \quad +8.57 \quad -8.57 \quad -8.57 \quad +8.57 \quad +11.43\}^t.$$

General Loading: When a member is loaded in a general form, it must be replaced by an equivalent loading. Such a loading can be found as the superposition of two cases; case 1 consists of the given loading, but the ends of the member are fixed. The fixed end forces (actions), denoted by FEA, can be found using tables from books on the strength of materials. Case 2 is the given structure subjected to the reverse of the fixed end actions only. Obviously, the sum of the loads and reactions of case 1 and case 2 will have the same effect as that of the given loading. This superposition process is illustrated in the following example:

Example 5: A two-span beam is considered as shown in Figure 3.5(a). The fixed end actions are provided in (b), and the equivalent forces are illustrated in Figure 3.5(c). The structure is twice indeterminate, and the primary structure is obtained by introducing two hinges as shown in (d). The applied nodal forces and redundants are depicted in Figure 3.5(e) and (f), respectively.

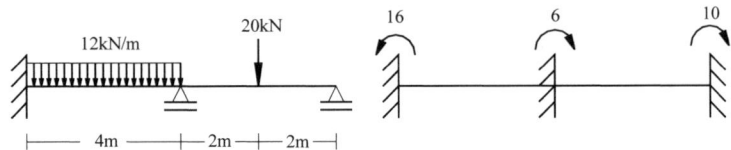

(a) A two-span beam. (b) Fixed end actions.

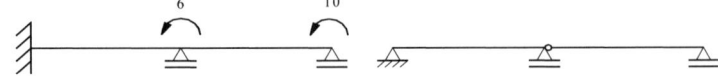

(c) The equivalent loading. (d) The selected primary structure.

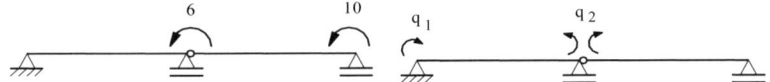

(e) Applied force on primary structure. (f) Redundants on primary structure.

Fig. 3.5 A two-span beam with general loading.

$\mathbf{B}_0$ and $\mathbf{B}_1$ matrices are formed as

$$\mathbf{B}_0 = \begin{bmatrix} -1 & 0 & 0 \\ 0 & +1 & 0 \\ 0 & 0 & 0 \\ 0 & 0 & +1 \end{bmatrix} \text{ and } \mathbf{B}_1 = \begin{bmatrix} -1 & 0 \\ 0 & +1 \\ 0 & -1 \\ 0 & 0 \end{bmatrix},$$

and the unassembled flexibility matrix of the structure is constructed as

$$\mathbf{F}_m = \frac{L}{6EI} \begin{bmatrix} 2 & -1 & & \\ -1 & 2 & & \\ & & 2 & -1 \\ & & -1 & 2 \end{bmatrix}.$$

Substituting in Eq. (3-21), we have

$$\mathbf{D}_{11} = \frac{L}{6EI}\begin{bmatrix} 2 & 1 \\ 1 & 4 \end{bmatrix} \text{ and } \mathbf{D}_{10} = \frac{L}{6EI}\begin{bmatrix} 2 & 1 & 0 \\ 1 & 2 & 1 \end{bmatrix}.$$

The inverse of $\mathbf{D}_{11}$ is computed as

$$\mathbf{D}_{11}^{-1} = -\frac{1}{7} \times \frac{6EI}{L}\begin{bmatrix} 4 & -1 \\ -1 & 2 \end{bmatrix},$$

and $\mathbf{Q}$ can be obtained as

$$\mathbf{Q} = -\mathbf{D}_{11}^{-1}\mathbf{D}_{10} = -\frac{1}{7}\begin{bmatrix} 4 & -1 \\ -1 & 2 \end{bmatrix}\begin{bmatrix} 2 & 1 & 0 \\ 1 & 2 & 1 \end{bmatrix} = -\frac{1}{7}\begin{bmatrix} 7 & 2 & -1 \\ 0 & 3 & 2 \end{bmatrix}.$$

Now $\mathbf{r}$ is computed as

$$\mathbf{r}' = \left(\begin{bmatrix} -1 & 0 & 0 \\ 0 & 1 & 0 \\ 0 & 0 & 0 \\ 0 & 0 & 1 \end{bmatrix} + \begin{bmatrix} 1 & 2/7 & -1/7 \\ 0 & -3/7 & -2/7 \\ 0 & 3/7 & 2/7 \\ 0 & 0 & 0 \end{bmatrix}\right)\begin{bmatrix} 0 \\ 6 \\ 10 \end{bmatrix} = \begin{bmatrix} 0.285 \\ 0.572 \\ 5.428 \\ 10.00 \end{bmatrix},$$

Adding the fixed end reaction, the final member forces are obtained as

$$\mathbf{r} = \{16.285 \; -15.428 \; 15.428 \; 0.00\}^t.$$

3.2.7 OPTIMAL FORCE METHOD

For an efficient force method, the matrix $\mathbf{G}$ should be

(a) sparse;

(b) well conditioned; and

(c) properly structured, that is, narrowly banded.

In order to provide the properties (a) and (b) for $\mathbf{G}$, the structure of $\mathbf{B}_1$ should be carefully designed since the pattern of $\mathbf{F}_m$ for a given discretisation is unchanged, that is, a suitable statical basis should be selected. This problem is treated in different forms by various methods. In the following, graph-theoretical methods are described for the formation of appropriate statical bases of different types of skeletal structures. The property (c) above has a totally combinatorial nature and is studied in Chapters 5 and 6.

Pattern Equivalence: Matrix $\mathbf{B}_1$, containing a statical basis, in partitioned form, is pattern equivalent to $\mathbf{C}^t$, where $\mathbf{C}$ is the cycle-member incidence matrix. Similarly, $\mathbf{B}_1^t \mathbf{F}_m \mathbf{B}_1$ is pattern equivalent to $\mathbf{CIC}^t$ or $\mathbf{CC}^t$. This correspondence transforms some structural problems associated with the characterisation of $\mathbf{G} = \mathbf{B}_1^t \mathbf{F}_m \mathbf{B}_1$ into combinatorial problems of dealing with $\mathbf{CC}^t$.

As an example, if a sparse matrix $\mathbf{G}$ is required, this can be achieved by increasing the sparsity of $\mathbf{CC}^t$. Similarly, for a banded $\mathbf{G}$, instead of ordering the elements of a statical basis (SESs), one can order the corresponding cycles. This transformation has many advantages, such as the following:

1. The dimension of $\mathbf{CC}^t$ is often smaller than that of $\mathbf{G}$. For example, for a space frame, the dimension of $\mathbf{CC}^t$ is sixfold smaller and for a planar frame, it is threefold smaller than that of $\mathbf{G}$. Therefore, the optimisation process becomes much simpler when combinatorial properties are used.

2. The entries of $\mathbf{C}$ and $\mathbf{CC}^t$ are elements of Z_2 and therefore easier to operate when compared to $\mathbf{B}_1$ and $\mathbf{G}$, which have real numbers as their entries.

3. The advances made in combinatorial mathematics and graph theory become directly applicable to structural problems.

4. A correspondence between algebraic and graph-theoretical methods becomes established.

3.3 FORCE METHOD FOR THE ANALYSIS OF FRAME STRUCTURES

In this section, frame structures are considered in their perfect conditions; that is, the joints of a frame are assumed to be rigid and connected to each other by elastic members and supported by a rigid foundation.

For this type of skeletal structure, a statical basis can be generated on a cycle basis of its graph model. The function representing the degree of static indeterminacy, $\gamma(S)$, of a rigid-jointed structure is directly related to the first Betti number $b_1(S)$ of its graph model,

$$\gamma(S) = \alpha b_1(S) = \alpha[M(S) - N(S) + b_0(S)], \qquad (3\text{-}51)$$

where $\alpha = 3$ or 6 depending on whether the structure is either a planar or a space frame.

For a frame structure, matrix $\mathbf{B}_0$ can easily be generated using the shortest route tree of its model, and $\mathbf{B}_1$ can be formed by constructing three or six SESs on each element of a cycle basis of S.

In order to obtain a flexibility matrix of maximal sparsity, special cycle bases should be selected as defined in the next section. Methods for the formation of a cycle basis can be divided into two groups, namely, (a) topological methods and (b) graph-theoretical approaches.

Topological methods useful for the selection of cycle bases by hand were developed by Henderson and Maunder [78] and Kaveh [94,95]; a complete description of these methods is presented in Kaveh [111]. Graph-theoretical methods suitable for computer applications were developed by Kaveh [96, 100,104].

3.3.1 MINIMAL AND OPTIMAL CYCLE BASES

A matrix is called *sparse* if many of its entries are zero. The interest in sparsity arises because its exploitation can lead to enormous computational saving and many large matrices that occur in the analysis of practical structures can be made sparse if they are not already so. A matrix can therefore be considered sparse if there is an advantage in exploiting its zero entries.

The *sparsity coefficient* χ of a matrix is defined as its number of non-zero entries. A cycle basis $C = \{C_1, C_2, C_3, \ldots, C_{b_1(S)}\}$ is called *minimal* if it corresponds to a minimum value of:

$$L(C) = \sum_{i=1}^{b_1(S)} L(C_i). \tag{3-52}$$

Obviously, $\chi(\mathbf{C}) = L(C)$ and a minimal cycle basis can be defined as a basis that corresponds to minimum $\chi(\mathbf{C})$. A cycle basis for which $L(C)$ is near minimum is called a *subminimal* cycle basis of S.

A cycle basis corresponding to maximal sparsity of the $\mathbf{CC}^t$ is called an *optimal* cycle basis of S. If $\chi(\mathbf{CC}^t)$ does not differ considerably from its minimum value, then the corresponding basis is termed *suboptimal*.

72   OPTIMAL STRUCTURAL ANALYSIS

The matrix intersection coefficient $\sigma_i(\mathbf{C})$ of row i of cycle-member incidence matrix $\mathbf{C}$ is the number of row j such that

(a) $j \in \{i+1, i+2, ..., b_1(S)\}$,

(b) $C_i \cap C_j \neq \varnothing$, that is, there is at least one k such that the column k of both cycles C_i and C_j (rows i and j) contains non-zero entries.

Now it can be shown that

$$\chi(\mathbf{C}^t\mathbf{C}) = b_1(S) + 2 \sum_{i=1}^{b_1(S)-1} \sigma_i(\mathbf{C}). \tag{3-53}$$

This relationship shows the correspondence of a cycle-member incidence matrix $\mathbf{C}$ and its cycle basis adjacency matrix. In order to minimise $\chi(\mathbf{CC}^t)$, the value of $\sum_{i=1}^{b_1(S)-1} \sigma_i(\mathbf{C})$ should be minimised since $b_1(S)$ is a constant for a given structure S, that is, γ-cycles with a minimum number of overlaps should be selected.

In the force method, an optimal cycle basis is needed corresponding to the maximum sparsity of the $\mathbf{CC}^t$ matrix. However, because of the complexity of this problem, most of the research has been concentrated on minimal cycle basis selection, except that of [105,108], which minimises the overlaps of the cycles rather than only their length.

3.3.2 SELECTION OF MINIMAL AND SUBMINIMAL CYCLE BASES

Cycle bases of graphs have many applications in various fields of engineering. The amount of work in these applications depends on the cycle basis chosen. A basis with shorter cycles reduces the time and storage required for some applications, that is, it is ideal to select a minimal cycle basis, and for some other applications minimal overlaps of cycles are needed, that is, optimal cycle bases are preferred. In this section, the formation of minimal and subminimal cycle bases is first discussed. Then the possibility of selecting optimal and suboptimal cycle bases is investigated.

Minimal cycle bases were considered first by Stepanec [212] and improved by Zykov [237]. Many practical algorithms for selecting subminimal cycle bases have been developed by Kaveh [94] and Cassell et al. [23]. Similar methods have been presented by Hubicka and Syslø [84], claiming the formation of a minimal cycle basis of a graph. Kolasinska [139] found a counterexample to the algorithm of Hubicka and Syslø. A similar conjecture was made by Kaveh [94] for planar graphs; however, a counterexample has been given by Kaveh and Roosta [129].

Recently, Horton [83] presented a polynomial time algorithm to find minimal cycle bases of graphs, which was improved by Kaveh and Mokhtar-zadeh [119].

In this section, the merits of the algorithms developed by different authors are discussed; a method is given for selection of minimal cycle bases, and efficient approaches are presented for the generation of subminimal cycle bases.

Formation of a Minimal Cycle on a Member: A minimal length cycle C_i on a member m_j, called its *generator*, can be formed by using the shortest route tree algorithm as follows:

Start the formation of two Shortest Route Trees (SRTs) rooted at the two end nodes n_s and n_t of m_j, and terminate the process as soon as the SRTs intersect each other (not through m_j itself) at say n_c. The shortest paths between n_s and n_c, and n_t and n_c, together with m_j, form a minimal cycle C_i on m_j. Using this algorithm, cycles of prescribed lengths can also be generated.

As an example, C_i is a minimal cycle on m_j in Figure 3.6. The SRTs are shown in bold lines. The generation of SRTs is terminated as soon as n_c has been found.

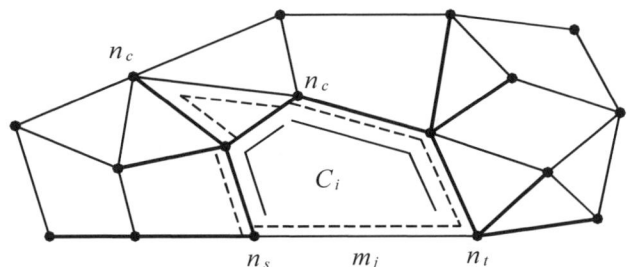

Fig. 3.6 A minimal cycle on a member.

A minimal cycle on a member m_j passing through a specified node n_k can similarly be generated. An SRT rooted at n_k is formed, and as soon as it hits the end nodes of m_j, the shortest paths are found by backtracking between n_k and n_s, and n_k and n_t. These paths together with m_j form the required cycle. As an example, a minimal cycle on m_j containing n_k is illustrated by dashed lines in Figure 3.6.

Different Cycle Sets for Selecting a Cycle Basis: It is obvious that a general cycle can be decomposed into its simple cycles. Therefore, it is natural to confine the considered set to only simple cycles of S. Even such a cycle set, which forms a subspace of the cycle space of the graph, has many elements and is therefore uneconomical for practical purposes.

74   OPTIMAL STRUCTURAL ANALYSIS

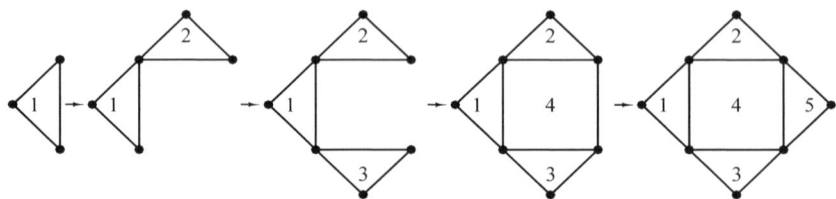

Fig. 3.7 A graph S and the selected cycles.

In order to overcome the above difficulty, Kaveh [94] used an expansion process, selecting the smallest admissible (independent with additional restriction) cycles, one at a time, until $b_1(S)$ cycles forming a basis had been obtained. In this approach, a very limited number of cycles were checked for being an element of a basis. As an example, the expansion process for selecting a cycle basis of S is illustrated in Figure 3.7.

Hubicka and Syslø [84] employed a similar approach, without the restriction of selecting one cycle at each step of expansion. In their method, when a cycle has been added to the previously selected cycles, increasing the first Betti number of the expanded part by "p", then p created cycles have been formed. As an example, in this method, Steps 4 and 5 will be combined into a single step, and addition of cycle 5 will require immediate formation of cycle 4. The above method is modified, and an efficient algorithm is developed for the formation of cycle bases by Kaveh and Roosta [129].

Finally, Horton [83] proved that the elements of a minimal cycle basis lie in between a cycle set consisting of the minimal cycles on each member of S that passes through each node of S, that is, each member is taken in turn and all cycles of minimal length on such a member passing through all the nodes of S are generated. Obviously, $M(S) \times M(S)$ such cycles will be generated.

Independence Control: Each cycle of a graph can be considered as a column vector of its cycle-member incidence matrix. An algebraic method such as the Gaussian elimination may then be used for checking the independence of a cycle with respect to the previously selected sub-basis. However, although this method is general and reduces the order dependency of the cycle selection algorithms, like many other algebraic approaches its application requires a considerable amount of storage space.

The most natural graph-theoretical approach is to employ a spanning tree of S and form its fundamental cycles. This method is very simple; however, in general, its use leads to long cycles. The method can be improved by allowing the inclusion of each used chord in the branch set of the selected tree. Further reduction in length may be achieved by generating an SRT from a centre node of a graph, and the use of its chords in ascending order of distance from the centre node; see Kaveh [96].

A third method, which is also graph-theoretical, consists of using admissible cycles. Consider the following expansion process, with S being a two-connected graph,

$$C_1 = C^1 \to C^2 \to C^3 \to \ldots \to C^{b_1(S)} = S,$$

where $C^k = \bigcup_{i=1}^{k} C_i$. A cycle C_{k+1} is called an *admissible* cycle if for $C^{k+1} = C^k \cup C_{k+1}$,

$$b_1(C^{k+1}) = b_1(C^k \cup C_{k+1}) = b_1(C^k) + 1. \tag{3-54}$$

It can easily be proved that the above admissibility condition is satisfied if any of the following conditions hold:

1. $A_{k+1} = C^k \cap C_{k+1} = \emptyset$, where $\emptyset$ is an empty intersection;

2. $\bar{b}_1(A_{k+1}) = r - s$, where r and s are the numbers of components of C^{k+1} and C^k, respectively;

3. $\bar{b}_1(A_{k+1}) = 0$ when C^k and C^{k+1} are connected ($r = s$).

In the above relations, $\bar{b}_1(A_i) = \bar{M}_i - \bar{N}_i + 1$, where $\bar{M}_i$ and $\bar{N}_i$ are the numbers of members and nodes of A_i, respectively.

As an example, the sequence of cycle selection in Figure 3.8 will be as specified by their numbers.

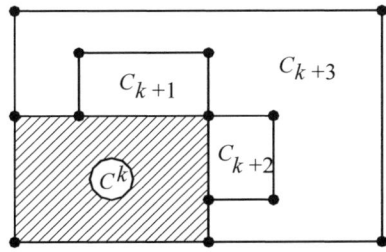

Fig. 3.8 A cycle and its bounded cycles.

76   OPTIMAL STRUCTURAL ANALYSIS

A different approach suggested by Hubicka and Sysło, in which

$$b_1(C^{k+1}) = b_1(C^k) + p, \qquad (3\text{-}55)$$

is considered to be permissible. However, a completion is performed for $p > 1$. As an example, when C_3 is added to C^k, its first Betti number is increased by 3 and therefore, cycles C_1 and C_2 must also be selected at that stage, before further expansion.

The mathematical concepts involved in the formation of a cycle basis having been discussed, three different algorithms are now described.

Algorithm 1 (Kaveh 1974)

Step 1: Select a pseudo-centre node of maximal degree O. Such a node can be selected manually or automatically using the graph or algebraic graph-theoretical methods discussed in Chapter 5.

Step 2: Generate an SRT rooted at O, form the set of its chords and order them according to their distance from O.

Step 3: Form one minimal cycle on each chord in turn, starting with the chord nearest to the root node. A corresponding simple path that contains members of the tree and the previously used chords is chosen, hence providing the admissibility of the selected cycle.

This method selects subminimal cycle bases, using the chords of an SRT. The nodes and members of the tree and consequently the cycles are partially ordered according to their distance from O. This is the combinatorial version of the Turn-back method to be discussed in Section 3.7.2 on algebraic force methods.

Algorithm 2 (Kaveh 1974)

Step 1: Select a centre or pseduo-centre node of maximal degree O.

Step 2: Use any member incident with O as the generator of the first minimal cycle. Take any member not used in C_1 and incident with O and generate the second minimal cycle on it. Continue this process until all the members incident with O are used as the members of the selected cycles. The cycles selected so far are admissible, since the intersection of each cycle with the previously selected cycles is a simple path (or a single node) resulting in an increase of the first Betti number by unity for each cycle.

Step 3: Choose a starting node O', adjacent to O, that has the highest degree. Repeat a step similar to Step 2, testing each selected cycle for admissibility. If the cycle formed on a generator m_k fails the test, then examine the other minimal cycles on m_k to find out if any such cycle exists. If no admissible minimal cycle can be found on m_k, then,

(i) form admissible minimal cycles on the other members incident with O'. If m_k does not belong to one of these subsequent cycles, then,

(ii) search for an admissible minimal cycle on m_k, since the formation of cycles on other previous members may now have altered the admissibility of this cycle. If no such cycle can be found, leave m_k unused. In this step, more than one member may be left unused.

Step 4: Repeat Step 3 using a node adjacent to O and/or O' having the highest degree as the starting node. Continue the formation of cycles until all the nodes of S have been tested for cycle selection. If all the members have not been used, select the shortest admissible cycle available for an unused member as generator. Then test the minimal cycles on the other unused members, in case the formation of the longer cycle has altered the admissibility. Each time a minimal cycle is found to be admissible, add to C^i and test all the minimal cycles on the other unused members again. Repeat this process, forming other shortest admissible cycles on unused members as generators, until S is re-formed and a subminimal cycle basis has been obtained.

Both of the above two algorithms are order-dependent, and various starting nodes may alter the result. The following algorithm is more flexible and less order-dependent, and in general leads to the formation of shorter cycle bases.

Algorithm 3 (Kaveh 1976)

Step 1: Generate as many admissible cycles of length 3 as possible. Denote the union of the selected cycles by C^n.

Step 2: Select an admissible cycle of length 4 on an unused member. Once such a cycle C_{n+1} is found, check the other unused members for possible admissible cycles of length 3. Again select an admissible cycle of length 4 followed by the formation of possible three-sided cycles. This process is repeated until no admissible cycles of length 3 and 4 can be formed. Denote the generated cycles by C^m.

Step 3: Select an admissible cycle of length 5 on an unused member. Then check the unused members for the formation of three-sided admissible cycles. Repeat Step 2 until no cycle of length 3 or 4 can be generated. Repeat Step 3 until no cycle of length 3, 4 or 5 can be found.

Step 4: Repeat Step 3 by considering higher-length cycles, until $b_1(S)$ admissible cycles forming a subminimal cycle basis are generated.

Remark: The cycle basis C formed by Algorithms 1 to 3 can be further improved by exchanging the elements of the selected basis. In each step of this process, a shortest cycle C_i' independent of the cycles of $C \setminus C_i$ is replaced by C_i if $L(C_i') < L(C_i)$. This process is repeated for $i = 1, 2, \ldots, b_1(S)$.

This additional operation increases the computational time and storage, and its use is recommended only when the formation of a minimal cycle basis is required.

Algorithm 4 (Horton 1987)

Step 1: Find a minimum path $P(n_i, n_j)$ between each pair of nodes n_i and n_j.

Step 2: For each node n_k and member $m_l = (n_i, n_j)$, generate the cycle having m_l and n_k as $P(n_k, n_i) + P(n_k, n_j) + (n_i, n_j)$ and calculate its length. Degenerate cases in which $P(n_k, n_i)$ and $P(n_k, n_j)$ have nodes other than n_k in common can be omitted.

Step 3: Order the cycles by their weight (or length).

Step 4: Use the Greedy Algorithm to find a minimal cycle basis from this set of cycles. This algorithm is given in Appendix B.

A simplified version of the above Algorithm can be designed as follows:

Step 1: Form a spanning tree of S rooted from an arbitrary node, and select its chords.

Step 2: Take the first chord and form $N(S) - 2$ minimal cycles, each being formed on the specified chord containing a node of S (except the two end nodes of this chord).

Step 3: Repeat Step 2 for the other chords, in turn, until $[M(S) - N(S) + 1] \times [N(S) - 2]$ cycles are generated. Repeated and degenerate cycles should be discarded.

Step 4: Order the cycles in ascending magnitude of their lengths.

Step 5: Using the above set of cycles, employ the Greedy Algorithm to form a minimal cycle basis of S.

The main contribution of Horton's Algorithm is the limit imposed on the elements of the cycle set used in the Greedy Algorithm. The use of matroids and the Greedy

Algorithm has been suggested by Kaveh [94,96], and they have been employed by Lawler [148] and Kolasinska [139].

3.3.3 EXAMPLES

Example 1: Consider a planar graph S, as shown in Figure 3.9, for which $b_1(S) = 18 - 11 + 1 = 8$. When Algorithm 3 is used, the selected basis consists of four cycles of length 3, three cycles of length 4 and one cycle of length 5, as follows:

$$C_1 = (1, 2, 3), C_2 = (1, 8, 9), C_3 = (2, 6, 3), C_4 = (2, 5, 6), C_5 = (1, 4, 5, 2)$$

$$C_6 = (1, 7, 5, 2), C_7 = (8, 6, 2, 1), C_8 = (10, 8, 6, 3, 11).$$

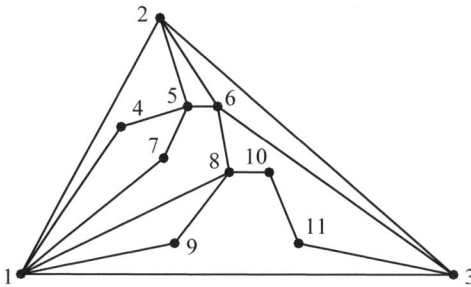

Fig. 3.9 A planar graph S.

The total length of the selected basis is $L(C) = 29$, which is a counter example for minimality of a mesh basis, since, for any such basis of S, $L(C) > 29$.

Example 2: In this example, S is the model of a space frame, considered as $S = \bigcup_{i=1}^{27} S_i$, where a typical S_i is depicted in Figure 3.10(a). For S_i there are 12 members joining eight corner nodes and a central node joined to these corner nodes. The model S is shown in Figure 3.10(b), in which some of the members are omitted for clarity of the diagram. For this graph, $b_1(S) = 270$.

80   OPTIMAL STRUCTURAL ANALYSIS

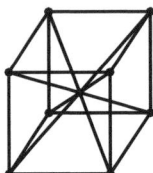
(a) A typical S_i ($i = 1, \ldots, 27$).

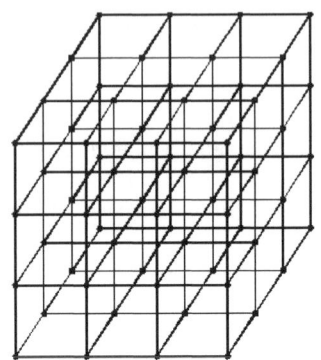
(b) S with some omitted members.

Fig. 3.10 A space frame S.

The selected cycle basis using any of the algorithms consists of 270 cycles of length 3, forming a minimal cycle basis of S. For Algorithm 3, the use of different starting nodes leads to a minimal cycle basis, showing the capability of this method.

Example 3: S is a planar graph with $b_1(S) = 9$, as shown in Figure 3.11. The application of Algorithm 3 results in the formation of a cycle of length 3 followed by the selection of five cycles of length 4. Then, member $\{1, 6\}$ is used as the generator of a six-sided cycle $C_7 = (1, 2, 3, 4, 5, 6, 1)$. Member $\{2, 10\}$ is then employed to form a seven-sided cycle $C_8 = (2, 11, 12, 13, 14, 15, 10, 2)$, followed by the selection of a five-sided cycle $C_9 = (10, 5, 4, 3, 2, 10)$. The selected cycle basis has a total length of $L(C) = 41$, and is not a minimal cycle basis. A shorter cycle basis can be found by Algorithm 4, consisting of one three-sided and five four-sided cycles, together with the following cycles:

$C_7 = (1, 2, 10, 5, 6, 1)$, $C_8 = (2, 3, 4, 5, 10, 2)$ and $C_9 = (2, 11, 12, 13, 14, 15, 10, 2)$,

forming a basis with the total length of 40. However, the computation time and storage for Algorithm 3 are far less than for Algorithm 4, as compared in [119].

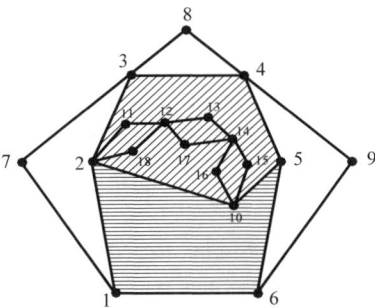

Fig. 3.11 A planar graph S.

3.3.4 OPTIMAL AND SUBOPTIMAL CYCLE BASES

In what follows, a direct method and an indirect approach, which often lead to the formation of optimal cycle bases, are presented. Much work is needed before the selection of an optimal cycle basis of a graph becomes feasible.

Suboptimal Cycle Bases: A Direct Approach

Definition 1: An *elementary contraction* of a graph S is obtained by replacing a path containing all nodes of degree 2 with a new member. A graph S contracted to a graph S' is obtained by a sequence of elementary contractions. Since in each elementary contraction k nodes and k members are reduced, the first Betti number does not change in a contraction, that is, $b_1(S) = b_1(S')$. The graph S is said to be homeo-morphic to S'; see Figure 3.12.

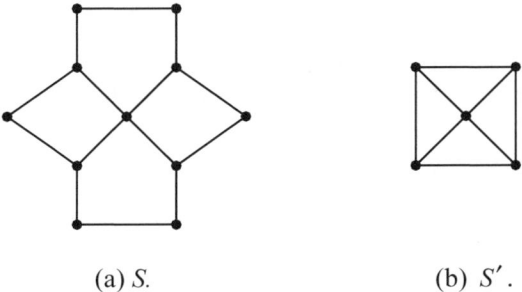

(a) S. (b) S'.

Fig. 3.12 S and its contracted graph S'.

This operation is performed in order to reduce the size of the graph and also because the number of members in an intersection of two cycles is unimportant; a single member is enough to render $C_i \cap C_j$ non-empty, and hence to produce a non-zero entry in $\mathbf{CC}^t$.

82   OPTIMAL STRUCTURAL ANALYSIS

Definition 2: Consider a member m_i of a graph S. On this member, p minimal cycles of length q can be generated. p is called the *incidence number* (IN) and q is defined as the *cycle length number* (CLN) of m_i. In fact, p and q are measures assigned to a member to indicate its potential as a member in the elements of a cycle basis. In the process of expansion for cycle selection, an artificial increase in p results in the exclusion of this element from a minimal cycle, keeping the number of overlaps as small as possible.

Space graphs need a special treatment. For these graphs, when a member has $p = 1$, then the next shortest length cycles with $q' = q + 1$ (1 being the next smallest possible integer) are also considered. Denoting the number of such cycles by p', the IN and CLN for this type of member are taken as

$$I_{jk} = p' + 1 \quad \text{and} \quad I^c_{jk} = (q + p'q') / (1 + p'), \qquad (3\text{-}56)$$

respectively. The end nodes of the considered member are j and k.

Definition 3: The *weight* of a cycle is defined as the sum of the incidence numbers of its members.

Algorithm A

Step 1: Contract S into S', and calculate the IN and CLN of all its members.

Step 2: Start with a member of the least CLN and generate a minimal weight cycle on this member. For members with equal CLNs, the one with the smallest IN should be selected. A member with these two properties will be referred to as "a member of the least CLN with the smallest IN".

Step 3: On the next unused member of the least CLN with the smallest IN, generate an admissible minimal weight cycle. In the case when a cycle of minimal weight is rejected due to inadmissibility, the next unused member should be considered. This process is continued as far as the generation of admissible minimal weight cycles is possible. After a member has been used as many times as its IN, before each extra usage, increase the IN of such a member by unity.

Step 4: On an unused member of the least CLN, generate one admissible cycle of the smallest weight. This cycle is not a minimal weight cycle, as otherwise it would have been selected at Step 3. Such a cycle is called a *subminimal weight cycle*. Again, update the INs for each extra usage. Now repeat Step 3, since the formation of the new subminimal weight cycle may have altered the admissibility condition of the other cycles and selection of further minimal weight cycles may now have become possible.

Step 5: Repeat Step 4, selecting admissible minimal and subminimal weight cycles, until $b_1(S')$ of these cycles are generated.

Step 6: A reverse process to that of the contraction of Step 1 transforms the selected cycles of S' into those of S.

This algorithm leads to the formation of a suboptimal cycle basis, and for many models encountered in practice, the selected bases have been optimal.

Suboptimal Cycle Bases: An Indirect Approach

Definition 1: The *weight* of a member in the following algorithm is taken as the sum of the degrees of its end nodes.

Algorithm B

Step 1: Order the members of S in ascending order of weight. In all the subsequent steps use this ordered member set.

Step 2: Generate as many admissible cycles of length α as possible, where α is the length of the shortest cycle of S. Denote the union of the selected cycles by C^m. When α is not specified, use the value $\alpha = 3$.

Step 3: Select an admissible cycle of length $\alpha + 1$ on an unused member (use the ordered member set). Once such a cycle C_{m+1} is found, control the other unused members for possible admissible cycles of length α. Again select an admissible cycle of length $\alpha + 1$, followed by the formation of possible α-sided cycles. This process is repeated until no admissible cycles of length α and $\alpha + 1$ can be found. Denote the generated cycles by C^n.

Step 4: Select an admissible cycle C_{n+1} of length $\alpha + 2$ on an unused member. Then check the unused members for the formation of α-sided cycles. Repeat Step 2 until no cycle of length α or $\alpha + 1$ can be generated. Repeat Step 3 until no cycles of length α, $\alpha + 1$ or $\alpha + 2$ can be found.

Step 5: Take an unused member and generate an admissible cycle of minimal length on this member. Repeat Steps 1, 2 and 3.

84   OPTIMAL STRUCTURAL ANALYSIS

Step 6: Repeat Step 4 until $b_1(S)$ admissible cycles, forming a suboptimal cycle basis, are generated.

The ordered member set affects the selection process in two ways:

1. Generators are selected in ascending weight order, hence increasing the possibility of forming cycles from the dense part of the graph. This increases the chance of cycles with smaller overlaps being selected.

2. From cycles of equal length formed on a generator, the one with smallest total weight (sum of the weights of the members of a cycle) is selected.

The cycle bases generated by this algorithm are suboptimal; however, the results are inferior to those of the direct method A.

Remark: Once a cycle basis C is formed by Algorithm A or Algorithm B, it can be further improved by exchanging the elements of C. In each step of this process, a cycle C_k is controlled for the possibility of being exchanged by the ring sum of C_k and a combination of the cycles of $C \setminus C_k$, in order to reduce the overlap of the cycles. The process is repeated until no improvement can be achieved. This additional operation increases the computational time and storage, and should only be used when the corresponding effort is justifiable, for example, this may be the case when a non-linear analysis or a design optimisation is performed using a fixed cycle basis.

3.3.5 EXAMPLES

In this section, examples of planar and space frames are studied. The cycle bases selected by Algorithms A and B are compared with those developed for generating minimal cycle bases (Algorithms 1–4). Simple examples are chosen in order to illustrate the process of the methods presented clearly. The models, however, can be extended to those containing a greater number of members and nodes of high degree to show the considerable improvements to the sparsity of matrix $\mathbf{CC}^t$.

Example 1: Consider a space frame as shown in Fig. 13(a) with the corresponding graph model S as illustrated in Fig. 13(b). For this graph $b_1(S) = 12$, and therefore 12 independent cycles should be selected as a basis. Algorithm B selects a minimal cycle basis containing the following cycles:

$C_1 = (1, 2, 3)$, $C_2 = (1, 2, 5)$, $C_3 = (1, 3, 4)$, $C_4 = (1, 5, 4)$, $C_5 = (2, 3, 6, 7)$, $C_6 = (3, 4, 7, 8)$, $C_7 = (4, 5, 8, 9)$, $C_8 = (6, 7, 8, 9)$, $C_9 = (7, 8, 11, 12)$, $C_{10} = (6, 7, 10, 11)$, $C_{11} = (9, 8, 12, 13)$, $C_{12} = (10, 11, 12, 13)$,

which corresponds to

$$\chi(\mathbf{C}) = 4 \times 3 + 8 \times 4 = 44$$

and $\quad \chi(\mathbf{CC^t}) = 12 + 2 \times 23 = 58.$

Use of Algorithm A leads to the formation of a similar basis, with the difference that $C'_8 = (6, 9, 10, 13)$ is generated in place of $C_8 = (6, 7, 8, 9)$, corresponding to

$$\chi(C') = 4 \times 3 + 8 \times 4 = 44,$$

$$\chi(\mathbf{C'C'^t}) = 12 + 2 \times 20 = 52.$$

The CLNs and Ins of the members used in this algorithm are illustrated in Figure 3.13(b).

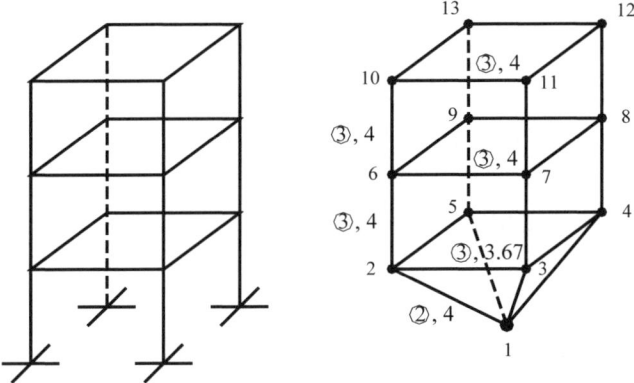

(a) A space structure. (b) The graph model S of the structure.

Fig. 3.13 A space frame, and CLNs and Ins of its members.

Example 2: In this example, S is a space structure with $b_1(S) = 33$, as shown in Figure 3.14(a). Both Algorithms 3 and A select 33 cycles of length 4, that is, a minimal cycle basis with $\chi(\mathbf{C}) = 4 \times 33 = 132$ is obtained.

The basis selected by Algorithm 3 contains (in the worst case) all four-sided cycles of S except those that are shaded in Figure 3.14(a), with $\chi(\mathbf{CC^t}) = 233$.

Algorithm A selects all three-sided cycles of S except those shaded in Figure 3.14(b), with $\chi(\mathbf{CC^t}) = 190$. It will be noticed that, for structures containing nodes of higher degrees, considerable improvement is obtained by the use of Algorithm A.

86 OPTIMAL STRUCTURAL ANALYSIS

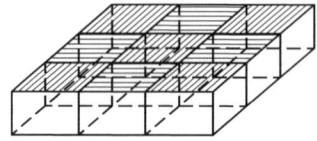 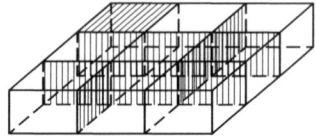

(a) A minimal cycle basis. (b) A suboptimal cycle basis.

Fig. 3.14 Minimal and suboptimal cycle bases of S.

Example 3: Consider a space frame as shown in Figure 3.15, for which $b_1(S) = 10$. The minimal cycle basis selected by Algorithm 3 consists of the following cycles:

$C_1 = (1, 2, 3)$, $C_2 = (4, 5, 6)$, $C_3 = (7, 8, 9)$, $C_4 = (10, 11, 12)$, $C_5 = (1, 2, 5, 4)$,
$C_6 = (2, 3, 6, 5)$, $C_7 = (4, 5, 8, 7)$, $C_8 = (5, 6, 9, 8)$, $C_9 = (7, 8, 11, 10)$,
$C_{10} = (8, 9, 12, 11)$,

corresponding to $\chi(\mathbf{C}) = 4 \times 3 + 6 \times 4 = 36$ and $\chi(\mathbf{CC}^t) = 10 + 2 [0 + 0 + 0 + 2 + 3 + 3 + 4 + 3 + 4] = 10 + 2 \times 19 = 48$.

However, the following non-minimal cycle basis has a higher $\chi(\mathbf{C})$, and leads to a more sparse $\mathbf{CC}^t$ matrix. The selected cycles are as follows:

$C_1 = (1, 2, 3)$, $C_2 = (1, 2, 5, 4)$, $C_3 = (2, 3, 6, 5)$, $C_4 = (1, 3, 6, 4)$, $C_5 = (4, 5, 8, 7)$,
$C_6 = (5, 6, 9, 8)$, $C_7 = (4, 6, 9, 7)$, $C_8 = (7, 8, 11, 10)$, $C_9 = (8, 9, 12, 11)$,
$C_{10} = (10, 11, 12)$,

for which $\chi(\mathbf{C'}) = 2 \times 3 + 8 \times 4 = 38$, corresponding to $\chi(\mathbf{C'C'}^t) = 10 + 2 [1 + 2 + 3 + 1 + 2 + 3 + 1 + 2 + 2] = 10 + 2 \times 16 = 42$.

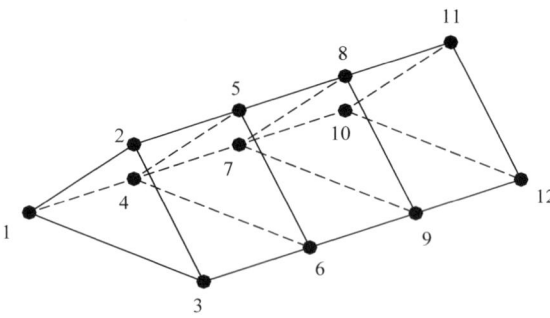

Fig. 3.15 A space frame S.

Therefore, the idea of having an optimal cycle basis in between minimal cycle bases is incorrect.

Example 4: Consider the skeleton of a structure S, comprising six flipped flags, as shown in Figure 3.16(a), for which $b_1(S) = 6$. After contraction, S' is obtained as illustrated in Figure 3.16(b). Obviously, this is a planar graph. The CLNs for the members are 3s, and the IN for member (1, 2) is 6, and for the remaining members it is equal to 1. Algorithm 3 selects a minimal cycle basis for S', consisting of six three-sided cycles, corresponding to

$$\chi(\mathbf{C}) = 6 \times 3 = 18 \text{ and } \chi(\mathbf{CC^t}) = 6 + 2\,[0 + 1 + 2 + 3 + 4 + 5] = 6 + 2 \times 15 = 36.$$

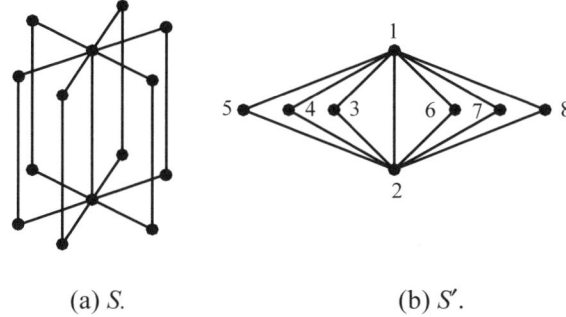

(a) S. (b) S'.

Fig. 3.16 A flipped flag before and after contraction.

However, the following non-minimal cycle basis has a higher $\chi(\mathbf{C'})$, and leads to a lower sparsity, $\chi(\mathbf{C'C'^t})$:

$$C_1 = (1, 3, 2, 4),\ C_2 = (1, 4, 2, 5),\ C_3 = (1, 2, 3),\ C_4 = (1, 2, 6),$$
$$C_5 = (1, 6, 2, 7),\ C_6 = (1, 7, 2, 8).$$

For this basis, $\chi(\mathbf{C'}) = 4 \times 4 + 2 \times 3 = 22$, corresponding to $\chi(\mathbf{C'C'^t}) = 6 + 2\,[0 + 1 + 1 + 1 + 1 + 1] = 6 + 2 \times 5 = 16$. After the back transformation from S' to S, we have $\chi(\mathbf{C}) = 4 \times 6 + 2 \times 4 = 32$, corresponding to $\chi(\mathbf{CC^t}) = 6 + 2\,[0 + 1 + 1 + 1 + 1 + 1] = 16$.

3.3.6 AN IMPROVED TURN-BACK METHOD FOR THE FORMATION OF CYCLE BASES

In this section, the combinatorial Turn-back method of Kaveh [94] is improved to obtain shorter cycle bases. This method covers all the counter examples, known for the minimality of the selected cycle bases.

Step 1: Generate an SRT rooted from an arbitrary node O. Identify its chords, and order them according to their distance numbers from O.

88   OPTIMAL STRUCTURAL ANALYSIS

Step 2: Select the shortest length cycle of the graph on a chord and add this chord (generator) to the tree members. Repeat this process with all the chords, forming cycles of the least length containing the tree members and the previously used chords only. The selected cycles are all admissible, that is, the addition of each cycle increases the first Betti number of the expanded part of the graph by unity. Store these cycles in C.

Step 3: Form all the new cycles of the same length on the remaining chords, allowing the use of more than one unused chord in their formation.

Step 4: Control the cycles formed in Step 3 to find only one cycle having a generator that is in none of the other connected cycles formed in Step 3. When such a chord is found, add the corresponding cycle to C and include its generator in the tree members. Repeat this control until no such cycle can be found.

Step 5: Select a cycle of the next higher length in the graph containing only one chord. Add the selected cycle to C and its generator to the tree members.

Step 6: Control the cycles formed in Step 3 to find a cycle containing only one unused chord. Add such a cycle to C and add its chord to the tree members. Repeat this control until no cycle of this property can be found.

Step 7: Repeat Step 4.

Step 8: Repeat Steps 5 and 6 and continue this repetition with the same length until no cycle in Step 5 can be found.

Step 9: Repeat Steps 3 to 8, until $b_1(S)$ cycles forming a cycle basis are included in C.

3.3.7 EXAMPLES

Example 1: A graph in the form of the one-skeleton of a torus-type structure is considered; see Figure 3.17. An SRT is selected, as shown in bold lines. The cycles selected in Step 2 are given below:

C = {(1, 2, 6), (1, 4, 5), (1, 5, 6), (1, 2, 13), (1, 4, 16), (1, 13, 16), (2, 3, 7), (2, 6, 7), (2, 3, 14), (2, 13, 14), (4, 5, 8), (4, 15, 16), (5, 6, 10), (5, 9, 10), (5, 8, 9), (12, 13, 16), (11, 12, 16), (11, 15, 16)}.

The execution of Step 3 results in the following cycles:

(3, 7, 8), (3, 4, 8), (7, 11, 12), (7, 8, 12), (8, 9, 12), (9, 13, 14), (9, 10, 14), (10, 14, 15), (10, 11, 15), (9, 12, 13), (3, 14, 15), (3, 4, 15).

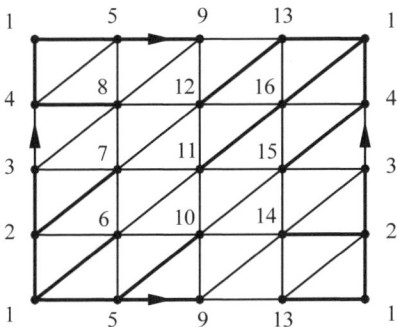

Fig. 3.17 Graph S and the selected SRT.

Twelve cycles are generated, increasing the first Betti number by 12. The control of Step 4 leads to generators {10, 11} and {7, 11}, corresponding to the cycles (10, 11, 15) and (7, 11, 12), respectively. Thus no cycle is selected.

In Step 5, a cycle of length 4 containing an unused chord is formed. On {3, 4}, cycle (1, 2, 3, 4) is generated and added to C. Then, in Step 6, the following cycles are added to C:

(3, 4, 8) for {3, 8}, (3, 7, 8) for {7, 8}, (3, 4, 15) for {3, 15}, (3, 14, 15) for {14, 15}.

In Step 7 no cycle is found, but in Step 8, the execution of Step 5 leads to cycle (1, 5, 9, 13) on {9, 13}, and Step 6 leads to the following cycles completing C and forming a minimal cycle basis of S:

(9, 12, 13) for {9, 12}, (9, 13, 14) for {9, 14}, (8, 9, 12) for {8, 12}, (7, 8, 12) for {7, 12}, (7, 11, 12) for {7, 11}, (9, 10, 14) for {10, 14}, (10, 14, 15) for {10, 15}, and (10, 11, 15) for {10, 11}.

Example 2: A space graph is considered as illustrated in Figure 3.18. An SRT is selected as shown in bold lines. The application of Step 2 leads to the following cycle set:

C = {(1, 2, 6, 7), (1, 5, 6, 10), (2, 3, 7, 8), (4, 5, 9, 10), (6, 7, 11, 12), (6, 10, 11, 15), (7, 8, 12, 13), (9, 10, 14, 15), (11, 12, 16, 17), (11, 15, 16, 20), (12, 13, 17, 18), (14, 15, 19, 20), (21, 22, 26, 27), (21, 25, 26, 30), (22, 23, 27, 28), (24, 25, 29, 30)}.

90   OPTIMAL STRUCTURAL ANALYSIS

In Step 3, the following cycles are generated:

(3, 4, 8, 9), (8, 9, 13, 14), (13, 14, 18, 19), (16, 17, 21, 22), (17, 18, 22, 23), (18, 19, 22, 23), (18, 19, 23, 24), (19, 20, 24, 25), (16, 20, 21, 25) ,(23, 24, 28, 29).

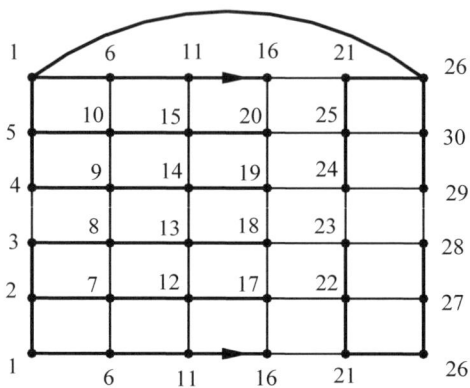

Fig. 3.18 A space graph and the selected SRT.

These cycles contain 11 unused chords. The control of Step 4 shows that {3, 4} and {28, 29} are included in one cycle, and therefore all the chords remain unused. In the next step, a cycle of length 5 including an unused chord is generated and added to C. Only with chord {3, 4}, the five-sided cycle (1, 2, 3, 4, 5) is generated, and in Step 6 the following three-sided cycles are selected:

(3, 4, 8, 9), (8, 9, 13, 14), and (13, 14, 18, 19).

Step 7 is carried out, and cycle (23, 24, 28, 29) on {28, 29} is found; repetition of this control leads to cycle (18, 19, 23, 24) on {23, 24}. In the next step, no cycle is selected. The execution of Steps 3 and 4 in Step 9 does not result in any cycle.

The execution of Step 5 in Step 9 forms cycle (1, 6, 11, 16, 21, 26) on chord {16, 21}, and the execution of Step 6 leads to the following cycles:

(16, 20, 21, 25) for {20, 25}, (19, 20, 24, 25) for {19, 24}, (16, 17, 21, 22) for {17, 22}, and (17, 18, 22, 23) for {18, 23}.

The selected cycles form a minimal cycle basis.

3.3.8. AN ALGEBRAIC GRAPH-THEORETICAL METHOD FOR CYCLE BASIS SELECTION

Consider a simple graph as shown in Figure 3.19, with $b_1(S) = 4$. Horton's algorithm forms the following cycle set Q:

$Q = \{C_1(1, 6, 9), C_2(2, 3, 7), C_3(7,8,9), C_4(4, 5, 8), C_5(1, 6, 7, 8), C_6(2, 3, 8, 9), C_7(4, 5, 7, 9), C_8(1, 2, 3, 6, 8), C_9(1, 4, 5, 6, 7), C_{10}(2, 3, 4, 5, 9), C_{11}(1, 2, 3, 4, 5, 6)\}$.

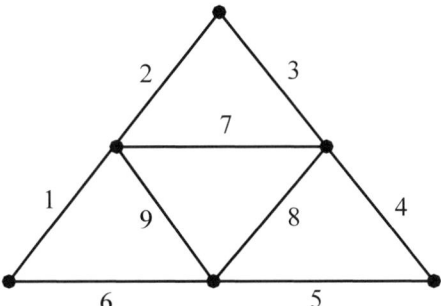

Fig. 3.19 A simple graph with $b_1(S) = 4$.

The cycle adjacency graph of S contains the nodes in a one-to-one correspondence with these cycles, and two nodes are connected to each other if the corresponding cycles have at least one member in common. Naturally, such a graph will not be simple and will have multiple members. The *weighted adjacency matrix* $\mathbf{A}^*$ of the new graph is constructed as

$$\mathbf{A}^* = \begin{array}{c} \\ 1 \\ 2 \\ 3 \\ 4 \\ 5 \\ 6 \\ 7 \\ 8 \\ 9 \\ 10 \\ 11 \end{array} \begin{array}{c} \begin{array}{ccccccccccc} 1 & 2 & 3 & 4 & 5 & 6 & 7 & 8 & 9 & 10 & 11 \end{array} \\ \left[\begin{array}{ccccccccccc} 3 & 0 & 1 & 0 & 2 & 1 & 1 & 2 & 2 & 1 & 2 \\ 0 & 3 & 1 & 0 & 1 & 2 & 1 & 2 & 1 & 2 & 2 \\ 1 & 1 & 3 & 1 & 2 & 2 & 2 & 1 & 1 & 1 & 0 \\ 0 & 0 & 1 & 3 & 1 & 1 & 2 & 1 & 2 & 2 & 2 \\ 2 & 1 & 2 & 1 & 4 & 1 & 1 & 3 & 3 & 0 & 2 \\ 1 & 2 & 2 & 1 & 1 & 4 & 1 & 3 & 0 & 3 & 2 \\ 1 & 1 & 2 & 2 & 1 & 1 & 4 & 0 & 3 & 3 & 2 \\ 2 & 2 & 1 & 1 & 3 & 3 & 0 & 5 & 2 & 2 & 4 \\ 2 & 1 & 1 & 2 & 3 & 0 & 3 & 2 & 5 & 2 & 4 \\ 1 & 2 & 1 & 2 & 0 & 3 & 3 & 2 & 2 & 5 & 4 \\ 2 & 2 & 0 & 2 & 2 & 2 & 2 & 4 & 4 & 4 & 6 \end{array} \right] \end{array}.$$

92   OPTIMAL STRUCTURAL ANALYSIS

Once $\mathbf{A}^*$ is formed, the largest eigenvalue λ_1 with the corresponding eigenvector having all positive entries can be easily calculated. $\mathbf{A}^*$ is real and symmetric, and it can be shown that all entries of $\mathbf{A}^{*k}$ are positive. Thus, it is primitive and, according to the Perron Frobenious theorem, λ_1 is real and positive and a simple root of the characteristic equation corresponds to a unique eigenvector $\mathbf{v}_1$ with all entries positive. Such an eigenvector can be obtained by the following simple algorithm:

Let $\mathbf{v} = \{1,1, \ldots, 1\}^t$. Then the components of $\mathbf{A}^{*t}\mathbf{v}$ are the number of walks of length k beginning at an arbitrary node of S and ending at n_i. If n_i is a good starting node, this number will be larger. Thus, for k, one should obtain some average number defined as the *accessibility index* by Gould [65], which indicates how many walks go through a node on an average. With a suitable normalisation, $\mathbf{A}^{*k}\mathbf{v}$ converges to the largest eigenvector $\mathbf{v}_1$ of $\mathbf{A}^*$; see Straffing [213].

As an example, for the cycle adjacency matrix discussed in Section 3.3.8, the largest eigenvalue is calculated as $\lambda_1 = 21.8815$, and the corresponding eigenvector $\mathbf{v}_1^*$ is obtained and its entries are reordered as follows:

$\mathbf{v}_1^* = \{0.1782\ \ 0.2124\ \ 0.2124\ \ 0.2124\ \ 0.2718\ \ 0.2718\ \ 0.2718\ \ 0.3654\ \ 0.3654\ \ 0.3654\ \ 0.4590\}^t$.
The entries of this vector correspond to the cycle numbers as

$$\mathbf{P} = \{3\ 1\ 2\ 4\ 5\ 6\ 7\ 9\ 8\ 10\ 11\}^t.$$

Algorithm (Kaveh and Rahami [120])

The algorithm is simple and consists of the following steps:

Step 1: Contract S to S'.

Step 2: Form the cycle subspace using Horton's approach.

Step 3: Form the cycle adjacency matrix $\mathbf{A}^*$.

Step 4: Calculate the largest eigenvector $\mathbf{v}_1$ of $\mathbf{A}^*$.

Step 5: Put the entries of $\mathbf{v}_1$ in ascending order to obtain $\mathbf{v}_1^*$ and construct a vector $\mathbf{P}$ containing the order of the cycles in $\mathbf{v}_1^*$.

Step 6: Choose the first entry of $\mathbf{P}$ as the first cycle, save it in $\mathbf{C}^*$ and remove it from $\mathbf{P}$.

Step 7: Select the next admissible cycle from the new **P**, starting from its first entry, save it in C* and remove it from **P**.

Step 8: Repeat Step 7 until $b_1(S)$ admissible cycles in C* forming a suboptimal cycle basis are constructed.

For the graph shown in Figure 3.18, the selected cycles are chosen sequentially from **P** = {3 1 2 4 5 6 7 9 8 10 11}^{t} as C_3, C_1, C_2, and C_4. These cycles with the corresponding sequence are found to be admissible and selected through steps 6 to 8 of the algorithm. The members of these cycles from the cycle set Q are (7, 8, 9), (1, 6, 9), (2, 3, 7) and (4, 5, 8), respectively, forming a cycle basis consisting of four three-sided cycles.

3.3.9 EXAMPLES

In this section, the selected cycle bases using the algorithm of Horton [83] and the present algorithm are compared. It should be noted that these two algorithms have different aims, namely, the first algorithm is designed for the formation of a minimal cycle basis, while the present algorithm aims at the selection of cycles with smallest possible overlaps leading to a suboptimal cycle basis.

Example 1: A planar graph is considered as shown in Figure 3.20. Use of Horton's algorithm leads to the formation of the following cycle basis:

$$C_1 = \{(1, 9, 18), (2, 3, 4), (8, 9, 10), (10, 11, 12), (11, 17, 16, 18),$$
$$(11, 14, 15, 18), (3, 13, 10, 18), (5, 6, 7, 8, 12)\}$$

corresponding to $\chi(\mathbf{C}) = 29$ and $\chi(\mathbf{D}) = 8 + 2 \times 16 = 40$.

The present algorithm leads to the following cycle basis:

$$C_1^* = \{(2, 3, 4), (1, 9, 18), (10, 11, 12), (11, 16, 17, 18), (8, 9, 10),$$
$$(14, 15, 16, 17), (1, 3, 5, 6, 7), (5, 6, 7, 8, 13)\}.$$

corresponding to $\chi(\mathbf{C}) = 30$ and $\chi(\mathbf{D}) = 8 + 2 \times 9 = 26$.

94   OPTIMAL STRUCTURAL ANALYSIS

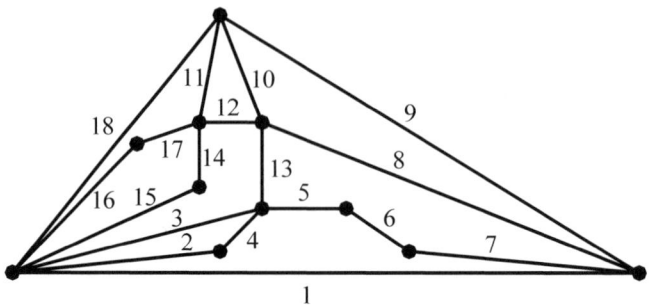

Fig. 3.20 A planar graph S with $b_1(S) = 8$.

Example 2: A simple graph is considered as shown in Figure 3.21. Use of Horton's algorithm leads to the formation of the following cycle basis:

$$C_2 = \{(1, 2, 3), (4, 5, 6), (7, 8, 9), (10, 11, 12), (1, 14, 4, 13), (2, 15, 5, 14),$$
$$(4, 17, 7, 16), (5, 18, 8, 17), (7, 20, 10, 19), (8, 21, 11, 2)\},$$

corresponding to $\chi(\mathbf{C}) = 36$ and $\chi(\mathbf{D}) = 10 + 2 \times 19 = 48$.

The present algorithm leads to the following cycle basis:

$$C_2^* = \{(1, 2, 3), (10, 11, 12), (1, 13, 4, 14), (2, 14, 5, 15), (3, 6, 13, 15), (8, 11, 20,$$
$$21), (9, 12, 19, 21), (7, 10, 19, 20), (4, 7, 16, 17), (5, 8, 17, 18)\},$$

corresponding to $\chi(\mathbf{C}) = 38$ and $\chi(\mathbf{D}) = 10 + 2 \times 17 = 44$.

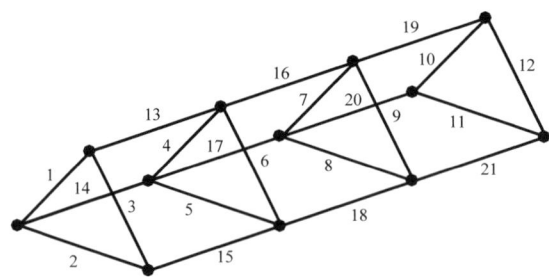

Fig. 3.21 A simple graph S with $b_1(S) = 10$.

OPTIMAL FORCE METHOD OF STRUCTURAL ANALYSIS   95

Example 3: A graph model in the form of a $3 \times 3 \times 1$ grid is considered, as shown in Figure 3.22. Use of Horton's algorithm leads to the formation of a minimal cycle basis, corresponding to $\chi(\mathbf{C}) = 132$ and $\chi(\mathbf{D}) = 33 + 2 \times 100 = 233$.

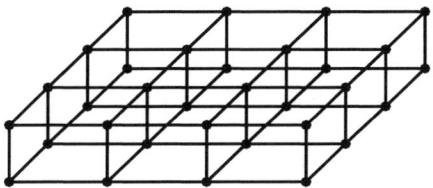

Fig. 3.22 A $3 \times 3 \times 1$ cube-type graph with $b_1(S) = 33$.

The cycle basis selected by the present algorithm corresponds to $\chi(\mathbf{C}) = 132$ and $\chi(\mathbf{D}) = 33 + 2 \times 77 = 187$.

Example 4: In this example, S is the model of a space frame, considered as $S = \bigcup_{i=1}^{b_1(S)} S_i$, where a typical S_i is depicted in Figure 3.23(a). For S_i there are 12 members joining eight corner nodes. The model is shown in Figure 3.23(b), in which some of the members are omitted for clarity of the diagram. For this graph, $b_1(S) = 270$.

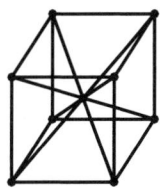 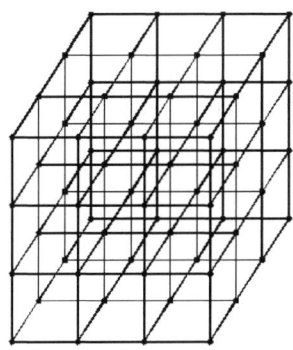

(a) A typical $S_i (i = 1, ..., 27)$.   (b) S with some omitted members.

Fig. 3.23 A space frame S.

Using Horton's algorithm, 324 three-sided and 108 four-sided cycles are generated. Higher-length cycles could also be generated; however, since an optimal cycle basis for a symmetric graph of Figure 3.23 does not seem to have any such cycle, higher-length cycles are not selected. The weighted adjacency matrix $\mathbf{A}^*$ is formed and, using the present algorithm, a suboptimal cycle basis is generated. This basis consists of 270 cycles of length 3 corresponding to $\chi(\mathbf{C}) = 270 \times 3 = 810$ and $\chi(\mathbf{D}) = 270 + 2 \times 617 = 1504$ non-zero entries.

A lower bound can be obtained for $\sigma(\mathbf{D})$ and hence $\chi(\mathbf{D})$. For calculating this bound every factor should be considered in its optimal condition. In a cycle list, if a member appears n times, $n(n-1)/2$ units will be added to $\sigma(\mathbf{D})$. As an example, if $n = 2$, then one unit will be added to $\sigma(\mathbf{D})$, and for $n = 3$, three units will be added to $\sigma(\mathbf{D})$.

Consider the graph model of Figure 3.23(a). Using the above argument, $\sigma(\mathbf{D})$ is obtained as follows:

For this graph, $M(S_i) = 20$, $N(S_i) = 9$ and $b_1(S_i) = 12$. If only cycles of length 3 are included in the selected basis, $12 \times 3 = 36$ members should be used. From the existing members, 12 members corresponding to the edges of the cube, each can be present only in one three-sided cycle, and therefore the remaining overlaps $36 - 29 = 16$ should be distributed between $20 - 12 = 8$ members. In the best case, for each member, two overlaps can be allocated, leading to an increase of $\sigma(\mathbf{D})$ by $8 \times \frac{3(3-1)}{2} = 24$ units. The present algorithm leads to a cycle basis with $\sigma(\mathbf{D}) = 24$, which is an optimal cycle basis.

For this example, a lower bound is calculated as follows:

The selected cycles containing $270 \times 3 = 810$ members should be formed on 360 members. The repeated members should be from $810 - 360 = 450$ remaining members. Each of the 36 members on the edges of the graph S can be present only in one 3-sided cycle. Thus $360 - 36 = 324$ members will be left on which repetition can be present. If these members are used twice, then $450 - 324 = 126$ members will be left for which 3 repetitions can be present. Therefore, the minimum value of $\sigma(\mathbf{D})$ will be:

$$\text{Min } \chi(\mathbf{D}) = 324 + [\frac{3(3-1)}{2} - 1] \times 126 = 576.$$

Such a bound may or may not be achievable, since the independence of the cycles should also be satisfied. The present algorithm leads to $\sigma(\mathbf{D}) = 617$ which is close to the ideal condition.

3.4 CONDITIONING OF THE FLEXIBILITY MATRICES

The use of the digital computer for problems in structural analysis requires the solution of a large system of algebraic equations of the form

$$\mathbf{Ax} = \mathbf{b}, \qquad (3\text{-}57)$$

as mentioned at the opening of Chapter 5. This is true for both the force method and the displacement approach. Sometimes the solution of Eq. (3-57) changes greatly on small perturbation in the matrix $\mathbf{A}$. Then we say $\mathbf{A}$ is ill-conditioned with respect to this solution. The accuracy of the solution of Eq. (3-57) can be sensitive to the characteristics of the matrix $\mathbf{A}$. Therefore, it is important to study these characteristics and their interrelationships with the source, propagation and distribution of possible errors. In doing so, better methods of problem formulation must be found, and techniques for predicting, detecting and minimising solution errors must be devised. The ill-conditioning of stiffness matrices for the displacement method of analysis was studied by Shah [197]. In his work, methods were suggested for improving the conditioning of the stiffness matrices. A mathematical investigation of matrix error analysis is due to Rosanoff and Ginsburg [188]. In their work, it was shown that numerically unstable equations may arise in physically stable problems. Thus, the need for routine measurement of matrix conditioning numbers associated with various patterns of formulation is emphasised. The effect of substructuring on the conditioning of stiffness matrices was investigated by Grooms and Rowe [68], who concluded that substructuring does not significantly influence the accuracy of solution of ill-conditioned systems. Filho [52] suggested an orthogonalisation method for the best conditioning of flexibility and stiffness matrices; however, this is an impractical approach for multi-member complex structures.

Optimisation of the conditioning of equilibrium equations when an algebraic force method is employed was studied by Robinson and Haggenmacher [187]. For the combinatorial force method, studies have been limited to increasing the sparsity of cycle basis incidence matrices; see Henderson [76] and Goodspeed and Martin [64] (see also Cassell [21] for a discussion on the latter reference). Recently, methods have been developed for selecting particular types of statical and kinematical bases, leading to flexibility and stiffness matrices that are better conditioned than classical ones; see Kaveh [109].

In structural engineering, one of the important sources of ill-conditioning is the use of members that have widely different stiffnesses (or flexibilities) in a structure. The application of standard statical or kinematical bases (though optimal) leads to ill-conditioned structural matrices. In this chapter, methods are developed for generating special cycle and cutset bases corresponding to statical and kinematical bases, which provide the best possible conditioning for flexibility and stiffness matrices, respectively.

3.4.1 CONDITION NUMBER

In order to measure the conditioning of a matrix, various numbers are defined and employed in practice; see Kaveh [94,113]. The most commonly used condition number is $|\lambda_{max}|/|\lambda_{min}|$, where λ_{max} is the eigenvalue of the largest modulus and λ_{min} is the eigenvalue of the least modulus as defined in the following text:

Eigenvalues and eigenvectors are related to the conditioning of matrices. The ratio of the extreme eigenvalues of a matrix $|\lambda_{max}|/|\lambda_{min}|$ can be taken as its condition number. It can easily be shown that the logarithm to the base 10 of this condition number is roughly proportional to the maximum number of significant figures lost in inversion or in the solution of simultaneous equations. Thus, the number of good digits, g, in the solution, is given by

$$g = p - \log(|\lambda_{max}|/|\lambda_{min}|) = p - PL. \qquad (3\text{-}58)$$

In this relationship, $PL = \log(|\lambda_{max}|/|\lambda_{min}|)$ and p is a number that varies from machine to machine. For example, the IBM/360 uses approximately 8 digits for single-precision and 16 digits for double-precision calculations. It should be mentioned that the above estimate is conservative, and experience shows that PL is one digit on the safe side. The importance of this condition number justifies more explanation and a simple numerical example.

Symmetric matrices can be written as a linear combination of rank-one matrices as

$$\mathbf{A} = \sum_{i=1}^{n} \lambda_i \mathbf{v}_i \mathbf{u}_i^t, \qquad (3\text{-}59)$$

and

$$\mathbf{A}^{-1} = \sum_{i=1}^{n} (1/\lambda_i) \mathbf{v}_i \mathbf{u}_i^t, \qquad (3\text{-}60)$$

with $\mathbf{v}_i^t \mathbf{u}_i = 1$ for $i = 1, ..., n$. In the above equations, λ_i is the ith eigenvalue and $\mathbf{v}_i$ is the corresponding eigenvector of $\mathbf{A}$, and $\mathbf{u}_i$ is the ith eigenvector of $\mathbf{A}^{-1}$. Equation (3-59) shows that the rank-one matrices of the eigenvectors enter the matrix $\mathbf{A}$ in amount proportional to their respective eigenvalues. The lower mode of $\mathbf{A}$ becomes weakly represented as the ratio of the extremal eigenvalues becomes large. Specifically, as a first approximation for each power of 10 in the ratio $|\lambda_{max}|/|\lambda_{min}|$, the lower mode will lose about 1 decimal digit in a finite computer number set representation of the matrix. On the other hand, the lower mode of $\mathbf{A}$ is the upper mode of $\mathbf{A}^{-1}$, because the coefficients of the linear combination (the eigenvalues) are inverted. Therefore, inverting matrices without some feel for their conditioning can lead to wrong solutions. Consider a 2 × 2 matrix such as

$$\mathbf{A} = \begin{bmatrix} 1/9 & 1/10 \\ 1/10 & 1/11 \end{bmatrix} = \begin{bmatrix} 0.11111111 & 0.10000000 \\ 0.10000000 & 0.09090909 \end{bmatrix}.$$

The eigenvalues and eigenvectors of **A** with eight digits are

$$\lambda_1 = 0.20151896 \quad \mathbf{v}_1 = \mathbf{u}_1 = \begin{Bmatrix} 0.74178794 \\ 0.67063452 \end{Bmatrix},$$

$$\lambda_2 = 0.0005012437 \quad \mathbf{v}_2 = \mathbf{u}_2 = \begin{Bmatrix} 0.67063452 \\ -0.74178794 \end{Bmatrix},$$

leading to $\lambda_1/\lambda_2 = 402.0379 = 10^{2.604}$. From Eq. (3-59), matrix **A** can be written as

$$\mathbf{A} = \lambda_1 \mathbf{v}_1 \mathbf{u}_1^t + \lambda_2 \mathbf{v}_2 \mathbf{u}_2^t$$

$$= 0.20151896 \begin{Bmatrix} 0.74178794 \\ 0.67063452 \end{Bmatrix} \{0.74178794 \quad 0.67063452\}$$

$$+ 0.0005012437 \begin{Bmatrix} 0.67063452 \\ -0.74178794 \end{Bmatrix} \{0.67063452 \quad -0.74178794\}$$

$$= \begin{bmatrix} 0.11088567 & 0.10024935 \\ 0.10024935 & 0.090633285 \end{bmatrix} + \begin{bmatrix} 0.00022543467 & -0.00024935298 \\ -0.00024935298 & 0.00027580902 \end{bmatrix}$$

$$= \begin{bmatrix} 0.11111111 & 0.099999997 \\ 0.099999997 & 0.090909094 \end{bmatrix}.$$

In forming this eight-digit approximation to the matrix, the component matrix $\lambda_2 \mathbf{v}_2 \mathbf{u}_2^t$, which has three leading zeros in its elements, is truncated to about five digits. Therefore, an eight-digit representation of the matrix **A** contains about five digits of information about the rank-one matrix $\mathbf{v}_2 \mathbf{u}_2^t$.

Similarly, consider $\mathbf{A}^{-1}$ formed as

$$\mathbf{A}^{-1} = (\frac{1}{\lambda_1}) \mathbf{v}_1 \mathbf{u}_1^t + (\frac{1}{\lambda_2}) \mathbf{v}_2 \mathbf{u}_2^t$$

$$= \begin{bmatrix} 2.7305090 & 2.4685944 \\ 2.4685944 & 2.2318030 \end{bmatrix} + \begin{bmatrix} 897.26951 & -992.46859 \\ -992.46859 & 1097.7681 \end{bmatrix}$$

$$= \begin{bmatrix} 900.00002 & -990.00000 \\ -990.000000 & 1099.9999 \end{bmatrix} \approx \begin{bmatrix} 900 & -990 \\ -990 & 1100 \end{bmatrix}.$$

Notice that the rank of matrix $\mathbf{v}_2 \mathbf{u}_2^t$, which was only available to about five digits in the approximation of $\mathbf{A}$, is the largest component of $\mathbf{A}^{-1}$. One should expect that five digits would be about the most one could obtain by numerically inverting the approximate matrix.

The true inverse can be obtained using rational number arithmetic, and is shown in the above equation to the right of the approximation sign. Using eight-digit arithmetic, the approximate matrix is inverted, yielding

$$\mathbf{A}^{-1} = \begin{bmatrix} 0.11111111 & 0.10000000 \\ 0.10000000 & 0.09090909 \end{bmatrix}^{-1} = \begin{bmatrix} 900.00089 & -990.00099 \\ -990.00099 & 1100.0011 \end{bmatrix}.$$

The poorest terms in this approximate inverse are the off-diagonal terms, which have barely six significant digits. For this matrix,

$$\log_{10} |\lambda_{max}| / |\lambda_{min}| = \log_{10} 402.0379 = 2.604.$$

Therefore, one should expect the approximate inverse to be limited to $8 - 2.6 = 5.4$ good digits. It should be mentioned that for positive definite and symmetric matrices, the calculation of $|\lambda_{max}| / |\lambda_{min}|$ can be carried out by the power method using Rayleigh's Quotient. Since a structural matrix $\mathbf{A}$ is symmetric and positive definite, the convergence of the procedure is ensured and the largest eigenvalue λ_{max} of $\mathbf{A}$ can be easily calculated. The largest eigenvalue of $\mathbf{A}^{-1}$ provides the smallest eigenvalue of $\mathbf{A}$. This method becomes especially simple if the inverse of the matrix is obtained as part of the calculation. However, the inversion of $\mathbf{A}$ can be avoided by using the fact that, if the eigenvalues of $\mathbf{A}$ are $\lambda_{min}, ..., \lambda_{max}$, then the eigenvalues of $c\mathbf{I} - \mathbf{A}$ are $c - \lambda_{min}, ..., c - \lambda_{max}$. Therefore, if constant c is greater than λ_{max}, then the largest eigenvalue of $c\mathbf{I} - \mathbf{A}$ will be $c - \lambda_{min}$. This provides a simple approach for evaluating λ_{min}. Simple computer programs for calculating λ_{min} and λ_{max} of a positive definite matrix are provided in [113].

3.4.2 WEIGHTED GRAPH AND AN ADMISSIBLE MEMBER

The relative stiffnesses (or flexibilities) of members of a structure can be considered as positive integers associated with the members of the graph model of a structure, resulting in a *weighted graph*.

Let S be the model of a frame structure and $\mathbf{k}_m$ denote the stiffness matrix of an element m_i in a global coordinate system selected for the structure. A *weight* can be defined for m_i, using the diagonal entries k_{ii} of $\mathbf{k}_m$, as

$$W(m_i) = \Sigma k_{ii} = 2(\alpha_1 + \alpha_4^z + \alpha_3^z), \quad (3\text{-}61)$$

where
$$\alpha_1 = \frac{EA}{L}, \quad \alpha_3^z = \frac{4EI}{L} \quad \text{and} \quad \alpha_4^z = \frac{12EI}{L^3}.$$

A different weight employing the square roots of the diagonal entries of $\mathbf{k}_{m_i}$ can also be used:

$$W(m_i) = \Sigma \sqrt{k_{ii}} = 2[(\alpha_1)^{1/2} + (\alpha_4^z)^{1/2} + (\alpha_3^z)^{1/2}]. \quad (3\text{-}62)$$

Other weight functions may be defined for representing the relative stiffnesses of the members of S, as appropriate.

Definition: Let the weight of members m_1, m_2, ..., $m_{M(S)}$ be defined by $W(m_1)$, $W(m_2)$, ..., $W(m_{M(S)})$, respectively. A member m_i is called *F-admissible* if

$$W(m_i) \geq \frac{1}{\alpha} \sum_{j=1}^{M(S)} \frac{W(m_j)}{M(S)}, \quad (3\text{-}63)$$

where α is an integer number that can be taken as 2, 3, ... We have used $\alpha = 2$; however, a complete study using other values of α is required. If a member is not F-admissible, it is called *inadmissible* or *S-admissible*.

3.4.3 OPTIMALLY CONDITIONED CYCLE BASES

In order to obtain optimally conditioned flexibility matrices, special statical bases, correspondingly cycle bases possessing particular properties, must be selected.

102 OPTIMAL STRUCTURAL ANALYSIS

A cycle basis is defined as an *optimally conditioned cycle basis* if

(a) it is an optimal cycle basis, that is, the number of non-zero entries of the corresponding cycle adjacency matrix is minimum, leading to a maximal sparsity of the flexibility matrix;

(b) the members of greatest weight of S are included in the overlaps of the cycles, that is, the off-diagonal terms of the corresponding flexibility matrix have the smallest possible magnitudes.

A weighted graph may have more than one optimal cycle basis. The one satisfying condition (b) is optimally conditioned. However, if no such cycle basis exists, then a compromise should be found in satisfying conditions (a) and (b). In other words, a basis should be selected that partially satisfies both conditions. Since there is no algorithm for the formation of an optimal cycle basis, one should look only for a suboptimally conditioned cycle basis.

Example: Consider a 3 × 3 grid as shown in Figure 3.24(a), with the relative weights of the members being encircled. An optimal cycle basis of S, as shown in Fig. 3.24(b), contains nine regional cycles (mesh basis) and corresponds to

$$L_T = \sum_{i=1}^{8} L(C^i \cap C_{i+1}) = 1+1+1+2+2+1+2+2 = 12 \cdot$$

The weight of the members contained in the overlaps is determined as

$$W_T = \sum_{i=1}^{8} W(C^i \cap C_{i+1}) = 2+2+10+12+12+1+3+3 = 45,$$

where L_T and W_T are the *length* and *weight* of the overlaps of the selected cycles, respectively.

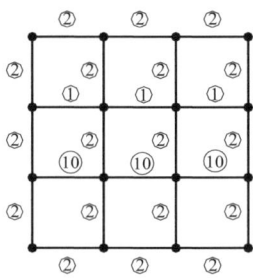

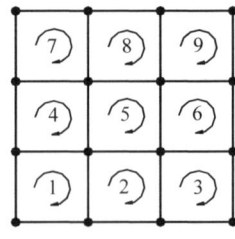

(a) A weighted graph S.   (b) An optimal cycle basis of S.

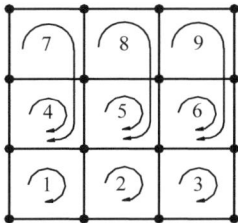

(c) A suboptimally conditioned cycle basis.

Fig. 3.24 A single-layer rigid-jointed grid S.

A suboptimal cycle basis of S is illustrated in Figure 3.24(c), for which

$$L'_{i=1} = \sum_{i=1}^{8} L(C^i \cap C_{i+1}) = 1+1+1+2+2+3+4+4 = 18.$$

The weights of the members contained in the overlaps are calculated as

$$W'_{i=1} = \sum_{i=1}^{8} W(C^i \cap C_{i+1}) = 2+2+10+12+12+14+16+16 = 84.$$

The weight of the overlaps of the selected cycles is considerably increased at the expense of some increase in their lengths, and hence some decrease in the sparsity of their cycle adjacency matrix. Obviously, W_T can be further increased; however, the decrease in sparsity will significantly influence the optimality of the cycle basis.

In this structure, the members of weight one are inadmissible according to the definition of the previous section, since $1 < \frac{1}{2} \times \frac{69}{24} = 1.43$.

3.4.4 FORMULATION OF THE CONDITIONING PROBLEM

The problem of selecting an optimally conditioned cycle basis can be stated in the following mathematical form:

$$\text{Min} \sum_{i=1}^{b_1(\bar{S})-1} L(C^i \cap C_{i+1}), \qquad (3\text{-}64)$$

104 OPTIMAL STRUCTURAL ANALYSIS

and

$$\text{Max} \sum_{i=1}^{b_1(\bar{S})-1} W(C^i \cap C_{i+1}), \qquad (3\text{-}65)$$

where $\bar{S}$ is a contracted S as defined in Section 3.3.4 and $C^i = \bigcup_{j=1}^{i} C_j$.

As can be seen, the problem is a multi-objective optimisation problem, and the following algorithms are designed such that both objective functions are partially satisfied simultaneously.

3.4.5 SUBOPTIMALLY CONDITIONED CYCLE BASES

In this section, three algorithms are developed for the selection of suboptimally conditioned cycle bases of a weighted graph. On each selected cycle, three or six SESs are formed, depending on S being a planar or a space frame, respectively. The condition number of the flexibility matrix corresponding to the selected statical basis is obtained using the methods of Section 3.2.

Algorithm A

This algorithm uses the chords of a special spanning tree to ensure the independence of the selected cycles. In order to avoid the inclusion of inadmissible chords in the intersections of the cycles, such chords are not added to the set of members to be used for generation of the cycles of S.

Step 1: Select the centre "O" of S with a graph or algebraic graph-theoretical method.

Step 2: Generate an SRT using the members of highest weights, that is,

> 2.1 take all members incident with O and assign "1" to the other ends;
>
> 2.2 find all members incident with nodes denoted by "1" and order them in ascending magnitude of their weights;
>
> 2.3 select the tree members from the above ordered members, and assign "2" to the other ends.

Step 3: Repeat Step 2 as many times as needed until all the nodes of S are spanned and an SRT is formed.

Step 4: Order the members incident with "1" in ascending magnitude of weight and use the members of maximal weight as the chord of the first minimal length cycle. If this chord is an F-admissible one, add it to the list of the tree members, and denote this list by T^c.

Step 5: Generate the second shortest length cycle on the second maximal weight member incident with "1" using the members of T^c. Again add the chord to T^c if it is F-admissible. Continue this process until all chords incident with the nodes labelled as "1" are used.

Step 6: Repeat Steps 4 and 5 for all the nodes labelled as "2". Repeat this process sequentially for all the nodes labelled as 3, 4, ..., k, until a basis is selected.

This algorithm generates suboptimally conditioned cycle bases, and has the following advantages compared with the algorithm for generating a fundamental cycle basis:

(a) Starting node at the centre of S: limits the length of the generated cycles.

(b) Employing the used chords in the formation of cycles: reduces the length of the selected cycles.

(c) Forbidding the addition of F-inadmissible chords: prevents the inclusion of weak members in the overlaps of the cycles.

(d) Using members of highest weight in each stage of generating an SRT: leaves the weaker members as chords, which can be excluded because of inadmissibility.

One can select a spanning tree of maximal weight employing the Greedy Algorithm (see Appendix B) in place of an SRT of maximal weight with respect to the centre node of S; however, in general, longer cycles will then be selected, corresponding to a cycle adjacency matrix of less sparsity.

An improvement may be achieved by comparison of the centre node (or nodes) and adjacent nodes to select a node of higher average weight as a starting node. The average weight of a node is taken as the sum of the weights of the members incident with $n_i/\deg n_i$. This improvement is due to the inclusion of all the members of the root node in T^c.

Example: In the following, a simple grid is considered, and the drawback of using a spanning tree of maximal weight compared with an SRT of maximal weight rooted at the centre node O is illustrated; see Figure 3.25. The inadmissible members are shown in dashed lines, and the selected trees are illustrated in bold lines. Use of a spanning tree results in much longer cycles, corresponding to a less sparse cycle adjacency matrix $\mathbf{CC}^t$. This in turn leads to a conditioning of $\mathbf{G}$, which in general is worse than the result obtained by an SRT.

106 OPTIMAL STRUCTURAL ANALYSIS

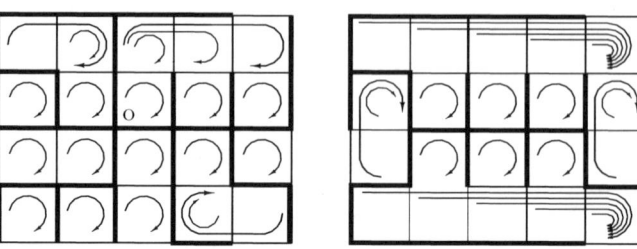

(a) A basis generated using an SRT.   (b) A basis using a spanning tree.

Fig. 3.25 Comparison of two different cycle bases.

Algorithm B

This algorithm is a modified version of the Algorithm 3 presented in Section 3.3.2 for selecting a suboptimal cycle basis of S, in which the relative stiffnesses of the members are also taken into account.

Step 1: Contract S into S' by replacing all paths with nodes of degree 2 by a single member. If a path contains an F-inadmissible member, then the replaced member will also be considered as F-inadmissible.

Step 2: Calculate the IN and CLN of the members of S.

Step 3: Start with a member of the least CLN and generate a minimal weight cycle C_1 on this member. The weight of a cycle in this algorithm is taken to be the sum of the INs of its members.

Step 4: Generate the second admissible cycle of minimal weight C_2 on the next member of the least CLN. If $C_1 \cap C_2$ contains an F-inadmissible member, and $C_1 \oplus C_2$ does not contain such a member, then exchange C_2 with $C_1 \oplus C_2$; otherwise take C_2 as the second cycle of the basis.

Step 5: Subsequently, select the kth admissible smallest weight cycle C_k on an unused member having the least CLN. If $C^{k-1} \cap C_k$ contains an F-inadmissible member, and $C_i \oplus C_j$ does not have such a member, then exchange C_k with $C_k \oplus C_j$; otherwise take C_k as the kth cycle. In the above relationship C_j are the generated cycles adjacent to C_k.

Step 6: The process of Step 5 should be continued as long as the generation of admissible minimal weight cycles is possible. After a member has been used as many times as its IN, before each additional usage, increase the IN of such a member by unity.

Step 7: On an unused member of the least length number, generate one admissible cycle of the smallest weight. This cycle is not a minimal weight cycle, otherwise it would have been selected at Step 4. Such a cycle is known as a subminimal weight cycle. Again a process similar to Step 5 should be performed for possible interchange of the cycle, and the INs should be updated for each additional usage. Now Step 6 should be repeated, since the formation of the new subminimal weight cycle may have altered the admissibility condition of the other cycles, and the selection of further minimal weight cycles may now have become possible.

Step 8: Repeat Step 7, selecting minimal and subminimal weight cycles with the process of combining for better conditioning, until $b_1(S') = b_1(S)$ cycles are generated.

Step 9: A process reverse to that of the contraction performed in Step 1 transforms the selected cycle basis of S' to that of S.

Remark: The idea of exchanging a cycle C_k with $C_k \oplus C_j$ at Step 5 of the above algorithm can be generalised to exchanging C_k with the ring sum of C_k and a linear combination of other cycles of $C^{k-1} \backslash C_k$.

Alternatively, instead of performing cycle exchange in the process of expansion, one may use a similar exchange process after the formation of a cycle basis, in order to increase the weight of the overlaps for the element of the basis to be selected.

The above two types of operations can be collectively performed; however, this approach requires additional computer time and storage and its use should be justified.

The algorithm B is implemented on a PC, and the improvements obtained on the conditioning of the flexibility matrices by using this method are studied through some examples.

3.4.6 EXAMPLES

Example 1: A three-storey frame is considered, as shown in Figure 3.26. Three cases are studied using two types of member properties:

108   OPTIMAL STRUCTURAL ANALYSIS

Type 1 $A_1 = 0.00106$ m^2 $I_1 = 0.00000171$ m^4
Type 2 $A_2 = 0.00970$ m^2 $I_2 = 0.0001961$ m^4.

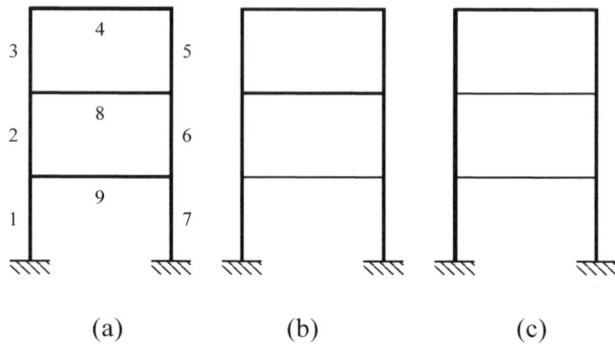

Fig. 3.26 Three-storey frames with different member properties.

The elastic modulus of the material is taken as $E = 2.1 \times 10^8$ kN/m^2, and all the members have $L = 3$ m. Type 1 members are shown in normal lines and type 2 members are illustrated in bold lines.

Algorithm 3 of Section 3.3.2 is applied to these frames, and in all the cases regional cycles are formed as an optimal (minimal) cycle basis. For each cycle, three SESs are generated, and $\mathbf{B}_1$ and the corresponding flexibility matrices $\mathbf{G}$ are formed. The condition numbers for these matrices are obtained as 1.971889, 3.611656, and 3.692658 for frame type a, b and c, respectively.

Algorithm B of Section 3.4.5 selected the following cycles as a suboptimally conditioned cycle basis:

For (a) $C_1 = (7, 9, 1)$, $C_2 = (4, 5, 8, 3)$ and $C_3 = (2, 8, 6, 9)$.
For (b) $C_1 = (7, 9, 1)$, $C_2 = (4, 5, 8, 3)$ and $C_3 = (2, 8, 6, 7, 1)$.
For (c) $C_1 = (7, 9, 1)$, $C_2 = (4, 5, 8, 3)$ and $C_3 = (2, 8, 6, 7, 1)$.

The corresponding flexibility matrices have condition numbers as 1.971889, 4.160444, and 3.883811 for frames type a, b and c, respectively.

Example 2: A two-storey frame with three bays is considered, as shown in Figure 3.27. The same member properties are used and three cases are studied. The calculated condition numbers are obtained as 2.95416, 4.504203 and 4.311532, for type a, b and c, respectively. Algorithm 3 of Section 3.3.2 is used, and the selected cycle bases are optimal for all three cases.

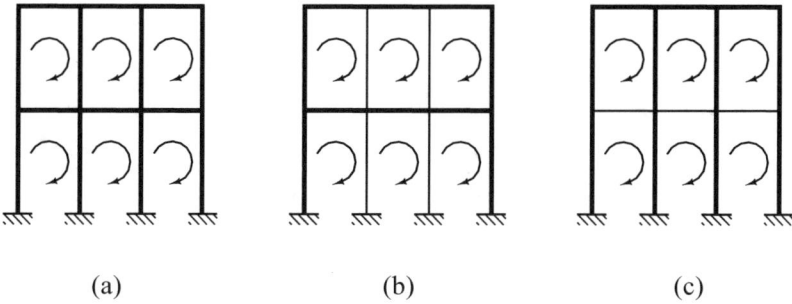

(a) (b) (c)

Fig. 3.27 Two-storey frame with different member types.

Algorithm B of Section 3.4.5 is applied, and the selected cycles for each case are illustrated in Figure 3.28. The corresponding flexibility matrices have the condition numbers as 2.942885, 3.770917 and 3.742143 for type a, b and c, respectively.

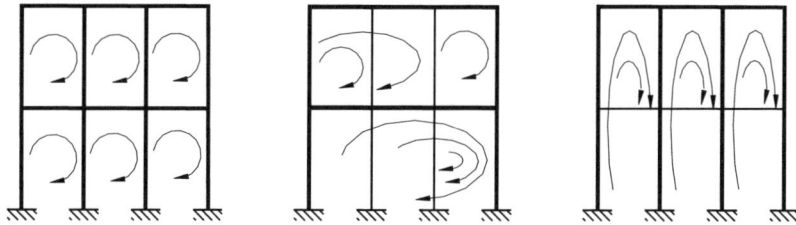

Fig. 3.28 Selected cycle bases using Algorithm B.

The considerable improvement is due to the formation of suboptimal cycle bases used in place of optimal cycle bases. It should be noted that these comparisons are made against the best existing algorithm, since the sparsity itself has a great influence on the conditioning of flexibility matrices.

3.4.7 FORMATION OF $\mathbf{B}_0$ AND $\mathbf{B}_1$ MATRICES

In order to generate the elements of a $\mathbf{B}_0$ matrix, a basic structure of S should be selected. For this purpose, a spanning forest consisting of $NG(S)$ SRTs is used, where $NG(S)$ is the number of ground (support) nodes of S. As an example, for S shown in Figure 3.29(a), two Shortest Route (SR) subtrees are generated; see Figure 3.29(b).

110   OPTIMAL STRUCTURAL ANALYSIS

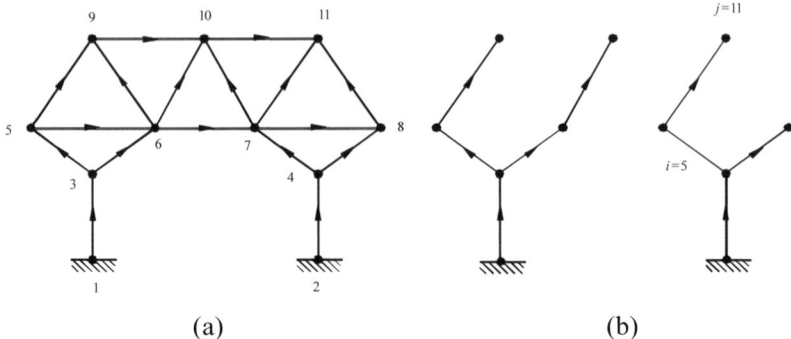

Fig. 3.29 S and two of its SR subtrees.

The orientation assigned to each member of S is from the lower-numbered node to its higher numbered end. For each SR subtree, the orientation is given in the direction of its growth from its support node.

MATRIX $\mathbf{B}_0$: This is a $6M(S) \times 6NL(S)$ matrix, where $M(S)$ and $NL(S)$ are the numbers of members and loaded nodes of S, respectively. If all the free nodes are loaded, then

$$NL(S) = N(S) - NG(S),$$

where $NG(S)$ is the number of support nodes.

For a member, the internal forces are represented by the components at the lower-numbered end. Obviously, the components at the other end can be obtained by considering the equilibrium of the member.

The coefficients of $\mathbf{B}_0$ can be obtained by considering the transformation of each joint load to the ground node of the corresponding subtree. $[\mathbf{B}_0]_{ij}$ for member i and node j is given by a 6×6 submatrix as

$$[\mathbf{B}_0]_{ij} = \alpha_{ij} \begin{bmatrix} 1 & 0 & 0 & 0 & 0 & 0 \\ 0 & 1 & 0 & 0 & 0 & 0 \\ 0 & 0 & 1 & 0 & 0 & 0 \\ 0 & -\Delta z & \Delta y & 1 & 0 & 0 \\ \Delta z & 0 & -\Delta x & 0 & 1 & 0 \\ -\Delta y & \Delta x & 0 & 0 & 0 & 1 \end{bmatrix}, \qquad (3\text{-}66)$$

in which Δx, Δy and Δz are the differences of the coordinates of node j with respect to the lower-numbered end of member i, in the selected global coordinate system, and α_{ij} is the orientation coefficient defined as

$$\alpha_{ij} = \begin{cases} +1 & \text{if member is positively oriented in the tree containing } j, \\ -1 & \text{if member is negatively oriented in the tree containing } j, \\ 0 & \text{if member is not in the tree containing node } j. \end{cases}$$

The $\mathbf{B}_0$ matrix can be obtained by assembling the $[\mathbf{B}_0]_{ij}$ submatrices as shown schematically in the following:

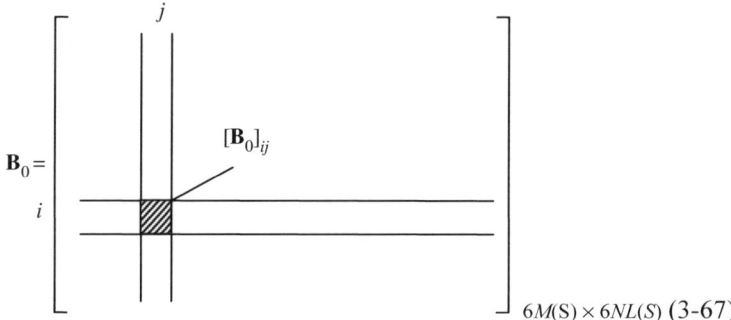

$$6M(S) \times 6NL(S) \quad (3\text{-}67)$$

MATRIX $\mathbf{B}_1$: This is a $6M(S) \times 6b_1(S)$ matrix, which can be formed using the elements of a selected cycle basis. For a space structure, six SESs can be formed on each cycle. Consider C_j and take a member of this cycle as its generator. Cut the generator in the neighbourhood of its beginning node and apply six bi-actions as illustrated in Figure 3.30.

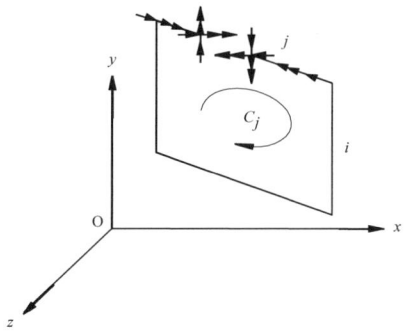

Fig. 3.30 A cycle and the considered bi-action at a cut.

The internal forces under the application of each bi-action are a SES. As for the matrix $\mathbf{B}_0$, a submatrix $[\mathbf{B}_1]_{ij}$ of $\mathbf{B}_1$ is a 6×6 submatrix, the columns of which show the internal forces at the lower-numbered end of member i under the application of six bi-actions at the cut of the generator j,

112   OPTIMAL STRUCTURAL ANALYSIS

$$[\mathbf{B}_1]_{ij} = \beta_{ij} \begin{bmatrix} 1 & 0 & 0 & 0 & 0 & 0 \\ 0 & 1 & 0 & 0 & 0 & 0 \\ 0 & 0 & 1 & 0 & 0 & 0 \\ 0 & -\Delta z & \Delta y & 1 & 0 & 0 \\ \Delta z & 0 & -\Delta x & 0 & 1 & 0 \\ -\Delta y & \Delta x & 0 & 0 & 0 & 1 \end{bmatrix}, \qquad (3\text{-}68)$$

in which Δx, Δy and Δz are the differences of the coordinates x, y and z of the beginning node of the generator j and the beginning node of the member i. The orientation coefficient β_{ij} is defined as

$$\beta_{ij} = \begin{cases} +1 & \text{if member } i \text{ has the same orientation of the cycle generated on } j, \\ -1 & \text{if member } i \text{ has the reverse orientation of the cycle generated on } j, \\ 0 & \text{if member is not in the cycle whose generator is } j. \end{cases}$$

The pattern of $\mathbf{B}_1$ containing $[\mathbf{B}_1]_{ij}$ submatrices is shown below:

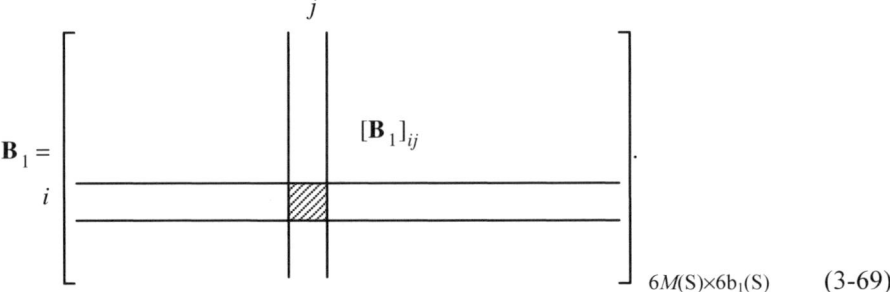

$$6M(S) \times 6b_1(S) \qquad (3\text{-}69)$$

Subroutines for the formation of $\mathbf{B}_0$ and $\mathbf{B}_1$ matrices are included in the program presented in [113].

Example 1: A four by four planar frame is considered as shown in Figure 3.31.

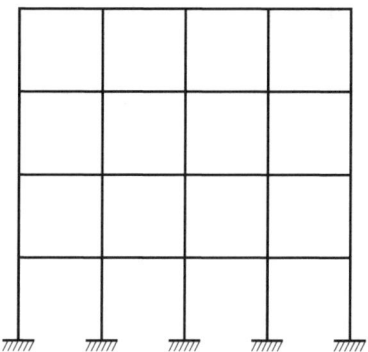

Fig. 3.31 A four by four planar frame S.

The patterns of $\mathbf{B}_1$ and $\mathbf{B}_1^t\mathbf{B}_1$ formed on the elements of the cycle basis selected by any of the methods of Section 3.3.4 are depicted in Figure 3.32, corresponding to $\chi(\mathbf{B}_1) = 241$ and $\chi(\mathbf{B}_1^t\mathbf{B}_1) = 388$.

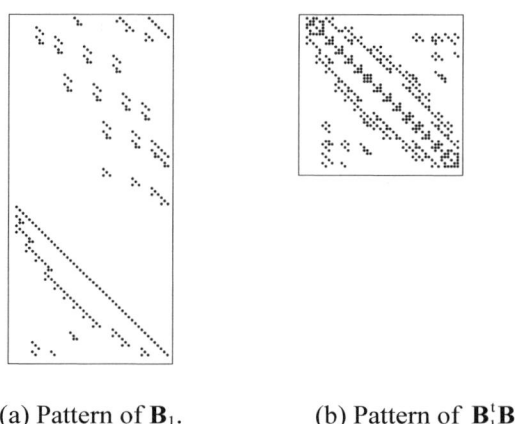

(a) Pattern of $\mathbf{B}_1$. (b) Pattern of $\mathbf{B}_1^t\mathbf{B}_1$.

Fig. 3.32 Patterns of $\mathbf{B}_1$ and $\mathbf{B}_1^t\mathbf{B}_1$ matrices for S.

114   OPTIMAL STRUCTURAL ANALYSIS

Example 2: A one-bay three-storey frame is considered as shown in Figure 3.33.

The patterns of $\mathbf{B}_1$ and $\mathbf{B}_1^t\mathbf{B}_1$ matrices formed on the elements of the cycle basis selected by any of the graph-theoretical algorithms of Section 3.3.4 are shown in Figure 3.34, corresponding to $\chi(\mathbf{B}_1) = 310$ and $\chi(\mathbf{B}_1^t\mathbf{B}_1) = 562$.

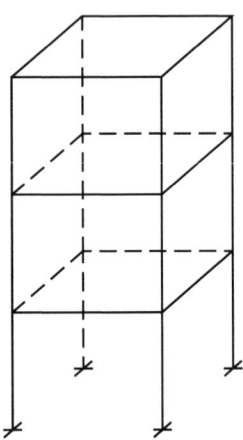

Fig. 3.33 A simple space frame S.

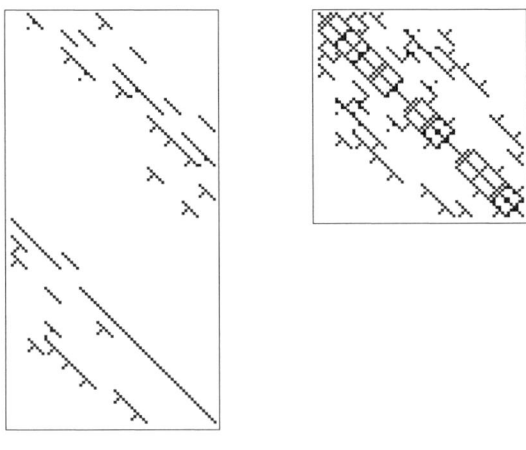

(a) Pattern of $\mathbf{B}_1$.   (b) Pattern of $\mathbf{B}_1^t\mathbf{B}_1$.

Fig. 3.34 Patterns of $\mathbf{B}_1$ and $\mathbf{B}_1^t\mathbf{B}_1$ matrices for S.

Once $\mathbf{B}_0$ and $\mathbf{B}_1$ are computed, the remaining steps of the analysis are the same as those presented in Section 3.2.4. The interested reader may also refer to standard textbooks such as those of McGuire and Gallagher [163], Przemieniecki [184], or Pestel and Leckie [176] for further information.

3.5 GENERALISED CYCLE BASES OF A GRAPH

In this section, S is considered to be a connected graph. For $\gamma(S) = aM(S) + bN(S) + c\gamma_0(S)$, the coefficients b and c are assumed to be integer multiples of the coefficient $a >; 0$. Only those coefficients given in Table 2.1 are of interest.

3.5.1 DEFINITIONS

Definition 1: A subgraph S_i is called an *elementary subgraph* if it does not contain a subgraph $S_i' \subseteq S_i$ with $\gamma(S_i') > 0$. A connected rigid subgraph T of S containing all the nodes of S is called a *γ-tree* if $\gamma(T) = 0$. For $\gamma(S_i) = b_1(S_i)$, a γ-tree becomes a tree in graph theory.

Obviously, a structure whose model is a γ-tree is statically determinate when $\gamma(S)$ describes the DSI of the structure. The ensuing stress resultants can uniquely be determined everywhere in the structure by equilibrium only. Examples of γ-trees are shown in Figure 3.35.

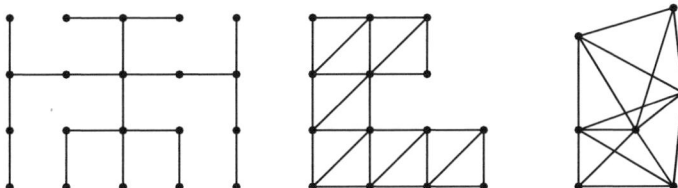

(a) $\gamma(S) = 3M - 3N + 3$.   (b) $\gamma(S) = M - 2N + 3$.   (c) $\gamma(S) = M - 3N + 6$.

Fig. 3.35 Examples of γ-trees.

Notice that $\gamma(T) = 0$ does not guarantee the rigidity of a γ-tree. For example, the graph models depicted in Figure 3.36 both satisfy $\gamma(T) = 0$; however, neither is rigid.

116 OPTIMAL STRUCTURAL ANALYSIS

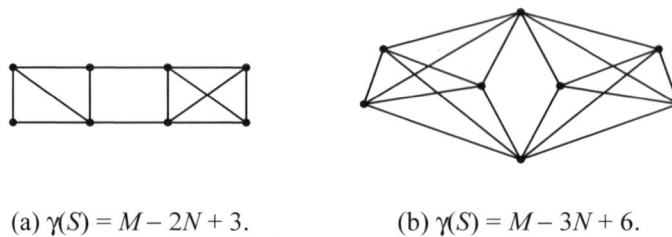

(a) $\gamma(S) = M - 2N + 3$.   (b) $\gamma(S) = M - 3N + 6$.

Fig. 3.36 Structures satisfying $\gamma(T) = 0$ that are not rigid.

Definition 2: A member of $S - T$ is called a γ-*chord* of T. The collection of all γ-chords of a γ-tree is called the γ-*cotree* of S.

Definition 3: A *removable subgraph* S_j of a graph S_i is the elementary subgraph for which $\gamma(S_i - S_j) = \gamma(S_i)$, that is, the removal of S_j from S_i does not alter its DSI. A γ-tree of S containing two chosen nodes, which has no removable subgraph, is called a γ-*path* between these two nodes.

As an example, the graphs shown in Figure 3.37 are γ-paths between the specified nodes n_s and n_t.

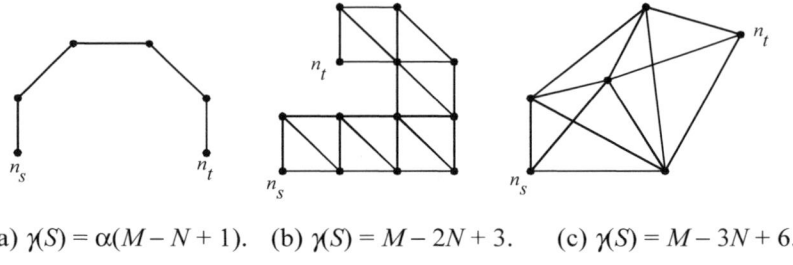

(a) $\gamma(S) = \alpha(M - N + 1)$.   (b) $\gamma(S) = M - 2N + 3$.   (c) $\gamma(S) = M - 3N + 6$.

Fig. 3.37 Examples of γ-paths.

Definition 4: A connected rigid subgraph of S with $\gamma(C_k) = a$, which has no removable subgraph, is termed a γ-*cycle* of S. The total number of members of C_k, denoted by $L(C_k)$, is called the *length* of C_k. Examples of γ-cycles are shown in Figure 3.38.

OPTIMAL FORCE METHOD OF STRUCTURAL ANALYSIS 117

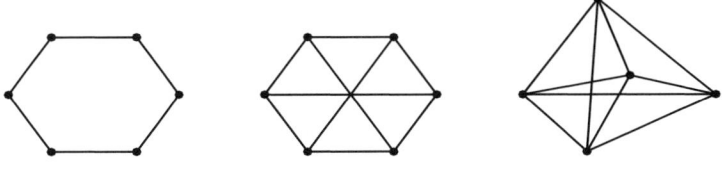

(a) $\gamma(S) = \alpha(M - N + 1)$. (b) $\gamma(S) = M - 2N + 3$. (c) $\gamma(S) = M - 3N + 6$.

Fig. 3.38 Examples of γ-cycles.

Definition 5: Let m_i be a γ-chord of T. Then $T \cup m_i$ contains a γ-cycle C_i, which is defined as a *fundamental γ-cycle* of S with respect to T. Using the Intersection Theorem of Chapter 2, it can easily be shown that

$$\gamma(T \cup m_i) = 0 + (a + 2b + c) - (2b + c) = a,$$

indicating the existence of a γ-cycle in $T \cup m_i$. For a rigid T, the corresponding fundamental γ-cycle is also rigid, since the addition of an extra member between the existing nodes of a graph cannot destroy the rigidity. A fundamental γ-cycle can be obtained by omitting all the removable subgraphs of $T \cup m_i$.

Definition 6: A maximal set of independent γ-cycles of S is defined as a *generalised cycle basis* (GCB) of S. A maximal set of independent fundamental γ-cycles is termed a *fundamental generalised cycle basis* of S. The dimension of such a basis is given be $\eta(S) = \gamma(S)/a$.

As an example, a GCB of a planar truss is illustrated in Figure 3.39.

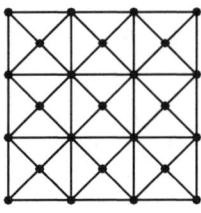

(a) A planar truss S.

118   OPTIMAL STRUCTURAL ANALYSIS

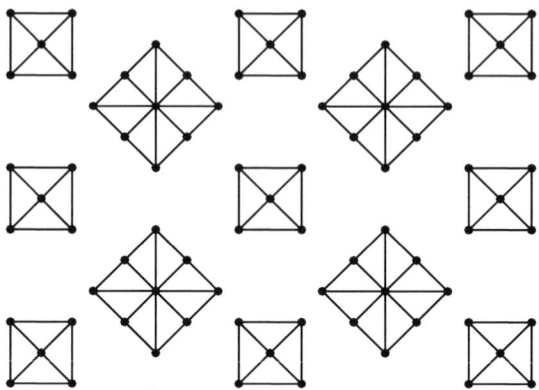

(b) A generalised cycle basis of S.

Fig. 3.39 A planar truss S, and the elements of a GCB of S.

Definition 7: A *generalised cycle basis-member incidence matrix* **C** is an $\eta(S) \times M$ matrix with entries -1, 0 and $+1$, where $c_{ij} = 1$ (or -1) if γ-cycle C_i contains positively (or negatively) oriented member m_j, and $c_{ij} = 0$ otherwise. The *generalised cycle adjacency matrix* is defined as **D**, which is an $\eta(S) \times \eta(S)$ matrix when undirected γ-cycles are considered; then the negative entries of **C** become positive.

3.5.2 MINIMAL AND OPTIMAL GENERALISED CYCLE BASES

A generalised cycle basis $C = \{C_1, C_2, ..., C_{\eta(S)}\}$ is called *minimal* if it corresponds to a minimum value of

$$L(C) = \sum_{i=1}^{\eta(S)} L(C_i). \tag{3-70}$$

Obviously, $\chi(\mathbf{C}) = L(C)$ and a minimal GCB can be defined as a basis that corresponds to minimum $\chi(\mathbf{C})$. A GCB for which $L(C)$ is near minimum is called a *subminimal* GCB of S.

A GCB corresponding to maximal sparsity of the GCB adjacency matrix is called an *optimal* GCB of S. If $\chi(\mathbf{CC}^t)$ does not differ considerably from its minimum value, then the corresponding basis is termed *suboptimal*.

The matrix intersection coefficient $\sigma_i(\mathbf{C})$ of row i of GCB incidence matrix $\mathbf{C}$ is the number of row j such that

(a) $j \in \{i+1, i+2, ..., \eta(S)\}$,

(b) $C_i \cap C_j \neq \emptyset$, that is, there is at least one k such that the column k of both γ-cycles C_i and C_j (rows i and j) contains non-zero entries.

Now it can be shown that

$$\chi(\mathbf{CC}^t) = \eta(S) + 2 \sum_{i=1}^{\eta(S)-1} \sigma_j(\mathbf{C}). \tag{3-71}$$

This relationship shows the correspondence of a GCB incidence matrix $\mathbf{C}$ and that of its GCB adjacency matrix. In order to minimise $\chi(\mathbf{CC}^t)$, the value of $\sum_{i=1}^{\eta(S)-1} \sigma_j(\mathbf{C})$ should be minimised, since $\eta(S)$ is a constant for a given structure S, that is, γ-cycles with a minimum number of overlaps should be selected.

3.6 FORCE METHOD FOR THE ANALYSIS OF PIN-JOINTED PLANAR TRUSSES

The methods described in Section 3.5 are applicable to the selection of generalised cycle bases for different types of skeletal structures. However, the use of these algorithms for trusses engenders some problems, which are discussed in [1]. In this section, two methods are developed for selecting suitable GCBs for planar trusses. In both methods, special graphs are constructed for the original graph model S of a truss, containing all the connectivity properties required for selecting a suboptimal GCB of S.

3.6.1 ASSOCIATE GRAPHS FOR SELECTION OF A SUBOPTIMAL GCB

Let S be the model of a planar truss with triangulated panels, as shown in Figure 3.40. The associate graph of S, denoted by $A(S)$, is a graph whose nodes are in a one-to-one correspondence with the triangular panels of S, and two nodes of $A(S)$ are connected by a member if the corresponding panels have a common member in S.

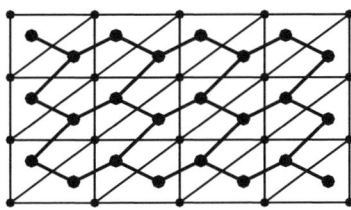

Fig. 3.40 A planar truss S and its associate graph $A(S)$.

120 OPTIMAL STRUCTURAL ANALYSIS

If S has some cut-outs, as shown in Figure 3.41, then its associate graph can still be formed, provided each cut-out is surrounded by triangulated panels.

For trusses containing adjacent cut-outs, a cut-out with cut-nodes in its boundary, or any other form violating the above-mentioned condition, extra members can be added to S. The effect of such members should then be included in the process of generating its SESs.

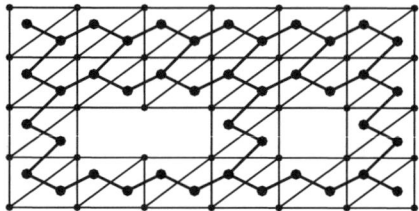

Fig. 3.41 S with two cut-outs and its $A(S)$.

Theorem A: For a fully triangulated truss (except for the exterior boundary), as in Figure 3.40, the dimension of a statical basis $\gamma(S)$ is equal to the number of its internal nodes, which is the same as the first Betti number of its associate graph, that is,

$$\gamma(S) = N_i(S) = b_1[A(S)]. \tag{3-72}$$

Proof: Let M' and N' be the number of members and nodes of $A(S)$, respectively. By definition,

$$N' = R(S) - 1,$$

and $\quad M' = M_i(S) = M(S) - M_e(S) = M(S) - N_e(S) = M(S) - [N(S) - N_i(S)].$

Thus, $b_1[A(S)] = M' - N' + 1 =$

$$M(S) - [N(S) - N_i(S)] - R(S) + 1 + 1 = 2 - R(S) + M(S) - N(S) + N_i(S).$$

By Euler's polyhedron formula, we have

$$2 - R(S) + M(S) - N(S) = 0.$$

Therefore,

$$b_1[A(S)] = N_i(S)$$

For trusses that are not fully triangulated, as described in Chapter 2, we have

$$\gamma(S) = N_i(S) - M_c(S).$$

A Cycle of $A(S)$ and the Corresponding γ-Cycle of S: In Figure 3.42(a), a triangulated truss and its associate graph, which is a cycle, are shown for which

$$\gamma(S_i) = N_i = 1 = b_1[A(S)].$$

Since C_1 of $A(S)$ corresponds to one γ-cycle of S, it is called a *type I cycle*, denoted by C_I. A γ-cycle of S is shown with continuous lines, and its γ-chords are depicted with dashed lines.

Figure 3.42(b) shows a truss unit with one cut-out. In general, if a cut-out is an m-polygon, then the completion of the triangulation requires $m - 3$ members. Instead, m internal nodes will be created, increasing the DSI by m. Hence, Eq. (3-72) yields

$$\gamma(S) = m - (m - 3) = 3,$$

while $\qquad b_1[A(S)] = 1.$

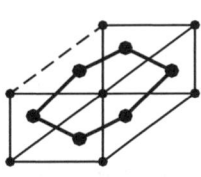 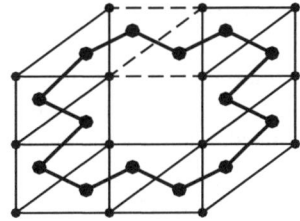

(a) A type C_I cycle. (b) A type C_{III} cycle.

Fig. 3.42 Two different types of cycles.

However, in this case, S contains three γ-cycles. A γ-path P and three γ-chords (dashed lines) are depicted in Figure 3.42(b). Obviously, $P \cup m_i$ ($i = 1, 2, 3$) forms three γ-cycles, which correspond to a cycle of type C_{III} of $A(S)$. Thus, two types of cycles, C_I and C_{III}, should be recognised in $A(S)$, and an appropriate number of γ-cycles will then be generated.

Algorithm AA

Step 1: Construct the associate graph $A(S)$ of S.

Step 2: Select a mesh basis of $A(S)$ using an appropriate cycle selection algorithm. For fully triangulated S, Algorithms 1 to 3 (Section 3.3.2) generate cycle bases with three-sided elements.

Step 3: Select the γ-cycles of S corresponding to the cycles of $A(S)$. One γ-cycle for each cycle of type C_I and three γ-cycles for each cycle of type C_III should be chosen.

Once a GCB is selected, on each γ-cycle one SES can easily be formed. Therefore, a statical basis with localised SESs will be obtained.

Example: Let S be the graph model of a planar truss, as shown in Figure 3.41, for which $\gamma(S) = 11$. For $A(S)$, five cycles of length 6 of type C_I, and two cycles of lengths 14 and 18 of type C_III are selected. Therefore, a total of $5 + 3 \times 2 = 11$ γ-cycles of S is obtained. On each cycle of type C_I, one SES and on each cycle of type C_III three SESs are constructed, and a statical basis consisting of localised SESs is thus obtained.

3.6.2 MINIMAL GCB OF A GRAPH

Theoretically, a minimal GCB of a graph can be found using the Greedy Algorithm developed for matroids. This will be discussed in Appendix B after matroids have been introduced, and only the algorithm is briefly outlined here.

Consider the graph model of a structure, and select all of its γ-cycles. Order the selected γ-cycles in ascending order of length. Denote these cycles by a set C. Then perform the following steps:

Step 1: Choose a γ-cycle C_1 of the smallest length, that is, $L(C_1) \leq L(C_i)$ for all $C_i \in C$.

Step 2: Select the second γ-cycle C_2 from $C - \{C_1\}$ that is independent of C_1 and $L(C_2) \leq L(C_i)$ for all γ-cycles of $C - \{C_1\}$.

Step 3: Subsequently, choose a γ-cycle C_k from $C - \{C_1, C_2, ..., C_{k-1}\}$ that is independent of $C_1, C_2, ..., C_{k-1}$ and $L(C_k) \leq (C_i)$ for all $C_i \in C - \{C_1, C_2, ..., C_{k-1}\}$.

After $\eta(S)$ steps, a minimal GCB will be selected by this process, a proof of which can be found in Kaveh [113].

3.6.3 SELECTION OF A SUBMINIMAL GCB: PRACTICAL METHODS

In practice, three main difficulties are encountered in an efficient implementation of the Greedy Algorithm. These difficulties are briefly mentioned in the following:

1. Selection of some of the γ-cycles for some $\gamma(S)$ functions
2. Formation of all of the γ-cycles of S
3. Checking the independence of γ-cycles.

In order to overcome the above difficulties, various methods are developed. The bases selected by these approaches correspond to very sparse GCB adjacency matrices, although these bases are not always minimal.

Method 1

This is a natural generalisation of the method for finding a fundamental cycle basis of a graph and consists of the following steps:

Step 1: Select an arbitrary γ-tree of S, and find its γ-chords.

Step 2: Add one γ-chord at a time to the selected γ-tree to form fundamental γ-cycles of S with respect to the selected γ-tree.

The main advantage of this method is the fact that the independence of γ-cycles is guaranteed by using a γ-tree. However, the selected γ-cycles are often quite long, corresponding to highly populated Generalised Cycle Basis (GCB) adjacency matrices.

Method 2

This is an improved version of Method 1, in which a special γ-tree has been employed and each γ-chord is added to γ-tree members after being used for the formation of a fundamental γ-cycle.

Step 1: Select the centre "O" of the given graph. Methods for selecting such a node will be discussed in Chapter 5.

Step 2: Generate an SR γ-tree rooted at the selected node O and order its γ-chords according to their distance from O. The *distance* of a member is taken as the sum of the shortest paths between its end nodes and O.

Step 3: Form a γ-cycle on the γ-chord of the smallest distance number and add the used γ-chord to the tree members, that is, form $T \cup m_1$.

124 OPTIMAL STRUCTURAL ANALYSIS

Step 4: Form the second γ-cycle on the next nearest γ-chord to O, by finding a γ-path in $T \cup m_1$ (not through m_2). Then add the second used γ-chord m_2 to $T \cup m_1$ obtaining $T \cup m_1 \cup m_2$.

Step 5: Subsequently form the kth γ-cycle on the next unused γ-chord nearest to O, by finding a γ-path in the $T \cup m_1 \cup m_2 \cup ... \cup m_{k-1}$ (not through m_k). Such a γ-path together with m_k forms a γ-cycle.

Step 6: Repeat Step 5 until $\eta(S)$ of γ-cycles are selected.

Addition of the used γ-chords to the γ-tree members leads to a considerable reduction in the length of the selected γ-cycles, while maintaining the simplicity of the independence check.

In this method, the use of an SRT orders the nodes and members of the graph. Such an ordering leads to fairly banded member-node incidence matrices. Considering the columns corresponding to tree members as independent columns, a base is effectively selected for the cycle matroid of the graph; see Kaveh [94,112].

Method 3

This method uses an expansion process, at each step of which one independent γ-cycle is selected and added to the previously selected ones. The independence is secured using an admissibility condition defined as follows:

A γ-cycle C_{k+1} added to the previous selected γ-cycles $C^k = C_1 \cup C_2 \cup ... \cup C_k$ is called *admissible* if

$$C^k \cup C_{k+1}) = \gamma(C^k) + a, \qquad (3\text{-}73)$$

where "a" is the coefficient defined in Table 2.1. The method can now be described as follows:

Step 1: Select the first γ-cycle of minimal length C_1.

Step 2: Select the second γ-cycle of minimal length C_2 that is independent of C_1, that is, select the second admissible γ-cycle of minimal length.

Step k: Subsequently, find the kth admissible γ-cycle of minimal length. Continue this process until $\eta(S)$ independent γ-cycles forming a subminimal GCB are obtained.

A γ-cycle of minimal length can be generated on an arbitrary member by adding a γ-path of minimal length between the two end nodes of the member (not through the member itself). The main advantage of this method is that of avoiding the formation of all γ-cycles of S and also the independence control, which becomes feasible by graph-theoretical methods.

The above methods are elaborated for specific $\gamma(S)$ functions in subsequent sections, and examples are included to illustrate their simplicity and efficiency.

3.7 FORCE METHOD OF ANALYSIS FOR GENERAL STRUCTURES

Combinatorial methods for the force method of structural analysis have been presented in previous sections. These methods are very efficient for skeletal structures and, in particular, for rigid-jointed frames. For a general structure, the underlying graph or hypergraph of a SES has not yet been properly defined, and much research is still to be done. Algebraic methods, on the other hand, can be formulated in a more general form to cover different types of structures such as skeletal structures and finite element models. The main drawbacks of these methods are the larger storage requirements and the higher number of operations.

These difficulties can be overcome partially by employing combinatorial approaches within the algebraic methods, whenever such tools are available and their use can lead to simplifications.

3.7.1 FLEXIBILITY MATRICES OF FINITE ELEMENTS

In this section, the force–displacement relationship is established for a family of finite elements, namely plane stress and plain strain problems. Triangular and rectangular elements are considered with constant and linearly varying stress fields, respectively.

Constant Stress Triangular Element: For this element, the nodal forces in a global coordinate system have six components, as shown in Figure 3.43(a). The element forces are taken as natural forces acting along the sides of the triangle, as shown in Figure 3.43(b); see Argyris et al. [6].

126 OPTIMAL STRUCTURAL ANALYSIS

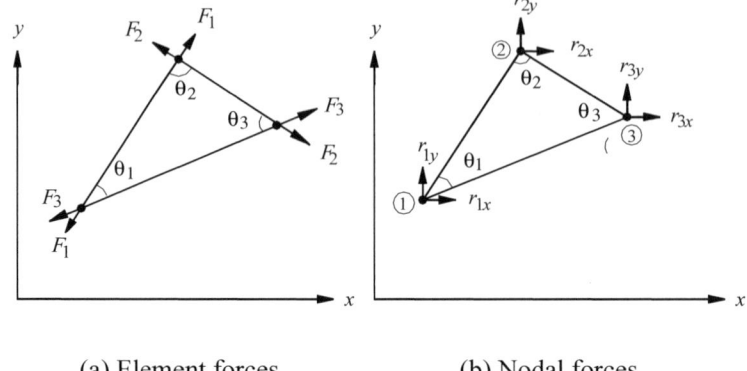

(a) Element forces.   (b) Nodal forces.

Fig. 3.43 A triangular element.

The nodal forces and element forces are related by projection as

$$\begin{bmatrix} r_{1x} \\ r_{1y} \\ r_{2x} \\ r_{2y} \\ r_{3x} \\ r_{3y} \end{bmatrix} = \begin{bmatrix} -l_{12} & 0 & l_{31} \\ -m_{12} & 0 & m_{31} \\ l_{12} & -l_{23} & 0 \\ m_{12} & -m_{23} & 0 \\ 0 & l_{23} & -l_{31} \\ 0 & m_{23} & -m_{31} \end{bmatrix} \begin{bmatrix} F_1 \\ F_2 \\ F_3 \end{bmatrix}, \qquad (3\text{-}74)$$

where l_{ij} and m_{ij} are the direction cosines of the side ij of the triangle.

The element forces are now related to stress resultants; see Figure 3.44. First F_1 is considered as the only natural force acting on the element and the internal stresses are calculated as follows:

$$y_{23}\sigma_x + x_{32}\sigma_{xy} = \frac{2l_{12}}{t}F_1 \qquad (3\text{-}75\text{a})$$

$$-y_{31}\sigma_x + x_{31}\sigma_{xy} = \frac{2l_{12}}{t}F_1 \qquad (3\text{-}75\text{b})$$

$$y_{31}\sigma_y + x_{31}\sigma_{xy} = \frac{2m_{12}}{t}F_1 \qquad (3\text{-}75\text{c})$$

OPTIMAL FORCE METHOD OF STRUCTURAL ANALYSIS   127

Solution of Eqs. (3-75) is obtained as follows:

$$\sigma_x = \frac{2l_{12}^2}{th_3}F_1, \sigma_y = \frac{2m_{12}^2}{th_3}F_1, \text{ and } \sigma_{xy} = \frac{2l_{12}m_{12}}{th_3}F_1, \quad (3\text{-}76)$$

where

$$\begin{cases} x_{ij} = x_i - x_j \\ y_{ij} = y_i - y_j \end{cases} \text{ for } i,j = 1,2,3.$$

In the above relations, h_3 is the height of the triangle corresponding to corner 3. Permutation of the indices results in the stresses produced by $\mathbf{F}_2$ and $\mathbf{F}_3$, and in matrix form these equations can be collectively written as

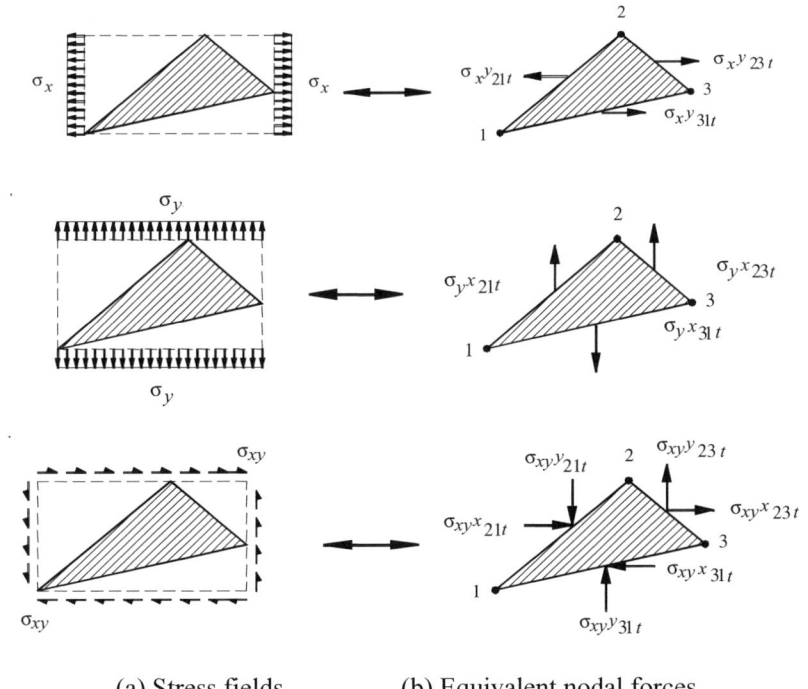

(a) Stress fields.   (b) Equivalent nodal forces.

Fig. 3.44 The stress fields and their equivalent nodal forces.

128 OPTIMAL STRUCTURAL ANALYSIS

$$\begin{bmatrix} \sigma_x \\ \sigma_y \\ \sigma_{xy} \end{bmatrix} = \frac{2}{t} \begin{bmatrix} \dfrac{l_{12}^2}{h_3} & \dfrac{l_{23}^2}{h_1} & \dfrac{l_{31}^2}{h_2} \\ \dfrac{m_{12}^2}{h_3} & \dfrac{m_{23}^2}{h_1} & \dfrac{m_{31}^2}{h_2} \\ \dfrac{m_{12} l_{12}}{h_3} & \dfrac{m_{23} l_{23}}{h_1} & \dfrac{m_{31} l_{31}}{h_2} \end{bmatrix} \begin{bmatrix} F_1 \\ F_2 \\ F_3 \end{bmatrix}, \qquad (3\text{-}77)$$

or
$$\sigma = \overline{c}\, F. \qquad (3\text{-}78)$$

The matrix $\overline{c}$ represents statically equivalent stress system due to unit force **F**. The flexibility matrix of the element can be written as

$$\mathbf{F}_m = \int_V \overline{c}^t \phi\, \overline{c}\, dV. \qquad (3\text{-}79)$$

The integration is taken over the volume of the element, where

$$\phi = \frac{1}{E}\begin{bmatrix} 1 & -v & 0 \\ -v & 1 & 0 \\ 0 & 0 & 2(1+v) \end{bmatrix}, \qquad (3\text{-}80)$$

is the matrix relating the stresses to strains, $\varepsilon = \phi\sigma$, in plane stress problems, and E and v are the Young's modulus and Poisson's ratio, respectively. The force–displacement relationship for a triangular element becomes

$$\mathbf{u}_m = \mathbf{f}_m\, \mathbf{r}_m, \qquad (3\text{-}81)$$

where $\mathbf{u}_m$ and $\mathbf{r}_m$ are the element displacements and element forces, respectively. The flexibility matrix of the element can now be written as

$$\mathbf{f}_m = \frac{2}{Et}\begin{bmatrix} A(\theta_3,\theta_2,\theta_1) & B(\theta_2) & B(\theta_1) \\ B(\theta_2) & A(\theta_{13},\theta_2,\theta_3) & B(\theta_3) \\ B(\theta_1) & B(\theta_3) & A(\theta_2,\theta_1,\theta_3) \end{bmatrix}, \qquad (3\text{-}82)$$

where t is the thickness of the element, and A and B are functions defined as follows:

$$A(\theta_i, \theta_j, \theta_k) = \frac{\sin\theta_i}{\sin\theta_j \sin\theta_k}, (i, j, k = \text{permutation of } 1,2,3), \quad (3\text{-}83a)$$

$$B(\theta_i) = \cos\theta_i \cot\theta_i - \nu\sin\theta_i, (i = 1, 2, 3), \quad (3\text{-}83b)$$

where θ_i, θ_j and θ_k are the angles of the triangle.

Linear Stress Rectangular Element: For this element, the nodal forces in a global coordinate system have eight components, as shown in Figure 3.45(a). The element forces are taken as natural forces along the sides and one diagonal, as shown in Figure 3.45(b). The nodal forces and element forces are related, similar to the triangular element, as,

$$\begin{bmatrix} r_{1x} \\ r_{1y} \\ r_{2x} \\ r_{2y} \\ r_{3x} \\ r_{3y} \\ r_{4x} \\ r_{4y} \end{bmatrix} = \begin{bmatrix} -1 & -\Omega & 0 & 0 & 0 \\ 0 & -\beta\Omega & -1 & 0 & 0 \\ 1 & 0 & 0 & 0 & 0 \\ 0 & 0 & 0 & 0 & -1 \\ 0 & 0 & 0 & -1 & 0 \\ 0 & 0 & 1 & 0 & 0 \\ 0 & \Omega & 0 & 1 & 0 \\ 0 & -\beta\Omega & 0 & 0 & 1 \end{bmatrix} \begin{bmatrix} F_1 \\ F_2 \\ F_3 \\ F_4 \\ F_5 \end{bmatrix}, \quad (3\text{-}84)$$

where: $\beta = \dfrac{b}{a}$ and $\Omega = \dfrac{1}{\sqrt{1+\beta^2}}$.

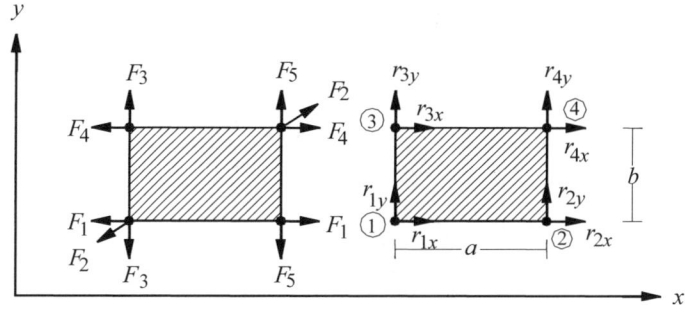

(a) Element forces. (b) Nodal forces.

Fig. 3.45 A rectangular element.

130 OPTIMAL STRUCTURAL ANALYSIS

Considering Figure 3.46, for this element the plane stresses are written as,

$$\begin{cases} \sigma_x = c_1 + c_2 \eta \\ \sigma_y = c_3 + c_4 \xi \\ \sigma_{xy} = c_5 \end{cases} \qquad (3\text{-}85)$$

where $c_1, c_2, \ldots, c_5$ are constants and,

$$\xi = \frac{x}{a} \quad \text{and} \quad \eta = \frac{y}{b},$$

a and b being the length and width of the element, respectively.

$$\mathbf{f}_m = \frac{1}{Et} \begin{bmatrix} 4\beta & -v & \dfrac{\beta^2 - v}{\sqrt{1+\beta^2}} & -2\beta & -v \\ -v & 4/\beta & \dfrac{1-v\beta^2}{\sqrt{1+\beta^2}} & -v & -2/\beta \\ \dfrac{\beta^2 - v}{\sqrt{1+\beta^2}} & \dfrac{1-v\beta^2}{\sqrt{1+\beta^2}} & \dfrac{1+\beta^2}{\beta} & \dfrac{\beta^2 - v}{\sqrt{1+\beta^2}} & \dfrac{1-v\beta^2}{\beta\sqrt{1+\beta^2}} \\ -2\beta & -v & \dfrac{\beta^2 - v}{\sqrt{1+\beta^2}} & 4\beta & -v \\ -v & -2/\beta & \dfrac{1-v\beta^2}{\beta\sqrt{1+\beta^2}} & -v & 4/\beta \end{bmatrix}. \qquad (3\text{-}86)$$

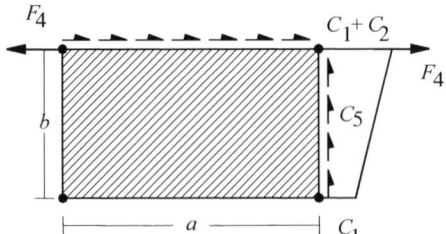

Fig. 3.46 The stress fields and their equivalent nodal forces.

The unassembled flexibility matrix of the structure can now be formed by using the above matrix for each element as block diagonal entries. This matrix is incorporated in the algebraic force method of the next section.

The element flexibility matrices for other elements, such as constant stress tetrahedron elements, higher-order plane stress and plane strain elements and triangular and rectangular plate bending elements, can be similarly formulated; see Przemieniecki [184].

3.7.2 ALGEBRAIC METHODS

Consider a discrete or discretised structure S, which is assumed to be statically indeterminate. Let $\mathbf{r}$ denote the m-dimensional vector of generalised independent element (member) forces, and $\mathbf{p}$ the n-vector of nodal loads. The equilibrium conditions of the structure can then be expressed as

$$\mathbf{Ar} = \mathbf{p}, \tag{3-87}$$

where $\mathbf{A}$ is an $n \times m$ *equilibrium matrix*. The structure is assumed to be rigid, and therefore $\mathbf{A}$ has a full rank, that is, $t = m - n > 0$, and rank $\mathbf{A} = n$.

The member forces can be written as,

$$\mathbf{r} = \mathbf{B}_0 \mathbf{p} + \mathbf{B}_1 \mathbf{q}, \tag{3-88}$$

where $\mathbf{B}_0$ is an $m \times n$ matrix such that $\mathbf{AB}_0$ is an $n \times n$ identity matrix and $\mathbf{B}_1$ is an $m \times t$ matrix such that $\mathbf{AB}_1$ is an $n \times t$ zero matrix. $\mathbf{B}_0$ and $\mathbf{B}_1$ always exist for a structure, and in fact many of them can be found for a structure. $\mathbf{B}_1$ is called a *self-stress matrix* as well as a *null basis matrix*. Each column of $\mathbf{B}_1$ is known as a *null vector*. Notice that the null space, null basis and null vectors correspond to complementary solution space, statical basis and SESs, respectively, when S is taken as a general structure.

Minimisation of the potential energy requires that $\mathbf{r}$ minimise the quadratic form,

$$\tfrac{1}{2} \mathbf{r}^t \mathbf{F}_m \mathbf{r}, \tag{3-89}$$

subject to the constraint as in Eq. (3-87). $\mathbf{F}_m$ is an $m \times m$ block diagonal element flexibility matrix. Using Eq. (3-88), it can be seen that $\mathbf{q}$ must satisfy the following equation:

$$(\mathbf{B}_1^t \mathbf{F}_m \mathbf{B}_1)\mathbf{q} = -\mathbf{B}_1^t \mathbf{F}_m \mathbf{B}_0 \mathbf{p}, \tag{3-90}$$

where $\mathbf{B}_1^t \mathbf{F}_m \mathbf{B}_1 = \mathbf{G}$ is the *overall flexibility matrix of the structure*. Computing the redundant forces $\mathbf{q}$ from Eq. (3-43), $\mathbf{r}$ can be found using Eq. (3-3). The structure of $\mathbf{G}$ is again important, and its sparsity, bandwidth and conditioning govern the efficiency of the force method. For the sparsity of $\mathbf{G}$, one can search for a sparse $\mathbf{B}_1$ matrix, which is often referred to as the sparse null basis problem.

Many algorithms exist for computing a null basis $\mathbf{B}_1$ of a matrix $\mathbf{A}$. For the moment, let $\mathbf{A}$ be partitioned so that

$$\mathbf{AP} = [\mathbf{A}_1, \mathbf{A}_2], \tag{3-91}$$

where $\mathbf{A}_1$ is $n \times n$ and non-singular, and $\mathbf{P}$ is a permutation matrix that may be required in order to ensure that $\mathbf{A}_1$ is non-singular. One can write,

$$\mathbf{B}_1 = \mathbf{P} \begin{bmatrix} -\mathbf{A}_1^{-1} \mathbf{A}_2 \\ \mathbf{I} \end{bmatrix}. \tag{3-92}$$

By simple multiplication it becomes obvious that

$$\mathbf{AB}_1 = [\mathbf{A}_1 \quad \mathbf{A}_2] \begin{bmatrix} -\mathbf{A}_1^{-1} \mathbf{A}_2 \\ \mathbf{I} \end{bmatrix} = \mathbf{0}.$$

A permutation $\mathbf{P}$ that yields a non-singular $\mathbf{A}_1$ matrix can be chosen purely symbolically, but this says nothing about the possible numerical conditioning of $\mathbf{A}_1$ and the resulting $\mathbf{B}_1$.

In order to control the numerical conditioning, pivoting must be employed. There are many such methods based on various matrix factorisations, including the Gauss–Jordan elimination, **QR**, **LU**, **LQ** and Turn-back methods. Some of these methods are briefly studied in the following text.

Gauss–Jordan Elimination Method: In this approach, one creates an $n \times n$ identity matrix $\mathbf{I}$ in the first columns of $\mathbf{A}$ by column changes and a sequence of n pivots. This procedure can be expressed as

$$\mathbf{G}_n \mathbf{G}_{n-1} \ldots \mathbf{G}_2 \mathbf{G}_1 \mathbf{AP} = [\mathbf{I}, \mathbf{M}], \tag{3-93}$$

where $\mathbf{G}_i$ is the ith pivot matrix and $\mathbf{P}$ is an $m \times m$ column permutation matrix (so $\mathbf{P}^t = \mathbf{P}$) and $\mathbf{I}$ is an $n \times n$ identity matrix, and $\mathbf{M}$ is an $n \times t$ matrix. Denoting $\mathbf{G}_n \mathbf{G}_{n-1} \ldots \mathbf{G}_2 \mathbf{G}_1$ by $\mathbf{G}$ we have,

$$\mathbf{GAP} = [\mathbf{I}, \mathbf{M}], \tag{3-94}$$

or

$$AP = G^{-1}[I, M] = [G^{-1}, G^{-1}M], \qquad (3-95)$$

which can be regarded as the Gauss–Jordan factorisation of **A**, and

$$B_0 = \bar{P} \begin{bmatrix} G \\ 0 \end{bmatrix} \quad \text{and} \quad B_1 = \bar{P} \begin{bmatrix} -M \\ I \end{bmatrix}. \qquad (3-96)$$

Example 1: The four by four planar frame of Figure 3.31 is reconsidered. The patterns of B_1 and $B_1^t B_1$ formed by the Gauss–Jordan elimination method are depicted in Figure 3.47, corresponding to $\chi(B_1) = 491$ and $\chi(B_1^t B_1) = 1342$.

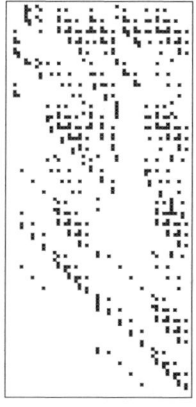

(a) Pattern of B_1. (b) Pattern of $B_1^t B_1$.

Fig. 3.47 Patterns of B_1 and $B_1^t B_1$ matrices for S.

Example 2: The three-storey frame of Figure 3.26 is reconsidered, and the Gauss–Jordan elimination method is used. The patterns of B_1 and $B_1^t B_1$ matrices formed are shown in Figure 3.48, corresponding to $\chi(B_1) = 483$ and $\chi(B_1^t B_1) = 1592$.

134   OPTIMAL STRUCTURAL ANALYSIS

(a) Pattern of $\mathbf{B}_1$.   (b) Pattern of. $\mathbf{B}_1^t\mathbf{B}_1$

Fig. 3.48 Patterns of $\mathbf{B}_1$ and $\mathbf{B}_1^t\mathbf{B}_1$ matrices for S.

LU Decomposition Method: Using the **LU** decomposition method, one obtains the **LU** factorisation of **A** as

$$\mathbf{PA} = \mathbf{LU} \text{ and } \mathbf{U}\overline{\mathbf{P}} = [\mathbf{U}_1, \mathbf{U}_2], \tag{3-97}$$

P and $\overline{\mathbf{P}}$ are again permutation matrices of order $n \times n$ and $m \times m$, respectively. Now $\mathbf{B}_0$ and $\mathbf{B}_1$ can be written as

$$\mathbf{B}_0 = \overline{\mathbf{P}}\begin{bmatrix} \mathbf{U}_1^{-1}\mathbf{L}^{-1}\mathbf{P} \\ \mathbf{0} \end{bmatrix} \text{ and } \mathbf{B}_1 = \overline{\mathbf{P}}\begin{bmatrix} -\mathbf{U}_1^{-1}\mathbf{U}_2 \\ \mathbf{I} \end{bmatrix}. \tag{3-98}$$

Example 1: The four by four planar frame of Figure 3.31 is reconsidered. The patterns of $\mathbf{B}_1$ and $\mathbf{B}_1^t\mathbf{B}_1$ formed by the **LU** factorisation method are depicted in Figure 3.49. The sparsity for the corresponding matrices are $\chi(\mathbf{B}_1) = 408$ and $\chi(\mathbf{B}_1^t\mathbf{B}_1) = 1248$.

OPTIMAL FORCE METHOD OF STRUCTURAL ANALYSIS   135

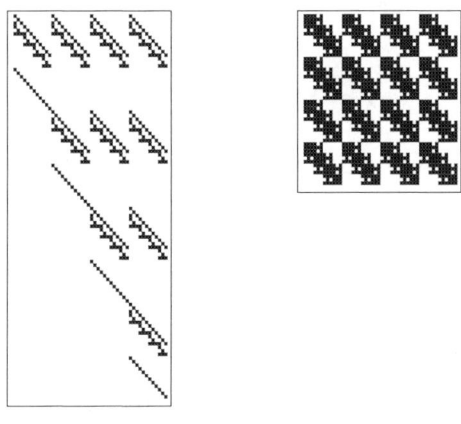

(a) Pattern of $\mathbf{B}_1$.   (b) Pattern of $\mathbf{B}_1^t\mathbf{B}_1$.

Fig. 3.49 Patterns of $\mathbf{B}_1$ and $\mathbf{B}_1^t\mathbf{B}_1$ matrices for S.

Example 2: The three-storey frame of Figure 3.33 is reconsidered, and the **LU** factorisation method is used. The patterns of $\mathbf{B}_1$ and $\mathbf{B}_1^t\mathbf{B}_1$ matrices formed are shown in Figure 3.50, corresponding to $\chi(\mathbf{B}_1) = 504$ and $\chi(\mathbf{B}_1^t\mathbf{B}_1) = 1530$.

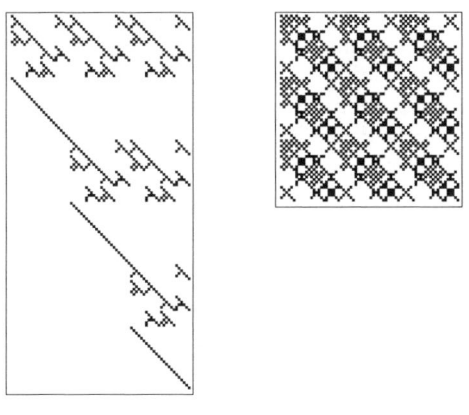

(a) Pattern of $\mathbf{B}_1$.   (b) Pattern of $\mathbf{B}_1^t\mathbf{B}_1$.

Fig. 3.50 Patterns of $\mathbf{B}_1$ and $\mathbf{B}_1^t\mathbf{B}_1$ matrices for S.

QR Decomposition Method: Using a **QR** factorisation algorithm with column pivoting yields

$$AP = Q\,[R_1, R_2], \qquad (3\text{-}99)$$

where **P** is again a permutation matrix and R_1 is an upper triangular matrix of order n. B_1 can be obtained as

$$B_1 = P\begin{bmatrix} -R_1^{-1}R_2 \\ I \end{bmatrix}. \qquad (3\text{-}100)$$

Turn-back LU Decomposition Method: Topçu developed a method, the so-called Turn-back **LU** procedure, which is based on **LU** factorisation and often results in highly sparse and banded B_1 matrices. Heath et al. [75] adopted this method for use with **QR** factorisation. Owing to the efficiency of this method, a brief description of their approach will be presented in the following text.

Write the matrix $A = (a_1, a_2, ..., a_n)$ by the columns. A *start column* is a column such that the ranks of $(a_1, a_2, ..., a_{s-1})$ and $(a_1, a_2, ..., a_s)$ are equal. Equivalently, a_s is a start column if it is linearly dependent on lower-numbered columns. The coefficients of this linear dependency give a null vector whose highest numbered non-zero is in positions. It is easy to see that the number of start columns is $m - n = t$, the dimension of the null space of **A**.

The start column can be found by performing a **QR** factorisation of **A**, using orthogonal transformations to annihilate the subdiagonal non-zeros. Suppose that in carrying out the **QR** factorisation we do not perform column interchanges but simply skip over any columns that are already zero on and below the diagonal. The result will then be a factorisation of the form

$$A = Q \begin{bmatrix} R \\ 0 \end{bmatrix} \qquad (3\text{-}101)$$

The start columns are those columns where the upper triangular structure jogs to the right, that is, a_s is a start column if the highest non-zero position in column s of **R** is no larger than the highest non-zero position in earlier columns of **R**.

The Turn-back method finds one null vector for each start column a_s, by "turning back" from column s to find the smallest k for which columns $a_s, a_{s-1}, ..., a_{s-k}$ are linearly dependent. The null vector has a non-zero only in position $s - k$ through s. Thus, if k is small for most of the start columns, then the null basis will have a small profile. Notice that the turn-back operates on **A**, and not on **R**. The initial **QR** factorisation of **A** is used only to determine the start columns and then discarded.

The null vector that the Turn-back method finds from start column a_S may not be non-zero in position s. Therefore, this method needs to have some way to guarantee that its null vectors are linearly independent. This can be accomplished by forbidding the left-most column of the dependency for each null vector from participating in any later dependencies. Thus, if the null vector for start column a_s has its first non-zero in position $s - k$, every null vector for a start column to the right of a_s will be zero in position $s - k$.

Although the term "Turn-back" is introduced in [219], the basic idea had also been used in [22,94]. Since this correspondence simplifies the understanding of the Turn-back method, it is briefly described in the following.

For the Algorithm 1 of Section 3.3.2, the use of an SRT orders the nodes and members of the graph simultaneously, resulting in a fairly banded member-node incidence matrix **B**. Considering the columns of **B** corresponding to tree members as independent columns, effectively a cycle is formed on each ordered chord (start column) by turning back in **B** and establishing a minimal dependency, using the tree members and previously used chords. The cycle basis selected by this process forms a base for the cycle matroid of the graph, as described in Kaveh [94,112]. Therefore, the idea used in Algorithm 1 and its generalisation for the formation of a GCB in [94,98,113] seem to constitute an idea similar to that of the algebraic Turn-back method.

Example 1: The four by four planar frame of Figure 3.31 is reconsidered. The patterns of $\mathbf{B}_1$ and $\mathbf{B}_1^t\mathbf{B}_1$ formed by the Turn-back LU factorisation method are depicted in Figure 3.51, corresponding to $\chi(\mathbf{B}_1) = 240$ and $\chi(\mathbf{B}_1^t\mathbf{B}_1) = 408$.

Example 2: The three-storey frame of Figure 3.33 is reconsidered, and the Turn-back LU factorisation method is used. The patterns of $\mathbf{B}_1$ and $\mathbf{B}_1^t\mathbf{B}_1$ matrices formed are shown in Figure 3.52, corresponding to $\chi(\mathbf{B}_1) = 476$ and $\chi(\mathbf{B}_1^t\mathbf{B}_1) = 984$.

A comparative study of various force methods has been made in [119].

Many algorithms have been developed for the selection of null bases, and the interested reader may refer to [28,29,179,181].

138   OPTIMAL STRUCTURAL ANALYSIS

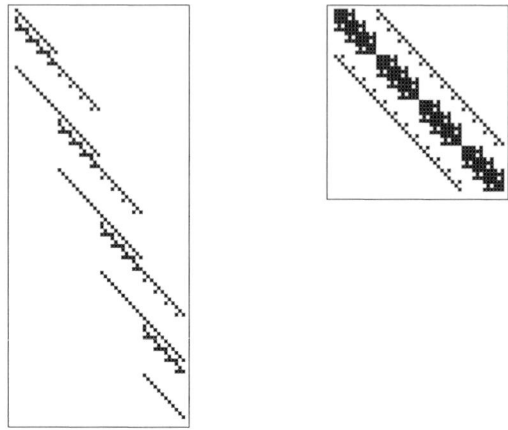

(a) Pattern of $\mathbf{B}_1$.   (b) Pattern of $\mathbf{B}_1^t \mathbf{B}_1$.

Fig. 3.51 Patterns of $\mathbf{B}_1$ and $\mathbf{B}_1^t \mathbf{B}_1$ matrices for S.

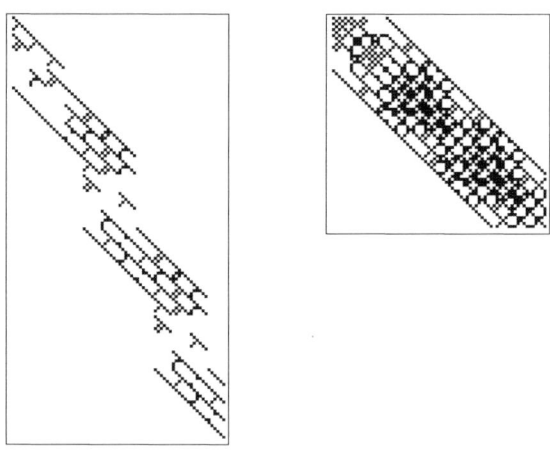

(a) Pattern of $\mathbf{B}_1$.   (b) Pattern of $\mathbf{B}_1^t \mathbf{B}_1$.

Fig. 3.52 Patterns of $\mathbf{B}_1$ and $\mathbf{B}_1^t \mathbf{B}_1$ matrices for S.

OPTIMAL FORCE METHOD OF STRUCTURAL ANALYSIS 139

EXERCISES

3.1   For each set of integers a, b and c of Table 2.1, draw an arbitrary γ-tree and a γ-cycle.

3.2   Construct a γ-tree for the following graph when it is viewed as the graph model of a planar truss:

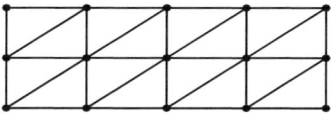

3.3   In Exercise 4.2, select a fundamental γ-cycle basis of S and form its γ-cycle adjacency matrix.

3.4   Find a graph for which Algorithm 3 fails to select a minimal cycle basis. Repeat this exercise for Algorithm 2.

3.5   Form $\mathbf{B}_0$ and $\mathbf{B}_1$ matrices by selecting a suitable SRT and cycle basis for the following planar frame:

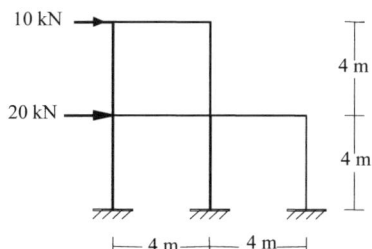

3.6   Form $\mathbf{B}_0$ and $\mathbf{B}_1$ for the planar truss of Exercise 3.2, when it is supported in a statically determinate fashion. Choose the support nodes arbitrarily.

140   OPTIMAL STRUCTURAL ANALYSIS

3.7   Perform a complete analysis for the following planar truss using the force method. EA = Constant:

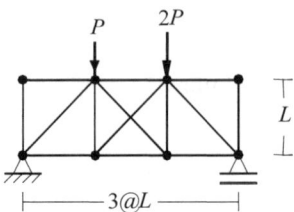

3.8   Perform a complete force method analysis for the following continuous beam:

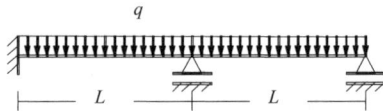

3.9   Prove the minimality of the cycle basis selected by Horton´s algorithm.

3.10   Why do the regional cycles of a planar graph form a cycle basis (mesh basis)?

3.11   Use Algorithms A, B and C to find suboptimally conditioned cycle bases for the following weighted graphs. The numbers 1 and 2 show the member types as given in Section 3.4.

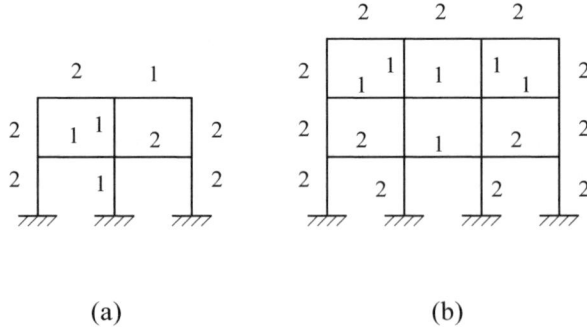

(a)   (b)

3.12   Write a computer program to calculate the largest and the smallest eigenvalues for adjacency matrices of graphs.

CHAPTER 4
Optimal Displacement Method of Structural Analysis

4.1 INTRODUCTION

In this chapter, the principles introduced in Chapter 1 are used for the formulation of the general displacement method of structural analysis. Computational aspects are discussed, and many worked examples are included to illustrate the concepts and principles being used. In order to show the generality of the methods introduced for the formation of the element stiffness matrices, the stiffness matrix of a simple finite element is also derived.

Special attention is paid to the graph theory aspects of the displacement method for rigid-jointed structures, where the pattern equivalence of structural and graph theory matrices is used. The standard displacement method employs cocycle bases of structural graph models; however, for general solutions, a cutset basis of the model should be employed. This becomes vital when solutions leading to well-conditioned stiffness matrices are required. Methods for the selection of such cutset bases are described in this chapter.

In the last half century, considerable progress has been made in the matrix analysis of structures; see for example, Argyris and Kelsey [6], Kardestuncer [93], Livesley [153], McGuire and Gallagher [163], Meek [164], Prezmieniecki [183], Vanderbilt [222], Ziegler [234] and Zienkiewicz [235]. The topic has been generalised to finite elements and extended to the stability and dynamic analysis of structures. This progress is due to the simplicity, modularity and flexibility of matrix methods.

Optimal Structural Analysis   A. Kaveh
© 2006 Research Studies Press Limited

142   OPTIMAL STRUCTURAL ANALYSIS

4.2 FORMULATION

In this section, a matrix formulation using the basic tools of structural analysis – equilibrium of forces, compatibility of displacements and force–displacement relationships – is provided. The notations are chosen from those most often encountered in structural mechanics.

4.2.1 COORDINATE SYSTEMS TRANSFORMATION

Consider a structure S with M members and N nodes, each node having α degrees of freedom (DOF). The kinematic indeterminacy (DKI) of S may then be determined as

$$\eta(S) = \alpha N - \beta, \qquad (4\text{-}1)$$

where β is the number of constraints due to the support conditions. As an example, $\eta(S)$ for the planar truss S depicted in Figure 4.1(a) is given by $\eta(S) = 7 \times 2 - 3 = 11$; for the plane frame illustrated in Figure 4.1(b), it is calculated as $\eta(S) = 8 \times 3 - 4 \times 3 = 12$; and for the space frame shown in Figure 4.1(c), it is calculated as $\eta(S) = 12 \times 6 - 6 \times 6 = 36$.

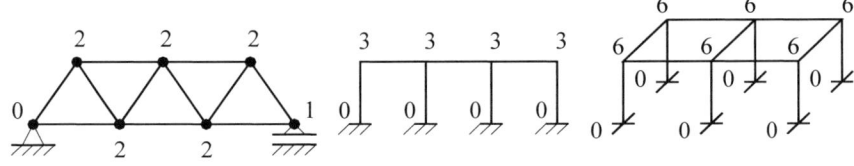

(a) A planar truss.   (b) A planar frame.   (c) A space frame.

Fig. 4.1 The degrees of freedom of the joints for three structures.

One can also calculate $\eta(S)$ by simple addition of the DOF of the joints of the structure, that is, for the truss S, $\eta(S) = 2 + 2 + 2 + 2 + 2 + 1 = 11$; for the planar frame, $\eta(S) = 4 \times 3 = 12$; and for the space frame, $\eta(S) = 6 \times 6 = 36$.

For a structure, the stiffness matrices of the elements should be prepared in a single coordinate system known as the *global coordinate system* in order to be able to perform the assembling process. However, the stiffness matrices of individual members are usually written first in coordinate systems attached to the members, known as *local coordinate systems*. Therefore a transformation is needed before the assembling process. Typical local and global coordinate systems are illustrated in Figure 4.2.

OPTIMAL DISPLACEMENT METHOD OF STRUCTURAL ANALYSIS   143

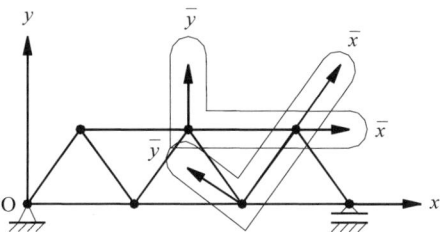

Fig. 4.2 Local $\bar{x}, \bar{y}$ system and global coordinate x, y system.

A global coordinate system can be selected arbitrarily; however, it may be advantageous to select this system such that the structure falls in the first quadrant of the plane in order to have positive coordinates for the nodes of the structure. On the other hand, a local coordinate system of a member is so chosen that it has one of its axes along the member and the second axis lies in its plane of symmetry (if it has one) and the third axis is chosen such that it results in a right-handed coordinate system.

The transformation from a local coordinate system to a global coordinate system can be performed as illustrated in Figure 4.3, in which x, y, z is the global system and x_2, y_2, z_2, often denoted by $\overline{xyz}$, is the local system.

For rotation about the y-axis, the relation between x_1, y_1, z_1 and x, y, z can be expressed as follows:

$$\begin{bmatrix} x_1 \\ y_1 \\ z_1 \end{bmatrix} = \begin{bmatrix} \cos\alpha & 0 & \sin\alpha \\ 0 & 1 & 0 \\ -\sin\alpha & 0 & \cos\alpha \end{bmatrix} \begin{bmatrix} x \\ y \\ z \end{bmatrix}. \qquad (4\text{-}2)$$

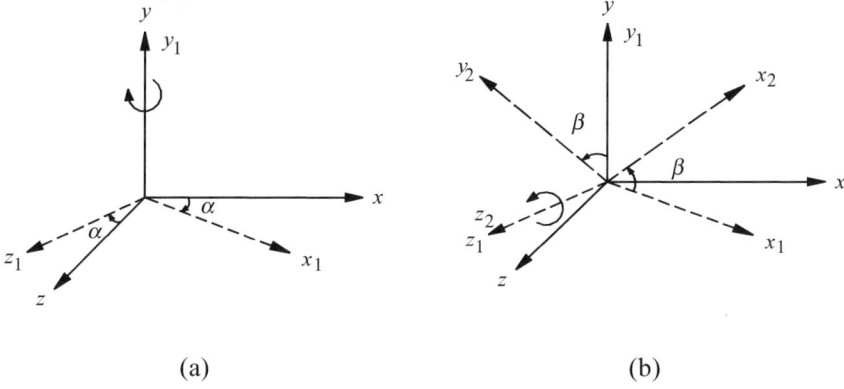

(a)   (b)

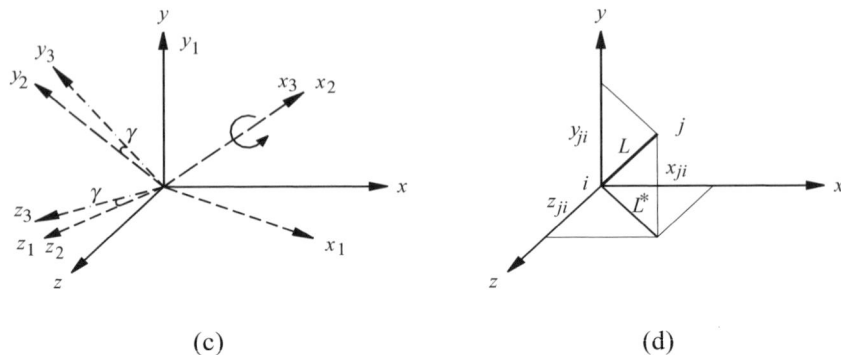

(c)   (d)

Fig. 4.3 Transformation from local coordinate system to global coordinate system.

Similarly, for rotation about the z_1-axis, x_2, y_2, z_2 and x_1, y_1, z_1 are related by

$$\begin{bmatrix} x_2 \\ y_2 \\ z_2 \end{bmatrix} = \begin{bmatrix} \cos\beta & \sin\beta & 0 \\ -\sin\beta & \cos\beta & 0 \\ 0 & 0 & 1 \end{bmatrix} \begin{bmatrix} x_1 \\ y_1 \\ z_1 \end{bmatrix}, \qquad (4\text{-}3)$$

and, for rotation about the x_2-axis, x_3, y_3, z_3 and x_2, y_2, z_2 are related as follows:

$$\begin{bmatrix} x_3 \\ y_3 \\ z_3 \end{bmatrix} = \begin{bmatrix} 1 & 0 & 0 \\ 0 & \cos\gamma & \sin\gamma \\ 0 & -\sin\gamma & \cos\gamma \end{bmatrix} \begin{bmatrix} x_2 \\ y_2 \\ z_2 \end{bmatrix}. \qquad (4\text{-}4)$$

Combining the above transformations we have

$$\mathbf{T} = \begin{bmatrix} (\cos\alpha\cos\beta) & (\sin\beta) & (\cos\beta\sin\alpha) \\ -(\sin\alpha\sin\gamma + \cos\alpha\sin\beta\cos\gamma) & (\cos\beta\cos\gamma) & (\sin\gamma\cos\alpha - \sin\alpha\sin\beta\cos\gamma) \\ -(\sin\alpha\cos\gamma - \cos\alpha\sin\beta\sin\gamma) & (-\cos\beta\sin\gamma) & (\cos\alpha\cos\gamma + \sin\alpha\sin\beta\sin\gamma) \end{bmatrix}.$$
(4-5)

where

$$\begin{bmatrix} x_3 \\ y_3 \\ z_3 \end{bmatrix} = [\mathbf{T}] \begin{bmatrix} x \\ y \\ z \end{bmatrix}. \qquad (4\text{-}6)$$

The representations of a vector in the local coordinate system $\overline{\Gamma}$ and the global coordinate system Γ are related as follows:

$$\overline{\Gamma} = \mathbf{T}\, \Gamma. \qquad (4\text{-}7)$$

It can easily be proved that **T** is an orthogonal matrix, that is,

$$[\mathbf{T}]^{-1} = [\mathbf{T}]^t. \qquad (4\text{-}8)$$

In the above transformation, γ represents the tilt of the member, which is quite often zero. Thus, **T** can be simplified as

$$\mathbf{T} = \begin{bmatrix} \cos\alpha \cos\beta & \sin\beta & \sin\alpha \cos\beta \\ -\cos\alpha \sin\beta & \cos\beta & -\sin\alpha \sin\beta \\ -\sin\alpha & 0 & \cos\alpha \end{bmatrix}, \qquad (4\text{-}9)$$

and for the two-dimensional case and "α equal to zero", **T** reduces to

$$\mathbf{T} = \begin{bmatrix} \cos\beta & \sin\beta \\ -\sin\beta & \cos\beta \end{bmatrix}. \qquad (4\text{-}10)$$

Equation (4-9) can easily be written in terms of the coordinates of the two ends of a vector. Considering Figure 4.3(b) and using simple trigonometry, Eq. (4-9) becomes

$$\mathbf{T} = \begin{bmatrix} x_{ji}/L & y_{ji}/L & z_{ji}/L \\ -x_{ji}y_{ji}/L*L & L*/L & y_{ji}z_{ji}/L*L \\ -z_{ji}/L* & 0 & x_{ji}/L* \end{bmatrix}, \qquad (4\text{-}11)$$

where

$$x_{ji} = x_j - x_i \qquad y_{ji} = y_j - y_i \qquad z_{ji} = z_j - z_i$$

$$L^* = (z_{ji}^2 + x_{ji}^2)^{\frac{1}{2}} \text{ and } L = (z_{ji}^2 + y_{ji}^2 + x_{ji}^2)^{\frac{1}{2}}. \qquad (4\text{-}12)$$

Notice that **T** transforms a three-dimensional vector from a global to a local coordinate system and $\mathbf{T}^t$ performs the reverse transformation. However, if the element forces or element displacements (distortions) consist of p vectors, the block diagonal matrix with p submatrices should be used. As an example, for a beam element

146 OPTIMAL STRUCTURAL ANALYSIS

of a space frame with each node having 6 DOF, the transformation matrix is a 12 × 12 matrix of the form

$$\mathbf{T} = \begin{bmatrix} \mathbf{T} & & & \\ & \mathbf{T} & & \\ & & \mathbf{T} & \\ & & & \mathbf{T} \end{bmatrix}. \tag{4-13}$$

4.2.2 ELEMENT STIFFNESS MATRIX USING UNIT DISPLACEMENT METHOD

Consider a general element, as shown in Figure 4.4, with n member forces,

$$\mathbf{r}_m = \{r_1\ r_2\ \ldots\ r_n\}^t \tag{4-14}$$

and n member displacements,

$$\mathbf{u}_m = \{u_1\ u_2\ \ldots\ u_n\}^t. \tag{4-15}$$

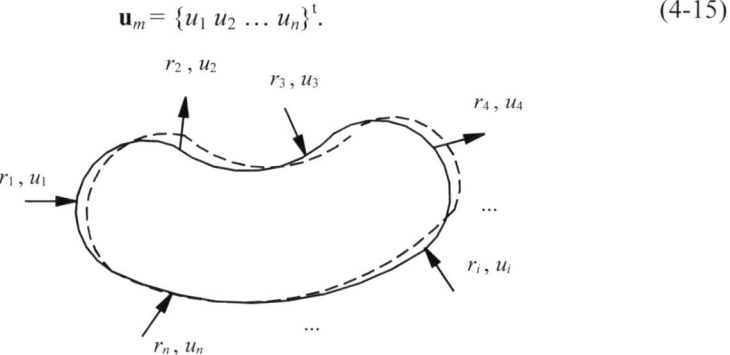

Fig. 4.4 A general element with its nodal loads and nodal displacements.

A typical force component r_i can be found by using the unit displacement method to be

$$r_i = \iiint_V \hat{\boldsymbol{\varepsilon}}_i^t \boldsymbol{\sigma}\ dV, \tag{4-16}$$

where $\hat{\boldsymbol{\varepsilon}}_i$ represents the matrix of compatible strains due to a unit displacement in the direction of r_i, and $\boldsymbol{\sigma}$ is the exact stress matrix due to the applied forces $\mathbf{r}_m$. The unit displacements can be used in turn for all the points where member forces are applied, and therefore,

OPTIMAL DISPLACEMENT METHOD OF STRUCTURAL ANALYSIS

$$\mathbf{r}_m = \iiint_V \hat{\varepsilon}^t \sigma \, dV, \tag{4-17}$$

where

$$\hat{\varepsilon} = \{\hat{\varepsilon}_1 \ \hat{\varepsilon}_2 \ ... \ \hat{\varepsilon}_n\}^t. \tag{4-18}$$

For a linear system the total strain,

$$\mathbf{e} = \{e_{xx} \ e_{yy} \ e_{zz} \ e_{xy} \ e_{yz} \ e_{xz}\}^t, \tag{4-19}$$

can be expressed as

$$\mathbf{e} = \mathbf{bu}, \tag{4-20}$$

where **b** is the exact strain due to the unit displacement **u**.

The stress–strain relationship can be written as

$$\sigma = \chi\mathbf{bu}, \tag{4-21}$$

where:

$$\chi = \frac{E}{(1+v)(1-2v)} \begin{bmatrix} 1-v & v & v & & & \\ v & 1-v & v & & \mathbf{0} & \\ v & v & 1-v & & & \\ & & & \frac{1-2v}{2} & 0 & 0 \\ & \mathbf{0} & & 0 & \frac{1-2v}{2} & 0 \\ & & & 0 & 0 & \frac{1-2v}{2} \end{bmatrix} \tag{4-22}$$

Substituting in Eq. (4-17) we have

$$\mathbf{r}_m = \iiint_V \hat{\varepsilon}^t \chi \mathbf{b} \, dV \, \mathbf{u}_m, \tag{4-23}$$

or

$$\mathbf{r}_m = \mathbf{k}_m \mathbf{u}_m, \tag{4-24}$$

where

$$\mathbf{k}_m = \iiint_V \bar{\varepsilon}^t \chi \mathbf{b} \, dV, \qquad (4\text{-}25)$$

represents the element stiffness matrix.

The evaluation of matrix **b**, representing the exact strain distributions, can often be difficult, if not impossible. Hence, in case there is no exact distribution, an approximate relationship may be used. Usually, matrix **b** is selected such that it satisfies at least the equations of compatibility. Denoting this approximate matrix by $\hat{\varepsilon}$ and using $\hat{\varepsilon} = \hat{\mathbf{b}}$, we have

$$\mathbf{k}_m = \iiint_V \hat{b}^t \chi \hat{\mathbf{b}} \, dV. \qquad (4\text{-}26)$$

This equation will be used for the derivation of the stiffness matrices of a finite element in Section 4.5.1.

As an example, consider the prismatic bar element shown in its local coordinate system in Figure 4.5. According to the definition of such an element, only axial forces are present.

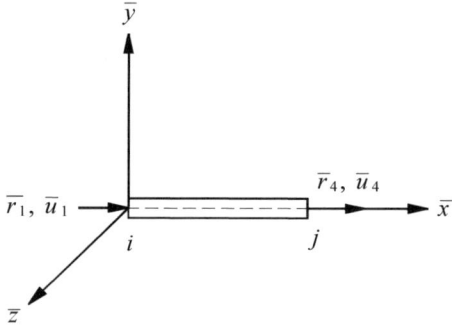

Fig. 4.5 A bar element in its local coordinate system.

From the theory of elasticity, the axial strain is expressed as

$$\varepsilon_{xx} = \frac{\partial u_x}{\partial x}. \qquad (4\text{-}27)$$

The displacement u_x along the longitudinal axis of the bar can be expressed as

$$u_x = A_1 x + A_2. \qquad (4\text{-}28)$$

OPTIMAL DISPLACEMENT METHOD OF STRUCTURAL ANALYSIS 149

From the boundary conditions,

$$u_x = \bar{u}_1 \text{ at } x = 0, \text{ and } u_x = \bar{u}_4 \text{ at } x = L. \tag{4-29}$$

Hence,

$$A_1 = \frac{\bar{u}_4 - \bar{u}_1}{L} \text{ and } A_2 = \bar{u}_1. \tag{4-30}$$

By substituting in Eq. (4-28), we have

$$u_x = \frac{\bar{u}_4 - \bar{u}_1}{L} x + \bar{u}_1. \tag{4-31}$$

Now axial strain can be evaluated as follows:

$$\varepsilon_{xx} = \frac{\partial u_x}{\partial x} = \frac{1}{L}(u_2 - u_1) = \frac{1}{L}[-1 \quad +1]\begin{bmatrix} u_1 \\ u_2 \end{bmatrix}. \tag{4-32}$$

The above strain distribution is exact, and

$$\hat{\mathbf{b}} = \mathbf{b} = \frac{1}{L}[-1 \quad +1]. \tag{4-33}$$

Since a bar element is one dimensional, χ is a 1×1 matrix defined as

$$\chi = E. \tag{4-34}$$

Substituting in Eq. (4-26), we have

$$\mathbf{k}_m = \int_0^L \frac{1}{L}\begin{bmatrix} -1 \\ 1 \end{bmatrix}\frac{E}{L}[-1 \quad 1]A\,dx, \tag{4-35}$$

and

$$\mathbf{k}_m = \frac{EA}{L}\begin{bmatrix} 1 & -1 \\ -1 & 1 \end{bmatrix}. \tag{4-36}$$

This method will also be used for the derivation of the finite element stiffness matrices in subsequent sections.

150 OPTIMAL STRUCTURAL ANALYSIS

4.2.3 ELEMENT STIFFNESS MATRIX USING CASTIGLIANO'S THEOREM

In this section, a different approach, using Castigliano's theorem, is described for the formation of element stiffness matrices. Consider a general element as shown in Figure 4.4. Suppose that loads are applied at certain points (specified as nodes) 1, 2, ..., n. Let v_i be the displacement of node i along the applied load p_i. The loads are applied in a pseudo-static manner, increasing gradually from zero. Assuming a linear behaviour, the work done by an external force $\mathbf{p} = \{p_1, p_2, ..., p_n\}$ through the displacement $\mathbf{v} = \{v_1, v_2, ..., v_n\}$ can be written as

$$W = \frac{1}{2}(p_1 v_1 + p_2 v_2 + ... + p_n v_n). \qquad (4\text{-}37)$$

According to the principle of conservation of energy,

$$W = U, \qquad (4\text{-}38)$$

and therefore

$$U = \frac{1}{2}(p_1 v_1 + p_2 v_2 + ... + p_n v_n). \qquad (4\text{-}39)$$

If a small variation is now given to v_i while keeping the other displacement components constant, then the variation of $\mathbf{v}$ with respect to v_i can be written as

$$\frac{\partial U}{\partial v_i} = \tfrac{1}{2}[p_i + \frac{\partial p_1}{\partial v_i} v_1 + \frac{\partial p_2}{\partial v_i} v_2 + ... + \frac{\partial p_n}{\partial v_i} v_n]. \qquad (4\text{-}40)$$

According to Castigliano's theorem,

$$\frac{\partial U}{\partial v_i} = p_i. \qquad (4\text{-}41)$$

Thus,

$$p_i = [\frac{\partial p_1}{\partial v_i} v_1 + \frac{\partial p_2}{\partial v_i} v_2 + ... + \frac{\partial p_n}{\partial v_i} v_n], \qquad (4\text{-}42)$$

or in a matrix form for all $i = 1, ..., n$, we have

OPTIMAL DISPLACEMENT METHOD OF STRUCTURAL ANALYSIS

$$\begin{bmatrix} p_1 \\ p_2 \\ \cdot \\ \cdot \\ p_n \end{bmatrix} = \begin{bmatrix} \frac{\partial p_1}{\partial v_1} & \frac{\partial p_2}{\partial v_1} & \cdot \cdot & \frac{\partial p_n}{\partial v_1} \\ \frac{\partial p_1}{\partial v_2} & \frac{\partial p_2}{\partial v_2} & \cdot \cdot & \frac{\partial p_n}{\partial v_2} \\ \cdot & \cdot & \cdot \cdot & \cdot \\ \cdot & \cdot & \cdot \cdot & \cdot \\ \frac{\partial p_1}{\partial v_n} & \frac{\partial p_2}{\partial v_n} & \cdot \cdot & \frac{\partial p_n}{\partial v_n} \end{bmatrix} \begin{bmatrix} v_1 \\ v_2 \\ \cdot \\ \cdot \\ v_n \end{bmatrix}. \qquad (4\text{-}43)$$

According to definition, the above coefficient matrix forms the stiffness matrix of the elastic body defined by its n nodes as illustrated in Figure 4.4.

A typical element of the stiffness matrix k_{ij} is given by

$$k_{ij} = \frac{\partial p_j}{\partial v_i}. \qquad (4\text{-}44)$$

Using Castigliano's first theorem,

$$k_{ij} = \frac{\partial}{\partial v_i}(\frac{\partial U}{\partial v_j}) = \frac{\partial^2 U}{\partial v_i \partial v_j}. \qquad (4\text{-}45)$$

Similarly,

$$k_{ji} = \frac{\partial p_i}{\partial v_j} = \frac{\partial^2 U}{\partial v_j \partial v_i}. \qquad (4\text{-}46)$$

Since the order of differentiation should not affect the result of our problems, we have

$$k_{ij} = k_{ji}, \qquad (4\text{-}47)$$

which is a proof of the symmetry of the stiffness matrices both for a structure and for an element.

As an example, consider the prismatic bar element as shown in its local coordinate system in Figure 4.5. According to the definition of such an element, only axial forces are present.

152   OPTIMAL STRUCTURAL ANALYSIS

The strain energy of this bar can be calculated as follows:

$$U = \frac{1}{2}\iiint \sigma_{xx}\varepsilon_{xx}\,dx\,dy\,dz = \frac{E}{2}\iiint \varepsilon_{xx}^2\,dx\,dy\,dz = \frac{EA}{2}\int \varepsilon_{xx}^2\,dx. \tag{4-48}$$

On the other hand,

$$\varepsilon_{xx} = \frac{\partial u_x}{\partial x}. \tag{4-49}$$

Using Eq. (4-31) and by substituting in Eq. (4-46), the strain energy of the bar is calculated as

$$U = \frac{EA}{2L}[\bar{u}_4^2 - 2\bar{u}_4\bar{u}_1 + \bar{u}_1^2]. \tag{4-50}$$

Hence

$$\bar{k}_{11} = \frac{\partial^2 U}{\partial \bar{u}_1^2} = \frac{EA}{L},$$

$$\bar{k}_{14} = \bar{k}_{41} = \frac{\partial^2 U}{\partial \bar{u}_1 \partial \bar{u}_4} = -\frac{EA}{L}, \tag{4-51}$$

$$\bar{k}_{44} = \frac{\partial^2 U}{\partial \bar{u}_4^2} = \frac{EA}{L}, \text{ and}$$

$$\bar{k}_{ij} = 0 \text{ for all other components.}$$

Therefore, the stiffness matrix of a bar element in the selected local coordinate system is obtained, and

$$\begin{bmatrix} \bar{r}_1 \\ \bar{r}_2 \\ \bar{r}_3 \\ \bar{r}_4 \\ \bar{r}_5 \\ \bar{r}_6 \end{bmatrix} = \frac{EA}{L} \begin{bmatrix} 1 & 0 & 0 & -1 & 0 & 0 \\ 0 & 0 & 0 & 0 & 0 & 0 \\ 0 & 0 & 0 & 0 & 0 & 0 \\ -1 & 0 & 0 & 1 & 0 & 0 \\ 0 & 0 & 0 & 0 & 0 & 0 \\ 0 & 0 & 0 & 0 & 0 & 0 \end{bmatrix} \begin{bmatrix} \bar{u}_1 \\ \bar{u}_2 \\ \bar{u}_3 \\ \bar{u}_4 \\ \bar{u}_5 \\ \bar{u}_6 \end{bmatrix}. \tag{4-52}$$

4.2.4 STIFFNESS MATRIX OF A STRUCTURE

Let **p** and **v** represent the joint loads and joint displacements of a structure. Then the force–displacement relationship for the structure can be expressed as

$$\mathbf{p} = \mathbf{Kv}, \tag{4-53}$$

where **K** is an $\alpha N \times \alpha N$ symmetric matrix, known as the *stiffness matrix* of the structure. Expanding the ith equation of the above system, the force p_i can be expressed in terms of the displacements $\{v_1, v_2, \ldots, v_{\alpha N}\}$ as

$$p_i = \mathrm{K}_{i1} v_1 + \mathrm{K}_{i2} v_2 + \ldots + \mathrm{K}_{i\alpha N} v_{\alpha N}. \tag{4-54}$$

A typical coefficient $\mathbf{K}_{ij}$ is the value of the force p_i required to be applied at the ith component of the structure in order to produce a displacement $v_j = 1$ at j and zero displacements at all the other components.

The member forces **r** can be related to nodal forces **p** by

$$\mathbf{p} = \mathbf{Br}. \tag{4-55}$$

Using the contragradient relationship, the joint displacements **v** can be related to member distortions **u** by

$$\mathbf{u} = \mathbf{B}^t \mathbf{v}. \tag{4-56}$$

For each individual member of the structure, the member forces can be related to member distortions by an element stiffness matrix $\mathbf{k}_m$. A block diagonal matrix containing these element stiffness matrices is known as the *unassembled stiffness matrix* of the structure, denoted by **k**. Obviously,

$$\mathbf{r} = \mathbf{ku}. \tag{4-57}$$

This equation together with Eqs (4-55) and (4-56) yields

$$\mathbf{p} = \mathbf{BkB}^t \mathbf{v}. \tag{4-58}$$

Therefore,

$$\mathbf{K} = \mathbf{BkB}^t \tag{4-59}$$

is obtained. The matrix **K** is singular since the boundary conditions of the structure are not yet applied. For an appropriately supported structure, the deletion of the rows and columns of **K** corresponding to the support constraints results in a positive definite matrix, known as the *reduced stiffness matrix* of the structure.

A symmetric matrix **S** is called *positive definite* if $\mathbf{x}^t\mathbf{S}\mathbf{x} > 0$ for every non-zero vector **x**. As shown earlier, the stiffness matrix **K** of a structure is symmetric. This matrix is also positive definite since

$$\mathbf{p}^t\mathbf{v} = (\mathbf{Kv})^t\mathbf{v} = \mathbf{v}^t\mathbf{K}^t\mathbf{v} = \mathbf{v}^t\mathbf{K}\mathbf{v} = 2W \tag{4-60}$$

and W is always positive.

Let us illustrate the stiffness method by means of a simple example. Consider a fixed end beam with a load P applied at its mid-span. This beam is discretised as two beam elements, as shown in Figure 4.6(a), with two DOF for each node (axial deformation is ignored for simplicity). The components of element forces and element distortions are depicted in Figure 4.6(b) and those of the entire structure are illustrated in Figure 4.6(c).

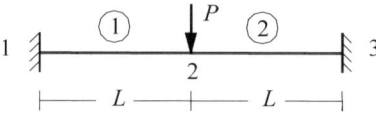

(a) A fixed ended beam S.

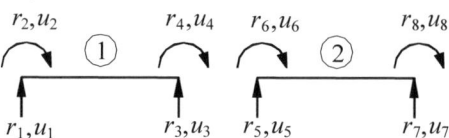

(b) Member forces and member distortions.

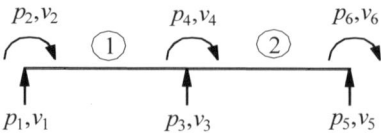

(c) Nodal forces and nodal displacements of the entire structure.

Fig. 4.6 Illustration of the analysis of a simple structure.

OPTIMAL DISPLACEMENT METHOD OF STRUCTURAL ANALYSIS   155

For each element such as element 1, the stiffness matrix can be written as

$$\begin{bmatrix} r_1 \\ r_2 \\ r_3 \\ r_4 \end{bmatrix} = \begin{bmatrix} k_{11} & k_{12} & k_{13} & k_{14} \\ k_{21} & k_{22} & k_{23} & k_{24} \\ k_{31} & k_{32} & k_{33} & k_{34} \\ k_{41} & k_{42} & k_{43} & k_{44} \end{bmatrix} \begin{bmatrix} u_1 \\ u_2 \\ u_3 \\ u_4 \end{bmatrix}, \qquad (4\text{-}61)$$

and for the entire structure we have

$$\begin{bmatrix} p_1 \\ p_2 \\ p_3 \\ p_4 \\ p_5 \\ p_6 \end{bmatrix} = \begin{bmatrix} K_{11} & K_{12} & K_{13} & K_{14} & K_{15} & K_{16} \\ K_{21} & K_{22} & K_{23} & K_{24} & K_{25} & K_{26} \\ K_{31} & K_{32} & K_{33} & K_{34} & K_{35} & K_{36} \\ K_{41} & K_{42} & K_{43} & K_{44} & K_{45} & K_{46} \\ K_{51} & K_{52} & K_{53} & K_{54} & K_{55} & K_{56} \\ K_{61} & K_{62} & K_{63} & K_{64} & K_{65} & K_{66} \end{bmatrix} \begin{bmatrix} v_1 \\ v_2 \\ v_3 \\ v_4 \\ v_5 \\ v_6 \end{bmatrix}. \qquad (4\text{-}62)$$

Element stiffness matrices $\mathbf{k}_1$ and $\mathbf{k}_2$ can be easily constructed using the definition of k_{ij}. For a beam element, ignoring its axial deformation, these terms are shown in Figure 4.7.

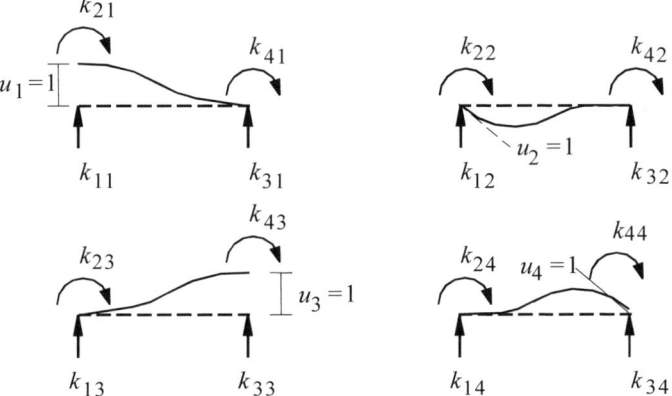

Fig. 4.7 Stiffness coefficients of a beam element ignoring its axial deformation.

The structure has a uniform cross section, and both elements have the same length. Therefore, using the force–displacement relationship from Chapter 1,

$$\mathbf{k}_1 = \mathbf{k}_2 = \frac{2EI}{L} \begin{bmatrix} 6/L^2 & -3/L & -6/L^2 & -3/L \\ -3/L & 2 & 3/L & 1 \\ -6/L^2 & 3/L & 6/L^2 & 3/L \\ -3/L & 1 & 3/L & 2 \end{bmatrix}. \tag{4-63}$$

The unassembled stiffness matrix is an 8×8 matrix of the form

$$\mathbf{k} = \begin{bmatrix} \mathbf{k}_1 & \mathbf{0} \\ \mathbf{0} & \mathbf{k}_2 \end{bmatrix}. \tag{4-64}$$

Now, consider the equilibrium of the joints of the structure, resulting in

$$p_1 = r_1, \; p_2 = r_2, \; p_3 = r_3 + r_5,$$
$$p_4 = r_4 + r_6, \; p_5 = r_7, \; p_6 = r_8 \tag{4-65}$$

or in a matrix form we have,

$$\begin{bmatrix} p_1 \\ p_2 \\ p_3 \\ p_4 \\ p_5 \\ p_6 \end{bmatrix} = \begin{bmatrix} 1 & \cdot & \cdot & \cdot & \cdot & \cdot & \cdot & \cdot \\ \cdot & 1 & \cdot & \cdot & \cdot & \cdot & \cdot & \cdot \\ \cdot & \cdot & 1 & \cdot & 1 & \cdot & \cdot & \cdot \\ \cdot & \cdot & \cdot & 1 & \cdot & 1 & \cdot & \cdot \\ \cdot & \cdot & \cdot & \cdot & \cdot & \cdot & 1 & \cdot \\ \cdot & \cdot & \cdot & \cdot & \cdot & \cdot & \cdot & 1 \end{bmatrix} \begin{bmatrix} r_1 \\ r_2 \\ r_3 \\ r_4 \\ r_5 \\ r_6 \\ r_7 \\ r_8 \end{bmatrix}, \tag{4-66}$$

and more compactly,

$$\mathbf{p} = \mathbf{Br}, \tag{4-67}$$

where

$$\mathbf{B} = \begin{bmatrix} 1 & \cdot & \cdot & \cdot & \cdot & \cdot & \cdot & \cdot \\ \cdot & 1 & \cdot & \cdot & \cdot & \cdot & \cdot & \cdot \\ \cdot & \cdot & 1 & \cdot & 1 & \cdot & \cdot & \cdot \\ \cdot & \cdot & \cdot & 1 & \cdot & 1 & \cdot & \cdot \\ \cdot & \cdot & \cdot & \cdot & \cdot & \cdot & 1 & \cdot \\ \cdot & \cdot & \cdot & \cdot & \cdot & \cdot & \cdot & 1 \end{bmatrix},$$

is known as the *equilibrium matrix*.

Consider now the compatibility of displacements:

$$u_1 = v_1, u_2 = v_2, u_3 = u_5 = v_3,$$

$$u_4 = u_6 = v_4, u_7 = v_5, u_8 = v_6. \qquad (4\text{-}68)$$

In a matrix form, we have

$$\begin{bmatrix} u_1 \\ u_2 \\ u_3 \\ u_4 \\ u_5 \\ u_6 \\ u_7 \\ u_8 \end{bmatrix} = \begin{bmatrix} 1 & \cdot & \cdot & \cdot & \cdot & \cdot \\ \cdot & 1 & \cdot & \cdot & \cdot & \cdot \\ \cdot & \cdot & 1 & \cdot & \cdot & \cdot \\ \cdot & \cdot & \cdot & 1 & \cdot & \cdot \\ \cdot & \cdot & 1 & \cdot & \cdot & \cdot \\ \cdot & \cdot & \cdot & 1 & \cdot & \cdot \\ \cdot & \cdot & \cdot & \cdot & 1 & \cdot \\ \cdot & \cdot & \cdot & \cdot & \cdot & 1 \end{bmatrix} \begin{bmatrix} v_1 \\ v_2 \\ v_3 \\ v_4 \\ v_5 \\ v_6 \end{bmatrix}, \qquad (4\text{-}69)$$

and in compact form we have

$$\mathbf{u} = \mathbf{E}\mathbf{v} = \mathbf{B}^t\mathbf{v}. \qquad (4\text{-}70)$$

where

$$\mathbf{E} = \begin{bmatrix} 1 & \cdot & \cdot & \cdot & \cdot & \cdot \\ \cdot & 1 & \cdot & \cdot & \cdot & \cdot \\ \cdot & \cdot & 1 & \cdot & \cdot & \cdot \\ \cdot & \cdot & \cdot & 1 & \cdot & \cdot \\ \cdot & \cdot & 1 & \cdot & \cdot & \cdot \\ \cdot & \cdot & \cdot & 1 & \cdot & \cdot \\ \cdot & \cdot & \cdot & \cdot & 1 & \cdot \\ \cdot & \cdot & \cdot & \cdot & \cdot & 1 \end{bmatrix},$$

is known as the *compatibility matrix*.

The reason for the matrix $\mathbf{E}$ being the transpose of the matrix $\mathbf{B}$ has already been discussed in Chapter 3; however, by using the principle of virtual work, a simple proof can be obtained. Consider

$$W = \text{work done by external loads} = \frac{1}{2}\mathbf{v}^t\mathbf{p},$$

158   OPTIMAL STRUCTURAL ANALYSIS

$$U = \text{strain energy} = \frac{1}{2}\mathbf{u}^t\mathbf{r}.$$

Then, equating W and U, we have $\mathbf{E} = \mathbf{B}^t$, which completes the proof. It should be mentioned that this equality holds for a general structure, and it is the result of the contragradient relationship introduced in Chapter 1.

The stiffness matrix of the entire structure is then obtained as

$$\mathbf{K} = \frac{2EI}{L}\begin{bmatrix} 6/L^2 & -3/L & -6/L^2 & -3/L & 0 & 0 \\ -3/L & 2 & 3/L & 1 & 0 & 0 \\ -6/L^2 & 3/L & 12/L^2 & 0 & -6/L^2 & -3/L \\ -3/L & 1 & 0 & 4 & 3/L & 1 \\ 0 & 0 & -6/L^2 & 3/L & 6/L^2 & 3/L \\ 0 & 0 & -3/L & 1 & 3/L & 2 \end{bmatrix}. \quad (4\text{-}71)$$

By applying the boundary conditions,

$$v_1 = v_2 = v_5 = v_6 = 0,$$

and deleting the displacements, the following reduced stiffness matrix is formed.

$$\begin{bmatrix} p_3 \\ p_4 \end{bmatrix} = \frac{2EI}{L}\begin{bmatrix} 12/L^2 & 0 \\ 0 & 4 \end{bmatrix}\begin{bmatrix} v_3 \\ v_4 \end{bmatrix}. \quad (4\text{-}72)$$

Since $p_4 = 0$ and $p_3 = -P$, $v_3 = \dfrac{p_3 L^3}{24EI} = \dfrac{-PL^3}{24EI}$.

4.2.5 STIFFNESS MATRIX OF A STRUCTURE: AN ALGORITHMIC APPROACH

From the above simple example, it can be seen that matrix $\mathbf{B}$ is a very sparse Boolean matrix, and the direct formation of $\mathbf{BkB}^t$ using matrix multiplication requires a considerable amount of storage. In the following, it is shown that one can form $\mathbf{BkB}^t$ with an assembling process (known also as *planting*), as follows:

Consider an element "a" of a structure, as shown in Figure 4.8, for which the element stiffness matrix can be written as

$$\mathbf{k}_a = \left[\begin{array}{c|c} \mathbf{k}_{ii} & \mathbf{k}_{ij} \\ \hline \mathbf{k}_{ji} & \mathbf{k}_{jj} \end{array}\right], \quad (4\text{-}73)$$

OPTIMAL DISPLACEMENT METHOD OF STRUCTURAL ANALYSIS 159

where i and j are the two end nodes of member a. Pre- and post-multiplication in the form of $\mathbf{BkB}^t$ has the following effect on $\mathbf{k}_a$:

$$\begin{bmatrix} 0 & 0 \\ 0 & 0 \\ 0 & 0 \\ \mathbf{I} & 0 \\ 0 & 0 \\ 0 & \mathbf{I} \\ 0 & 0 \\ 0 & 0 \end{bmatrix} \begin{bmatrix} \mathbf{k}_{ii} & \mathbf{k}_{ij} \\ \mathbf{k}_{ji} & \mathbf{k}_{jj} \end{bmatrix} \begin{bmatrix} 0 & 0 & 0 & \mathbf{I} & 0 & 0 & 0 & 0 \\ 0 & 0 & 0 & 0 & 0 & \mathbf{I} & 0 & 0 \end{bmatrix} = \begin{bmatrix} 0 & 0 \\ 0 & 0 \\ 0 & 0 \\ \mathbf{I} & 0 \\ 0 & 0 \\ 0 & \mathbf{I} \\ 0 & 0 \\ 0 & 0 \end{bmatrix} \begin{bmatrix} 0 & 0 & 0 & \mathbf{k}_{ii} & 0 & \mathbf{k}_{ij} & 0 & 0 \\ 0 & 0 & 0 & \mathbf{k}_{ji} & 0 & \mathbf{k}_{jj} & 0 & 0 \end{bmatrix}$$

(4-74)

$$= \begin{array}{c} 1 \\ 2 \\ 3 \\ 4 \\ 5 \\ 6 \\ 7 \\ 8 \end{array} \begin{bmatrix} 0 & 0 & 0 & 0 & 0 & 0 & 0 & 0 \\ 0 & 0 & 0 & 0 & 0 & 0 & 0 & 0 \\ 0 & 0 & 0 & 0 & 0 & 0 & 0 & 0 \\ 0 & 0 & 0 & \mathbf{k}_{ii} & 0 & \mathbf{k}_{ij} & 0 & 0 \\ 0 & 0 & 0 & 0 & 0 & 0 & 0 & 0 \\ 0 & 0 & 0 & \mathbf{k}_{ji} & 0 & \mathbf{k}_{jj} & 0 & 0 \\ 0 & 0 & 0 & 0 & 0 & 0 & 0 & 0 \\ 0 & 0 & 0 & 0 & 0 & 0 & 0 & 0 \end{bmatrix}$$

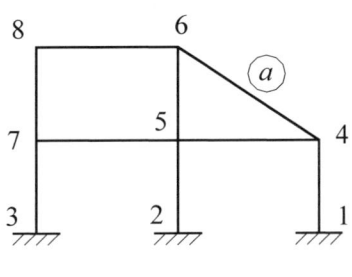

Fig. 4.8 A structural model S.

The adjacency matrix of S is also an 8×8 matrix, and the effect of node 4 being adjacent to node 6 is the existence of unit entries in the same locations as the sub-matrices of the element "a". One can build up the adjacency matrix of a graph by the addition of the effect of one member at a time. In the same way, one can also form the overall stiffness matrix of the structure by the addition of the contribution of every member in succession. As an example, for the graph shown in Figure 4.8, the overall stiffness matrix has the following pattern:

160   OPTIMAL STRUCTURAL ANALYSIS

$$\begin{array}{c} \ 1\ 2\ 3\ 4\ 5\ 6\ 7\ 8 \\ \begin{array}{c}1\\2\\3\\4\\5\\6\\7\\8\end{array}\begin{bmatrix} 1 & \cdot & \cdot & 1 & \cdot & \cdot & \cdot & \cdot \\ \cdot & 1 & \cdot & \cdot & 1 & \cdot & \cdot & \cdot \\ \cdot & \cdot & 1 & \cdot & \cdot & \cdot & 1 & \cdot \\ 1 & \cdot & \cdot & 1 & 1 & 1 & \cdot & \cdot \\ \cdot & 1 & \cdot & 1 & 1 & 1 & 1 & \cdot \\ \cdot & \cdot & \cdot & 1 & 1 & 1 & \cdot & 1 \\ \cdot & \cdot & 1 & \cdot & 1 & \cdot & 1 & 1 \\ \cdot & \cdot & \cdot & \cdot & \cdot & 1 & 1 & 1 \end{bmatrix} \end{array}$$ (4-75)

Non-zero entries are shown as "1". For a stiffness matrix, each of these non-zero entries is an $\eta \times \eta$ submatrix, where η is the DOF of each node of the structure. For example, for a planar truss, $\eta = 2$, and for a space frame, $\eta = 6$. The formation of the stiffness matrix by the above process is known as the *assembling* or *planting* of the stiffness matrix of a structure.

4.3 TRANSFORMATION OF STIFFNESS MATRICES

Methods for the formation of element stiffness matrices have been presented in Section 4.2. In the following, the stiffness matrices for bar and beam elements are transformed to global coordinate systems using the transformation described in Section 4.2.1.

From Eq. (4-7), we have

$$\bar{\mathbf{r}} = \mathbf{T}\mathbf{r},$$ (4-76)

$$\bar{\mathbf{u}} = \mathbf{T}\mathbf{u}.$$ (4-77)

From the definition of an element stiffness matrix in a local coordinate system,

$$\bar{\mathbf{r}} = \bar{\mathbf{k}}\bar{\mathbf{u}}.$$ (4-78)

By the substitution of Eqs (4-76) and (4-77) into the above equation, we have

$$\mathbf{r} = \mathbf{T}^{-1}\bar{\mathbf{k}}\mathbf{T}\mathbf{u} = \mathbf{T}^{t}\bar{\mathbf{k}}\mathbf{T}\mathbf{u}.$$ (4-79)

By definition of a stiffness matrix in a global coordinate system, we have

$$\mathbf{r} = \mathbf{k}\mathbf{u}.$$ (4-80)

OPTIMAL DISPLACEMENT METHOD OF STRUCTURAL ANALYSIS 161

A comparison of Eq. (4-79) and Eq. (4-80) results in

$$\mathbf{k} = \mathbf{T}^t \bar{\mathbf{k}} \mathbf{T}. \tag{4-81}$$

4.3.1 STIFFNESS MATRIX OF A BAR ELEMENT

Equation (4-52) provides the stiffness matrix of a bar element in its local coordinate system. This matrix in the global system, as shown in Figure 4.9, can be written as

$$\mathbf{k} = \begin{bmatrix} \mathbf{T} & \\ & \mathbf{T} \end{bmatrix}^t [\bar{\mathbf{k}}] \begin{bmatrix} \mathbf{T} & \\ & \mathbf{T} \end{bmatrix}. \tag{4-82}$$

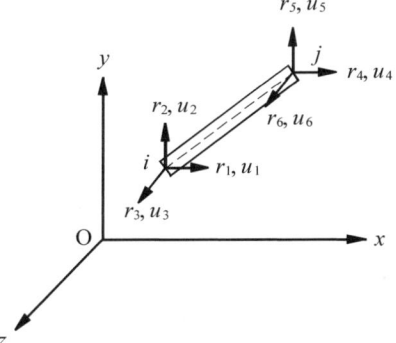

Fig. 4.9 A bar element of a space truss.

Denoting **T** in Eq. (4-32) by

$$\mathbf{T} = \begin{bmatrix} T_{11} & T_{12} & T_{13} \\ T_{21} & T_{22} & T_{23} \\ T_{31} & T_{32} & T_{33} \end{bmatrix}, \tag{4-83}$$

162   OPTIMAL STRUCTURAL ANALYSIS

$\mathbf{k}_m$ can be written as

$$\mathbf{k} = \frac{EA}{L} \begin{bmatrix} T_{11}^2 & T_{11}T_{12} & T_{11}T_{13} & -T_{11}^2 & -T_{11}T_{12} & -T_{11}T_{13} \\ T_{11}T_{12} & T_{12}^2 & T_{12}T_{13} & -T_{11}T_{12} & -T_{12}^2 & -T_{12}T_{13} \\ T_{11}T_{13} & T_{12}T_{13} & T_{13}^2 & -T_{11}T_{13} & -T_{12}T_{13} & -T_{13}^2 \\ -T_{11}^2 & -T_{11}T_{12} & -T_{11}T_{13} & T_{11}^2 & T_{11}T_{12} & T_{11}T_{13} \\ -T_{11}T_{12} & -T_{12}^2 & -T_{12}T_{13} & T_{11}T_{12} & T_{12}^2 & T_{12}T_{13} \\ -T_{11}T_{13} & -T_{12}T_{13} & -T_{13}^2 & T_{11}T_{13} & T_{12}T_{13} & T_{13}^2 \end{bmatrix}. \quad (4\text{-}84)$$

The entries of the above matrix can be found using the T_{ij} from Eq. (4.11). As an example, the stiffness matrix of bar 1 in the planar truss shown in Figure 4.10 can be obtained as

$$T_{11} = \frac{x_{21}}{(x_{12}^2 + y_{12}^2 + z_{12}^2)^{\frac{1}{2}}} = \frac{1}{\sqrt{2}} = \frac{\sqrt{2}}{2},$$

$$T_{12} = \frac{y_{21}}{(x_{12}^2 + y_{12}^2 + z_{12}^2)^{\frac{1}{2}}} = -\frac{1}{\sqrt{2}} = -\frac{\sqrt{2}}{2}.$$

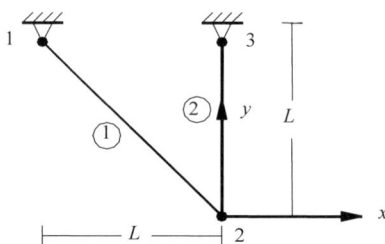

Fig. 4.10 A planar truss and the selected global coordinate system.

Therefore,

$$\mathbf{k}_1 = \frac{EA}{L\sqrt{2}} \begin{bmatrix} 0.5 & -0.5 & -0.5 & 0.5 \\ -0.5 & 0.5 & 0.5 & -0.5 \\ -0.5 & 0.5 & 0.5 & -0.5 \\ 0.5 & -0.5 & -0.5 & 0.5 \end{bmatrix}.$$

4.3.2 STIFFNESS MATRIX OF A BEAM ELEMENT

Consider a prismatic beam element as shown in Figure 4.11. The element forces and element distortions are defined by the following vectors:

$$\bar{\mathbf{r}} = \{r_1, r_2, r_3, ..., r_{12}\}^t,$$

and

$$\bar{\mathbf{u}} = \{u_1, u_2, u_3, ..., u_{12}\}^t,$$

where r_1 to r_3 are the force components at end i and r_4 to r_6 are moment components at end i. Also r_7 to r_9 are the force and r_{10} to r_{12} are the moment components, respectively, at the end j, and u_i ($i = 1, ..., 12$) are correspondingly the translations and rotations at the ends i and j of the element.

Using one of the methods presented in Section 4.2.2, the stiffness matrix of the beam element, in the local coordinate system defined in Figure 4.11, can be obtained from Eq. (4-83) as

$$\bar{\mathbf{k}} = \frac{E}{L}\begin{bmatrix} A & 0 & 0 & 0 & 0 & 0 & -A & 0 & 0 & 0 & 0 & 0 \\ 0 & 12I_z/L^2 & 0 & 0 & 0 & 6I_z/L & 0 & 0 & -12I_z/L^2 & 0 & 0 & 6I_z/L \\ 0 & 0 & 12I_y/L^2 & 0 & -6I_y/L & 0 & 0 & 0 & -12I_y/L^2 & 0 & -6I_y/L & 0 \\ 0 & 0 & 0 & J/2(1+v) & 0 & 0 & 0 & 0 & 0 & -J/2(1+v) & 0 & 0 \\ 0 & 0 & -6I_y/L & 0 & 4I_y & 0 & 0 & 0 & -6I_y/L & 0 & 2I_y & 0 \\ 0 & 6I_z/L & 0 & 0 & 0 & 4I_z & 0 & -6I_z/L & 0 & 0 & 0 & 2I_z \\ -A & 0 & 0 & 0 & 0 & 0 & A & 0 & 0 & 0 & 0 & 0 \\ 0 & -12I_z/L^2 & 0 & 0 & 0 & -6I_z/L & 0 & 12I_y/L^2 & 0 & 0 & 0 & -6I_z/L \\ 0 & 0 & -12I_y/L^2 & 0 & 6I_y/L & 0 & 0 & 0 & 12I_y/L^2 & 0 & 6I_y/L & 0 \\ 0 & 0 & 0 & -J/2(1+v) & 0 & 0 & 0 & 0 & 0 & J/2(1+v) & 0 & 0 \\ 0 & 0 & -6I_y/L & 0 & 2I_y & 0 & 0 & 0 & 6I_y/L & 0 & 4I_y & 0 \\ 0 & 6I_z/L & 0 & 0 & 0 & 2I_z & 0 & -6I_z/L & 0 & 0 & 0 & 4I_z \end{bmatrix} \quad (4\text{-}85)$$

In this matrix, I_y, I_z and J are the moments of inertia with respect to the $\bar{y}$ and $\bar{z}$ axes, and J is the polar moment of inertia of the section. E specifies the elastic modulus and v is the Poisson ratio. L denotes the length of the beam.

164   OPTIMAL STRUCTURAL ANALYSIS

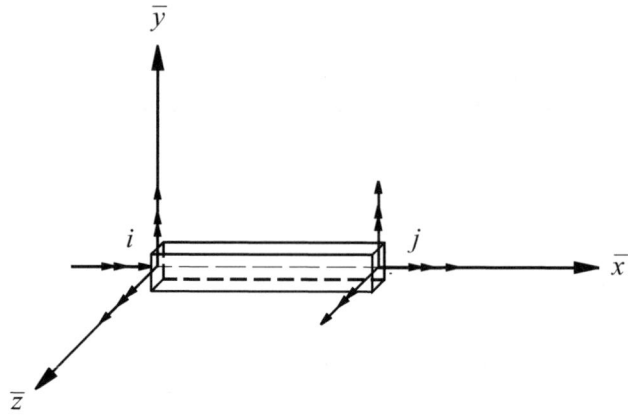

Fig. 4.11 A beam element in the local coordinate system.

For the two-dimensional case, the columns and rows corresponding to the third dimension can easily be deleted to obtain the stiffness matrix of an element of a planar frame.

The stiffness matrix in a global coordinate system can be written as

$$\mathbf{k} = \begin{bmatrix} \mathbf{T} & & \\ & \mathbf{T} & \\ & & \mathbf{T} \\ & & & \mathbf{T} \end{bmatrix}^t [\bar{\mathbf{k}}] \begin{bmatrix} \mathbf{T} & & \\ & \mathbf{T} & \\ & & \mathbf{T} \\ & & & \mathbf{T} \end{bmatrix}. \qquad (4\text{-}86)$$

For the two-dimensional case,

$$\mathbf{k} = \begin{bmatrix} \mathbf{T} & \\ & \mathbf{T} \end{bmatrix}^t [\bar{\mathbf{k}}] \begin{bmatrix} \mathbf{T} & \\ & \mathbf{T} \end{bmatrix}. \qquad (4\text{-}87)$$

The entries of **k** are as follows:

$k_{11} = T_{11}^2 \alpha_1 + T_{21}^2 \alpha_4^z$

$k_{21} = T_{11}T_{12}\alpha_1 + T_{21}T_{22}\alpha_4^z \quad k_{22} = T_{12}^2 \alpha_1 + T_{22}^2 \alpha_4^z$

$k_{31} = T_{21}\alpha_2^z \quad k_{32} = T_{22}\alpha_2^z \quad k_{33} = \alpha_3^z$

OPTIMAL DISPLACEMENT METHOD OF STRUCTURAL ANALYSIS 165

$k_{41} = -T_{11}^2 \alpha_1 + T_{21}^2 \alpha_4^z \quad k_{42} = -T_{21}T_{22}\alpha_4^z - T_{12}T_{11}\alpha_1 \quad k_{43} = -T_{21}\alpha_2^z \quad k_{44} = -T_{21}\alpha_2^z$

$k_{52} = -T_{21}^2 \alpha_4^z - T_{12}^2 \alpha_1$

$k_{54} = T_{21}T_{22}\alpha_4^z + T_{12}T_{11}\alpha_1 \quad k_{55} = T_{22}^2 \alpha_4^z + T_{12}^2 \alpha_1$

$k_{61} = T_{21}\alpha_2^z \quad k_{62} = T_{22}\alpha_2^z \quad k_{63} = \alpha_6^z \quad k_{64} = -T_{21}\alpha_2^z \quad k_{65} = -T_{22}\alpha_2^z \quad k_{66} = \alpha_3^z \qquad (4\text{-}88)$

in which

$$\alpha_1 = \frac{EA}{L}, \quad \alpha_2^z = \frac{6EI_z}{L^2}, \quad \alpha_3^z = \frac{4EI_z}{L}, \quad \alpha_4^z = \frac{12EI_z}{L^3}, \text{ and } \alpha_6^z = \frac{2EI_z}{L}.$$

As an example, for element 1 of the planar frame shown in Figure 4.12, we have,

$$T_{11} = 0 \quad T_{12} = 1 \quad T_{21} = -1 \quad T_{22} = 0,$$

and the stiffness matrix of the element is obtained as

$$\mathbf{k}_1 = 10^6 \begin{bmatrix} 1.25 & 0 & -0.75 & -1.25 & 0 & -0.75 \\ 0 & 200 & 0 & 0 & -200 & 0 \\ -0.75 & 0 & 6 & 0.75 & 0 & 3 \\ -1.25 & 0 & 0.75 & 1.25 & 0 & 0.75 \\ 0 & -200 & 0 & 0 & 200 & 0 \\ -0.75 & 0 & 3 & 0.75 & 0 & 6 \end{bmatrix}.$$

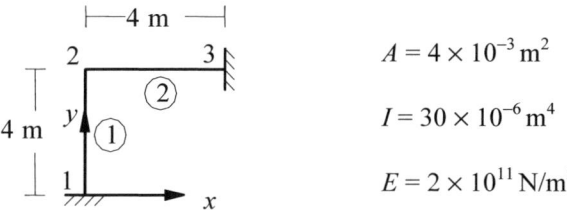

$A = 4 \times 10^{-3} \text{ m}^2$

$I = 30 \times 10^{-6} \text{ m}^4$

$E = 2 \times 10^{11} \text{ N/m}^2$

Fig. 4.12 A planar frame.

4.4 DISPLACEMENT METHOD OF ANALYSIS

Once the stiffness matrix of an element is obtained in the selected global coordinate system, it can be planted in the specified and initialised overall stiffness matrix of the structure **K**, using the process described in Section 4.2.5.

Example: Let S be a planar truss with an arbitrary nodal and element numbering, as shown in Figure 4.13. The entries of the transformation matrices of the members are calculated using Eq. (4-11) and Eq. (4-12) as follows:

For bar 1,

$$T_{11} = \frac{x_2 - x_1}{2} = \frac{1-0}{2} = \frac{1}{2} \text{ and } T_{12} = \frac{y_2 - y_1}{2} = \frac{\sqrt{3}-0}{2} = \frac{\sqrt{3}}{2}.$$

Similarly, for bar 2,

$$T_{11} = \frac{1}{2}, \quad T_{12} = -\frac{\sqrt{3}}{2},$$

and for bar 3,

$$T_{11} = 1, \quad T_{12} = 0.$$

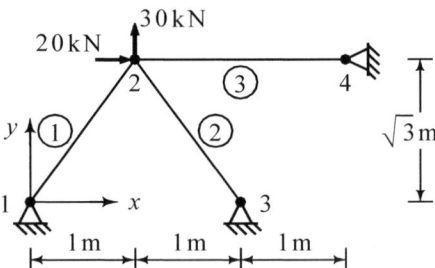

Fig. 4.13 A planar truss and the selected global coordinate system.

Using the relationship

$$\begin{bmatrix} F_i^x \\ F_i^y \\ F_j^x \\ F_j^x \end{bmatrix} = \frac{EA}{L} \begin{bmatrix} T_{11}^2 & T_{11}T_{12} & -T_{11}^2 & -T_{11}T_{12} \\ T_{11}T_{12} & T_{12}^2 & -T_{11}T_{12} & -T_{12}^2 \\ -T_{11}^2 & -T_{11}T_{12} & T_{11}^2 & T_{11}T_{12} \\ -T_{11}T_{12} & -T_{12}^2 & T_{11}T_{12} & T_{12}^2 \end{bmatrix} \begin{bmatrix} \delta_i^x \\ \delta_i^y \\ \delta_j^x \\ \delta_j^y \end{bmatrix}, \quad (4-89)$$

OPTIMAL DISPLACEMENT METHOD OF STRUCTURAL ANALYSIS 167

the stiffness matrices of the members are computed directly in the selected global coordinate system.

Now the stiffness matrices can be formed using Eq. (4-89):

For bar 1: $\quad \mathbf{k}_1 = \dfrac{EA}{2} \left[\begin{array}{cc|cc} 0.25 & 0.433 & -0.25 & -0.433 \\ 0.433 & 0.75 & -0.433 & -0.75 \\ \hline -0.25 & -0.433 & 0.25 & 0.433 \\ -0.433 & -0.75 & 0.433 & 0.75 \end{array} \right].$

For bar 2: $\quad \mathbf{k}_2 = \dfrac{EA}{2} \left[\begin{array}{cc|cc} 0.25 & -0.433 & -0.25 & 0.433 \\ -0.433 & 0.75 & 0.433 & -0.75 \\ \hline -0.25 & 0.433 & 0.25 & -0.433 \\ 0.433 & -0.75 & -0.433 & 0.75 \end{array} \right].$

For bar 3: $\quad \mathbf{k}_3 = \dfrac{EA}{2} \left[\begin{array}{cc|cc} 1 & 0 & -1 & 0 \\ 0 & 0 & 0 & 0 \\ \hline -1 & 0 & 1 & 0 \\ 0 & 0 & 0 & 0 \end{array} \right].$

The overall stiffness matrix of the structure is an 8×8 matrix, which can easily be formed by planting the three member stiffness matrices as follows:

$$\mathbf{K} = \dfrac{EA}{2} \begin{bmatrix} 0.250 & 0.433 & -0.250 & -0.433 & 0 & 0 & 0 & 0 \\ 0.433 & 0.750 & -0.433 & -0.750 & 0 & 0 & 0 & 0 \\ -0.250 & -0.433 & 1.500 & 0 & -0.250 & 0.433 & -1.00 & 0 \\ -0.433 & -0.750 & 0 & 1.500 & 0.433 & -0.750 & 0 & 0 \\ 0 & 0 & -0.250 & 0.433 & 0.250 & -0.433 & 0 & 0 \\ 0 & 0 & -0.433 & -0.750 & -0.433 & 0.750 & 0 & 0 \\ 0 & 0 & -1.00 & 0 & 0 & 0 & 1.00 & 0 \\ 0 & 0 & 0 & 0 & 0 & 0 & 0 & 1.00 \end{bmatrix}.$$

By partitioning $\mathbf{K}$ into 2×2 submatrices, it can easily be seen that it is pattern equivalent to the node adjacency matrix of the graph model of the structure as follows:

168   OPTIMAL STRUCTURAL ANALYSIS

$$\mathbf{C} * \mathbf{C}^{*t} = \begin{bmatrix} * & * & 0 & 0 \\ * & * & * & * \\ 0 & * & * & 0 \\ 0 & * & 0 & * \end{bmatrix}.$$

This pattern equivalence simplifies certain problems in structural mechanics, such as ordering the variables for bandwidth or profile reduction. Methods for increasing the sparsity, using special cutset bases, and improving the conditioning of structural matrices will be discussed in Section 4.7.

4.4.1 BOUNDARY CONDITIONS

The matrix **K** is singular, since the boundary conditions have to be applied. Consider,

$$\mathbf{p} = \mathbf{K}\mathbf{v},$$

and partition it for free and constraint DOF as follows:

$$\begin{bmatrix} \mathbf{p}_f \\ \mathbf{p}_c \end{bmatrix} = \begin{bmatrix} \mathbf{K}_{ff} & \mathbf{K}_{fc} \\ \mathbf{K}_{cf} & \mathbf{K}_{cc} \end{bmatrix} \begin{bmatrix} \mathbf{v}_f \\ \mathbf{v}_c \end{bmatrix}. \qquad (4\text{-}90)$$

This equation has a mixed nature; $\mathbf{p}_f$ and $\mathbf{v}_c$ have known values and $\mathbf{p}_c$ and $\mathbf{v}_f$ are unknowns. $\mathbf{K}_{ff}$ is known as the *reduced stiffness matrix* of the structure, which is non-singular for a rigid structure.

For boundary conditions such as $\mathbf{v}_c = \mathbf{0}$, it is easy to delete the corresponding rows and columns to obtain

$$\mathbf{p}_f = \mathbf{K}_{ff}\mathbf{v}_f, \qquad (4\text{-}91)$$

from which $\mathbf{v}_f$ can be obtained by solving the above set of equations. In a computer, this can be done by multiplying the diagonal entries of $\mathbf{K}_{cc}$ by a large number such as 10^{20}. An alternative approach is possible by equating the diagonal entries of $\mathbf{K}_{cc}$ to unity and all the other entries of these rows and columns to zero. If $\mathbf{v}_c$ contains some specified values, $\mathbf{p}_c$ will have corresponding $\mathbf{v}_c$ values. A third method, which is useful when a structure has more constraint DOF (such as many supports), consists of the formation of element stiffness matrices considering the corresponding constraints, that is, to form the reduced stiffness matrices of the elements in place of their complete matrices. This leads to some reduction in storage, and is also at the expense of additional computational effort.

OPTIMAL DISPLACEMENT METHOD OF STRUCTURAL ANALYSIS

As an example, the reduced stiffness matrix of the structure shown in Figure 4.13 can be obtained from **K** by deleting the rows and columns corresponding to the three supports 1, 3 and 4:

$$\begin{bmatrix} 20 \\ 30 \end{bmatrix} = \frac{EA}{2} \begin{bmatrix} 1.5 & 0 \\ 0 & 1.5 \end{bmatrix} \begin{bmatrix} u_{2x} \\ u_{2y} \end{bmatrix}.$$

Solving for the joint displacements, we have

$$u_{2x} = \frac{40}{1.5EA} \text{ and } u_{2y} = \frac{40}{EA}.$$

The member distortions can easily be extracted from the displacement vector, and multiplication by the stiffness matrix of each member results in its member forces in the global coordinate system. As an example, for member 3 we have,

$$\begin{bmatrix} r_{2x} \\ r_{2y} \\ r_{4x} \\ r_{4y} \end{bmatrix} = \frac{EA}{2} \begin{bmatrix} 1 & 0 & -1 & 0 \\ 0 & 0 & 0 & 0 \\ -1 & 0 & 1 & 0 \\ 0 & 0 & 0 & 0 \end{bmatrix} \begin{bmatrix} 40/1.5EA \\ 40/EA \\ 0 \\ 0 \end{bmatrix} = \begin{bmatrix} 13.33 \\ 0 \\ -13.33 \\ 0 \end{bmatrix}.$$

A transformation yields the member forces in the local coordinate systems, $\mathbf{r}_1 = \{-23.99 \ 23.99\}^t$, $\mathbf{r}_2 = \{-10.659 \ 10.65\}^t$ and $\mathbf{r}_3 = \{13.33 \ -13.33\}^t$

4.4.2 GENERAL LOADING

The joint load vector of a structure can be computed in two parts. The first part comes from the external concentrated loads and/or moments, which are applied to the joints defined as the nodes of S. The components of such loads are most easily specified in a global coordinate system and can be assembled to the joint load vector **p**.

The second part comes from the loads, which are applied to the spans of the members. These loads are usually defined in the local coordinate system of a member. For each member, the fixed end actions (FEA) can be calculated using existing classical formulae or tables. A simple computer program can be prepared for this purpose. The FEA should then be expressed in the global coordinate system using the transformation matrix given by Eq. (4-11). The FEA should then be reversed and applied to the end nodes of the members. These components can be added to **p** to form the final joint load vector. After **p** has been assembled and the boundary conditions imposed, the corresponding equations should be solved to obtain the

170   OPTIMAL STRUCTURAL ANALYSIS

joint displacements of the structure. Member distortions can then be extracted for each member in the reverse order to that used in assembling the **p** vector.

Example 1: A two-span continuous beam is considered as shown in Figure 4.14(a). EI is taken to be constant along the beam.

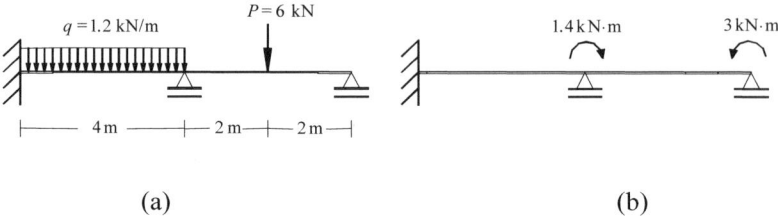

(a)   (b)

Fig. 4.14 A continuous beam and its equivalent loading.

For continuous beams, the transformation matrix **T** from local coordinate to global coordinate is identity, and therefore $\mathbf{k}_m = \overline{\mathbf{k}}_m$, that is, no transformation is required. Ignoring the axial deformation and using Eq. (4-63), the stiffness matrices of the elements are obtained as follows:

$$k_1 = k_2 = \frac{64}{4}\begin{bmatrix} 0.75 & 1.5 & -0.75 & 1.5 \\ 1.5 & 4 & -1.5 & 2 \\ -0.75 & -1.5 & 0.75 & -1.5 \\ 1.5 & 2 & -1.5 & 4 \end{bmatrix}.$$

Assembling the overall stiffness matrix and imposing the boundary conditions, the reduced stiffness matrix of the entire beam is obtained and the force–displacement relationship for the beam is written as

$$\begin{bmatrix} -1.40 \\ 3 \end{bmatrix} = 16 \begin{bmatrix} 8 & 2 \\ 2 & 4 \end{bmatrix} \begin{bmatrix} \theta_2^z \\ \theta_3^z \end{bmatrix}.$$

Solving the equations, we have

$$\begin{bmatrix} \theta_2^z \\ \theta_3^z \end{bmatrix} = \frac{1}{448} \begin{bmatrix} 4 & -2 \\ -2 & 8 \end{bmatrix} \begin{bmatrix} -1.4 \\ 3 \end{bmatrix} = \begin{bmatrix} -0.0259 \\ 0.0598 \end{bmatrix}.$$

OPTIMAL DISPLACEMENT METHOD OF STRUCTURAL ANALYSIS   171

Member forces are calculated as follows:

$$\begin{bmatrix} V_1 \\ M_1 \\ V_2 \\ M_2 \end{bmatrix} = 16 \begin{bmatrix} 0.75 & 1.5 & -0.75 & 1.5 \\ 1.5 & 4 & -1.5 & 2 \\ -0.75 & -1.5 & 0.75 & -1.5 \\ 1.5 & 2 & -1.5 & 4 \end{bmatrix} \begin{bmatrix} 0 \\ 0 \\ 0 \\ -0.0259 \end{bmatrix} + \begin{bmatrix} 2.4 \\ 1.6 \\ 2.4 \\ -1.6 \end{bmatrix} = \begin{bmatrix} 1.779 \\ 0.772 \\ 3.021 \\ -3.256 \end{bmatrix},$$

and

$$\begin{bmatrix} V_2 \\ M_2 \\ V_3 \\ M_3 \end{bmatrix} = 16 \begin{bmatrix} 0.75 & 1.5 & -0.75 & 1.5 \\ 1.5 & 4 & -1.5 & 2 \\ -0.75 & -1.5 & 0.75 & -1.5 \\ 1.5 & 2 & -1.5 & 4 \end{bmatrix} \begin{bmatrix} 0 \\ -0.0259 \\ 0 \\ +0.0598 \end{bmatrix} + \begin{bmatrix} 3 \\ 3 \\ 3 \\ -3 \end{bmatrix} = \begin{bmatrix} 3.814 \\ 3.258 \\ 2.186 \\ 0 \end{bmatrix}.$$

Example 2: A portal frame is considered as shown in Figure 4.15. The members are made of sections with $A = 150$ cm^2 and $I_z = 2 \times 10^4$ cm^4 and $E = 2 \times 10^4$ kN/cm^2. Calculate the joint rotations and displacements.

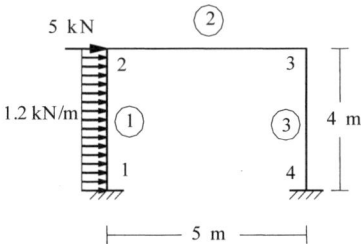

Fig. 4.15 A portal frame and its loading.

The equivalent joint loads are illustrated in Figure 4.16.

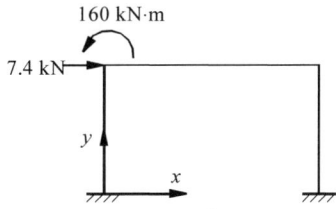

Fig. 4.16 Equivalent joint loads.

172   OPTIMAL STRUCTURAL ANALYSIS

Employing Eq. (4-88), the stiffness matrices for the members are obtained as follows:

For member 1,

$$\mathbf{k}_1 = 10^4 \left[\begin{array}{ccc|ccc} 0.008 & 0 & -1.5 & 0.008 & 0 & -1.5 \\ 0 & 0.75 & 0 & 0 & -0.75 & 0 \\ -1.5 & 0 & 400 & 1.5 & 0 & 200 \\ \hline 0.008 & 0 & 1.5 & 0.008 & 0 & 1.5 \\ 0 & -0.75 & 0 & 0 & 0.75 & 0 \\ -1.5 & 0 & 200 & 1.5 & 0 & 400 \end{array} \right],$$

and for member 2,

$$\mathbf{k}_2 = 10^4 \left[\begin{array}{ccc|ccc} 0.6 & 0 & 0 & -0.6 & 0 & 0 \\ 0 & 0.004 & 0.96 & 0 & -0.004 & 0.96 \\ 0 & 0.96 & 320 & 0 & -0.96 & 160 \\ \hline -0.6 & 0 & 0 & 0.6 & 0 & 0 \\ 0 & -0.004 & -0.96 & 0 & 0.004 & -0.96 \\ 0 & 0.96 & 160 & 0 & -0.96 & 320 \end{array} \right]$$

For member 3,

$$\mathbf{k}_3 = 10^4 \left[\begin{array}{ccc|ccc} 0.008 & 0 & 1.5 & -0.008 & 0 & 1.5 \\ 0 & 0.75 & 0 & 0 & -0.75 & 0 \\ 1.5 & 0 & 400 & -1.5 & 0 & 200 \\ \hline -0.008 & 0 & -1.5 & 0.008 & 0 & -1.5 \\ 0 & -0.75 & 0 & 0 & 0.75 & 0 \\ 1.5 & 0 & 200 & -1.5 & 0 & 400 \end{array} \right]$$

By assembling the stiffness matrices and imposing the boundary conditions, the following equations are obtained:

OPTIMAL DISPLACEMENT METHOD OF STRUCTURAL ANALYSIS 173

$$\begin{bmatrix} 7.4 \\ 0 \\ 160 \\ 0 \\ 0 \\ 0 \end{bmatrix} = 10^4 \begin{bmatrix} 0.608 & 0 & 1.5 & -0.6 & 0 & 0 \\ 0 & 0.754 & 0.96 & 0 & -0.004 & 0.96 \\ 1.5 & 0.96 & 720 & 0 & -0.96 & 160 \\ -0.6 & 0 & 0 & 0.608 & 0 & 1.5 \\ 0 & -0.004 & -0.96 & 0 & 0.754 & -0.96 \\ 0 & 0.96 & 160 & 1.5 & -0.96 & 720 \end{bmatrix} \begin{bmatrix} \delta_2^x \\ \delta_2^y \\ \theta_2^z \\ \delta_3^x \\ \delta_3^y \\ \theta_3^z \end{bmatrix}$$

Solving these equations, we obtain

$$\delta_2^x = 0.0659167, \quad \delta_2^y = 2.617764\text{E}{-}04, \quad \theta_2^z = -8.983453\text{E}{-}05,$$

$$\delta_3^x = 0.0653377, \quad \delta_3^y = -2.617704\text{E}{-}04 \text{ and } \theta_3^z = -1.16855\text{E}{-}04.$$

The final member forces can be found using the stiffness of the members, superimposed by the FEA.

4.5 STIFFNESS MATRIX OF A FINITE ELEMENT

In this section, a simple element is introduced from finite element methods, in order to show the capability of the method presented in Section 4.2.2, for the formation of element stiffness matrices.

4.5.1 STIFFNESS MATRIX OF A TRIANGULAR ELEMENT

For plane stress and plane strain problems, the displacements of a node can be specified by two components, and therefore, for each node of the triangular element, two DOF are considered, as shown in Figure 4.17.

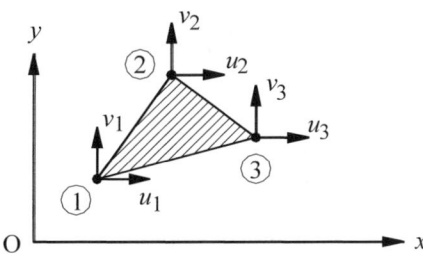

Fig. 4.17 A triangular element.

Element forces and displacements are defined by the following vectors:

$$\mathbf{r}_m = \{r_1 \quad r_2 \quad \ldots \quad r_6\}^t \text{ and } \mathbf{u}_m = \{u_1 \quad u_2 \quad \ldots \quad u_6\}^t. \quad (4\text{-}92)$$

A triangular element has its boundary attached continuously to the surrounding medium, and therefore no exact stiffness matrix can be derived. Therefore, an approximate solution should be sought.

The following displacement functions can be considered for the variation of the displacements:

$$u = \alpha_1 x + \alpha_2 y + \alpha_3 \text{ and } v = \alpha_4 x + \alpha_5 y + \alpha_6, \quad (4\text{-}93)$$

where $\alpha_1, \alpha_2, \ldots, \alpha_6$ are arbitrary constants that can be found from the displacements of the three nodes of the element. From the boundary conditions,

at node $i(x_i, y_i)$, $u = u_i$ and $v = v_i$,

at node $j(x_j, y_j)$, $u = u_j$ and $v = v_j$, $\quad (4\text{-}94)$

at node $k(x_k, y_k)$, $u = u_k$ and $v = v_k$,

the constants can be evaluated. Substituting in Eq. (4-93), we get

$$u = 1/2A\{[y_{kj}(x-x_j)-x_{kj}(y-y_j)]u_1 + [-y_{ki}(x-x_k)-x_{ki}(y-y_k)]u_3 + [y_{ji}(x-x_i)-x_{ji}(y-y_i)]u_5\},$$

$$v = 1/2A\{[y_{kj}(x-x_j)-x_{kj}(y-y_j)]u_2 + [-y_{ki}(x-x_k)-x_{ki}(y-y_k)]u_4 + [y_{ji}(x-x_i)-x_{ji}(y-y_i)]u_6\},$$

$$(4\text{-}95)$$

where

$$2A = 2(\text{area of the triangle}) = x_{kj} y_{ji} - x_{ji} y_{kj}, \quad (4\text{-}96)$$

and

$$x_{mn} = x_m - x_n \text{ and } y_{mn} = y_m - y_n. \quad (4\text{-}97)$$

From Eq. (4-95), it is obvious that both u and v vary linearly along each edge of the element and they depend only on the displacements of the two nodes on a particular edge. Therefore, the compatibility of displacements on two adjacent elements with a common boundary is satisfied.

By the theory of elasticity, the nodal displacements $\mathbf{u}_m^t = \{u_1, u_2, \ldots, u_6\}$ are related to total strains $e^t = \{e_{xx}, e_{yy}, e_{xy}\}$ by the following:

$$\mathbf{e} = \begin{bmatrix} e_{xx} \\ e_{yy} \\ e_{xy} \end{bmatrix} = \begin{bmatrix} \dfrac{\partial u}{\partial x} \\ \dfrac{\partial v}{\partial y} \\ \dfrac{\partial u}{\partial y} + \dfrac{\partial v}{\partial x} \end{bmatrix} = \frac{1}{2A} \begin{bmatrix} y_{kj} & 0 & -y_{ki} & 0 & y_{ji} & 0 \\ 0 & -x_{kj} & 0 & x_{ki} & 0 & -x_{ji} \\ -x_{kj} & y_{kj} & x_{ki} & -y_{ki} & -x_{ji} & y_{ji} \end{bmatrix} \begin{bmatrix} u_1 \\ u_2 \\ u_3 \\ u_4 \\ u_5 \\ u_6 \end{bmatrix}. \quad (4\text{-}98)$$

This relationship can be written in matrix notation as

$$\mathbf{e} = \hat{\mathbf{b}}\mathbf{u}, \quad (4\text{-}99)$$

where

$$\hat{\mathbf{b}} = \frac{1}{2A} \begin{bmatrix} y_{kj} & 0 & -y_{ki} & 0 & y_{ji} & 0 \\ 0 & -x_{kj} & 0 & x_{ki} & 0 & -x_{ji} \\ -x_{kj} & y_{kj} & x_{ki} & -y_{ki} & -x_{ji} & y_{ji} \end{bmatrix}. \quad (4\text{-}100)$$

The above equation indicates that, for a linearly varying displacement field, the strains are constant, and by Hooke's law it also leads to constant stresses. Substituting the total strain $\mathbf{e}$ in Eq. (4-96), we obtain the stress–displacement relationship,

$$\begin{bmatrix} \sigma_{xx} \\ \sigma_{yy} \\ \sigma_{xy} \end{bmatrix} = \frac{E}{2A(1-v^2)} \begin{bmatrix} y_{kj} & -v y_{kj} & -y_{ki} & v x_{ki} & y_{ji} & -v y_{ji} \\ v y_{kj} & -x_{kj} & -v y_{ki} & x_{ki} & v y_{ji} & -x_{ji} \\ -\Psi x_{kj} & \Psi y_{kj} & \Psi x_{ki} & -\Psi y_{ki} & -\Psi x_{ji} & \Psi y_{ji} \end{bmatrix} \begin{bmatrix} u_1 \\ u_2 \\ u_3 \\ u_4 \\ u_5 \\ u_6 \end{bmatrix}, \quad (4\text{-}101)$$

where v is the Poisson ratio and

$$\Psi = \frac{1-v}{2}.$$

The stiffness matrix is then calculated using Eq. (4-26), and for convenience it is presented in two separate parts as

$$\mathbf{k} = \mathbf{k}_n + \mathbf{k}_s, \quad (4\text{-}102)$$

where $\mathbf{k}_n$ represents the stiffness due to normal stresses and $\mathbf{k}_s$ represents the stiffness due to shearing stresses. Thus

176   OPTIMAL STRUCTURAL ANALYSIS

$$\mathbf{k}_n = \frac{Et}{4A(1-v^2)} \begin{bmatrix} y_{32}^2 & & & & & \\ -v y_{32} x_{32} & x_{32}^2 & & & \text{sym.} & \\ -y_{32} y_{31} & v x_{32} y_{31} & y_{31}^2 & & & \\ v y_{32} x_{31} & -x_{32} x_{31} & -v y_{31} x_{31} & x_{31}^2 & & \\ y_{32} y_{21} & -v x_{32} y_{21} & -y_{31} y_{21} & v x_{31} y_{21} & y_{21}^2 & \\ -v y_{32} x_{21} & x_{32} x_{21} & v y_{31} x_{21} & -x_{31} x_{21} & -v y_{21} x_{21} & x_{21}^2 \end{bmatrix},$$

and

$$\mathbf{k}_s = \frac{Et}{4A(1+v)} \begin{bmatrix} x_{32}^2 & & & & & \\ -x_{32} y_{32} & y_{32}^2 & & & \text{sym.} & \\ -x_{32} x_{31} & y_{32} x_{31} & x_{31}^2 & & & \\ x_{32} y_{31} & -y_{32} y_{21} & -x_{31} y_{31} & y_{31}^2 & & \\ x_{32} x_{21} & -y_{32} x_{21} & -x_{31} x_{21} & y_{31} x_{21} & x_{21}^2 & \\ -x_{32} y_{21} & y_{32} y_{21} & x_{31} y_{21} & -y_{31} y_{21} & -x_{21} y_{21} & y_{21}^2 \end{bmatrix}. \quad (4\text{-}103)$$

Using the same method, the stiffness matrices for other elements can be derived. Since there are many excellent books on finite element methods, no further details are provided here, and the interested reader may refer to Zienkiewicz [235], McGuire and Gallagher [163], and Bathe and Wilson [9] among many others.

4.6 COMPUTATIONAL ASPECTS OF THE MATRIX DISPLACEMENT METHOD

The main advantage of the displacement method is its simplicity for use in computer programming. This is due to the existence of a simple kinematical basis formed on a special cutset basis known as the *cocycle* basis of the graph model S of the structure. Such a basis does not correspond to the most sparse stiffness matrix; however, the sparsity is generally so good that there is usually no need to look further. However, if an optimal cutset basis of S is to be used in the displacement method, then all the problems involved in the force method, described in Chapter 3, still exist. The algorithm for the displacement method is summarised below. The coding for such an algorithm may be found in textbooks such as those of Vanderbilt [222] and Meek [164].

4.6.1 ALGORITHM

Step 1: Select a global coordinate system and number the nodes and members of the structure. An appropriate nodal ordering algorithm will be discussed in Chapter 5.

OPTIMAL DISPLACEMENT METHOD OF STRUCTURAL ANALYSIS

Step 2: After initialisation of all the vectors and matrices, read or generate the data for the structure and its members.

Step 3: For each member of the structure,

    (a) compute L, L^*, $\sin \alpha$, $\sin \beta$, $\sin \gamma$, $\cos \alpha$, $\cos \beta$, and $\cos \gamma$;

    (b) compute the rotation matrix **T**;

    (c) form the member stiffness matrix $\bar{\mathbf{k}}$ in its local coordinate system;

    (d) form the member stiffness matrix **k** in the selected global coordinate system;

    (e) plant **k** in the overall stiffness matrix **K** of the structure.

Step 4: For each loaded member,

    (a) read the FEA;

    (b) transform the FEA to the global coordinate system and reverse it to apply at joints;

    (c) store these joint loads in the specified overall joint load vector.

Step 5: For each loaded joint,

    (a) read the joint number and the applied joint loads;

    (b) store it in the overall joint load vector.

Step 6: Apply boundary conditions to the structural stiffness matrix **K** to obtain the reduced stiffness matrix $\mathbf{K}_{ff}$. Repeat the same for the overall joint load vector.

Step 7: Solve the corresponding equations to obtain the joint displacements.

Step 8: For each member,

    (a) extract the member distortions from the joint displacements;

    (b) rotate the member distortions to the local coordinate system;

    (c) compute the member stiffness matrix;

    (d) compute the member forces and FEA.

178 OPTIMAL STRUCTURAL ANALYSIS

Step 9: Compute the joint displacements and the member forces.

The application of the above procedure is now illustrated by a simple example so that the reader can use it to fully understand the computational steps.

4.6.2 EXAMPLE

Consider a planar truss, as shown in Figure 4.18. Member 1 has a uniform load of intensity 0.6 kN/m, and at joint 2 a concentrated load of magnitude 1.05 kN is applied. The cross-sectional areas for members are $2A$ and $1.8A$, respectively.

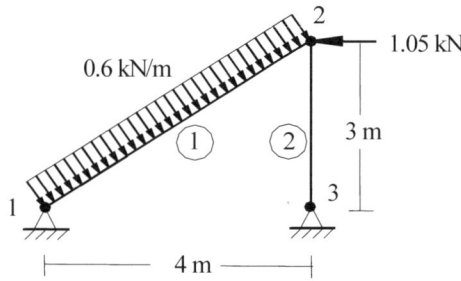

Fig. 4.18 A planar truss with general loading.

The selected global coordinate system and the equivalent nodal forces are illustrated in Figure 4.19. The stiffness matrices are formed as follows:

For member 1:

$$\mathbf{k}_1 = \frac{2EA}{5} \begin{bmatrix} 0.64 & 0.48 & -0.64 & -0.48 \\ 0.48 & 0.36 & -0.48 & -0.36 \\ -0.64 & -0.48 & 0.64 & 0.48 \\ -0.48 & -0.36 & 0.48 & 0.36 \end{bmatrix}.$$

For member 2:

$$\mathbf{k}_2 = \frac{1.8EA}{3} \begin{bmatrix} 0 & 0 & 0 & 0 \\ 0 & +1 & 0 & -1 \\ 0 & 0 & 0 & 0 \\ 0 & -1 & 0 & +1 \end{bmatrix}.$$

OPTIMAL DISPLACEMENT METHOD OF STRUCTURAL ANALYSIS 179

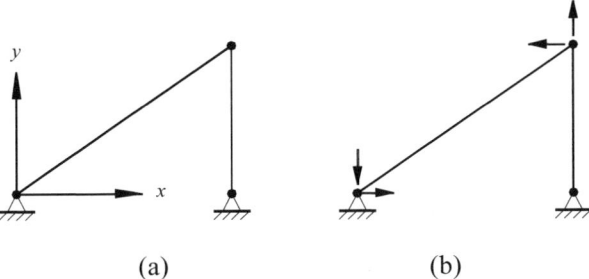

Fig. 4.19 The selected coordinate system and the equivalent nodal loads.

The overall stiffness matrix is then obtained as follows:

$$\mathbf{K} = EA \begin{bmatrix} 0.256 & 0.192 & -0.256 & -0.192 & 0 & 0 \\ 0.192 & 0.144 & -0.192 & -0.144 & 0 & 0 \\ -0.256 & -0.192 & 0.256 & 0.192 & 0 & 0 \\ -0.192 & -0.144 & 0.192 & 0.744 & 0 & -0.6 \\ 0 & 0 & 0 & 0 & 0 & 0 \\ 0 & 0 & 0 & -0.6 & 0 & 0.6 \end{bmatrix}.$$

The FEA are shown in Figure 4.19(b), and calculated for member 1 as follows:

$$\mathbf{FEA}_1 = \begin{bmatrix} 0 \\ 1.5 \\ 0 \\ 1.5 \end{bmatrix}.$$

These forces are reversed and transformed into the global coordinate system as follows:

$$\mathbf{T}_1^t(-\mathbf{FEA}_1) = \begin{bmatrix} 0.8 & -0.6 & 0 & 0 \\ 0.6 & 0.8 & 0 & 0 \\ 0 & 0 & 0.8 & -0.6 \\ 0 & 0 & 0.6 & 0.8 \end{bmatrix} \begin{bmatrix} 0 \\ -1.5 \\ 0 \\ -1.5 \end{bmatrix} = \begin{bmatrix} 0.9 \\ -1.2 \\ 0.9 \\ -1.2 \end{bmatrix}.$$

180 OPTIMAL STRUCTURAL ANALYSIS

By superimposing the concentrated force at node 2, the final vector of external forces is obtained as follows:

$$\mathbf{p} = \{0.9 \quad -1.2 \quad -0.15 \quad -1.2 \quad 0 \quad 0\}^t.$$

By substituting a large number such as 1.E + 30 for the diagonal entries corresponding to the zero displacement boundary conditions, we have

$$\begin{bmatrix} 0 \\ 0 \\ -0.15 \\ -1.2 \\ 0 \\ 0 \end{bmatrix} = EA \begin{bmatrix} 1.E+30 & 0.192 & -0.256 & -0.192 & 0 & 0 \\ 0.192 & 1.E+30 & -0.192 & -0.256 & 0 & 0 \\ -0.256 & -0.192 & 0.256 & 0.192 & 0 & 0 \\ -0.192 & -0.256 & 0.192 & 0.714 & 0 & -0.6 \\ 0 & 0 & 0 & 0 & 1.E+30 & 0 \\ 0 & 0 & 0 & -0.6 & 0 & 1.E+30 \end{bmatrix} [\mathbf{v}].$$

By solving these equations, we have

$$\mathbf{v} = \frac{1}{EA} \{0 \quad 0 \quad 0.845 \quad -1.907 \quad 0 \quad 0\}^t.$$

The member forces are now computed as follows:

$$\mathbf{r}_1 = \frac{2}{5} \begin{bmatrix} 1 & 0 & -1 & 0 \\ 0 & 0 & 0 & 0 \\ -1 & 0 & 1 & 0 \\ 0 & 0 & 0 & 0 \end{bmatrix} \begin{bmatrix} 0.8 & 0.6 & 0 & 0 \\ -0.6 & 0.8 & 0 & 0 \\ 0 & 0 & 0.8 & 0.6 \\ 0 & 0 & -0.6 & 0.8 \end{bmatrix} \begin{bmatrix} 0 \\ 0 \\ 0.845 \\ -1.907 \end{bmatrix} + \begin{bmatrix} 0 \\ 1.5 \\ 0 \\ 1.5 \end{bmatrix} = \begin{bmatrix} 0.179 \\ 1.5 \\ -0.179 \\ 1.5 \end{bmatrix},$$

and

$$\mathbf{r}_2 = \frac{3}{5} \begin{bmatrix} 1 & 0 & -1 & 0 \\ 0 & 0 & 0 & 0 \\ -1 & 0 & 1 & 0 \\ 0 & 0 & 0 & 0 \end{bmatrix} \begin{bmatrix} 0 & -1 & 0 & 0 \\ 1 & 0 & 0 & 0 \\ 0 & 0 & 0 & -1 \\ 0 & 0 & 1 & 0 \end{bmatrix} \begin{bmatrix} 0.845 \\ -1.907 \\ 0 \\ 0 \end{bmatrix} + \begin{bmatrix} 0 \\ 0 \\ 0 \\ 0 \end{bmatrix} = \begin{bmatrix} 1.091 \\ 0 \\ -1.091 \\ 0 \end{bmatrix}.$$

4.7 OPTIMALLY CONDITIONED CUTSET BASES

For an efficient displacement analysis of a structure, special considerations such as structuring its stiffness matrix and improving its conditioning should to be taken into account. The former will be discussed in Chapter 5 and the latter is studied in this section.

OPTIMAL DISPLACEMENT METHOD OF STRUCTURAL ANALYSIS   181

In order to optimise the conditioning of the stiffness matrices, special cutset bases must be used in the formation of kinematical bases.

A cutset basis with the following properties is defined as an *optimally conditioned cutset basis*:

(a) It is an optimal cutset basis, that is, the number of non-zero entries of its cutset adjacency matrix and the corresponding number of non-zero entries of its stiffness matrix are a minimum.

(b) The members of the lowest weight of S are included in the overlaps of the cutsets, that is, the off-diagonal terms of the corresponding stiffness matrix have the smallest possible magnitudes.

A weighted graph may or may not have an optimally conditioned cutset basis. However, if such a basis does not exist or cannot be found, then a compromise should be found to satisfy the above two conditions, that is, a basis that partially satisfies both conditions should be selected.

4.7.1 MATHEMATICAL FORMULATION OF THE PROBLEM

The cardinality of a cutset basis for a connected graph is given by

$$\rho(S) = N(S) - 1. \tag{4-104}$$

The problem of finding an optimally conditioned cutset basis can be stated as follows:

Select a cutset basis $\{C_1^*, C_2^*, ..., C_{\rho(S)}^*\}$ such that

$$L_s = \text{Min} \bigcup_{i=1}^{\rho(S)-1} W(C^{*^i} \cap C_{i+1}^*),$$

and

$$W_s = \text{Min} \bigcup_{i=1}^{\rho(S)-1} W(C^{*^i} \cap C_{i+1}^*), \tag{4-105}$$

with $(C^{*^i} = \bigcup_{j=1}^{i} C_j^*)$, L denoting the length and W indicating the weight of the members of $(C^{*^i} \cap C_{i+1}^*)$, respectively.

Again we have a multi-objective optimisation problem, whose solution is not obvious. Therefore, we design an algorithm that is practical and satisfies the required conditions partially.

4.7.2 SUBOPTIMALLY CONDITIONED CUTSET BASES

A fundamental cutset basis of a graph can easily be generated using each branch of a spanning tree as the generator of a cutset. A more common cutset basis, employed in the displacement method of structural analysis, is a cocycle basis of S. In this basis, each element simply isolates a node of S, excepting the ground node.

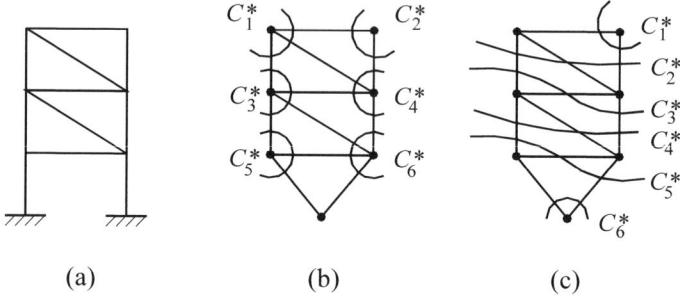

(a)   (b)   (c)

Fig. 4.20 A planar frame S, a cocycle basis and a cutset basis of S.

Although a cocycle basis corresponds to a rather sparse cutset adjacency matrix, other cutset bases corresponding to more sparse cutset adjacency matrices, leading to more sparse stiffness matrices, can be generated. As an example, consider a frame model S as depicted in Figure 4.20(a) for which a cocycle basis and a cutset basis are selected, as illustrated in Figures 4.20(b and c), respectively. The patterns of the corresponding cutset adjacency matrices are shown below using * for non-zero entries:

$$\mathbf{C}_1^* \mathbf{C}_1^{*t} = \begin{matrix} & 1 & 2 & 3 & 4 & 5 & 6 \\ 1 & * & * & * & * & & \\ 2 & * & * & & * & & \\ 3 & * & & * & * & & * \\ 4 & * & * & * & * & & * \\ 5 & & * & & * & * & \\ 6 & & * & * & * & & * \end{matrix} \qquad \mathbf{C}_2^* \mathbf{C}_2^{*t} = \begin{matrix} & 1 & 2 & 3 & 4 & 5 & 6 \\ 1 & * & * & & & & \\ 2 & * & * & * & & & \\ 3 & & * & * & * & & \\ 4 & & & * & * & * & \\ 5 & & & & * & * & * \\ 6 & & & & & * & * \end{matrix}$$

$$\chi(\mathbf{C}_1^* \mathbf{C}_1^{*t}) = 24 \qquad \chi(\mathbf{C}_2^* \mathbf{C}_2^{*t}) = 16$$

It will be realised that sparser stiffness matrices can be generated using suitable cutset bases rather than by employing the traditional cocycle basis [118].

In order to keep the off-diagonal terms small, the members of the overlaps of the cutsets should be as flexible as possible, that is, the lower weight members should

OPTIMAL DISPLACEMENT METHOD OF STRUCTURAL ANALYSIS

be included in the overlaps. In the following, three algorithms are designed for the formation of suboptimal cutset bases of the graph model of the structures.

4.7.3 ALGORITHMS

The formation of a cocycle basis of a graph model S is simple and straightforward. For this purpose, the members incident with each free node (except the selected datum node) are taken as an element of the basis. By repeating this operation for all the free nodes, the process of the generation is completed.

Algorithm A

Step 1: Generate a spanning tree of maximal weight. Order its members (branches) in ascending magnitudes of weight.

Step 2: Use a branch of the least weight, form the selected tree and form the first fundamental cutset on this branch.

Step 3: Form the next fundamental cutset on the unused branch of least weight.

Step k: Repeat Step 3 for the other unused branches until $\rho(S) = N(S) - 1$ independent cutsets forming a basis are generated.

Algorithm B

Step 1: Form a cocycle basis; denote the selected cocycles by C^{*1}.

Step 2: Take the first cocycle C_1^* of C^{*1} and combine with the remaining cocycles of C^{*1}. For each cocycle C_j^* (j = 2, ..., $\rho(S)$) satisfying the following condition, replace C_j^* with $C_i^* \oplus C_j^*$.

$$\text{Condition: } (L_{s2} < L_{s1}) \text{ or } (L_{s2} = L_{s1} \text{ and } W_{s2} < W_{s1}),$$

where $L_{s1}(W_{s1})$ and $L_{s2}(W_{s2})$ indicate the lengths (weights) before and after the application of the combining process, and $\oplus$ denotes the modulus 2 addition. The new set of cocycles and/or cutsets is denoted by C^{*2}.

Step 3: Take C_2^* of C^{*2} and repeat a process similar to that of Step 2.

Step k: Take C_k^* of C^{*k-1} and combine with the elements of C^{*k-1}. The process terminates when k becomes equal to $\rho(S)$.

184   OPTIMAL STRUCTURAL ANALYSIS

Algorithm C

This algorithm is the same as Algorithm B except that the corresponding condition is replaced by the following one:

Condition: $(W_{s2} < W_{s1})$ or $(W_{s2} = W_{s1}$ and $L_{s2} < L_{s1})$.

The selected bases are suboptimal and contain elements with lower weight members leading to kinematical bases corresponding to small off-diagonal terms for stiffness matrices.

4.7.4 EXAMPLE

A one-bay four-storey planar truss is considered as shown in Figure 4.21, with cross sections being designated by A_i. Typical member cross sections are

$A_1 = 20$ cm^2, $A_2 = 10$ cm^2, $A_3 = 5$ cm^2, $A_4 = 4$ cm^2 and $E = 2.1 \times 10^4$ kN/cm^2.

The patterns of the cutset bases adjacency matrices are illustrated in the following:

	1	2	3	4	5	6	7	8			1	2	3	4	5	6	7	8
1	*	*	*							1	*					*	*	
2	*	*	*	*						2		*		*		*		
3	*	*	*	*	*					3			*	*			*	*
4		*	*	*	*	*				4		*	*	*		*	*	*
5			*	*	*	*	*			5	*					*	*	
6				*	*	*	*	*		6	*	*			*	*	*	
7					*	*	*	*		7			*	*			*	*
8						*	*	*		8			*	*			*	*

Pattern of $C*C*^t$ by a cocycle basis.   Pattern of $C*C*^t$ by Algorithm A.

$$\begin{array}{c}1\ 2\ 3\ 4\ 5\ 6\ 7\ 8\\\begin{array}{c}1\\2\\3\\4\\5\\6\\7\\8\end{array}\begin{bmatrix}*&*&&&&&&\\ *&*&*&&&&&\\ &*&*&*&&&&\\ &&*&*&*&&&\\ &&&*&*&*&&\\ &&&&*&*&*&\\ &&&&&*&*&*\\ &&&&&&*&*\end{bmatrix}\end{array}\qquad\begin{array}{c}1\ 2\ 3\ 4\ 5\ 6\ 7\ 8\\\begin{array}{c}1\\2\\3\\4\\5\\6\\7\\8\end{array}\begin{bmatrix}*&*&&&&&&\\ *&*&*&&&&&\\ &*&*&*&&&&\\ &&*&*&*&&&\\ &&&*&*&*&&\\ &&&&*&*&*&*\\ &&&&&*&*&*\\ &&&&&*&*&*\end{bmatrix}\end{array}$$

Pattern of $\mathbf{C}*\mathbf{C}^{*t}$ by Algorithm B.   Pattern of $\mathbf{C}*\mathbf{C}^{*t}$ by Algorithm C.

The condition numbers of stiffness matrices, the sparsity and the magnitudes of L_s and W_s for the selected cutset bases are illustrated in Table 4.1.

Table 4.1 Comparison of the condition numbers and sparsities.

Algorithm	PL	$\chi(\mathbf{C}*\mathbf{C}^{*t})$	L_s	W_s
Cocycles	2.720131	34	13	75, 936.5
A	2.145762	32	12	48, 048.7
B	2.502612	22	7	46, 200.0
C	2.245613	24	8	36, 400.0

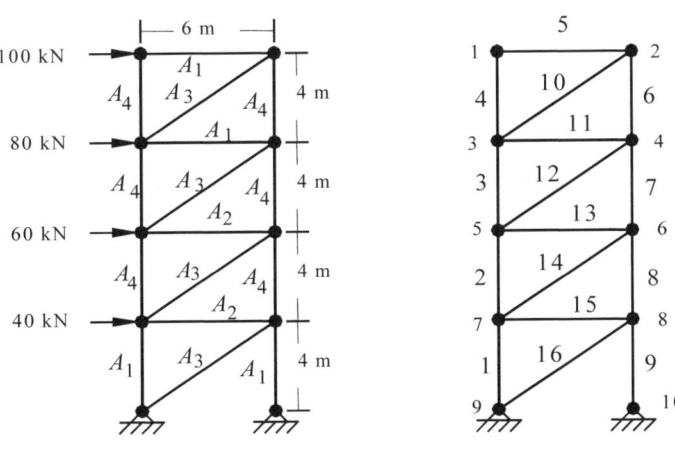

(a) A planar truss.   (b) The graph model S.

Fig. 4.21 A planar truss and its graph model S.

The execution time for the formation of the selected cutset bases (T_C) and the corresponding stiffness matrices (T_K) are presented in Table 4.2.

Table 4.2 Comparison of the computational time.

Time	Cocycle basis	A	B	C
T_C	0.00	0.88	0.76	0.88
T_K	0.43	0.65	0.48	0.60

Although the sparsity of stiffness matrices **K** for frame structures can be improved by the formation of special cutset bases in place of cocycle bases, the improvements, in general, are not significant. On the other hand, the conditioning of **K** can be improved by employing appropriate cutset bases. Algorithm B improves the conditioning of the stiffness matrices, maintaining the sparsity of the stiffness matrices. This improvement is more significant for Algorithms A and C, although the sparsity of **K** is not maintained.

EXERCISES

4.1 Compute the transformation matrix **T** of the element "*a*" for the following truss:

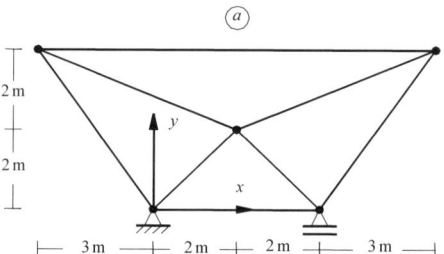

4.2 For the following planar frame, compute the joint displacements, ignoring the effect of axial deformations of the members. *EI* is considered constant for all the members.

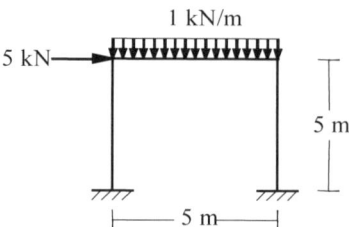

4.3 Compute the stiffness matrix of member "a" for the following pitched-roof portal frame, in the selected coordinate system:

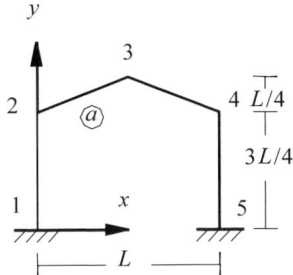

4.4 Consider a single-layer grid in the xz-plane loaded in the y-direction. Derive the stiffness matrices of a member of this grid in local and global coordinate systems.

4.5 Find the joint displacements of the space truss in the following diagram, where EA is considered to be the same for all the members:

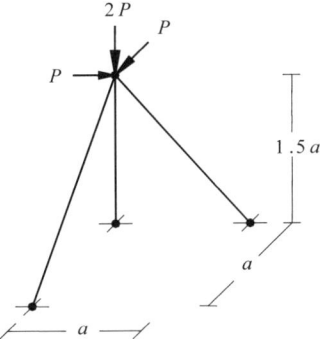

4.6 Derive the stiffness matrix of a beam element of a planar frame that has one end hinged.

4.7 Determine the patterns of the stiffness matrix and reduced stiffness matrix of the following pitched-roof frame using two different node numberings chosen arbitrarily:

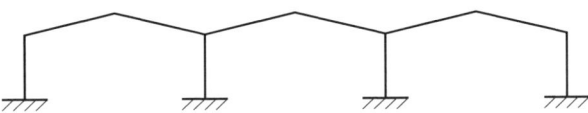

188   OPTIMAL STRUCTURAL ANALYSIS

4.8   Compute the joint load vector of the following frame:

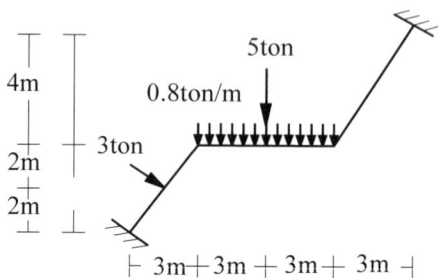

4.9   Perform the matrix stiffness analysis for the following truss, where $P = 100$ kN, $EA = 40 \times 10^6$ N and $L = 4$ m:

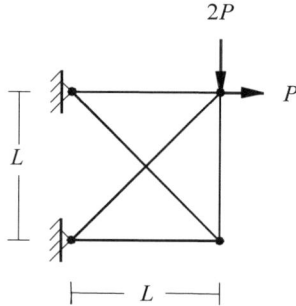

4.10   Derive the overall stiffness matrix of the following grid and calculate its joint displacements:

$EI = 60 \times 10^5$ N·m², $GJ = 100 \times 10^5$ N·m², $L = 4.0$ m and $P = 1.0 \times 10^5$ N.

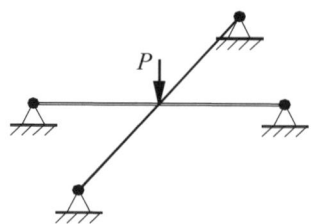

4.11 Perform the matrix stiffness analysis for the following grid:

$P = 30$ kN, $E = 2.1 \times 10^4$ kN/cm^2, $G = 1.05 \times 10^4$ kN/cm^2, $I_x = I_y = 25{,}170$ cm^4,

$I_z = 8560$ cm^4, $J = 33{,}730$ cm^4 and $A = 149$ cm^2.

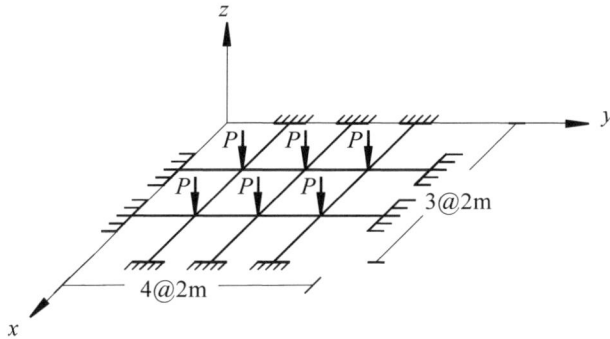

4.12 Find the condition numbers of the stiffness matrix of the following planar frame. For member 1, the section properties are $A_1 = 10.6 \times 10^{-4}$ m^2 and $I_1 = 17.1 \times 10^7$ m^4. For member 2, $A_2 = 9.7 \times 10^{-3}$ m^2 and $I_2 = 19.61 \times 10^{-5}$ m^4. Modulus of elasticity is equal to $E = 2.1 \times 10^4$ kN/cm^2.

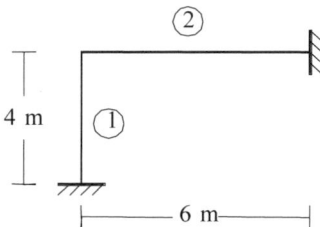

4.13 Study the effect of bandwidth reduction on the conditioning of the stiffness matrices of structures, and find out whether this effect is significant. Illustrate this fact numerically using a simple matrix.

CHAPTER 5

Ordering for Optimal Patterns of Structural Matrices: Graph Theory Methods

5.1 INTRODUCTION

In this chapter, ordering methods are presented for forming the elements of sparse structural matrices into special patterns. Such a transformation reduces the storage and the number of operations required for the solution, and leads to more accurate results. Graph-theory methods are presented for different approaches to reorder equations to preserve their sparsity, leading to predefined patterns. Alternative, objective functions are considered, and heuristic algorithms are presented to achieve these objectives. The three main methods for the solution of structural equations require the optimisation of bandwidth, profile and frontwidth, especially for those encountered in finite element (FE) analysis. Methods are presented for reducing the bandwidth of flexibility matrices. Bandwidth optimisation of rectangular matrices is presented for its use in the formation of sparse flexibility matrices.

Entries of the stiffness and flexibility matrices are provided with the most appropriate specified patterns for the solution of the corresponding equations. Realisation of these patterns (or not) affects the formulation of the mathematical models and the efficiency of the solution. Many patterns can be designed, depending on the solution scheme being used. Figure 5.1 shows some of the popular ones that are encountered in practice.

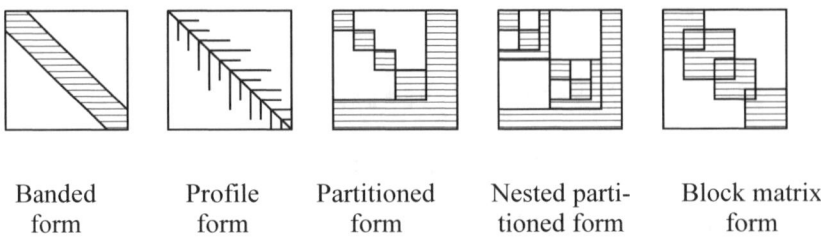

| Banded form | Profile form | Partitioned form | Nested partitioned form | Block matrix form |

Fig. 5.1 Different matrix forms.

The pattern equivalence of the stiffness matrix of a structure and the cutset basis matrix $\mathbf{C}^*\mathbf{C}^{*t}$ of its graph model and the pattern equivalence of the flexibility matrix of a structure with that of a generalised cycle basis matrix $\mathbf{CC}^t$ of its graph model reduce the size of the problem β-fold, β being the degrees of freedom of the nodes of the model for the displacement method, and $\beta = 1$–6 depending on the type of structure being studied using the force method.

5.2 BANDWIDTH OPTIMISATION

The analysis of many problems in structural engineering involves the solution of a set of linear equations of the form

$$\mathbf{Ax} = \mathbf{b}, \tag{5-1}$$

where $\mathbf{A}$ is a symmetric, positive-definite and usually very sparse matrix. For large structures encountered in practice, 30–50% of the computer execution time may be devoted for solving these equations. This figure may rise to about 80% in non-linear, dynamic or structural optimisation problems.

Different methods can be used for the solution of the system of equations, of which the Gaussian elimination is the most popular among structural analysts, since it is simple, accurate and practical, and produces some very satisfactory error bounds.

In the forward course of elimination, new non-zero entries may be created, but the back substitution does not lead to any new non-zero elements. It is beneficial to minimise the total number of such non-zero elements created during the forward course of the Gaussian elimination to reduce the round-off errors and the computer storage. Matrix $\mathbf{A}$ of Eq. (5-1) can be transformed by means of row and column operations to a form that leads to the creation of a minimum number of non-zero entries during the forward course of the elimination. This is equivalent to the "a priori" determination of the permutation matrices $\mathbf{P}$ and $\mathbf{Q}$, such that

$$\mathbf{PAQ} = \mathbf{G}. \tag{5-2}$$

When **A** is symmetric and positive-definite, it is advantageous to have **G** also symmetric so that only the non-zero elements on and above the diagonal of **G** need to be stored, and only about half as many arithmetic operations are needed in the elimination. The diagonal elements of **A** and **G** are the same, though their positions are different. In order to preserve symmetry, **P** is taken as **Q** so that Eq. (5-2) becomes

$$Q^t A Q = G. \qquad (5\text{-}3)$$

For transforming a symmetric matrix **A** into the forms depicted in Figure 5.1, various methods are available, some of which are described in this chapter. However, owing to the simplicity of the banded form, most of the material presented is confined to optimising the bandwidth of the structural matrices and other forms are only introduced briefly.

In the Gaussian elimination method, the time required to solve the resulting equations by the banded matrix technique is directly proportional to the square of the bandwidth of **A**. As mentioned earlier, the solution of these equations forms a large percentage of the total computational effort needed for the structural analysis. Therefore, it is not surprising that a lot of attention is being paid to the optimisation of the bandwidth of these sparse matrices. A suitable ordering of the elements of a kinematical basis for a structure reduces the bandwidth of **A**, hence decreasing the solution time, storage and round-off errors. Similarly, ordering the elements of a statical basis results in the reduction of the bandwidth of the corresponding flexibility matrix of the structure.

Iterative methods using different criteria for the control of the process of interchanging rows and columns of **A** are described by many authors, for example, see Rosen [189] and Grooms [67]. For these methods, in general, the required storage and CPU time can be high, making them uneconomical.

The first direct method for bandwidth reduction was recognised by Harary [73] in 1967, who posed the following question:

For a graph S with N(S) nodes, how can labels 1, 2, ..., N(S) be assigned to nodes in order to minimise the maximum absolute value of the difference between the labels of all pairs of adjacent nodes?

For a graph labelled in such an optimum manner, the corresponding adjacency matrix will have unit entries concentrated as closely as possible to its main diagonal.

In structural engineering, Cuthill and McKee [33] developed the first graph-theoretical approach for reducing the bandwidth of stiffness matrices. In their work, a level structure is used, which is called a *spanning tree* of a structure. The author's interest in bandwidth reduction was initially motivated by an interest in generating and ordering the elements of cycle bases and generalised cycle bases of a graph, as defined in Chapter 3, in order to reduce the bandwidth of the flexibility matrices [94,96]. For this purpose, a *shortest route tree* (SRT) has been used. The application of this approach has been extended to the elements of a kinematical basis (cutset basis) to reduce the bandwidth of stiffness matrices. Subsequently, it has been noticed that there is a close relation between Cuthill–McKee's *level structure* and the author's SRT. However, the SRT contains additional information about the connectivity properties of the corresponding structure.

Further improvements have been made by employing special types of SRTs such as the longest and narrowest ones [97]. The generation of a suitable SRT depends on an appropriate choice of starting node. Kaveh [94] used an end node of an arbitrary SRT having the least valency, which was chosen from its last counter (level). Gibbs et al. [59] employed a similar node and called it a *pseudo-peripheral node*. Cheng [26] used an algebraic approach to select a single node or a set of nodes as the root of an SRT. Kaveh employed two simultaneous SRTs for selecting a pseudo-peripheral node. A comparison of six different algorithms was made in [101]. Algebraic graph theory has also been used for finding a starting node; see Kaveh [113] and Grimes et al. [66]. Paulino et al. [174] used another type of algebraic graph-theoretical approach employing the Laplacian matrix of a graph for nodal ordering.

5.3 PRELIMINARIES

A matrix **A** is called *banded* when all its non-zero entries are confined within a band, formed by diagonals parallel to the main diagonal. Therefore, $A_{ij} = 0$, when $|i - j| > b$, and $A_{k,k-b} \neq 0$ or $A_{k,k+b} \neq 0$ for at least one value of k. b is the half-bandwidth and $2b + 1$ is known as the *bandwidth* of **A**. For example, for

$$\mathbf{A} = \begin{bmatrix} 1 & 6 & . & . & . \\ 6 & 2 & 7 & 9 & . \\ . & 7 & 3 & 8 & . \\ . & 8 & 9 & 4 & . \\ . & . & . & . & 5 \end{bmatrix}, \tag{5-4}$$

the bandwidth of **A** is $2b + 1 = 2 \times 2 + 1 = 5$.

A banded matrix can be stored in different ways. The *diagonal storage* of a symmetric banded $n \times n$ matrix **A** is an $n \times (b + 1)$ matrix **AN**. The main diagonals are stored in the last column, and lower co-diagonals are stored down-justified in the remaining columns. For example, **AN** for the above matrix is

$$\mathbf{AN} = \begin{bmatrix} \cdot & \cdot & 1 \\ \cdot & 6 & 2 \\ 0 & 7 & 3 \\ 9 & 8 & 4 \\ 0 & 0 & 5 \end{bmatrix}. \qquad (5\text{-}5)$$

When **A** is a sparse matrix, this storage scheme is very convenient, since it provides direct access, in the sense that there is a simple one-to-one correspondence between the position of an entry in the matrix $\mathbf{A}(i,j)$ and its position in $\mathbf{AN}(i, j - i + b + 1)$.

Obviously, the bandwidth depends on the order in which the rows and columns of **A** are arranged. This is why iterative techniques seek permutations of the rows and columns to make the size of the resulting bandwidth small. For symmetric matrices, identical permutations are needed for both the rows and the columns. When a system of linear equations has a banded matrix of coefficients and the system is solved by Gaussian elimination, with pivots being taken from the diagonals, all the operations are confined to the band and no new non-zero entries are generated outside the band. Therefore, the Gaussian elimination can be carried out in place, since a memory location is already reserved for any new non-zeros that might be introduced within the band.

For each row i of a symmetric matrix **A** define,

$$b_i = i - j_{\min}(i), \qquad (5\text{-}6)$$

where $j_{\min}(i)$ is the minimum column index in row i for which $A_{ij} \neq 0$. Therefore, the first non-zero of row i lies b_i positions to the left of the diagonal, and b is defined as:

$$b = \max(b_i). \qquad (5\text{-}7)$$

In Chapter 4, it is shown that the stiffness matrix **K** of a structure is pattern equivalent to the cutset basis matrix $\mathbf{C^*C^{*t}}$, where $\mathbf{C^*}$ is the cutset basis–member incidence matrix of the structural model S. Similarly, the flexibility matrix **G** is pattern equivalent to the cycle basis matrix $\mathbf{CC^t}$, where **C** is the cycle basis–member incidence matrix of S.

Reducing the bandwidths of $\mathbf{C^*C^{*t}}$ and $\mathbf{CC^t}$ directly influences those of **K** and **G**, respectively. Notice that the dimensions of $\mathbf{C^*C^{*t}}$ and $\mathbf{CC^t}$, for general space structures, are sixfold smaller than those of **K** and **G**, and therefore simpler to optimise.

For the displacement method of analysis, there exists a special cutset basis whose elements correspond to the stars of its nodes except for the ground node (cocycle basis). The adjacency matrix of such a basis naturally is the same as that of the node adjacency matrix of S, with the row and column corresponding to the datum node being omitted. In this chapter, such a special cutset basis is considered, and the nodes of S are ordered such that the bandwidth of its node adjacency matrix is reduced to the smallest possible amount.

Let **A** be the adjacency matrix of a graph S. Let i and j be the nodal numbers of member k, and let $\alpha_k = |i - j|$. Then the bandwidth of **A** can be defined as

$$b(\mathbf{A}) = 2\text{Max}\{\alpha_k: k = 1, 2, ..., M(S)\} + 1, \qquad (5\text{-}8)$$

where $M(S)$ is the number of members of S. To minimise the bandwidth of **A**, the value of $b(\mathbf{A})$ should be minimised. The bandwidth of the stiffness matrix **K** of a structure is related to that of **A** by

$$b(\mathbf{K}) = \beta b(\mathbf{A}), \qquad (5\text{-}9)$$

where β is the number of degrees of freedom of a typical node of the structure.

Papademetrious [170] has shown that the bandwidth minimisation problem is an NP-complete problem. Therefore, any approach to it is primarily of interest because of its heuristic value.

5.4 A SHORTEST ROUTE TREE AND ITS PROPERTIES

The main tool for most of the ordering algorithms using graph-theoretical approaches is the SRT of its model or its associate model. An SRT rooted at a node O, called the *starting node* (root) of the tree, is denoted by SRT_O and has the following properties.

The path from any node to the root through the tree is a shortest path. An algorithm for generating an SRT is given in Appendix A and therefore, only its properties relevant to the nodal number are discussed here.

An SRT decomposes (partitions) the node set of S into subsets according to their distance from the root. Each subset is called a *contour* (level) of the SRT, denoted by C_i. The contours of an SRT have the following properties:

$$\text{Adj}(C_i) \subseteq C_{i-1} \cup C_{i+1}, \qquad 1 < i < m$$

$$\text{Adj}(C_1) \subseteq C_2, \qquad (5\text{-}10)$$

$$\text{Adj}(C_m) \subseteq C_{m-1}.$$

The number of nodes in each contour is called the *width* of that contour, and the largest width of the contours of an SRT is called the *width of the SRT* rooted at the starting node O, denoted by $w(SRT_O)$. This number is known as the *width number* of O. The number of contours of an SRT (except the starting node contour) is the *height* of the tree denoted by $h(SRT_O)$. The *longest* SRT is the one with maximal height and the *narrowest* SRT is the one with minimal width.

For example, an SRT of S shown in Figure 5.2(a), rooted at O, denoted by SRT_O, has the following identities:

$w(C_1) = 1$, $w(C_2) = 2$, $w(C_3) = 3$, $w(C_4) = 4$, $w(C_5) = 5$, $w(C_6) = 5$, $w(C_7) = 4$, $w(C_8) = 3$, $w(C_9) = 2$ and $w(C_{10}) = 4$. Hence $w(SRT_O) = 5$ and $h(SRT_O) = 9$.

For the same graph model, an SRT rooted at O', which is shown in Figure 5.2(b), leads to $w(SRT_{O'}) = 9$ and $h(SRT_{O'}) = 5$.

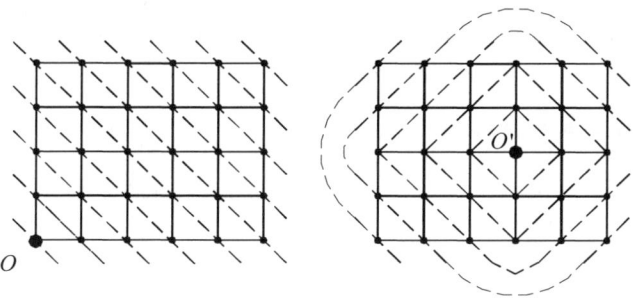

(a) An SRT rooted at O.   (b) An SRT rooted at O'.

Fig. 5.2 A graph S and two of its SRTs.

This simple example shows the importance of selecting an appropriate starting node. This is discussed in some detail in subsequent sections.

5.5 NODAL ORDERING FOR BANDWIDTH REDUCTION

The following four-step algorithm is employed for nodal ordering of graphs leading to banded node adjacency matrices. This method can be directly used for nodal ordering of skeletal structures resulting in banded stiffness matrices.

1. Find a suitable starting node.

2. Decompose the node set of S into ordered subsets (contours).

198 OPTIMAL STRUCTURAL ANALYSIS

3. Select a connected path (transversal) containing one representative node from each contour.

4. Order the nodes within each contour, to obtain the final nodal numbering of S.

All the above steps require the use of an SRT algorithm of Appendix A, known as the *breadth-first-search algorithm*. Therefore, a nodal ordering process may be considered as a multiple application of the SRT algorithm.

The node set of S can be decomposed into ordered subsets by means of a breadth-first-search algorithm. The quality of the results depends upon the choice of an appropriate starting node as the root of this tree. The results corresponding to the ordering within each contour, however, also depend upon the use of a suitable transversal containing one representative node from each contour.

Methods for finding suitable starting nodes have been developed by Cheng [26], Kaveh [111,121,124], Gibbs et al. [59] and Grimes et al. [66]. In the following text, various graph-theoretical methods are presented for finding good starting nodes and selecting suitable transversals.

5.5.1 A GOOD STARTING NODE

The *distance* $d(n_i,n_j)$ between two nodes n_i and n_j is defined to be the length of the shortest path between these nodes. The *eccentricity* of a node n_i is defined as

$$e(n_i) = \text{Max } d(n_i,n_j) \text{ for } j = 1, ..., N(S). \qquad (5\text{-}11)$$

The *diameter* of S is defined as

$$\delta(S) = \text{Max } e(n_i) \text{ for } i = 1, ..., N(S). \qquad (5\text{-}12)$$

For example, the eccentricity of n_2 in Figure 5.3 is $e(n_2) = 3$, and the diameter of S is $\delta(S) = 4$.

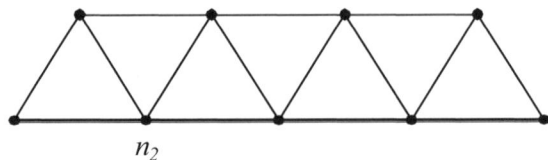

Fig. 5.3 A graph S

A node n_i of S is called *peripheral* if its eccentricity is the same as the diameter of S, that is, $\delta(S) = e(n_i)$. If the eccentricity is close to the diameter, then n_i is called a *pseudo-peripheral* node or a *good starting node*.

In this section, three algorithms are described for the selection of a good starting node or nodes for nodal numbering. Other algorithms have been developed, details of which can be found in Kaveh [113].

Algorithm A

Step 1: Start from an arbitrary node of S. Construct an SRT on this node and take a node of least valency from its last contour.

Step 2: Form a new SRT from the selected node, and record all the nodes of the last contour of the selected SRT.

Step 3: Form SRTs rooted at each of the recorded nodes and choose the one that corresponds to the narrowest SRT. The process of constructing an SRT is terminated as soon as the width of one of its contours exceeds the width of the previously selected SRT.

This algorithm is similar to the Gibbs et al. [59] algorithm, where the starting node O and another node of minimum valency from its last contour are selected as *pseudo-peripheral* or *diameteral* nodes.

Algorithm B

Step 1: Start with an arbitrary node, form an SRT on this node and take a node n_i of least valency from its last contour.

Step 2: Generate an SRT on n_i and find all nodes contained in its even, first and last contours.

Step 3: Generate an SRT on each node of these contours, and find the narrowest one. The process of formation of an SRT is terminated as soon as the width of one of its contours exceeds the width of the previously selected SRT. Denote the selected node by n_j.

Step 4: Check adjacent nodes to n_j for possible reduction in width, to decide the final starting node.

Algorithm C

Algorithms A and B may search for a good starting node in a single direction of a graph and do not meet nodes lying in other directions. Algorithm C overcomes this

200   OPTIMAL STRUCTURAL ANALYSIS

problem. In this method, the control of overall connectivity properties of the graph becomes feasible. The following example clearly illustrates this point.

Step 1: From an arbitrary node generate an SRT, and from its last contour select a node X_1 of minimal valency. Observe the width of the selected SRT.

Step 2: Generate an SRT from X_1, and select X_2 of the least valency from its last contour, and observe the width.

Step 3: Generate two SRTs simultaneously rooted at X_1 and X_2, and find the node X_3, which is the last node of S included in one of the SR subtrees. Once X_3 is found, terminate the process of forming SRTs. Generate an SRT from X_3 and observe its width. X_1 and X_2 are called the *generators* of X_3.

Step 4: Repeat the process of Step 3, using the pairs (X_1, X_3) and (X_2, X_3) as the generators to find X_4 and X_5, respectively. Construct the corresponding SRTs and observe their widths.

Step 5: Repeat the process of Step 3 for X_i (i = 3, 4, ...), along with the corresponding generator, until no further improvement in width is observed. The narrowest SRT should be selected for nodal decomposition of S.

An example of the application of this algorithm is depicted in Figure 5.4, where a cross-shaped grid S is considered.

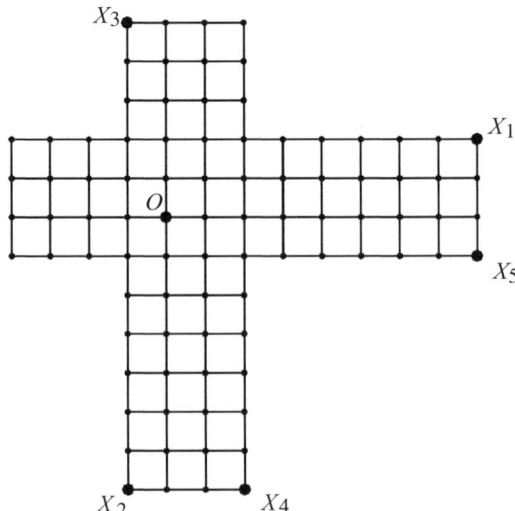

Fig. 5.4 A cross-shaped grid and the selected X_i (i = 1, ..., 5) by Algorithm C.

Starting from an arbitrary node "O", an SRT is generated and X_1 is obtained from its last contour. Generating a new SRT from X_1, node X_2 is chosen from its last contour. X_3 is the result of generating two simultaneous SRTs from X_1 and X_2. Using (X_1, X_3) and (X_2, X_3), nodes X_4 and X_5 are obtained, respectively. The widths of the selected SRTs rooted at X_1, X_2, X_3, X_4 and X_5 are 8, 8, 8, 11 and 10, respectively. Therefore, the process is terminated, and X_3 is taken as a good starting node of S.

5.5.2 PRIMARY NODAL DECOMPOSITION

Once a good starting node is selected, an SRT is constructed and its contours $\{C_1, C_2, ..., C_m\}$ are obtained. These subsets are now ordered according to their distances from the selected starting node. Obviously, many SRTs can be constructed on a node. Although all lead to the same nodal decompositions, different transversals are obtained for different SRTs. Thus, in the generation process, the nodes of each contour C_i are considered in ascending order of their valencies for selecting the nodes in C_{i+1}, to provide the conditions for the possibility of generating a minimal (or optimal) transversal as defined in the next section. Finding an optimal transversal before an SRT is fixed, seems to be a time-consuming problem. However, for most of the models encountered in practice, an optimal transversal lies between the minimal ones. In the following text, an algorithm is given for selecting a suboptimal transversal of an SRT.

5.5.3 TRANSVERSAL P OF AN SRT

A *transversal* of an SRT is defined as a connected path P containing one distinct node N_i from each contour C_i of an SRT. A *minimal transversal* is the one for which $\sum_{i=1}^{m} \deg(N_i)$ is minimum. An *optimal transversal* is the one leading to the best nodal numbering, that is, a numbering corresponding to the smallest bandwidth for the selected decomposition. The weight of a node is defined as its degree.

Algorithm

Step 1: Take a node N_m of minimal weight from the last contour C_m of the selected SRT.

Step 2: Find N_{m-1} from C_{m-1}, which is connected to N_m by a branch of the SRT.

Step 3: Repeat Step 2, selecting nodes N_{m-2}, N_{m-3}, ..., N_1, as the representative nodes of the contours C_{m-2}, C_{m-3}, ..., C_1, respectively.

202   OPTIMAL STRUCTURAL ANALYSIS

The above algorithm is a backtracking process from a node of minimal weight in the last contour C_m, and selects a transversal $P = \{N_1, N_2, ..., N_m\}$, which can now be used for ordering the nodes of the contours of the corresponding SRT.

5.5.4 NODAL ORDERING

Step 1: Number N_1 as "1".

Step 2: N_2 is a given number "2", and a SR subtree is generated from N_2, numbering the nodes of C_2 in the order of their occurrence in this SR subtree.

Step 3: Step 2 is repeated for numbering the nodes of C_3, C_4, ... , C_m, sequentially, using N_3, N_4, ... , N_m as the starting nodes of SR subtrees, until all the nodes of S are numbered.

Now the numbering can be reversed, in a way similar to that of the Reverse Cuthill–McKee algorithm, for possible reduction of fill-ins in the process of Gaussian elimination, which is discussed in Section.

5.5.5 EXAMPLE

The following simple example is chosen to illustrate the steps of the approaches presented, but the applications are not limited to such simple cases.

Let S be the graph model of a truss structure, as shown in Figure 5.5(a). Using one of the algorithms of Section 5.5.1, a good starting node A is found, and the corresponding SRTs are depicted in Figure 5.5(b). A transversal is selected as shown by bold lines in Figure 5.5(c). Then nodes are numbered contour by contour, employing the representative nodes as the starting nodes of SR subtrees as shown in Figure 5.5(d).

In order to cast the concepts developed for nodal ordering in a mathematical form, a connectivity coordinate system is defined for nodal numbering of S. A separate study of planar and space graphs results in clarifications about nodal numbering of space structures, as described in Kaveh [110].

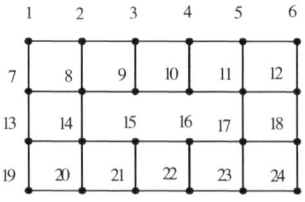
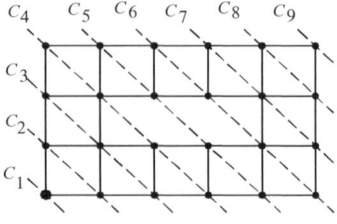

(a) Initial numbering of S.   (b) The selected SRT.

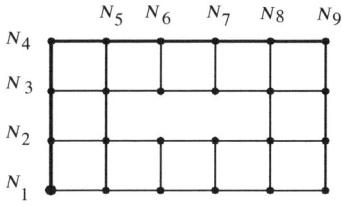 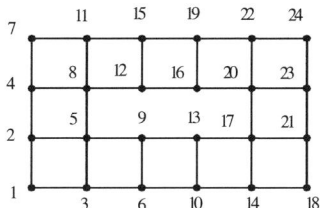

(c) The selected transversal *P*.   (d) Final nodal numbering of *S*.

Fig. 5.5 Graph model *S* and its nodal numbering.

5.6 FINITE ELEMENT NODAL ORDERING FOR BANDWIDTH OPTIMISATION

Extensions and applications of the nodal numbering algorithms to element ordering are due to Kaveh [101,102], Akhras and Dhatt [2], Everstine [43], Razzaque [185], Pina [177], Sloan and Randolph [206], Sloan [203], and Burgess and Lai [18].

For FE nodal ordering, different methods are developed. The application of a natural associate graph (NAG) in a two-step approach, was suggested by Kaveh [99], and later by Fenves and Law [50]. A corner-node method is developed by Kaveh [94], Cassell et al. [23] and Kaveh and Ramachandran [128]. The application of an element clique graph is due to Sloan [204] and Livesley and Sabin [153]. A comparative study of the application of these graphs was made by Kaveh and Behfar [115]. Additional graphs for transforming the information concerning the connectivity of the FE mesh to those of different simple graphs, are introduced and employed in efficient finite element nodal numbering by Kaveh and Roosta [133]. Excellent books on these topics are written by Duff et al. [39] and Pissanetsky [178].

In this section, the connectivity properties of FE models are embedded in the topological properties of nine different graphs. A nodal ordering is then performed on these graphs, leading to the element ordering of the corresponding FEMs, followed by their final nodal ordering. This process is summarised in the flow chart given in the following text.

For the sake of clarity, the nodes of the constructed graphs are referred to as *vertices*.

The complexities of the methods presented are given for a logical comparison of their efficiency. The interested reader may refer to Baase [8] for an analysis of the algorithms. The efficiency of the methods is also tested by some two-dimensional and three-dimensional FE models. The computational time and the bandwidth obtained for these models are presented for comparison.

204   OPTIMAL STRUCTURAL ANALYSIS

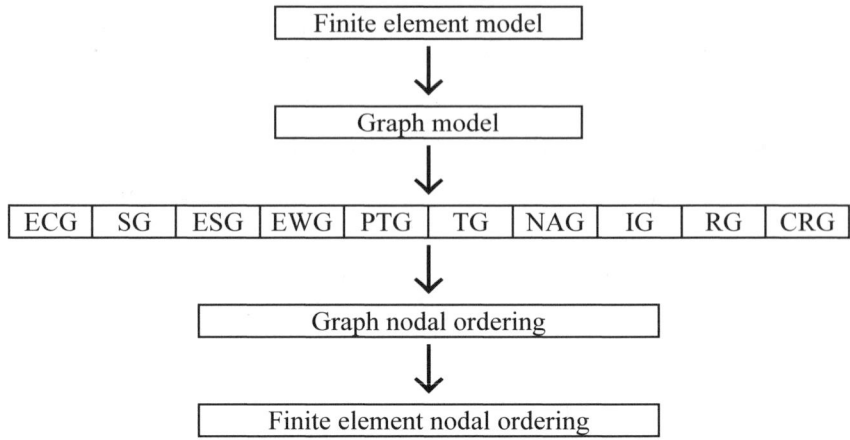

Notations: Element Clique Graph (ECG); Skeleton Graph (SG); Element Star Graph (ESG); Element Wheel Graph (EWG); Partially Triangulated Graph (PTG); Triangulated Graph (TG); Natural Associate Graph (NAG); Incidence Graph (IG); Representative Graph (RG); Complete Representative Graph (CRG).

5.6.1 ELEMENT CLIQUE GRAPH METHOD (ECGM)

Definition: The *element clique graph S* of an FEM is a graph whose vertices are the same as those of the FEM, and two vertices n_i and n_j of S are connected with a member if n_i and n_j belong to the same element in the FEM. The element clique graph (ECG) of the FEM shown in Figure 5.6(a) is illustrated in Figure 5.6(b).

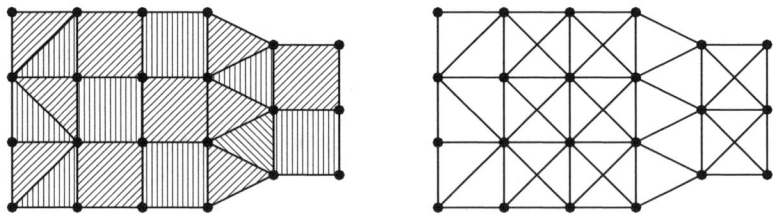

(a) An FEM.   (b) The element clique graph of the FEM.

Fig. 5.6 An FEM and its element clique graph.

In order to generate the ECG of an FEM, all pairs of nodes of each element of the FEM should be connected by members if such pairs are not connected in the previous steps. Let λ, δ and θ denote the number of elements of the FEM, the maximum degree of a vertex of the ECG and the maximum number of nodes of an element of the FEM, respectively. The formation of the ECG takes $O(\lambda\delta\theta^2)$ operations in the worst case. For this formation, the element-node list of the FEM is sufficient; however, since $M(ECG)$ is a high integer, especially for FEMs with higher-order elements, the compact adjacency list will be very large. This is a disadvantage of the ECG.

Let us consider an $m \times n$ grid of 4-node quadrilaterals; the compact adjacency list is a vector of length α, where $\alpha = (m + 1)(n + 1)$ denotes the number of nodes of the FEM. The diameter of the ECG of an FEM is small. For example, the diameter of the considered $m \times n$ grid is equal to d, where d is the same as m if $m \geq n$, otherwise d is equal to n. This property is very useful when multiple roots are used for a process. This is because more than one pair of vertices are the end vertices of the diameter of the ECG for most of the FEMs. For example, for an $m \times n$ grid, if $m > n$, then there will be $(n + 1)^2$ pairs of peripheral vertices. This graph model has another advantage, namely, when two nodes of the FEM are contained in an element, their corresponding vertices in the ECG are adjacent. This is useful since, for the computational aspects of the FEM, we explicitly or implicitly consider an FEM as a hypergraph. This graph is particularly suitable for bandwidth optimisation, since in this graph each vertex corresponds to a node of the FEM, and a single step is needed for direct nodal numbering of the considered FEM.

Algorithm

Step 1: Construct the element clique graph S of the considered FEM.

Step 2: Use a nodal numbering algorithm that is available (e.g. the algorithm presented in Section 5.5.4).

In this method, all the nodes of an element will be contained in at most two adjacent contours of an SRT; hence the bandwidth becomes dependent on the width of the SRT.

ANALYSIS OF ELEMENT CLIQUE GRAPH METHOD

Step 1: This step has time complexity $O(\lambda\delta\theta^2)$.

Step 2: This step has time complexity $O(\alpha^2\delta)$. This complexity corresponds to the complexity of the critical step of the nodal ordering algorithm.

5.6.2 SKELETON GRAPH METHOD (SGM)

Definition: The *1-skeleton graph S* of an FEM is a graph whose vertices are the same as the nodes of the FEM, and its members are the edges of the FEM. Figure 5.7 illustrates the skeleton graph of the FEM shown in Figure 5.6(a).

206   OPTIMAL STRUCTURAL ANALYSIS

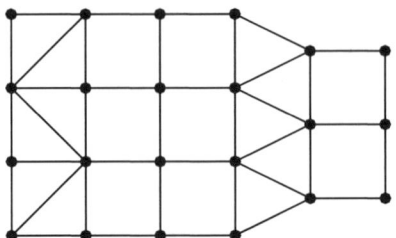

Fig. 5.7 The skeleton graph of the FEM of Figure 5.6(a).

In order to generate the skeleton graph (SG) of an FEM, one should connect all pairs of nodes of each element *i* of the FEM by a member that is not previously connected. The time complexity of the process of formation of the SG of an FEM is $O(\lambda\delta\tau)$, where τ is the same as the maximum number of members of the SG of an element. Since the bound of τ is $\theta(\theta-1)/2$, the complexity of this process is $O(\lambda\delta\theta^2)$. As shown, the time complexities for the formation of the SG and the ECG are the same for the worst case. However, the number of members of the SG is less than that of the ECG in FEMs containing elements with 4 or more nodes. For example, in the grid of Section 5.6.1, the compact adjacency list is a vector of length $(4mn + 2m + 2n)$. It is clear that the lengths of the index vectors of the compact adjacency lists of the SG and the ECG are always the same. The diameter of the SG of the FEM is large. For example, the diameter of the considered grid is the same as $(m + n)$. This property is very efficient when a single good starting vertex is needed. This is because of the existing low number of pairs of vertices being the peripheral nodes. For example, in the grid there are always only two pairs of peripheral vertices.

Simultaneous application of the ECG and the SG provides very efficient tools. For example, consider the small FE shown in Figure 5.8. Suppose an SRT is rooted from vertex 1 in the ECG of the FEM to find a good starting node with minimum degree from its last contour. Vertices 17 and 19 are found. They are the farthest from vertex 1 and have the same degree as 3 (in the ECG). However, vertex 19 is better than vertex 17, since $W(SRT_{19}) < W(SRT_{17})$. Instead of generating two SRTs from vertices 17 and 19, one can choose 19 by generating SRT_1 in the SG, because $d_{SG}(1,19) > d_{SG}(1,17)$, where $d_{SG}(i,j)$ denotes the distance between vertices *i* and *j* in the SG.

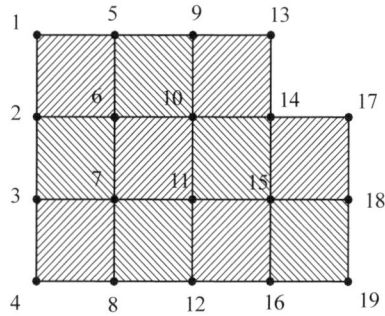

Fig. 5.8 A small finite element model.

Algorithm

Step 1: Construct the skeleton graph S of the considered FEM. For each element i of the FEM, connect two end nodes of each edge of element by a member. Such nodes should be connected only once.

Step 2: Order the vertices of S using any nodal numbering algorithm that is available (e.g. one of the algorithms presented in Section 5.5.1), thus obtaining a nodal ordering of S.

In order to generate the SG of an FEM, it is necessary to list the nodes of each element in a suitable order. In this method, the number of members of S is less than those of the Element Clique Graph Method (ECGM); however, in FEMs with triangular elements, the number of the members are the same. Therefore, this method takes less computer storage for keeping the connectivity of S. Generating an SRT in a SG may lead to the allocation of the nodes of an element in three or more adjacent contours. Therefore, the width of the SRT being used, along with the number of contours containing the nodes of an element of the FEM, specify the bandwidth.

ANALYSIS OF SKELETON GRAPH METHOD

Step 1: The running time for this step is $O(\lambda \delta \tau)$, where τ is the same as the maximum number of edges of an element. Since the bound of τ is $\theta(\theta-1)/2$, the time complexity of this step is $O(\lambda \delta \theta^2)$.

Step 2: This step requires $O(\alpha^2 \delta)$ time.

208 OPTIMAL STRUCTURAL ANALYSIS

5.6.3 ELEMENT STAR GRAPH METHOD (ESGM)

Definition: The *element star graph S* of an FEM has two sets of vertices, namely, the main set containing the same nodes as those of the FEM and a virtual set consisting of the virtual vertices associated in a one-to-one correspondence with the elements of the FEM. The member set of S is constructed by connecting the virtual vertex of each element i to all the nodes of the element i. The element star graph (ESG) of the FEM shown in Figure 5.6(a) is illustrated in Figure 5.9. The virtual vertices are shown by larger-sized dots.

In order to generate the ESG of an FEM, one should assign a vertex to each element and to each node, and then connect the vertex corresponding to an element to all vertices corresponding to the nodes of the element by a member. This process takes only $O(\lambda\theta)$ time for the worst case. For this process, the element-node list of the considered FEM is sufficient; however, since $N(\text{ESG})$ and $M(\text{ESG})$ are large integers, the compact adjacency list will be large. This is a disadvantage of the ESG. For the previously considered grid of 4-node quadrilateral elements, the compact adjacency list and its index vector take ($8mn$) and ($2mn + m + n + 1$) integer words of computer storage, respectively.

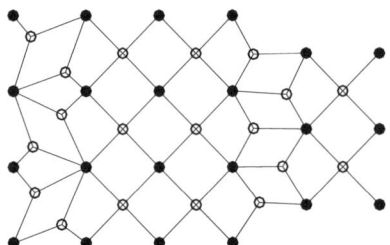

Fig. 5.9 The element star graph of the FEM of Figure 5.6(a).

Note that in this case, there are (mn) virtual vertices. The diameter of the ESG of an FEM is large. It can be easily shown that the diameter of the ESG of an FEM is twice the diameter of the ECG of the FEM; that is, $d_{\text{ESG}} = 2d_{\text{ECG}}$.

In the ESG of an FEM, the distance between each pair of vertices corresponding to two nodes of the FEM that share an element is equal to 2, while in the ECG it is equal to 1. This difference does not cause the ESG to lose the previously discussed property, which is the existence of more than one pair of peripheral nodes in most of the FEMs. Hence, this graph model is efficient for algorithms in which multiple roots need to be found. In this graph, the degree of each vertex corresponding to a node i of the FEM is the same as that of the number of elements of the FEM incident to node i.

Algorithm

Step 1: Construct the element star graph S of the considered FEM. For each element i, generate a virtual vertex labelling with $i + \alpha$, and connect the nodes of i to the vertex $i + \alpha$, where α is the total number of nodes of the FEM.

Step 2: Order the main vertices of S using a nodal numbering algorithm that is available, for example, one of the methods presented in Section 5.5.1. This step is similar to the previous methods, but virtual vertices need not be labelled in the process of the numbering of the nodes. The virtual vertices can easily be identified because their labels are above α.

In order to generate the ESG of an FEM, it is not necessary to list the nodes of each element in a specific order. In this method, $M(S)$ is higher than the skeleton graph method (SGM) and can also be higher than the ECGM (e.g. for an FEM with triangular elements). $N(S)$ of the star graph is equal to $\lambda + \alpha$, where λ denotes the number of elements of the FEM. Therefore this method requires more computer time than ECGM, for most of the cases, and is always longer than the SGM. Generation of an SRT in an ESG forces the nodes of an element to be contained at the most in three adjacent contours.

ANALYSIS OF ELEMENT STAR GRAPH METHOD

Step 1: The running time for this step is $O(\lambda\theta)$.

Step 2: This step requires $O(\beta^2\delta)$ time, where $\beta = \lambda + \alpha$.

5.6.4 ELEMENT WHEEL GRAPH METHOD (EWGM)

Definition: The *element wheel graph* S of an FEM is the union of the element star graph and the skeleton graph of the FEM. The element wheel graph (EWG) of the FEM shown in Figure 5.6(a) is illustrated in Figure 5.10. The virtual vertices are shown by larger-sized dots.

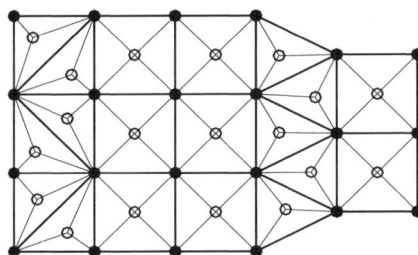

Fig. 5.10 The element wheel graph of the FEM of Figure 5.6(a).

In order to generate the EWG of an FEM, one should generate the SG of the FEM and then apply the process for the formation of the ESG. This procedure takes $O(\lambda\delta\theta^2 + \lambda\theta) = (\lambda\delta\theta^2)$ operations. It should be noted that δ denotes the maximum degree of a vertex of the SG. For the formation of the EWG like that of the SG, the element-node list along with the member list or other lists for a typical element should be given and the list of nodes for the elements should be provided in a suitable order. Since N(EWG) and M(EWG) are large integers, the compact adjacency list requires $(12mn + 2m + 2n)$ integer words. This is the same as the total computer storage needed for the SG and the ESG. The index vector of the compact adjacency list is like that of the ESG, which is a vector of length $(2mn + m + n + 1)$. The diameter of the SG, the ESG and the element wheel graph of an FEM have the following relation

$$d_{EWG} \leq d_{dSG} \text{ and } d_{EWG} \leq d_{ESG}. \tag{5-13}$$

This is due to the existence of the members of the SG and the ESG in the member set of the EWG. Clearly, the diameter of the ECG is less than or equal to that of the EWG. When an FEM contains higher-order elements, its EWG may contain several pairs of peripheral vertices, since the distance between the vertices of the EWG corresponding to the corner nodes of each element is the same as that of the ESG.

Algorithm

Step 1: Construct the element wheel graph S of the considered FEM. This can be done by generating the union of the ESG and SG.

Step 2: Order the main vertices of S using a nodal numbering algorithm that is available, for example, one of the methods presented in Section 5.5.1. This step should be carried out like Step 2 in the ESGM.

In order to generate the EWG of an FEM, it is necessary to list the nodes of each element in a suitable order. In this method, $M(S)$ is higher than that of the ESGM, and therefore it needs more computer storage than the ESGM. The nodes of an element of FEM are at most contained in three contours of the generalised SRT of the EWG.

ANALYSIS OF ELEMENT WHEEL GRAPH METHOD

Step 1: The running time for this step is $O(\lambda\delta\theta^2)$.

Step 2: This step requires $O(\beta^2\theta)$ time.

ORDERING: GRAPH THEORY METHODS   211

5.6.5 PARTIALLY TRIANGULATED GRAPH METHOD (PTGM)

Definition: The *partially triangulated graph S* of an FEM is a graph whose vertices are the same as the nodes of the FEM and an artificial vertex assigned to each element i is connected to all the original nodes of i. The selected nodes of the elements are found by generating all SR subtrees from a good starting vertex in the SG of the FEM and taking the first node of an element included in the SRT during the process of the generation. For example, for the FEM shown in Figure 5.6(a), an SR subtree is rooted from n_0 and shown in Figure 5.11(a), and the selected nodes of the elements are shown by larger-sized dots. The partially triangulated graph (PTG) of the FEM is shown in Figure 5.11(b).

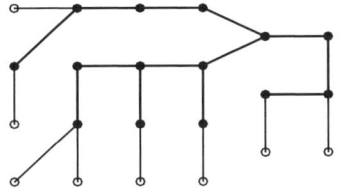
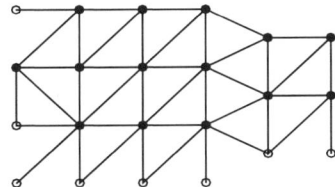

(a) The skeleton graph and an SR subtree of the FEM.

(b) The partially triangulated graph of the FEM.

Fig. 5.11 The skeleton, an SR subtree and the partially triangulated graph of the FEM of Figure 5.6(a).

In order to generate the PTG of an FEM, the following steps can be executed.

1. Generate the SG of the FEM.

2. Form an SRT rooted from an arbitrary node n_0, and select a node n_1 from the last contour of SRT_{n_0} with the minimum degree.

3. Form an SRT rooted from n_1, and select a node n_2 from the last contour of SRT_{n_1}, with minimum degree.

4. Form an SRT routed from n_2 and take n_s from n_0, n_1 and n_2 whose corresponding SRT has the least width.

5. Calculate the distance between each vertex of the SG and n_s.

6. For each element i, select a vertex that is the nearest node to n_s;

7. Form the PTG by connecting the vertex corresponding to the selected node of each element i to the vertices corresponding to other nodes of i; previously connected nodes should not be connected again.

The above process has time complexity $O(\lambda\delta\theta^2 + 4\alpha\delta + \lambda\theta + \lambda\delta\theta) = O(\lambda\delta\theta^2)$. Steps 2, 3 and 4 of this process are to find a good starting vertex n_s; however, other good starting node selection algorithms can be employed for these steps. For the formation of the PTG like that of the EWG, the element-node list or other lists for a typical element should be given, and the nodes of elements should be listed in a suitable order. The number of members of a PTG is low and is always less than or equal to $\lambda(\theta - 1)$. For example, for the considered $m \times n$ grid, the compact adjacency list is a vector of length ($6\,mn$). Since the PTG of an FEM has the same number of nodes as the FEM, the index vector of the compact adjacency list is a vector of length α, and for the considered grid $\alpha = (m + 1)(n + 1)$. The diameter of the PTG of an FEM is a high integer; for example, the diameter of the grid is $(m + n)$. Clearly, the diameter of the PTG of an FEM is greater than or equal to that of the ECG of the FEM. This is because, in this graph, the distance between two vertices corresponding to the two nodes of the FEM that are contained in an element is equal to 1 or 2.

Algorithm

Step 1: Construct the partially triangulated graph S of the considered FEM.

Step 2: Order the vertices of S using an available nodal numbering algorithm, for example, an algorithm of Section 5.5.1.

For generating the PTG of an FEM, it is necessary to list the nodes of each element in a suitable order. In this method, $M(S)$ may or may not be higher than that of the SGM. In the process of forming an SRT in a PTG, the nodes of an element may lie in one, two or three adjacent contours.

ANALYSIS OF PARTIALLY TRIANGULATED GRAPH METHOD

Step 1: The running time for this step is $O(\lambda\delta\theta^2 + 4\alpha\delta + \lambda\theta + \lambda\delta\theta) = O(\lambda\delta\theta^2)$.

Step 2: This step requires $O(\alpha^2\theta)$ time.

5.6.6 TRIANGULATED GRAPH METHOD (TGM)

Definition: The *triangulated graph* S of an FEM is the union of the partially triangulated graph and the skeleton graph of the FEM. The triangulated graph (TG) of the FEM shown in Figure 5.6(a) is illustrated in Figure 5.12. The selected vertices of the elements are the same as those of Figure 5.6(a).

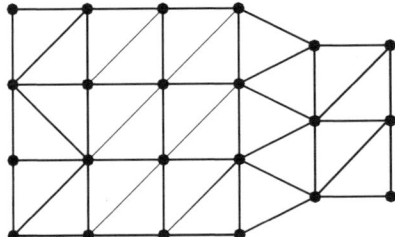

Fig. 5.12 The triangulated graph of the FEM of Figure 5.6(a).

In order to generate the TG of an FEM, one should generate the PTG of the FEM and then connect each pair of disconnected vertices that are adjacent in the SG by a member. This process of formation also has time complexity $O(\lambda \delta \theta^2)$ for the worst case. It is obvious that the formation of the TG of an FEM is similar to that of SG, EWG and PTG. The element list along with a data connectivity list, such as the member list of a typical element, should be given and the list of nodes of the elements should be provided in a suitable order. The number of members of a TG is always higher than or equal to that of the SG. It is interesting to note that the TG, SG and ECG become the same graph when the considered FEM contains bar and/or 3-node triangular elements. If the FEM contains only the bar elements, then the PTG is also included in this set.

The compact adjacency list is a comparatively long vector. For example, for the $m \times n$ grid, it is a vector of length ($6mn + 2m + 2n$). Since the TG of an FEM has the same number of nodes as the FEM, the index vector of the compact adjacency list takes $\alpha = (m + 1)(n + 1)$ words of memory. The diameter of the TG of an FEM is an integer between the diameters of ECG and the SG of the FEM. Clearly, its diameter is less than or equal to that of the PTG.

Algorithm

Step 1: Construct the TG S of the considered FEM. This step can be carried out by generating the PTG and the SG.

Step 2: Order the vertices of S using a nodal numbering algorithm.

In this method, the number of members is higher than that of the Partially triangulated graph method (PTGM). For an SRT in a TG, the nodes of an element of an FEM are contained in at most three adjacent contours.

214 OPTIMAL STRUCTURAL ANALYSIS

ANALYSIS OF TRIANGULATED GRAPH METHOD

The time complexity of this method is the same as that of the PTGM.

5.6.7 NATURAL ASSOCIATE GRAPH METHOD (NAGM)

Definition: The *natural associate graph* S of an FEM has its vertices in a one-to-one correspondence with the elements of the FEM, and two vertices of S are connected by a member if the corresponding elements have a common boundary. The NAG of the FEM shown in Figure 5.6(a) is illustrated in Figure 5.13.

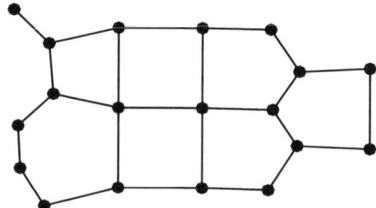

Fig. 5.13 The natural associate graph of the FEM of Figure 5.6(a).

In order to generate the NAG of an FEM, one of the following two methods can be employed. The first is a direct scheme that requires more computer time but less computer storage. In this case, only the element-node list should be provided. The second requires less computational time but uses larger memory. In this case, the node-element list along with the element-node list are provided as input data.

Method 1

Check each pair of elements i and j of the FEM for a common boundary. If i and j have such a boundary, then the vertices corresponding to i and j should be connected by a member in the NAG.

The time complexity of this method is $O(\lambda^2 \theta^2)$; however, if the maximum difference Δ between the labels of the two elements with a common boundary is given, then the time complexity reduces to $O(\lambda \Delta \theta^2)$. Hence, ordering of the elements should be performed in the process of mesh generation. However, this method is not efficient because of the high time complexity or dependency on the data. In the following text, a different method is presented that does not depend on the data order and requires far fewer operations, at the expense of greater computer storage, than Method 1.

Method 2

Step 1: Generate the node-element list of the considered FEM.

Step 2: Take each pair of elements incident at a node, and note whether they have more than one corner node in common.

Step 3: When two elements of equal or different dimensions have common corner nodes equal to or more than the smallest dimension of the elements, then the corresponding vertices in the NAG are connected by a member.

This method takes $O(\alpha\theta^2\varepsilon^2)$ operations, where ε is the same as the maximum number of elements containing a specified node. The element-node list and the node-element list have the same length; however, the index vector of the node-element list takes α words in place of λ words in the element-node list.

The number of members of the NAG of an FEM is a relatively small integer; hence its compact adjacency list uses a small amount of memory, which is to the advantage of the NAG. For example, in the $m \times n$ grid, the compact adjacency list is a vector of length $(4mn - 2m - 2n)$, and its index vector takes (mn) words. However, although the list for keeping the data connectivity of the NAG of an FEM uses low computer storage, some difficulties arise in the process of the formation of the graph for FEMs containing elements with mid-side nodes.

The NAG of the $m \times n$ grid is the same as that of the SG of an $(m - 1) \times (n - 1)$ grid. Hence, like the SG of the grid, the NAG has a relatively high diameter length of $m + n - 2$.

ALGORITHM

Step 1: Construct the natural associate graph S of the considered FEM.

Step 2: Order the vertices of S using a nodal numbering algorithm, to obtain an ordering for the elements of the FEM.

Step 3: Order the nodes of the FEM, element by element, in the same sequence as decided in Step 2. Within each element, priority is given to mid-nodes, passive and active nodes, respectively. A node is called *passive* if it has no incident new element; otherwise it is *active*.

Step 3 of this method can be carried out using the following process:

(I) Generate a matrix **NE** with a rows and ε columns, in which its ith row contains the labels of the elements containing node i, where ε is the same as the maximum number of elements incident to a specified node.

(II) For each element j ($j=1, ..., l$) execute the following steps, in turn.

(a) If j has a mid-node, label it first.

(b) Detect the active and passive nodes j using the matrix **NE**. It should be noted that using **NE** makes the process fast; however, instead one can check a node of j for incidence with a new element.

(c) Form a multiple root SR subtree from the active node of j.

(d) Label the passive nodes of j when they are selected in the multiple root SR subtree.

(e) Label the active nodes of j that are adjacent to the previously labelled nodes.

(f) Repeat Step (e) until all the active nodes of j are labelled.

In order to generate the NAG of an FEM, it is necessary to list the nodes of each element in a suitable order. In this algorithm, $M(S)$ has the least value among the methods presented so far; therefore, it takes less computer storage for keeping the connectivity data of S.

However, this method has a disadvantage in terms of programming. In order to check two elements having a common boundary, the nodes of each boundary of the elements or the dimension of the elements should be provided as input data. If the dimensions of the elements are given, it should be noted that, in three-dimensional models, the elements may have mid-side nodes. Hence, having three or more common nodes does not guarantee the existence of a common boundary. So the mid-side nodes in an irregular configuration, or the number of mid-side nodes of an element in a regular configuration of elements, should be provided as input data.

ANALYSIS OF NATURAL ASSOCIATE GRAPH METHOD

Step 1: The running time for this step is $O(\alpha\delta\varepsilon^2)$.

Step 2: This step requires $O(\lambda^2\delta)$ time.

Step 3: This step has time complexity $O(\lambda\sigma\theta)$, where σ is the maximum value of θ^2 and ε.

5.6.8 INCIDENCE GRAPH METHOD (IGM)

Definition: The *incidence graph* S of an FEM has its vertices in a one-to-one correspondence with the elements of the FEM, and two vertices of S are connected by a member, if the corresponding elements have a common node. Figure 5.14 shows the incidence graph of the FEM shown in Figure 5.6(a).

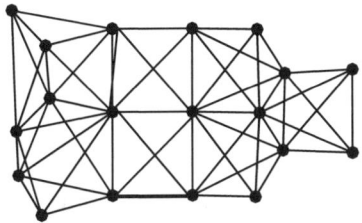

Fig. 5.14 The incidence graph of the FEM of Figure 5.6(a).

To generate the incidence graph (IG) of an FEM, one of the following two methods can be employed. The first is a direct approach, which requires more computational time but less words of memory, for which only the element-node list should be provided. The second scheme requires less computational time but more computer storage; the node-element list along with the element-node list should be provided.

Method 1

Check each pair of elements i and j of the FEM for a common node, and if they have such a node, connect it with a member to the corresponding vertices i and j in the IG.

The time complexity of this method is $O(\lambda^2\theta^2)$; however, as stated for the NAG generation, if the maximum difference Δ between the labels of two elements with a common node is given, the complexity reduces to $O(\lambda\Delta\theta)$.

Method 2

Step 1: Generate the node-element list of the *considered FEM*.

Step 2: Connect the representative vertices of each pair of elements that contain a common node.

218   OPTIMAL STRUCTURAL ANALYSIS

This method takes $O(\alpha\delta\varepsilon^2)$ operations, which is more efficient than Method 1, especially for FEMs with more number of elements.

The number of members $M(IG)$ of the IG of an FEM is relatively high, and its compact adjacency list takes a large amount of memory. For an arbitrary FEM, $M(IG) \geq M(NAG)$. The equality holds when every two elements have a common boundary or a common node. FEMs with only bar elements belong to this category. For example, for the grid of Section 5.6.7, $M(NAG) = 2mn - m - n$ and $M(IG) = 4mn - 3m - 3n + 2$ and the compact adjacency list of the IG takes ($8mn - 6m - 6n + 4$) integer words of memory. Its index vector has the same length as that of the NAG.

Algorithm

Step 1: Construct the incidence graph S of the considered FEM.

Step 2: Order the vertices of S using a nodal numbering algorithm, to obtain an ordering for the elements of the FEM.

Step 3: Order the nodes of the FEM, element by element, in the same sequence as decided in Step 2. Within each element, priority is given to mid-nodes, passive and active nodes, respectively.

ANALYSIS OF INCIDENCE GRAPH METHOD

Step 1: The running time for this step is $O(\alpha\delta\varepsilon^2)$.

Step 2 and Step 3 of this method have the same complexities as Steps 2 and 3 of the NAGM.

5.6.9 REPRESENTATIVE GRAPH METHOD (RGM)

Definition: Consider the skeleton graph of an FEM, and select an appropriate starting vertex, using any available algorithm. The nearest corner node of each element of the FEM is taken as the representative node of that element. The SR subtree of the SG of the FEM containing all the representative nodes of the elements is called a *representative graph S* of the FEM. The representative graph (RG) of the FEM shown in Figure 5.6(a) is illustrated in Figure 5.15.

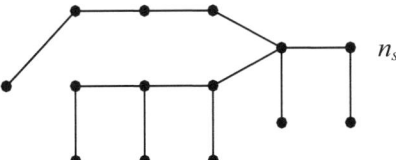

Fig. 5.15 The representative graph of the FEM in Figure 5.6(a).

In order to generate the RG of an FEM, the following steps should be executed:

Step 1: Execute Steps 1–4 of the algorithm for the formation of the PTG.

Step 2: Form an SR subtree step by step from n_s until each element of the FEM has a node whose corresponding vertex in SG is contained in the SR subtree. The first selected vertex (in the SR subtree) corresponding to the nodes of each element i should be taken as the representative node of i.

The generated SR subtree is the RG of an FEM. The first step takes $O(\lambda \delta \theta^2 + 3\alpha\delta) = O(\lambda \delta \theta^2)$ operations. For the execution of this step, the element-node list, along with the member list or other connectivity lists for a typical element, should be given and lists of nodes of elements should be provided in a suitable order. This is because of the need for the formation of the SG of the FEM.

The RG of an FEM is a tree; hence all theorems and properties of trees hold for this graph. The number of members of this graph is a very small integer and its upper bound is $(\alpha - 1)$. Thus, its compact adjacency list occupies a small amount of memory. For example, for the considered $m \times n$ grid, the compact adjacency list of the RG is a vector of length $(2mn - 2)$. This property of the RG of an FEM makes it an efficient model.

Algorithm

Step 1: Construct the RG of the FEM and number its vertices, which results in the ordering of the elements of the considered FEM.

Step 2: Use Step 3 of the NAGM to number the nodes of the FEM.

This method is the most efficient approach from the computational time and storage points of view for most of the practical models.

ANALYSIS OF THE REPRESENTATIVE GRAPH METHOD

Step 1: The running time for this step is $O(\lambda\delta\theta^2 + \alpha\delta + \lambda\theta + \alpha\delta\varepsilon) = O(\gamma)$, where ε is the maximum value of the elements of the $\{\lambda\sigma\theta^2, \alpha\delta\varepsilon\}$.

Step 2: This step requires $O(\lambda\sigma\theta)$ time.

5.6.10 DISCUSSION OF THE ANALYSIS OF ALGORITHMS

It can be concluded from the above complexity analyses that the RGM occupies the least number of operations for the worst case. In most of the practical models, it has been the fastest algorithm. ECGM, SGM, PTGM and TGM have the same complexity. The critical steps of these have time complexity $O(\alpha^2\delta)$, where α is a constant for an FEM but δ differs from one graph to another. However, for the practical models studied here, the following results are observed:

(a) The difference between the times required for ECGM and SGM is small.

(b) In general, TGM uses slightly more time than PTGM.

(c) For an FEM with low-order elements, ECGM and SGM, in general, take less time than, or nearly the same time as, PTGM and TGM. However, for FEMs with high-order elements, PTGM and TGM take far less time than ECGM and SGM.

The time complexity of the Element Wheel Graph Method (EWGM) for the worst case is the highest. For the practical models studied, the following results are obtained:

(a) In two-dimensional models with elements having less than 10 nodes, EWGM generally occupies the highest computational time, but in models with higher-order elements, SGM uses the highest computer time.

(b) In three-dimensional models, NAGM requires the highest computational time.

Excluding the RGM, the following results can be derived considering the speed for the practical models being studied:

(i) In two-dimensional models with low-order elements, ECGM and SGM may be the fastest methods; however, in FEMs with high-order elements, the incidence graph method (INGM) is generally the fastest approach.

(ii) In three-dimensional models, ESGM is generally the most economical algorithm.

5.6.11 COMPUTATIONAL RESULTS

The presented algorithms are implemented on a PC and many examples are examined, some of which are included in this section. The bandwidth of **D** and the relative computational time for the algorithms are provided.

Example 1: A planar FEM with three types of elements consisting of 4-node, 8-node and 12-node elements is considered as shown in Figure 5.16. This model contains 1959 nodes and 2250 elements. The combination of elements may not be practical; however, it is purposely chosen to illustrate the generality of the methods in dealing with the presence of different elements of a model. The results are presented in Table 5.1.

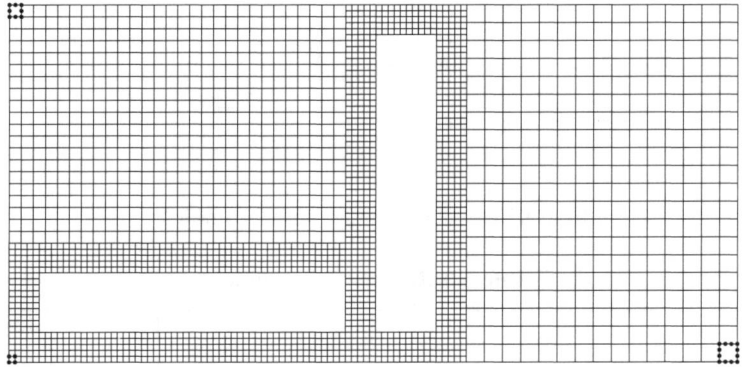

Fig. 5.16 A planar FEM.

Table 5.1 Results of Example 1.

Method	ECGM	SGM	ESGM	EWGM	PTGM	TGM	NAGM	INGM	RGM
b(**D**)	313	497	313	457	513	515	447	451	491
Time	29.77	27.02	21.92	36.09	20.65	22.03	18.29	15.71	9.72

Example 2: A three-dimensional FE model consisting of 480 (5812) 20-node cubic elements (each edge of the elements has a mid-side node) is considered, having a total of 2559 nodes. The results are depicted in Table 5.2.

Table 5.2 Results of Example 2.

Method	ECGM	SGM	ESGM	EWGM	PTGM	TGM	NAGM	INGM	RGM
b(**D**)	843	1173	843	787	1103	1103	1185	845	1195
Time	18.62	7.08	5.93	8.12	7.47	7.85	38.67	7.75	5.93

222 OPTIMAL STRUCTURAL ANALYSIS

Example 3: A planar FEM with two holes is considered as shown in Figure 5.17. Six FEMs with 1000 elements are studied with elements having 4 nodes, 4 nodes and a mid-node, 8 nodes, 8 nodes and a mid-node, 12 nodes, and 12 nodes and a mid-node. These models contain 1134, 2134, 3269, 4204 and 6404 nodes, respectively. The results are depicted in Table 5.3.

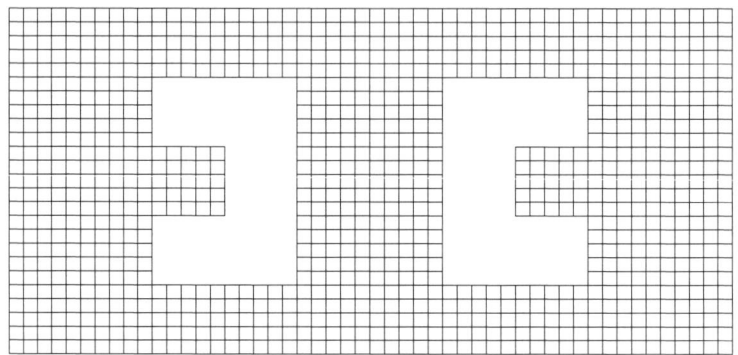

Fig. 5.17 A planar FEM with two holes.

Table 5.3 Results of Example 3.

Element		4 nodes	4 nodes + mid-node	8 nodes	8 nodes + mid-node	12 nodes	12 nodes + mid-node
ECGM	b(**D**)	111	217	333	437	553	657
	Time	4.12	7.25	15.32	19.55	30.49	37.90
SGM	b(**D**)	95	179	269	347	439	519
	Time	4.29	7.14	15.93	20.32	33.23	39.71
EWGM	b(**D**)	97	185	313	417	541	639
	Time	7.31	10.60	17.13	20.82	26.04	29.94
PTGM	b(**D**)	159	309	477	633	807	963
	Time	4.29	6.59	10.60	12.97	16.70	19.45
TGM	b(**D**)	167	327	479	619	791	945
	Time	4.17	6.98	11.15	13.90	17.74	21.09
NAGM	b(**D**)	95	177	271	353	447	529
	Time	5.22	6.87	10.28	12.31	16.03	18.84
INGM	b(**D**)	113	225	341	455	569	687
	Time	4.39	5.77	7.91	9.23	11.10	12.96
RGM	b(**D**)	95	177	271	353	447	529
	Time	2.70	4.18	6.48	8.07	10.05	12.24

Example 4: The FE model of a buttress dam is considered, the section of which is illustrated in Figure 5.18, consisting of 480 nodes and 603 elements. This model contains three layers of prismatic members, and each element contains six nodes.

The results are depicted in Table 5.4. The patterns of the node adjacency matrices are presented in the figure.

Table 5.4 Results of Example 4.

Method	ECGM	SGM	ESGM	EWGM	PTGM	TGM	NAGM	INGM	RGM
b(**D**)	125	221	125	221	229	213	175	125	187
Time	1.70	1.32	1.70	2.42	1.76	1.76	6.43	4.45	1.54

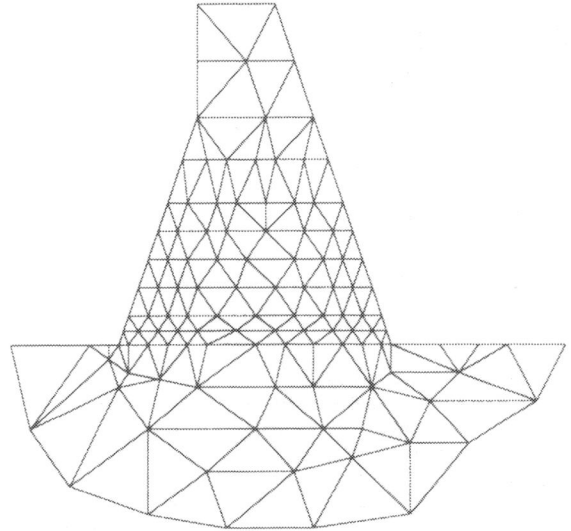

Fig. 5.18 A planar FEM.

5.6.12 DISCUSSIONS

The algorithms presented in this section transform the connectivity of the FEMs into the topological properties of different graphs. Then a nodal ordering algorithm undertakes numbering the nodes of the graphs, resulting in nodal numbering of the FEMs. All the methods presented are low-order polynomial time algorithms. Analyses are considered for the worst cases and compared. Such an analysis is the most logical way of comparing the algorithms, since most of the combinatorial optimisation algorithms are configuration dependent. Each algorithm presented has advantages and disadvantages, which become manifest when the algorithm is employed for models with different types of elements and connectivity properties. It should be noted that the relative performance of the algorithms also depends on the starting node selection algorithm and the nodal ordering algorithm being employed.

Finally, it should be mentioned that the simultaneous use of two graphs out of the nine graphs presented in this section for nodal ordering may lead to a combined model more informative than individual models; see Kaveh and Roosta [131].

5.7 FINITE ELEMENT NODAL ORDERING FOR PROFILE OPTIMISATION

5.7.1 INTRODUCTION

When a banded matrix of high order has a wide band and a large number of zeros inside it, the diagonal storage scheme may become wasteful. Then a *profile* (variable band) scheme of Jennings [88], the so-called *skyline scheme* (Felippa [49]), may be used.

Nodal numbering algorithms can also be applied to profile reduction. As mentioned earlier, after nodal numbering for bandwidth reduction, by reversing the ordering, a numbering corresponding to a much smaller profile can be found. This has been found by George [56] and proved by Liu and Sherman [151]. The method is known as the *Reverse Cuthill–McKee algorithm*. For the Cuthill–McKee type of ordering, the bandwidth remains unchanged when the order is reversed; however, the profile can never increase.

For example, consider a nodal numbering for a graph as shown in Figure 5.19(a) with corresponding adjacency matrix **A** in Figure 5.19(b). Reversing the nodal numbers as in Figure 5.19(c) leads to a matrix **A'** as depicted in Figure 5.19(d), with a reduction of the profile from 15 to 13.

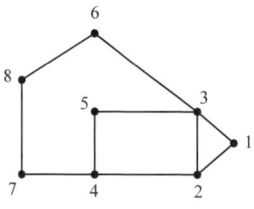
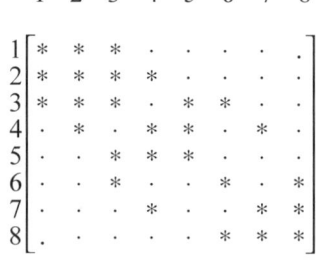

(a) A nodal numbering.　　　　　　　　(b) Matrix **A**.

ORDERING: GRAPH THEORY METHODS   225

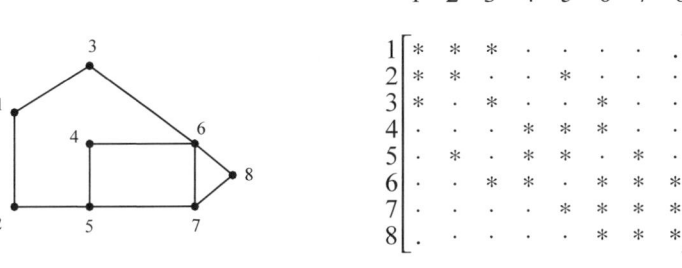

(c) Reverse of the nodal numbering of (a).   (d) Matrix **A**′.

Fig. 5.19 A Reverse Cuthill–McKee for nodal numbering.

There are many algorithms for profile and frontwidth reduction, which can be categorised in different ways. In this section the general algorithm of Souza and Murray [210] is adopted for nodal ordering of all the graph models presented in the previous section, to reduce the profile of sparse matrices with symmetric structures. This algorithm incorporates the algorithm for the selection of peripheral nodes, the re-sequencing scheme of Sloan [204] and the algorithm of Gibbs–King [58].

To proceed with the main algorithms for profile reduction, some definitions are stated in the following text.

The profile of an $n \times n$ matrix **A** is defined as,

$$P = \sum_{i=1}^{N} b_i, \qquad (5\text{-}14)$$

where the row bandwidth, b_i, for row i is defined as the number of inclusive entries from the first non-zero element in the row to the $(i + 1)$th entry. The efficiency of any given ordering for the profile solution scheme is related to the number of active equations during each step of the factorisation process. Formally, row j is defined to be active during the elimination of column i if $j \geq i$ and there exists $a_{ik} = 0$ with $k \leq i$. Hence, at the ith stage of the factorisation, the number of active equations is the number of rows of the profile that intersect column i, ignoring those rows already eliminated. Let f_i denote the number of equations that are active during the elimination of the variable x_i. It follows from the symmetric structures of **A** that

$$P = \sum_{i=1}^{N} f_i = \sum_{i=1}^{N} b_i, \qquad (5\text{-}15)$$

where f_i is commonly known as the *wavefront* or *frontwidth*. Assuming that N and the average value of f_i are reasonably large, it can be shown that a complete profile or front factorisation requires approximately $O(Nf^2)$ operations, where F is the root-mean-square wavefront, which is defined as

$$F = (\frac{1}{N}\sum_{i=1}^{N} f_i^2)^{0.5}. \tag{5-16}$$

Everstine [43] has shown that $P/N \leq F \leq W_{max} \leq B$, where W_{max} is the maximum wavefront. Hence, to minimise the storage requirement and solution time, it is imperative to reduce the profile and root-mean-square wavefront, respectively. As both P and F are related, any algorithm that seeks to minimise either inevitably tends to reduce the other as well. We call an algorithm efficient if it results in significant profile reduction in a reasonable computer time.

In the storage scheme due to Jennings, all elements that belong to the envelope are stored row by row including zeros, in a one-dimensional array, say **AN**. Diagonal elements are stored at the end of each row. The length of **AN** is equal to Profile (**A**) + n. An array of pointers **IN**, the entries of which are pointers to the locations of the diagonal elements in **AN**, is also necessary. Thus, the elements of row i, when $i > 1$, are in positions $\mathbf{IN}(i-1) + 1$ to $\mathbf{IN}(i)$. The only element of row 1 is A_{11}, stored in $\mathbf{AN}(1)$. The elements have consecutive, easily calculable column indices.

For example, the matrix of Eq. (5-4) has a profile equal to 4, and its envelope storage is

Position = 1 2 3 4 5 6 7 8 9

AN = $[1\ 6\ 2\ 7\ 3\ 9\ 8\ 4\ 5]$

IN = $[1\ 3\ 5\ 8\ 9]$

A variant of Jennings's scheme is obtained when the transpose of the lower envelope is stored. In this case, elements are stored column-wise, and since the columns of the matrix retain their length, the scheme is often termed *skyline storage*. The profile of a matrix also changes if the rows and columns are permuted.

5.7.2 GRAPH NODAL NUMBERING FOR PROFILE REDUCTION

Graph models defined in the previous section are incorporated in a general algorithm of Souza and Murray [209] to obtain ten approaches for profile reduction.

ORDERING: GRAPH THEORY METHODS

This algorithm is based on Sloan's algorithm, using priorities to control the selection of nodes from a priority queue. Some of its features are adapted in the following algorithms.

The numbering and control of nodes in the priority queue are carried out through the assignment of status, based on the numbering strategy of King [137], which operates as follows:

Take a node of minimum valency and number it "1". The set of nodes is now divided into three subsets, A, B and C. The subset A consists of nodes already numbered. The subset B is defined as $B = \text{Adj}(A)$; that is, it consists of all nodes adjacent to any node of A. The subset C contains the remaining nodes. Then, at each step, number the node of subset B that causes the smallest number of nodes of subset C to be transferred to subset B, and redefine A, B and C, accordingly.

For example, consider a graph S with original nodal numbering as shown in Figure 5.20(a).

Take node "5" as a starting node and number it as "1". Then:

$$A = \{5\}, B = \{1,8\} \text{ and } C = \{\text{the remaining nodes}\}.$$

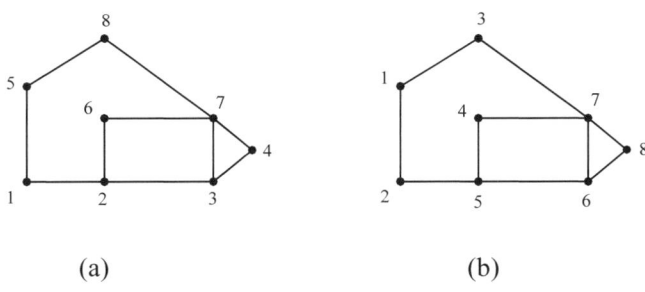

(a)   (b)

Fig. 5.20 An example of numbering by King's algorithm.

At this stage, 1, 8 are the next candidates. If 1 is taken to A, then 2 will come to B; and for 8, node 7 will join B. Therefore, arbitrarily, 1 is taken to A and numbered as "2". Now we have:

$$A = \{5,1\}, B = \{8,2\} \text{ and } C = \{\text{the remaining nodes}\}.$$

From new candidates 8 and 2, naturally 8 will be selected because it brings only 7 to B, while 2 brings 3 and 6. Therefore, 8 is numbered as 3. This process is continued until the nodal numbering of Figure 5.20(b) is obtained, which corresponds to a profile equal to 14.

228 OPTIMAL STRUCTURAL ANALYSIS

The nodes in the above strategy can be categorised more formally as follows:

Prior to the numbering, all the nodes of a graph model G of the considered FEM are assigned *inactive status*. When a node of G is inserted in the priority queue, it is assigned *pre-active status*. After a node is numbered, it is assigned *post-active status*. Nodes that are adjacent to a post-active node and do not have post-active status are defined as having *active status*, as shown in Figure 5.21. King's algorithm is generalised by Sloan [203] by introducing a priority queue to control the order to be followed in the numbering of the nodes. This algorithm consists of the following two phases:

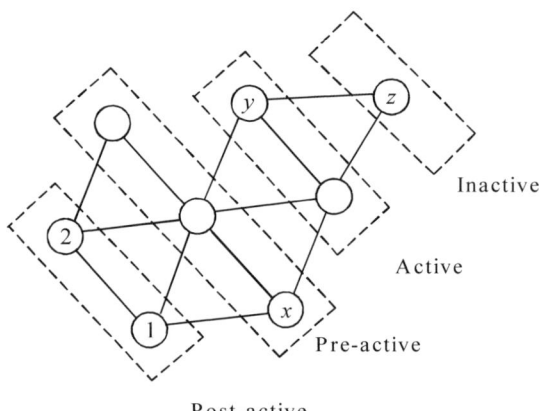

Fig. 5.21 Nodes in different status.

Phase 1: Selection of pseudo-peripheral nodes

The pair of starting nodes is determined according to the following steps:

Step 1: Choose an arbitrary node v of minimum degree.

Step 2: Generate an $\mathrm{SRT}_v = \{C_1^v, C_2^v, ..., C_d^v\}$ rooted from v. Let S be the list of the nodes of C_d^v, which is stored in the order of increasing degree.

Step 3: Decompose S into subsets S_j of cardinality $|S_j|$, $j = 1, 2, ..., \Delta$, where Δ is the maximum degree of any node of S, such that all nodes in S_j have degree j. Generate an SRT from each node y in S, for the first $1 \leq m_j \leq \Delta$. If $d(\mathrm{SRT}_y) > d(\mathrm{SRT}_v)$, then set $v = y$ and go to Step 2.

Step 4: Let u be the root of the longest SRT that has the smallest width. When the algorithm terminates, v and u are end points of a pseudo-diameter.

Step 1 of the above algorithm uses $O(\psi)$ operations, where ψ is the number of nodes of the graph employed for the connectivity of the considered FEM. The time complexity of Step 2 is $O(\psi\delta)$. The execution of Step 3 uses $O(\psi^2\delta)$ for the worst case. The time complexity of Step 4 is $O(\lambda)$. The critical step of the method is Step 3; its time complexity shows that the elapsed time for execution of this algorithm grows proportionally to the square of the number of nodes of the graph model.

Phase 2: Numbering

The general algorithm for nodal numbering of an arbitrary graph associated with an FEM consists of the following steps:

Step 1: The priority queue denoted by Q is initialised with a starting node s, that is, $Q_1 = s$. Set $n = 1$, where n is the length of the queue. The node s is assigned a pre-active status. Let k be the node count, which is initially set equal to zero or equal to the last number being used, in the case of disconnected graph models.

Step 2: Assign initial status and priorities to all the nodes.

Step 3: Select the node $u \in Q$ that has the maximum priority. Let i be the index of node u in the queue such that $Q_i = u$.

Step 4: Update queue, priority and status. Delete u from Q by setting $Q_i = Q_n$ and $n \leftarrow n - 1$. Insert nodes in queue: for each node x adjacent to u, whose status is inactive, set $n \leftarrow n + 1$ and $Q_n = x$. Assign node x a pre-active status and update priorities.

Step 5: Increment the node count by setting $k \leftarrow k + 1$, and label node u by *label*$(u) \leftarrow k$, where *label*$(.)$ contains the new labels of the nodes of the graph model. The node u is assigned a post-active status.

Step 6: If $n > 0$, that is, there are still nodes in the queue, then update priorities and status and go back to Step 3.

Step 7: Exit; that is, the new ordering is now completed, and the number of each node u is obtained as *label*(u).

5.7.3 NODAL ORDERING WITH ELEMENT CLIQUE GRAPH (NOECG)

In this method, Sloan's criteria and definition for profile reduction are adapted and the general algorithm of the previous section along with the element clique graph of the considered FEM are employed for ordering. In Sloan's algorithm, a quantity is defined and used as the current degree. The initial priority for each node is set to

$$P_v = W_1 \times d(e, v) - W_2 \times cd(v), \qquad (5\text{-}17)$$

where W_1 and W_2 are integers (set to $W_1 = 1$ and $W_2 = 2$ in the original algorithm of Sloan [203]), $d(e, v)$ is the distance of node v from the end node e, and $cd(v)$ is the current degree of v.

In Step 4 of the general algorithm, if u has pre-active status, then each node x that is adjacent to it has its priority incremented according to $p_x \leftarrow p_x + W_2$. This is equivalent to decreasing the current degree of node x by unity.

In Step 6, each node x that is adjacent to the node u has its priority and status updated if it is pre-active. Then it is assigned an active status and its priority is increased by setting $p_x \leftarrow p_x + W_2$. Each node y that is adjacent to x is examined next, according to the following conditions:

(i) if y is not post-active, its priority is incremented by setting $p_y \leftarrow p_x + W_2$;

(ii) else, if y is inactive, then it is assigned pre-active status and increased in the priority queue by setting $n \leftarrow n + 1$ and $Q_n = y$. The time complexity of this method is $O(\alpha^2)$ for the worst case.

5.7.4 NODAL ORDERING WITH SKELETON GRAPH (NOSG)

The method for ordering the nodes of the SG of an FEM, to reduce the profile differs in the following two ways from the method of Nodal Ordering with Element Clique Graph (NOECG) (i.e. Sloan's method):

1. The distance between each node of SG and s (not e) is considered.

2. The initial priorities of nodes are calculated in a different manner.

The steps of the algorithm are outlined as follows:

Step 1: Form an SRT from S and compute the distance $d(s, v)$ between each node v of the SG and the starting node s.

Step 2: Assign each node in the graph an inactive status and compute its initial priority, p_v, according to

$$P_v = -d(s, v) - 3 \times \deg(v), \qquad (5\text{-}18)$$

where $\deg(v)$ is the degree of node v.

Step 3: Initialise the priority queue Q with the starting node s, that is, $Q_1 = s$. Set $n = 1$, where n is the length of the queue. The node s is assigned pre-active status. Let k be the node count.

Step 4: While the priority queue is not empty, which is signified by $n > 0$, execute Steps 5–8.

Step 5: Select node $u \in Q$ that has the maximum priority. Let i be the index of the node u in the queue such that $Q_i = u$.

Step 6: Delete node u from the priority queue by setting $Q_i = Q_n$ and decreasing n according to $n \leftarrow n - 1$. If node u is not pre-active, go to Step 7. Otherwise, examine each node w that is adjacent to node u and increment its priority according to $p(w) = p(w) + 2$. If node w is inactive, then insert it in the priority queue with a pre-active status by setting $n \leftarrow n + 1$ and $Q_i = w$.

Step 7: Label node u with its new number by incrementing the node count according to $k \leftarrow k + 1$ and setting $label(u) \leftarrow k$. Assign node u a post-active status.

Step 8: Examine each node w that is adjacent to node u. If node w is pre-active, assign node w an active status, set $p(w) = p(w) + 2$ and examine each node x that is adjacent to node w. If node x is not post-active, increment its priority to $p(x) = p(x) + 2$. If node x is inactive, insert it in the priority queue with a pre-active status by setting $n \leftarrow n + 1$ and $Q_n = x$.

Once the above steps are carried out, the new label of each node v will be $label(v)$. The time complexity of this method is the same as that of the NOECG method, for the worst case. However, it is interesting to note that the Nodal Ordering with Skeleton Graph (NOSG) must be executed faster than the NOECG in average cases, since the value of n NOSG is mostly less than that in the NOECG. This is because, for FEMs containing elements with four or more nodes, the degree of nodes of the SG is less than those of the ECG. These two methods need the same lists for nodal ordering of the considered graph; however, note that the compact adjacency list of the SG occupies usually less memory than that of the ECG.

5.7.5 NODAL ORDERING WITH ELEMENT STAR GRAPH (NOESG)

The profile reduction algorithm that employs the ESG of an FEM is the same as that of NOECG with the following modifications being imposed:

If a virtual node u (a node whose old label is more than λ) is selected for being labelled, it should be labelled with its new number by λ plus another node count without incrementing according to $k' \leftarrow k' + 1$ and setting $label(u) = \lambda + k'$.

This modification enables the numbering of the elements of the main set to be varied continuously from 1 to α.

This method uses $O(\gamma^2)$ operations in worst case, where $\gamma = \lambda + \alpha$, and needs the same lists as the NOECG and the NOSG. Clearly, the lists needed for the Nodal Ordering with Element Star Graph (NOESG) take large amounts of memory, since ESG has λ nodes more than those of the ECG and SG.

5.7.6 NODAL ORDERING WITH ELEMENT WHEEL GRAPH (NOEWG)

The same method as that of Section 5.7.5 is employed for ordering the nodes of the EWG of an FEM. The time complexity of the Nodal Ordering with Element Wheel Graph (NOEWG) is the same as that of the NOESG for the worst case. However, the NOESG is executed faster than the NOEWG in average cases, since the value of n in the process of the NOESG is, in general, less than that of the NOEWG, since for all FEMs, the degrees of the nodes of the ESG are, in general, less than those of the EWG. These two methods require the same lists to be provided for nodal ordering of the considered graph model. However, the compact adjacency list of the ESG uses fewer words of memory than that of the EWG.

5.7.7 NODAL ORDERING WITH PARTIALLY TRIANGULATED GRAPH (NOPTG)

Ordering the nodes of the PTG of an FEM for profile reduction does not require the selection of a pair of pseudo-peripheral nodes. The same good starting node used for the formation of the PTG (the node found in the SG for the formation of the PTG) can be used again in the Nodal Ordering with Partially Triangulated Graph (NOPTG) as the starting nodes s. The following two steps together with Steps 3–8 of the NOSG presented in Section 5.7.4 complete the process of the NOPTG.

Step 1: Form an SRT from the good starting node s used for the formation of the PTG and compute the distance $d(s, v)$ between each node v of the PTG and the starting node s.

ORDERING: GRAPH THEORY METHODS 233

Step 2: Assign each node in the graph an inactive status and compute its initial priority, p_v, according to

$$P_v = -d(s, v) - 2 \times (\deg(v) + 1). \tag{5-19}$$

The time complexity of this method is clearly the same as that of the NOECG and NOSG methods for worst case. In the NOPTG method, the same lists needed for the previous four methods should be provided. However, some of these lists such as the compact adjacency list do not take the same number of words of memory in different graph models.

5.7.8 NODAL ORDERING WITH TRIANGULATED GRAPH (NOTG)

In order to number the nodes of the TG of an FEM for profile reduction, it is not necessary to find a pair of pseudo-peripheral nodes. The same good starting node used for the formation of the TG is employed again in the Nodal Ordering with Triangulated Graph (NOTG) as the starting node s. The following steps along with Steps 2–8 of the NOSG method complete the process of NOTG.

Step 1: Form an SRT from the good starting node s used in the formation of the TG and compute the distance $d(s, v)$ between each node v of the TG and the starting node s.

The time complexity of this method is the same as that for methods NOECG, NOSG and NOPTG for the worst case.

The value of n in the NOTG process is mostly greater than that in NOSG and NOPTG, since the degrees of the nodes of the TG are mostly more than those of the PTG and the SG. Thus, NOTG is executed more slowly than NOSG and NOPTG in average cases. An advantage of NOTG, like NOPTG, is that no pseudo-peripheral nodes are needed.

5.7.9 NODAL ORDERING WITH NATURAL ASSOCIATE GRAPH (NONAG)

The profile reduction algorithm that employs the NAG of the FEM consists of two phases. In the first phase, which is the same as NOECG, the nodes of the NAG are ordered. In the second phase, the nodes of the considered FEM are ordered on the basis of the new labels of the nodes of the NAG. This process consists of the following steps:

Step 1: For each node i of the graph model set $n(label(i)) = i$.

Step 2: For each element e corresponding to the node u, $u = n(j)$, $j = 1, 2, ..., a$, label the unlabelled nodes of e, in turn.

The time complexity of the first phase of this algorithm is $O(\lambda^2)$ and the second phase uses $O(\alpha + \lambda\theta) = O(\lambda\theta)$ operations.

This algorithm requires the same lists as the previous methods; however, the number of nodes of the graph model is equal to λ. Therefore, it is very efficient for FEMs containing higher-order elements. In the second phase of this method, an additional list with λ integer words of memory is needed, which is denoted by $n(.)$ in the steps of the process. However, this list can be created when most of the lists needed for the first phase are not required, and can be erased from the working memory.

5.7.10 NODAL ORDERING WITH INCIDENCE GRAPH (NOIG)

The profile heuristic that employs the IG of an FEM consists of two parts, as in the Nodal Ordering with Natural Associate Graph (NONAG) method. These phases are the same as those of NONAG, with IG being employed in place of NAG.

Time and memory complexities of the Nodal Ordering with Incidence Graph (NOIG) are the same as those of NONAG. However, the value of n is higher than that of the NONAG, since degrees of the IG are more than those of the NAG. Therefore, the NOIG should have slower execution than NONAG in average cases.

5.7.11 NODAL ORDERING WITH REPRESENTATIVE GRAPH (NORG)

This method consists of two parts. The first part orders the nodes of the RG, that is, the representative nodes of the elements of the considered FEM. The second phase orders the nodes of the considered FEM based on the new labels of the representative nodes of the elements of the FEM.

The first part consists of the following steps:

Step 1: Form an SRT from a good starting node s used for the formation of the RG and compute the distance $d(s, v)$ between each node v of the RG and the starting node s.

Step 2: Assign an inactive status to each node in the graph and compute its initial priority p_v, according to

$$P_v = -3 \times d(s, v) - \varepsilon(v), \tag{5-20}$$

where $\varepsilon(v)$ denotes the number of elements incident to node v.

Step 3: Initialise the priority queue Q with the starting node s used for the formation of the RG, that is, $Q_1 = s$. Set $n = 1$, where n is the length of the queue. The node s is assigned a pre-active status. Let k be the node count.

Step 4: While the priority queue is not empty, signified by $n > 0$, execute Steps 5–8.

Step 5: Select node $u \in Q$ that has the maximum priority. Let i be the index of the node u in the queue such that $Q_i = u$.

Step 6: Delete node u from the priority queue by setting $Q_i = Q_n$ and decrementing n according to $n \leftarrow n - 1$. If node i is not pre-active, go to Step 7, otherwise examine each node w that is adjacent to node u and increment its priority according to $p(w) = p(w) + 1$. If node w is inactive, then insert it in the priority queue with a pre-active status by setting $n \leftarrow n + 1$ and $Q_n = w$.

Step 7: Label node u with its new number by incrementing the node count according to $k \leftarrow k + 1$, and setting $label(u) \leftarrow k$. Assign node u a post-active status.

Step 8: Examine each node w that is adjacent to node u. If node w is pre-active, assign node w an active status, set $p(w) = p(w) + 1$ and examine each node x that is adjacent to node w. If node x is not post-active, increment its priority to $p(x) = p(x) + 1$. If node x is inactive, insert it in the priority queue with a pre-active status by setting $n \leftarrow n + 1$ and $Q_n = x$.

When the above steps are completely performed, the new label of each node v is $label(v)$. In this method, there is no need to find any pseudo-peripheral, and the same good starting node used for generating the RG is employed again in the process of numbering.

The second phase of the algorithm consists of the following steps:

Step 1: For each node i of the graph model set $n(label(i)) = i$.

Step 2: Set $k = 0$. Check each element e containing node u, $u = n(j)$, $j = 1,2, ..., \alpha$, in turn; if e does not contain a node v corresponding to $n(1)$ and $i < j$, then set $k \leftarrow k + 1$ and $m(k) = e$.

Step 3: Set $label(i) = 0$, where $i = 1,2,..., \alpha$.

Step 4: Set $l = 0$. Check each node w of element e, $e = m(j)$, $j = 1,2, ..., \alpha$, in turn; if $label(w) = 0$, then set $i = 0 + 1$ and $label(w) = 1$.

The time complexity of the first part of this algorithm is $O(\alpha^2)$, and the second part uses $O(\lambda \theta^2)$ operations. The time complexity of the second phase can be reduced by using an additional list in Step 2 to show whether element e has been previously detected. This procedure uses $O(\lambda\theta)$ operations. The first phase of the algorithm Nodal Ordering with Representative Graph (NORG) requires the same lists as the previous methods of profile reduction.

Complete Representative Graph (CREG)

This graph is the same as the REG with additional members connecting each pair of nodes in the Complete Representative Graph (CREG), if their corresponding nodes in the FEM are contained in the same element.

5.7.12 NODAL ORDERING WITH ELEMENT CLIQUE REPRESENTATIVE GRAPH (NOECRG)

The profile reduction of this method consists of two steps, as in the NONAG, NOIG and NORG methods. The first process is the same as that of Sloan's algorithm (NOECG), and the second step is similar to the second step of the NORG approach.

The time complexity and memory complexity of the Nodal Ordering with Element Clique Representative Graph (NOECRG) method are the same as those of the NORG, but the magnitude of n in the process of NOECRG is, in general, far higher than that of the NORG, since the degrees of the nodes of ECRG are generally much greater than those of RG. Therefore, NOECRG should be slower in execution than RG.

5.7.13 COMPUTATIONAL RESULTS

A program is developed for implementing the algorithms as shown in Tables 5.5–5.8, and many FEMs are studied. Four examples are presented here. For each problem a table is provided for illustrating the new profile obtained from the new labels of the nodes, and the elapsed time for executing the program is also given for each case. The numbers of nodes α and elements λ of each FEM are provided in the captions of the Figures 5.22–5.25.

ORDERING: GRAPH THEORY METHODS 237

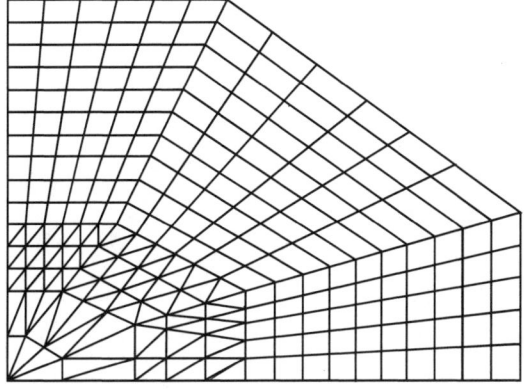

Fig. 5.22 $\alpha = 240$ and $\lambda = 499$.

Table 5.5

Algorithm	Profile	Elapsed time
NOECG	3207	0.22
NOSG	3236	0.22
NOESG	3367	0.44
NOEWG	3465	0.66
NOPTG	3194	0.28
NOTG	3237	0.27
NONAG	3365	0.71
NOIG	3365	0.60
NORG	3460	0.33
NOECRG	3185	0.44

238 OPTIMAL STRUCTURAL ANALYSIS

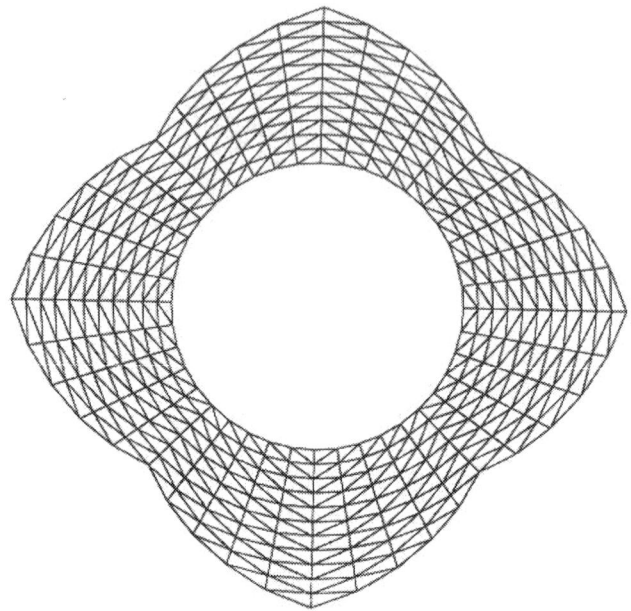

Fig. 5.23 $\alpha = 748$ and $\lambda = 1236$.

Table 5.6

Algorithm	Profile	Elapsed time
NOECG	7444	0.39
NOSG	8436	0.39
NOESG	8336	0.87
NOEWG	8256	1.27
NOPTG	8527	0.65
NOTG	8514	0.66
NONAG	7320	0.93
NOIG	7204	1.32
NORG	9388	0.66
NOECRG	7818	0.88

ORDERING: GRAPH THEORY METHODS 239

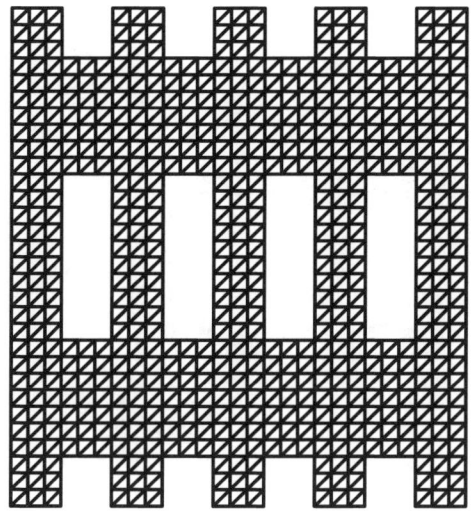

Fig. 5.24 $\alpha = 936$ and $\lambda = 1640$.

Table 5.7

Algorithm	Profile	Elapsed time
NOECG	12,248	0.72
NOSG	13,142	0.71
NOESG	13,016	1.37
NOEWG	13,049	2.03
NOPTG	13,282	1.16
NOTG	13,113	1.21
NONAG	12,631	1.54
NOIG	12,665	1.98
NORG	16,055	1.16
NOECRG	12,894	1.65

240 OPTIMAL STRUCTURAL ANALYSIS

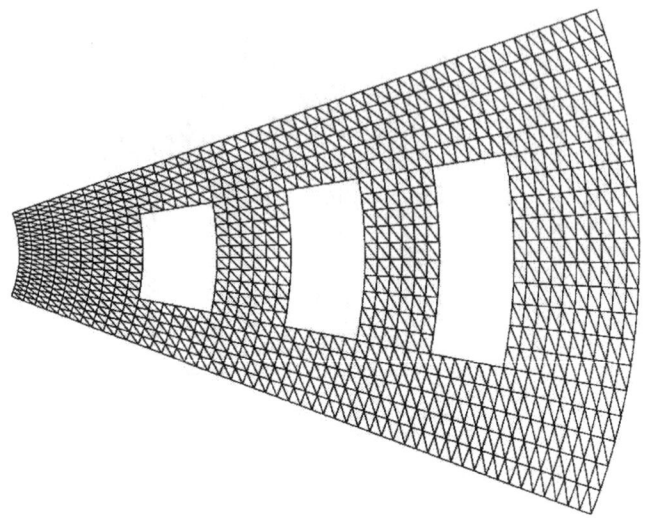

Fig. 5.25 $\alpha = 936$ and $\lambda = 1640$.

Table 5.8

Algorithm	Profile	Elapsed time
NOECG	15,223	0.88
NOSG	16,217	0.93
NOESG	16,008	1.87
NOEWG	15,852	2.63
NOPTG	15,391	1.48
NOTG	16,204	1.60
NONAG	15,482	2.15
NOIG	15,345	2.69
NORG	17,474	1.42
NOECRG	15,343	2.09

5.7.14 DISCUSSIONS

The algorithms presented for the profile reduction of sparse matrices with symmetric structures are analysed for the worst case to show their time and memory complexities.

The programs developed for these algorithms have been tested on many examples, and Table 5.9 is obtained, which illustrates the average computational time (in seconds) of the methods.

Table 5.9

Algorithm	Average of the computational time
NOECG	0.99
NOSG	0.77
NOESG	1.39
NOEWG	1.96
NOPTG	1.25
NOTG	1.32
NONAG	1.33
NOIG	1.45
NORG	1.17
NOECRG	1.90

5.8 ELEMENT ORDERING FOR FRONTWIDTH REDUCTION

For the solution of sparse systems of simultaneous equations arising from the FEM, the frontal methodology due to Irons [87] and the profile method described by George [56], as well as band-matrix techniques, are commonly used. These methods exploit the sparsity of the coefficient matrices generated by the FE approximation. They differ, however, in one significant respect: the band and profile methods first construct the coefficient matrix explicitly, while the frontal method arranges for the elimination of variables as it assembles the matrix.

The most suitable ordering of the equation is dependent on the type of equation-solving scheme adopted (i.e. whether a band, profile or frontal solver is used). In FE analysis, in the case of one degree of freedom per node, performing nodal ordering is equivalent to reordering the equations. In a more general problem with β degrees of freedom per node, there are β-coupled equations produced by each node. In this case re-sequencing is usually performed on the nodal numbering to reduce the bandwidth, profile or frontwidth, because the size of this problem is β times less than that for degree of freedom numbering.

In this section, a graph-theoretical approach is designed for element renumbering of the FE meshes for frontwidth reduction of sparse matrices with symmetric structures. In this heuristic, the problem is transformed into that of a graph nodal ordering. The IG of a mesh is used for representing its connectivity. The efficiency of the method is illustrated by worst-case analysis and examples of unstructured FE models.

5.8.1 DEFINITIONS

Neighbouring nodes of a subgraph S_i of S are the nodes contained in $N(S) - N(S_i)$ that are adjacent to the nodes of S_i.

A tree is rooted from a given node n; it may be denoted by T_n. A *spanning tree* is a tree containing all the nodes of S. A *shortest root tree* (SRT_{n_0}) rooted from a specified node (starting node) n_0, is a spanning tree for which the distance between every node n_j of T and n_0 is minimum, where the *distance* between the two nodes is taken as the number of members in the shortest path between these nodes. A *multiple root SRT* ($MRSRT_R$) is an SRT, but rooted from a set of nodes R.

A contour $C_{n_0}^k$ of an SRT_{n_0} contains all nodes with equi-distance k from n_0. The number of contours "d" of an SRT is known as its *depth* or *length*, and the highest number of nodes in a contour "w" specifies the *width* of an SRT. The last contour of an SRT_{n0} is denoted by $C_{n_0}^i$. A contour $C_{n_0}^d$ of an SRT_{n_0} can be disconnected, that is, there may be no path included in $C_{n_0}^i$ between all pairs of its nodes. Each component of a disconnected contour is called a *subcontour*. A *heeled SRT* ($HSRT_c^{n_0}$) is a MRSRT rooted from a subcontour C of a contour $C_{n_0}^i$ but expanded on the part that contains a component of $C_{n_0}^{i-1}$. The following algorithm is employed for generating an $HSRT_c^{n_0}$ in which C is a component of $C_{n_0}^i$ with a desired property, for example, the component with the smallest number of nodes.

1. Form SRT_{n_0} and check $C_{n_0}^i$ in order to find a subcontour c with the desired property.

2. Designate the nodes of c as C_c^0.

3. Select the nodes of $C_{n_0}^{i-1}$ that are adjacent to the nodes of C_c^0 as C_c^1.

4. Select unselected nodes adjacent to C_c^1 as C_c^2. Repeat this process to form $C_c^3, C_c^4, \ldots, C_c^j$, where C_c^j is a contour for which there is no unselected node.

5. If all nodes are selected, the formation of $HSRT_c^{n_0}$ is completed; otherwise, select unselected nodes adjacent to C_c^0 as C_c^{j+1}. Then take the unselected nodes adjacent to C_c^{j+1} as C_c^{j+2} and repeat this process until all the nodes are selected.

The *IG* of an FEM has its nodes in a one-to-one correspondence with the elements of the FEM, and two nodes are connected with an edge if the corresponding elements have at least one common node.

A *level structure* $L(r)$ of an FEM rooted from an element r (as the root) is defined as a partitioning of the set of elements into levels $l_1(r), l_2(r), ..., l_d(r)$ such that:

1. $l_1(r) = \{r\}$;

2. all elements adjacent to the elements in level $l_i(r)$ ($1 < i < d$) are in levels $l_{i-1}(r)$, $l_i(r)$ and $l_{i+1}(r)$;

3. all elements adjacent to the elements in level $l_d(r)$ are in levels $l_{d-1}(r)$ and $l_d(r)$.

The overall level structure may be shown as $L(r) = \{l_1(r), l_2(r), ..., l_d(r)\}$, where d is the depth of the level structure rooted at element r and is simply the total number of levels, and two elements are adjacent if they share a common node.

The *element adjacency list* of an FEM contains the list of elements adjacent to each element. The element-node list of an FEM contains the list of nodes of each element and is generally employed as input for data connectivity of FEMs. The node-element list contains the list of elements containing each node of the FEM.

Consider the solution of sparse linear systems of equations,

$$\mathbf{Ax = b}, \qquad (5\text{-}21)$$

where the $n \times n$ matrix $\mathbf{A}$ is a sum of elemental matrices:

$$\mathbf{A} = \sum_{i=1}^{m} \mathbf{A}^{[l]}, \qquad (5\text{-}22)$$

and the right-hand side vector $\mathbf{b}$ is of the form:

$$\mathbf{b} = \sum_{i=1}^{m} \mathbf{b}^{[l]}. \qquad (5\text{-}23)$$

In Eq. (5-22), each matrix $\mathbf{A}^{[l]}$ has entries only in the principal submatrix corresponding to the variables in element l and represents contributions from this element. This principal submatrix is assumed to be dense (any zeros are stored explicitly). The matrix $\mathbf{A}$ may be unsymmetric but the form of Eq. (5-22) implies that it has a symmetric pattern.

With reference to Eq. (5-21), column j is said to be *active* at stage i if $j \geq i$ and there is a non-zero entry in column j with a row index, k, such that $k \leq i$. Letting f_i denote the number of columns that are active at stage i, the *maximum frontwidth* of **A** is given by

$$F_{max} = \max \{f_i\}. \qquad 1 \leq i \leq n \qquad (5\text{-}24)$$

The *root-mean-squared frontwidth* is defined as

$$\tilde{F} = (\frac{1}{n}\sum_{i=1}^{n} f_i^2)^{\frac{1}{2}} \qquad (5\text{-}25)$$

5.8.2 DIFFERENT STRATEGIES FOR FRONTWIDTH REDUCTION

There are many algorithms for profile and frontwidth reduction, such as those of Kaveh and Ramachandran [128], Duff et al. [39], Razzaque [185], Pina [177], and many others, which can be categorised in different ways. Akin and Purdue [3] have divided these methods into two groups: direct and indirect approaches. In direct methods, elements are renumbered directly to minimise the profile and frontwidth. Indirect methods first renumber the nodes and then reorder the elements on the basis of the new labels of their nodes. When attempting to develop schemes for minimising the profile and frontwidth of sparse matrix equations, it is fruitful to consider schemes that are aimed at minimising the bandwidth. This is because the maximum frontwidth must always be less than or equal to the corresponding bandwidth (if the variables are eliminated in the same order). Thus one method of reducing the frontwidth is to resequence the nodes first to minimise the bandwidth and then to relabel the elements so that the new order of elimination is presented as closely as possible [3,185]. The effectiveness of this strategy is obviously dependent on the performance of the bandwidth minimisation procedure and thus suffers from the disadvantage of being indirect.

Another way to categorise the renumbering schemes is to consider how the algorithms use the connectivity of an FEM. In the general case, these algorithms can be grouped into two categories: engineering-based and graph theory–based heuristics. The first applies the element-node list, exclusive of other lists generated using this list such as the node-element list, to improve the efficiency of the considered method; for example, see Webb and Froncioni [225]. However, there are engineering-based methods that use the concepts of graph theory to form the auxiliary lists; for instance, Pina [177] has cryptically employed the NAG and the IG. Graph-theoretical heuristics can be found in [128,203–206] among many others. In these methods, the connectivity properties of FEMs are transformed into different graph models [133]. These transformations lessen the time complexities of the renumbering procedures at the expense of computer storage and time for keeping and generating the data connectivity of the graph models employed.

In this section, the renumbering procedures are divided into five main categories as follows:

1. Level (contour) expansion methods;

2. Sublevel (subcontour) expansion methods;

3. Nodal (elemental) expansion methods;

4. Divide-and-conquer methods; and

5. Node (element) shuffling methods.

It should be noted that in the rest of this section the term "node" is used as the unit of a model that should be renumbered. This unit can be a node, an element or a super element.

Level expansion procedures start the process of renumbering from one or more nodes as the first level of a level structure or the first contour of an SRT and renumber unlabelled nodes connected to the previous level contour containing renumbered nodes. This process is continued until all nodes are relabelled. Level expansion schemes are efficient for bandwidth reduction and are very easily implemented. However, it is simple to show that these methods can easily lead to inefficient profiles and frontwidths [43,205].

Sublevel expansion methods are similar to level expansion methods, but the process of renumbering expands from a sublevel. The process of renumbering is continued by renumbering the unrenumbered nodes of a sublevel connected to a sublevel with relabelled nodes; see, for example Plesek [180].

Nodal expansion methods for profile and frontwidth reduction are the most popular schemes. In these algorithms, each unrelabelled node is assigned a single number as its weight or priority number, and in each step one node with the highest priority is selected. Some of these methods such as the one discussed in [204] benefit from the global properties (such as pseudo-diameter) as well as local properties (such as degrees of nodes) of FEMs.

The divide-and-conquer strategy divides the set of nodes into two or more subsets and then treats each subset as a new set, and the process of division is continued until a specified condition (e.g. that a final subset should contain a single node) is fulfilled. The process of renumbering is carried out synchronously or when the process of division is completed [199].

The most important character of the node shuffling methods is the multiple modification of the new labels of two or more or even all nodes to improve the results. Such modifications may be made where the frontwidth or bandwidth has the highest magnitude, or may be carried out randomly. These methods suffer from high computer time requirement and benefit from very efficient results. It should be noted that the methods that apply the non-deterministic heuristics used in combinatorial optimisation such as Simulated Annealing belong to this category. In the classification by Lim et al. [150], where the profile and frontwidth reduction algorithms are categorised into five classes, these methods are considered as exhaustive search algorithms.

5.8.3 EFFICIENT ROOT SELECTION

A large number of algorithms for reducing the bandwidth, profile or frontwidth can be found in the literature. The most efficient algorithms are based on graph theory and require one or more starting nodes. The success of these algorithms is dependent on the selection of the starting nodes. Gibbs et al. [59] have presented an algorithm for finding a pair of starting nodes that are located at nearly maximal distance apart. These nodes are known as *pseudo-periphal nodes*. It is demonstrated through extensive tests available in the literature that this strategy provides good starting nodes. However, in some problems of FE analysis, namely, in models with square or annular meshes, there may be several candidates for pseudo-peripheral nodes. The selection of the pair of nodes, however, is not always indifferent. This is because of the heuristic nature of the algorithms and the parameters adopted for the selection of the end nodes. Consequently, if additional criteria were taken into account, either to limit the eligible nodes or to establish a new rank for the nodes, then fewer pairs would be equally rated. Modifications to the original pseudo-peripheral node finder strategy given by Gibbs et al. [59] have been proposed by Kaveh [101], George and Liu [57], Sloan and Randolph [206] and Sloan [204,203]. Souza and Murray [209] have added an additional parameter that is checked for the selection of the second end nodes of pseudo-peripheral nodes. Their modification generally improves the results of the renumbering algorithms of Gibbs et al. [59] and Gibbs–King [58], but at the expense of more computational time. This modified algorithm uses more computer time than that of Sloan [204] and seldom improves on its results (see, [209]). In terms of the computer storage needed for keeping connectivity data of the considered FEM, these two algorithms are similar; they apply the ECG.

There are also very simple methods in which the starting nodes are found using only the local properties of an FEM [177]. In these methods, the renumbering process generally suffers from the inefficiency of the selected starting nodes.

An efficient method for finding a root is presented in the following text. We do not say that an efficient root should contain a single node or two nodes. A selected root may contain one or several nodes. Removing this restriction from root selection schemes can lead to very efficient results [131], provided that the global properties of the considered FEM are contemplated.

For FE models composed of high-order elements, Sloan and Randolf [206] have noted that it is necessary to consider the corner nodes only when renumbering is carried out. This follows from the observation that an element ordering that is efficient for a model of low-order elements is also efficient for an equivalent model of high-order elements. Since an FEM of high-order elements may have a small number of corner nodes, but a large number of nodes in total, this approach leads to considerable economy in the ordering phase.

In this method, only the corner nodes of the element are essential; however, this does not mean that interior nodes and mid-side nodes make the method inefficient. In the efficient root selection procedure presented in this section, an FEM can contain meshes of different types and dimensions.

Algorithm

Step 1: Use the IG of the considered FEM and form a SRT_{n_0} rooted from a node n_0 with the minimum degree.

Step 2: Form an SRT_{n_1} rooted from a node n_1 of minimum degree from $C_{n_1}^d$.

Step 3: Check $C_{n_1}^1, C_{n_1}^2, ..., C_{n_1}^d$ to find a contour with at least two components for which a subcontour R' has a minimum number of neighbouring nodes.

Step 4: If R' is found, then generate a heeled SRT from R'.

Step 5: Select R as the desired root from n_0, n_1 and R' according to the minimum width of SRT_{n_0}, SRT_{n_1} and $HSRT_{R'}^{n_1}$.

Example: Consider the small multiconnected FEM shown in Figure 5.26(a). The IG of the FEM is depicted in Figure 5.26(b). The execution of the steps of the above algorithm leads to the following results.

Step 1: An SRT is formed from node 1, since its degree is minimum. The width of SRT_1 is equal to 15, and C_1^d contains nodes 46 and 58.

Step 2: The degree of node 58 is less than that of node 50; thus SRT_{58} is formed, since its width is equal to 13.

248 OPTIMAL STRUCTURAL ANALYSIS

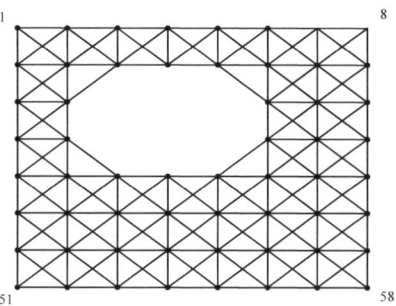

(a) A small multiconnected FEM.

(b) The incidence graph of the FEM.

Fig. 5.26 AN FEM and its graph model.

Step 3: The contours of SRT_{58} are checked; the subcontour containing nodes 4, 5, 6, 7, 8 and 12 is selected as R' because the number of its neighbouring nodes is minimum and equal to 6.

Step 4: $HSRT_{R'}^{58}$ is formed; its contours are as follows:

$C_{R'}^{0} = \{4, 5, 6, 7, 8, 12\}$, $C_{R'}^{1} = \{13, 14, 15, 16\}$, $C_{R'}^{2} = \{19, 20, 21\}$, $C_{R'}^{3} = \{24, 25, 26\}$, $C_{R'}^{4} = \{31, 32, 33, 34\}$, $C_{R'}^{5} = \{30, 38, 39, 40, 41, 42\}$, $C_{R'}^{6} = \{29, 37, 45, 46, 47, 48, 49, 50\}$, $C_{R'}^{7} = \{23, 28, 36, 44, 52, 53, 54, 55, 56, 57, 58\}$ $C_{R'}^{8} = \{17, 18, 22, 27, 35, 43, 51\}$, $C_{R'}^{9} = \{9, 10, 11\}$, $C_{R'}^{10} = \{1, 2, 3\}$.

The width of $\text{HSRT}_{R_1^5}^{58}$ is equal to 11.

Step 5: The width of $\text{HSRT}_{R_1^5}^{58}$ is the least one; thus $R' = \{4, 5, 6, 7, 8, 12\}$ is selected as R.

5.8.4 ALGORITHM FOR FRONTWIDTH REDUCTION

The frontal method can be simply viewed as a variation of the variable band solver that is tuned for matrices generated by the FEM. At the heart of most FE problems, one has to solve one or more linear systems of the form:

$$\mathbf{Kv} = \mathbf{p}, \qquad (5\text{-}26)$$

where
$$\mathbf{K} = \sum_{i=1}^{M} \mathbf{k}_i, \qquad (5\text{-}27)$$

and M spans the entire set of elements. $\mathbf{k}_i$ is the stiffness matrix of each individual element expressed in a global coordinate system, and $\mathbf{v}$ and $\mathbf{p}$ are the generalised displacement and force vectors, respectively. The matrix $\mathbf{K}$ is symmetric and positive-definite and usually very sparse. The sparseness property must be exploited if results are to be obtained in a reasonable time. One approach for solving such systems is to reorder the equations and to build a skyline or profile of the matrix to prevent the fill-in resulting from the factorisation. Next, the summation of the element contributions to the global stiffness matrix is performed, which defines the assembly phase. It is only then that the factorisation process can proceed, followed by forward and backward substitutions. Unfortunately, for large-scale three-dimensional problems, the memory space required for storing the values within the band or skyline representation of $\mathbf{K}$ often exceeds the available random access memory (RAM), so that the triangular factors of $\mathbf{K}$ have to be moved to auxiliary storage. This entails the design and implementation of rather critical algorithms for the out-of-core assembly and elimination processes; see, for example, Wilson and Dovey [232].

The frontal method is an alternative to band or skyline solvers that does not require the global stiffness matrix to be explicitly assembled and therefore significantly reduces the I/O time for out-of-core systems. Instead, assembly and elimination processes are merged into a single one. The elemental stiffness matrices are formed and assembled until a row of $\mathbf{K}$ is completed, and then eliminated. This process is carried out in a matrix $\mathbf{F}^{(k)}$ similar to the global matrix $\mathbf{K}$, which can be referred to as the *front matrix* or simply the front. This matrix needs to provide entries only for the degrees of freedom that are connected to those that have already been introduced in the front (i.e. assembled) and not yet eliminated (for most problems $\mathbf{F}^{(k)}$ will fit in the central memory of a supercomputer). The introduction of the ith element in the front can be written as

$$\mathbf{F}^{(k+1)} = \mathbf{F}^{(k)} + \mathbf{k}_i. \tag{5-28}$$

Once a row (equation) is assembled, the system is partitioned as

$$\mathbf{F}^{(k+1)}\mathbf{v} = \begin{bmatrix} \mathbf{d} & \mathbf{l} \\ \mathbf{l}^t & \mathbf{F}^{(k+1)}_{rr} \end{bmatrix} \begin{bmatrix} \mathbf{v}_k \\ \mathbf{v}_r \end{bmatrix} = \begin{bmatrix} \mathbf{p}_k \\ \mathbf{p}_r \end{bmatrix}, \tag{5-29}$$

where [$\mathbf{d}$ $\mathbf{1}$] is the pivotal row k and $\mathbf{F}^{(k+1)}_{rr}$ corresponds to the remaining degrees of freedom of $\mathbf{F}^{(k)}$. The elimination step can now be written as

$$\begin{aligned}\mathbf{F}^{(k+1)}_{rr} &= \mathbf{F}^{(k+1)}_{rr} - \mathbf{l}^t \mathbf{d}^{-1}\mathbf{l}, \\ \mathbf{p}_r &= \mathbf{p}_r - \mathbf{l}^t \mathbf{d} - \mathbf{1}\mathbf{p}_k. \end{aligned} \tag{5-30}$$

The pivotal row [$\mathbf{d}$ $\mathbf{1}$] and $\mathbf{p}_k$ are stored in a buffer area or an auxiliary storage for later use in the back-substitution step,

$$\mathbf{v}_k = \mathbf{d}^{-1}(\mathbf{p}_k - \mathbf{l}\mathbf{v}_r), \tag{5-31}$$

which is performed in the reverse order of the elimination. Further details can be found in Irons [87], Razzaque [185], and Duff et al. [39]. In order to employ this method in a vector computer, one may refer to Lesoinne et al. [149], Duff [40], Brusa and Riccio [17], Löhner [154] and references cited therein.

The average number of arithmetic operations in a single elimination step in a frontal algorithm is proportional to the mean-squared frontwidth, and the maximum amount of storage required for the frontal matrix during the Gaussian elimination is dependent upon the maximum frontwidth. Moreover, the total storage required and the amount of work involved in the back-substitution stage depend on the profile of the matrix. Thus the elements are numbered in such a way as to reduce F, $\tilde{F}$ and P. On the other hand, it may be shown that the total number of operations required for a profile or frontal elimination is $O(NF^2)$, where N denotes the dimension of the considered sparse symmetric matrix. Thus, to minimise the total storage, we minimise the profile, whereas to minimise the total computing time the root-mean-square frontwidth has to be reduced. For reducing the maximum amount of working memory during the elimination process, the minimisation of the maximum frontwidth is needed; see, for example, Carey and Oden [20].

ORDERING: GRAPH THEORY METHODS 251

The renumbering procedure of this section is targeted at minimising the maximum amount of working memory during the elimination process; on the other hand, attention is paid to reducing the maximum frontwidth F. However, optimising the computational time and total storage are also contemplated cryptically.

In the present algorithm when a node i is renumbered as x, a priority number is calculated for each unlabelled node adjacent to i. If there is no unlabelled node connected to i, then unlabelled nodes connected to the nodes with new labels $x - 1$, $x - 2$, ..., 1 are checked, in turn, until a node j is renumbered as $x + 1$. Then the priority numbers of the unlabelled nodes adjacent to $x + 1$ are calculated, and the one with the maximum priority number is renumbered as $x + 2$. This process is continued until all the nodes have been renumbered.

The priority number of a node i is calculated as,

$$PN_i = c_1 \times \frac{\text{dis}(i)}{d} - (c_2 + c_3) \times \frac{c\deg(i)}{md} + c_3 \times \frac{\text{dis}(i)}{md}, \qquad (5\text{-}32)$$

where, c_1, c_2 and c_3 are defined as follows:

if R is a subcontour, then $c_1 = 0.05$, $c_2 = 0.05$ and $c_3 = 0.9$,

else, $c_1 = 0.9$, $c_2 = 0.04$ and $c_3 = 0.06$.

d denotes the depth of the HSRTC_R^d, and $\text{dis}(i)$ denotes the distance between C_R^d and i. $\deg(i)$, $c\deg(i)$ and md denote the degree of i, the current degree of i and the highest degree of the nodes of the graph model respectively, where, going by Sloan and Randolph [206], the current degree of a node is the same as the number of unlabelled nodes connected to the node.

Algorithm

Step 1: Generate the incidence graph S of the considered FEM.

Step 2: Find an efficient root R using the method of Section 5.8.3.

Step 3: Form a (multiple root) SRT from R. Then form an MRSRT from C_R^d and calculate the distance between C_R^d and each node of S.

Step 4: Label the node (nodes) of R (in the same order as they are selected).

Step 5: Label an unlabelled node adjacent to the last labelled node j with the maximum priority number. If there is no unlabelled node connected to j, check the previously labelled nodes, in turn. Repeat this process until all nodes are labelled.

The above method consists of a nodal expansion process and is efficient from two significant viewpoints: (1) the priority numbers of a few number of nodes will be calculated in each process of renumbering of a node with the highest priority (compare with the existing nodal expansion methods), (2) the corner nodes only can be given.

5.8.5 COMPLEXITY OF THE ALGORITHM

Algorithms may be efficient in several ways. They may be economical in terms of computer memory, or computer time, or they may be easy for humans to read, write, or understand. Of these virtues, only the first two can easily be quantified.

The main problem in measuring the time and storage required by a computer program or subprogram lies in isolating the properties of the algorithm from the properties of the particular machine and language. In the following text, the frontwidth reduction heuristic is analysed for the worst case.

Step 1: The element-node list of an FEM is usually employed as input for data connecting of elements, and here such a list is used. This data structure requires $\lambda + 1 + \sum_{i=1}^{\lambda} \theta_i$ words of memory, where λ and θ_i denote the number of elements of the FEM and the number of nodes of element i. Clearly, when only the corner nodes are given, this data structure occupies far less storage. However, for an efficient programming of algorithms like the present one, this list should not be used exclusively. Another data structure should be employed along with the element-node list because finding elements connected to a specified element takes $O(\lambda\theta^2)$ operations, where

$$\theta = \max \theta_i \qquad 1 \le i \le \lambda \qquad (5\text{-}33)$$

The direct formation of the IG of an FEM has time complexity $O(\lambda^2\theta^2)$, which is inefficient. In order to improve this, the node-element list should be formed. This list requires $\alpha + 1 + \sum_{i=1}^{\alpha} \varepsilon_i$ words of memory, where α and ε_i denote the number of nodes of the FEM and the number of elements connected to node i. The formation of the node-element list takes $O(\theta\lambda)$ operations. When the node-element list is formed, the element-node list can be erased from working memory. Using the node-element list, the formation of the IG takes $O(\alpha\delta\varepsilon^2)$ operations, where δ denotes the maximum number of elements connected to an element and

$$\varepsilon = \max \varepsilon_i \qquad 1 \leq i \leq \lambda \qquad (5\text{-}34)$$

The list that keeps the data of connecting of the IG needs $\lambda + 1 + \sum_{i=1}^{\lambda} \delta_i$ words of memory. When the IG is generated, the node-element list can be erased.

Step 2: Scanning the nodes of the IG to find a node with the minimum degree takes $O(\lambda)$ operations; note that λ denotes the number of nodes in the IG or the number of elements in the FEM. Formation of an SRT, MRSRT or heeled SRT (HSRT) takes $O(\lambda\delta)$ operations and requires nearly $3 \times \lambda$ words of memory (for worst case). Controlling the contours to find the desired subcontour requires $O(\lambda\delta)$ operations.

Step 3: Execution of this step requires $O(\lambda\delta)$ operations and $3 \times \lambda$ words of memory.

Step 4: This step has the time complexity $O(1)$.

Step 5: The time complexity of this step is $O(\lambda\delta^2)$ and, excluding the data structure needed for keeping the data of connectivity of the IG, $3 \times \lambda$ words of memory are required for the efficient execution of this step. λ words are needed for keeping the distance, λ words are needed for keeping the new labels of the nodes and λ words are also needed to define which node has a new label.

5.8.6 COMPUTATIONAL RESULTS

A computer program is developed for the heuristic of this section and many models are studied; some of them are included in this section, and the maximum frontwidth F, root-mean-square $\tilde{F}$ and profile P, along with the computational time T using a PC (80486DXII), are provided for each model. In these examples, all corner and mid-side nodes are considered.

Example 1: An FEM with 240 nodes and 499 bar elements is considered as shown in Figure 5.27. For this model, the application of the present algorithm results in an ordering corresponding to $F = 19$, $\tilde{F} = 14.14$, $P = 6796$ and $T = 0.72$ (sec.).

254 OPTIMAL STRUCTURAL ANALYSIS

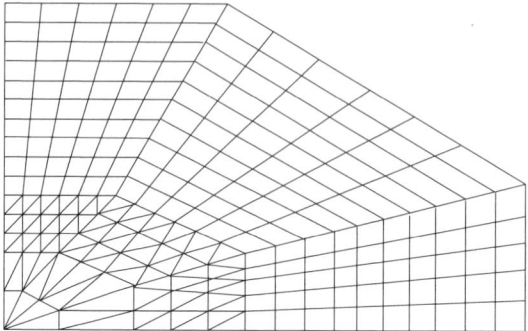

Fig. 5.27 An FEM with bar elements.

Example 2: An H-shaped FEM with 2096 nodes and 3900 triangular elements is considered as depicted in Figure 5.28. For this model, the results are $F = 71$, $\tilde{F} = 32.42$, $P = 11{,}676$ and $T = 12.36$.

Fig. 5.28 An H-shaped FEM with triangular elements.

Example 3: A multiconnected FEM is considered as illustrated in Figure 5.29. This model contains 1248 nodes and 1152 quadrilateral elements. The results are $F = 38$, $\tilde{F} = 33.02$, $P = 37{,}582$ and $T = 2.25$.

ORDERING: GRAPH THEORY METHODS 255

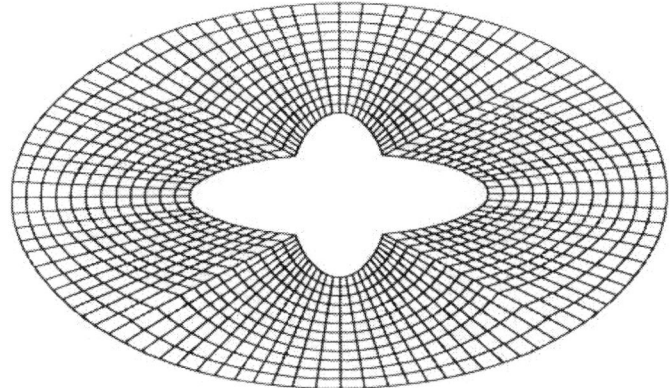

Fig. 5.29 A multiconnected FEM with quadrilateral elements.

Example 4: A multiconnected FEM with 1762 nodes and 910 4-node, 8-node and 12-node quadrilateral elements is considered as shown in Figure 5.30. The results are $F = 62$, $\tilde{F} = 40.46$, $P = 34{,}879$ and $T = 1.93$.

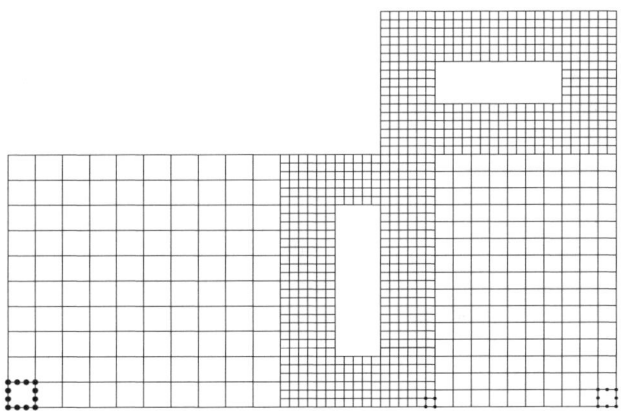

Fig. 5.30 A multiconnected FEM with different types of elements.

Example 5: An FEM with 2392 nodes and 1800 brick elements is considered, as illustrated in Figure 5.31. The execution of this method leads to $F = 199$, $\tilde{F} = 143.85$, $P = 251{,}929$ and $T = 16.98$.

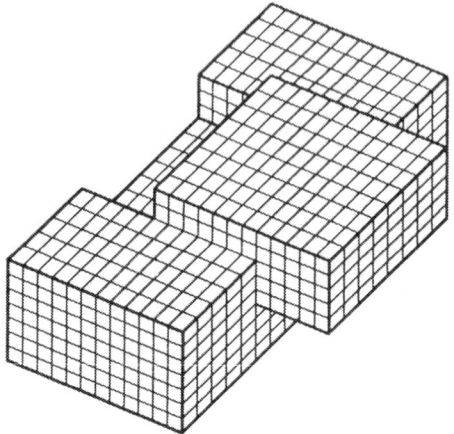

Fig. 5.31 An FEM with brick elements.

5.8.7 DISCUSSIONS

In this section, an algorithm is developed to renumber the elements of FEMs for frontwidth reduction of sparse matrices with symmetric structures. The present method requires less computer storage, has less time complexity and leads to small maximum frontwidth. In this method, the number of candidates for the next labelling stage is very less in comparison with the existing methods, and only the corner nodes should be given as input. An additional feature of the present procedure is the use of multiple node roots, leading to more efficient results than those when a single node root is employed. This heuristic is a graph theory–based method that contains a nodal expansion process. In the process of expansion, both local and global properties of the graph model of the considered FEM are used. The algorithm is applicable to one- to three-dimensional models containing meshes of different types and dimensions.

5.9 ELEMENT ORDERING FOR BANDWIDTH OPTIMISATION OF FLEXIBILITY MATRICES

The elements of a generalised cycle basis (GCB), as defined in Chapter 3, must be ordered to obtain a banded flexibility matrix **G**. This is similar to ordering the elements of a cutset basis (nodal numbering) for reducing the bandwidth of the stiffness matrix **K**. This problem can be transferred to a nodal ordering algorithm by defining appropriate mathematical structures for the transformation of the connectivity properties; see Kaveh [113]. Two approaches for this problem are developed in the following text.

5.9.1 AN ASSOCIATE GRAPH

An *associate graph* $A(B(S))$ of a GCB $B(S)$ of S is a graph whose nodes are in a one-to-one correspondence with the elements of $B(S)$, and two nodes are connected if two elements of $B(S)$ have at least one common member. For example, the associate graph of the mesh basis in Figure 5.32(a) is depicted in Figure 5.32(b).

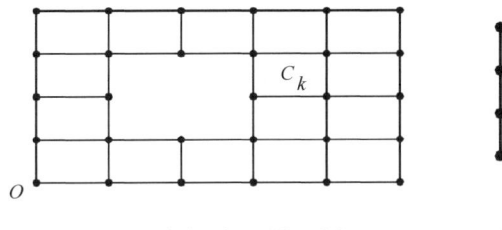 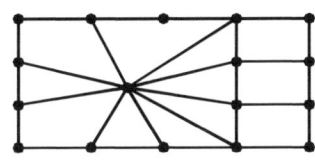

(a) A mesh basis $B(S)$ of S.   (b) The associate graph of $B(S)$.

Fig. 5.32 A mesh basis and its associate graph.

A *weighted associate graph* can be similarly defined. For this graph, the nodes and members are assigned integer numbers. The *weight* of a node in $A(B(S))$ is taken as the number of members of the corresponding cycle in S, and the weight of a member $m_k = (n_i, n_j)$ in $A(B(S))$ is taken as the number of members of $C_i \cap C_j$, where C_i and C_j are the cycles of S corresponding to the nodes n_i and n_j of $A(B(S))$, respectively.

5.9.2 DISTANCE NUMBER OF AN ELEMENT

The *distance* d_i of a node n_i of S from a selected node O is the length of the shortest path connecting n_i to O. The *distance number* of a cycle or a γ-cycle or an element C_k from O is defined as one of the following:

(a) The distance of the nearest node of C_k from O, denoted by d_k^n.

(b) The distance of the furthest node of C_k from O, denoted by d_k^f.

(c) The mean value of d_k^n and d_k^f; that is, $|(\frac{1}{2})(d_k^n + d_k^f)|$, where $|.|$ means the integer part of the number.

(d) The sum of $d_k^n + |(\frac{1}{2})L(C_k)|$, where $L(C_k)$ is the length of C_k.

(e) The mean value of the distance of the nodes of C_k; that is, $\sum_{i=1}^{L(C_k)} |d_i / L(C_k)|$.

258   OPTIMAL STRUCTURAL ANALYSIS

For example, the values defined above for a cycle C_k are shown in bold lines in Figure 5.32(b), and with respect to a reference node O are 5,6,5,7 and 5, respectively. For simplicity, only the integer parts of the divisions are considered.

Any of the definitions (a)–(e) can be used as the distance number of a cycle, a γ-cycle or an element of a FE model.

5.9.3 ELEMENT ORDERING ALGORITHMS

In the following text, two algorithms are presented for ordering the elements of a cycle basis, a GCB, n FEM or the substructures of a structure. However, for simplicity, we will refer to a GCB only.

Algorithm A

Step 1: Order the nodes of S with a nodal numbering algorithm.

Step 2: Use the same starting node as in Step 1 to form an SRT and find the distance numbers of the elements of the GCB.

Step 3: Assign these distance numbers to the nearest (furthest or any other appropriate intermediate) nodes of the elements of the GCB. In this process, a node may become the representative node of p elements. Then p independent distance numbers are assigned to the representative nodes.

Step 4: Order these nodes in ascending order of distance number. A node representing p elements receives p different (independent) numbers. For equidistant nodes, the same sequence as the nodal numbering of Step 1 should be used, to effect the connectivity properties of S.

Step 5: Order the elements of the GCB with the same numbers received by their representative nodes. This provides an efficient ordering for the elements of the GCB.

Algorithm B

Step 1: Construct the associate graph $A(B(S))$ of the GCB.

Step 2: Generate an SRT of S, starting from an appropriate node O, and find the distance numbers of the elements of the GCB.

Step 3: Assign these numbers to the nodes of $A(B(S))$, and order its nodes by a nodal numbering algorithm, with a starting node that corresponds to an element containing O.

ORDERING: GRAPH THEORY METHODS 259

Step 4: Reorder the nodes of $A(B(S))$ in ascending order of their distance numbers obtained in Step 2. For equidistant nodes, the same sequence as that obtained by the nodal numbering algorithm of Step 3 should be used.

Step 5: Number the elements in the same order as that obtained for their representative nodes in $A(B(S))$. This leads to an efficient numbering of the elements of the considered GCB.

Example: Let S be the model of a rigid-jointed planar frame. Suppose that the selected cycle basis consists of the boundaries of the bounded regions of S (a mesh basis) as shown in Figure 5.33(a).

For Algorithm A, an SRT starting from O is generated as in Figure 5.33(a), and the distance numbers of the cycles corresponding to definitions (a) and (e) of Section 5.9.2 are calculated and assigned to the representative nodes of the cycles. The nearest node of a cycle to O is taken as its representative node, as shown in Figures 5.33(b) and (c). These nodes are then ordered, leading to an ordered cycle basis. The bandwidths of the cycle adjacency matrices for these orderings are 15 and 13. The latter result can further be reduced to 11 by imposing additional restrictions in the process of ordering. Since the frame is planar, the bandwidths of the corresponding flexibility matrices will be 45 and 39, respectively.

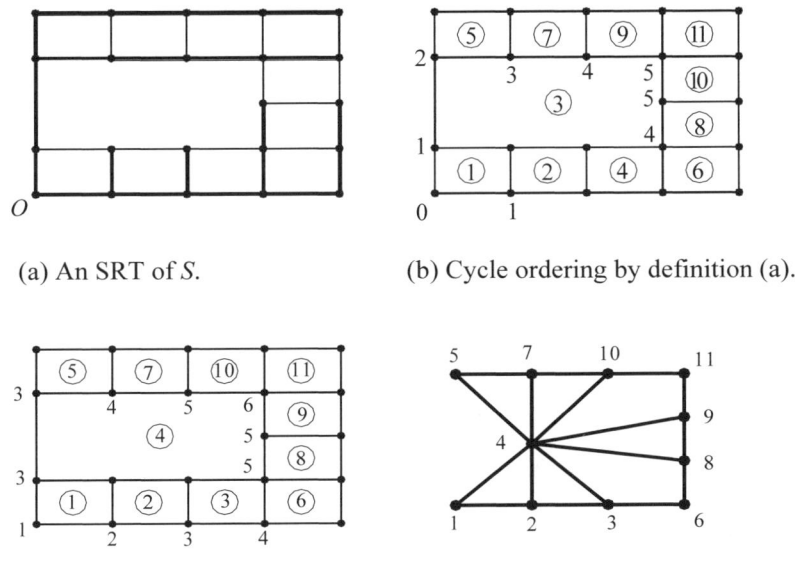

(a) An SRT of S. (b) Cycle ordering by definition (a).

(c) Cycle ordering by definition (e). (d) $A(B(S))$ and its nodal ordering.

Fig. 5.33 S and ordering the elements of its cycle basis.

Algorithm B is also applied to this example. The associate graph $A(B(S))$ of the mesh basis is formed, as shown in Figure 5.33(d). Using definition (e) for the distance number of the elements, the order of the nodes of $A(B(S))$ is obtained. The numbering of the cycles is shown in Figure 5.33(d), which corresponds to a bandwidth of 13 for its cycle adjacency matrix, and 39 for its flexibility matrix.

5.10 BANDWIDTH REDUCTION FOR RECTANGULAR MATRICES

In previous sections, the bandwidth optimisation of square matrices has been discussed. In structural analysis, it may also be desirable to reduce the bandwidth of some sparse rectangular matrices. For example, it may be beneficial to reduce the bandwidth of the equilibrium equations of a structure; see Kaneko et al. [92]. This can be done by optimising the bandwidth of the corresponding cutset basis incidence matrix **L**. Similarly, for compatibility equations, one can optimise the bandwidth of **C**.

In this section, a K-total graph is defined and two algorithms are presented for the bandwidth reduction of rectangular matrices.

5.10.1 DEFINITIONS

Let **B** be a rectangular matrix with m rows and n columns, whose entries are denoted by b_{ij}. For each row like i (except the first and the last row, where $i_d = 1$ and $i_d = n$, respectively), the integer part of the real number $i(n/m)$ is defined as i_d. Therefore, the entry of **B** at position (i, i_d) is considered as the ith diagonal entry. For square matrices, $m = n$, and $i = i_d$. The bandwidth of **B** is then defined as

$$b(\mathbf{B}) = m_r + m_l + 1, \qquad (5\text{-}35)$$

where

$$m_r = \max\{k - i_d \,|\, b_{ik} \neq 0, \; k > i_d\},$$
$$1 \leq i \leq n$$

and

$$m_l = \max\{i_d - k \,|\, b_{ik} \neq 0, \; k < i_d\}. \qquad (5\text{-}36)$$
$$1 \leq i \leq n$$

If **B** is a symmetric square matrix, then $m_r = m_l$ and $b(\mathbf{B})$ reduces to the conventional definition of square matrices. A rectangular matrix is called *banded* if $b(\mathbf{B})$ is small compared to m.

Matrix **B** in block submatrix form has the same pattern as **L**, that is, each non-zero entry of **L** corresponds to a $\eta \times \eta$ submatrix in **B**, where η is the degree of freedom of a node of the structure. Obviously, reduction of the bandwidth of **L** leads to a banded matrix **B**.

The terms "nodes" and "members" have been used for a graph S, and now we use "vertices" and "edges" for the elements of a *K-total graph*, which is defined as follows:

Associate one vertex with each member and each element of the selected cutset basis or a cycle (γ-cycle) basis of S. Connect two vertices with an edge if

(a) the corresponding members are incident;

(b) the corresponding cutsets (cycles or γ-cycles) are adjacent;

(c) the corresponding member and cutset (cycle or γ-cycle) are incident.

When a cutset or cycle is changed to a node of S, then the K-total graph becomes a total graph as defined in the graph theory (see, Behzad [10]).

Examples of K-$T(S)$ are shown in Figures 5.34 and 5.35, when the cocycle basis and the cycle basis are considered, respectively. In these figures small squares are used to represent members, and circles are employed to show the elements of the considered basis.

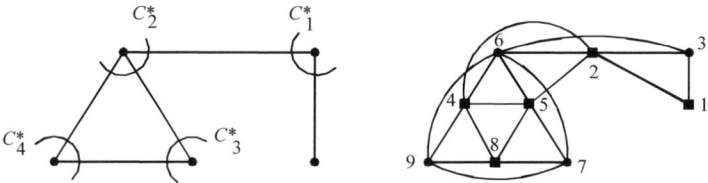

(a) S and the considered cocycle basis.   (b) K-$T(S)$ and its nodal ordering.

Fig. 5.34 Reduction of bandwidth for a cutset basis incidence matrix.

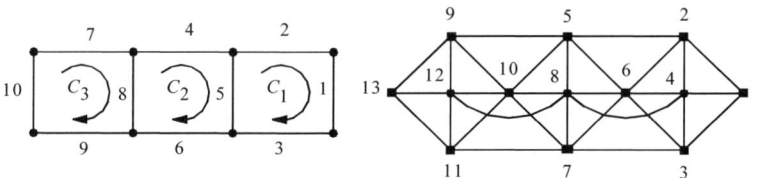

(a) S and the considered cycle basis.   (b) K-$T(S)$ and its nodal ordering.

Fig. 5.35 Reduction of bandwidth for a cycle basis incidence matrix.

262 OPTIMAL STRUCTURAL ANALYSIS

5.10.2 ALGORITHMS

Algorithm A

Construct the K-total graph of S and order its vertices. The corresponding sequence leads to a favourable order of cutsets (nodes) and members of S, to reduce the bandwidth of **L**, which is pattern equivalent to the coefficient matrix of the equilibrium equations. A similar approach reduces the bandwidth of **C**, when cycles (γ-cycles) are considered in place of cutsets.

This algorithm will now be applied to the examples of Figures 5.34 and 5.35, from which the corresponding orders for the elements of the bases and members of S are obtained.

Algorithm B

Order the nodes of S. Then order the unnumbered members of the stars of the nodes in the selected sequence, to obtain a reasonably banded **L** matrix.

In general, Algorithm A leads to a better result than Algorithm B, at the expense of additional computer time.

5.10.3 EXAMPLES

Consider a graph S as shown in Figure 5.36 with the corresponding member and cutset orders.

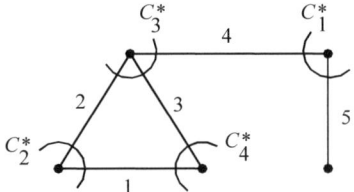

Fig. 5.36 S with arbitrarily ordered members and cutsets.

The cutset basis incidence matrix of S can be written as,

$$\mathbf{C}^* = \begin{array}{c} \\ C_1^* \\ C_2^* \\ C_3^* \\ C_4^* \end{array} \begin{array}{c} {\scriptstyle m_1 \; m_2 \; m_3 \; m_4 \; m_5} \\ \begin{bmatrix} \cdot & * & \cdot & \cdot & 1 & 1 \\ 1 & 1* & \cdot & \cdot & \cdot \\ \cdot & 1 & 1* & 1 & \cdot \\ 1 & \cdot & 1 & \cdot & * \end{bmatrix} \end{array} \qquad b(\mathbf{L}) = 4+4+1 = 9,$$

where artificially defined diagonal entries are shown with "*" sign. Using the ordering obtained by $K\text{-}T(S)$, the cutset basis incidence matrix becomes

$$C^* = \begin{array}{c} \\ C_1^* \\ C_2^* \\ C_3^* \\ C_4^* \end{array} \begin{array}{cccccc} m_1 & m_2 & m_3 & m_4 & m_5 \\ \left[\begin{array}{ccccc} 1^* & 1 & \cdot & \cdot & \cdot \\ \cdot & 1^* & 1 & 1 & \cdot \\ \cdot & \cdot & \cdot^* & 1 & 1 \\ \cdot & \cdot & 1 & \cdot & 1^* \end{array}\right] \end{array} \qquad b(L) = 2+2+1 = 5,$$

in which the non-zero entries are clustered to the diagonal of the matrix.

As a second example, consider S as shown in Figure 5.37, in which the regional cycles and members are arbitrarily numbered.

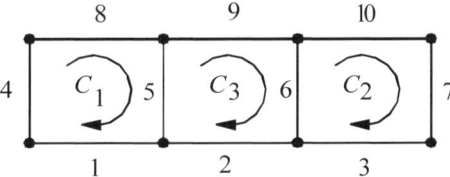

Fig. 5.37 S with arbitrarily numbered members and cycles.

The cycle basis incidence matrix for S is given as

$$C = \begin{array}{c} \\ C_1 \\ C_2 \\ C_3 \end{array} \begin{array}{cccccccccc} m_1 & m_2 & m_3 & m_4 & m_5 & m_6 & m_7 & m_8 & m_9 & m_{10} \\ \left[\begin{array}{cccccccccc} 1 & 0 & 0 & 1 & 1 & 0 & 0 & 1 & 0 & 0 \\ 0 & 0 & 1 & 0 & 0 & 1 & 1 & 0 & 0 & 1 \\ 0 & 1 & 0 & 0 & 1 & 1 & 0 & 0 & 1 & 0 \end{array}\right] \end{array}.$$

For this matrix, $b(\mathbf{C}) = 7 + 8 + 1 = 16$. By ordering the cycles and members simultaneously, using Algorithm A, the following cycle basis incidence matrix is obtained:

$$C = \begin{array}{c} \\ C_1 \\ C_2 \\ C_3 \end{array} \begin{array}{cccccccccc} m_1 & m_2 & m_3 & m_4 & m_5 & m_6 & m_7 & m_8 & m_9 & m_{10} \\ \left[\begin{array}{cccccccccc} 1 & 1 & 1 & 0 & 1 & 0 & 0 & 0 & 0 & 0 \\ 0 & 0 & 0 & 1 & 1 & 1 & 0 & 1 & 0 & 0 \\ 0 & 0 & 0 & 0 & 0 & 0 & 1 & 1 & 1 & 1 \end{array}\right] \end{array}.$$

The bandwidth for this matrix is obtained as $b(\mathbf{C}) = 4 + 3 + 1 = 8$.

For the force method of frames, the coefficient matrix of the equilibrium equations can be made banded by reducing the bandwidth of its member-cycle incidence matrix. After an algebraic force method is employed, a repeated application of the developed method makes the null basis matrix a banded one for subsequent applications. Similarly, if a combinatorial approach is used, the bandwidth reduction algorithm makes the cycle basis incidence matrix banded, leading to a banded statical basis (null basis) matrix.

5.10.4 BANDWIDTH REDUCTION OF FINITE ELEMENT MODELS

The algorithms presented in the previous section can also be applied to FE models, for their analysis by the algebraic force method; see Kaveh and Mokhtar-zadeh [119]. For such models, the K-total graph of an FEM is defined as follows:

Associate one vertex with each side and each element of the FEM, and connect two vertices with an edge if any of the following conditions hold:

1. sides are adjacent;

2. elements are adjacent;

3. a side and an element are incident.

The Algorithm A can now be adapted to FEMs as follows:

Step 1: Generate the K-total graph of the FE mesh S.

Step 2: Order the vertices of K-$T(S)$ by any nodal ordering algorithm that is available.

Step 3: Assign numbers to the members of K-$T(S)$ and to the elements of the considered FEM, in the order of their occurrence in the sequence selected in Step 2.

Example: Four groups of examples are considered as shown in Figure 5.38(a–d). In these figures, Ω is the aspect ratio of the element numbers in two perpendicular directions (x- and y-directions), which is taken as unity. The ratio of the length of the elements side in the x-direction to that of the y-direction is taken as 1.2. S is the refinement index of a group. In the group UT, Ω_1, Ω_2 are the aspect ratios of the element numbers in the two sides of the general configuration with respect to the central part of the model.

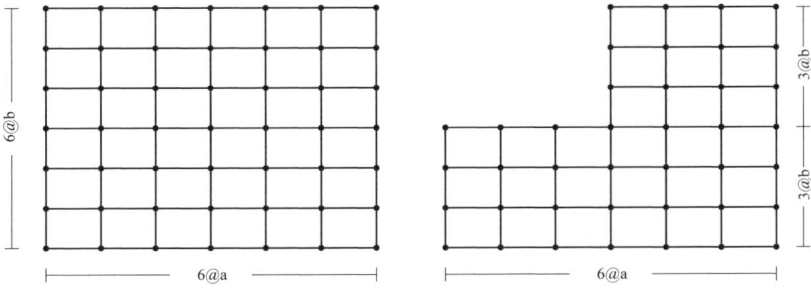

(a) Group RQ-Ω-S ($\Omega = 1, S = 6$).   (b) Group LQ-Ω-S ($\Omega = 1, S = 3$).

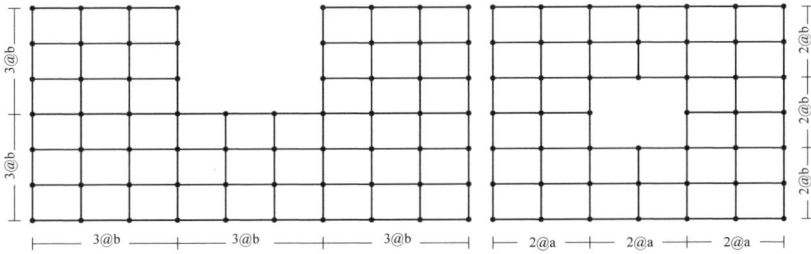

(c) Group UQ-Ω_1-$\Omega_{2\text{-}S}$ ($\Omega_1 = \Omega_2 = 1, S = 3$). (d) Group HQ-$\Omega$-S ($\Omega = 1, S = 2$).

Fig. 5.38 Test group examples.

The sparsity of the self-stress and flexibility matrices of the LQ and HQ groups is illustrated in Figure 5.39(a–d).

266 OPTIMAL STRUCTURAL ANALYSIS

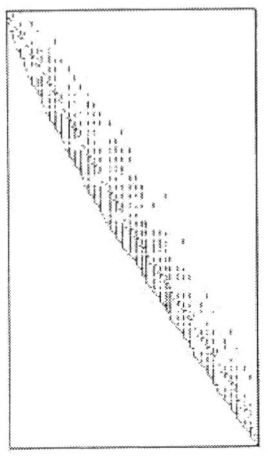

(a) Self-stress matrix of LQ-1-4.

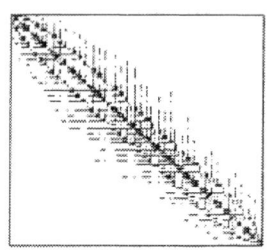

(b) Flexibility matrix of LQ-1-4.

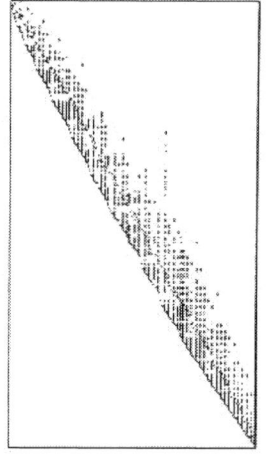

(c) Self-stress matrix of HQ-1-4.

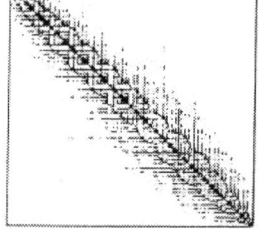

(d) Flexibility matrix of HQ-1-4.

Fig. 5.39 Self-stress and flexibility matrices.

5.11 GRAPH-THEORETICAL INTERPRETATION OF GAUSSIAN ELIMINATION

In this section, a simple graph-theoretical interpretation of the Gaussian elimination is presented, in order to establish a closer link between matrix algebra on the one hand and graph-theoretical concepts on the other hand.

ORDERING: GRAPH THEORY METHODS   267

Let **A** be a symmetric sparse matrix of order N, and let S be the corresponding graph. Suppose that Gaussian elimination by columns is performed on **A** until the factorisation $\mathbf{A} = \mathbf{U}^t\mathbf{D}\mathbf{U}$ is obtained. At the beginning of the kth step, all non-zeros in columns $1, 2, \ldots, k-1$ below the diagonal have been eliminated. Multiples of the kth row are then subtracted from all rows that have a non-zero in column k below the diagonal. On performing this operation, new non-zero entries may be introduced in row $k+1, \ldots, N$ to the right of column k. Cancellations may also occur, producing new zeros, but this is rare in practice and will be neglected. Consider the active submatrix at the kth step (an active submatrix contains all elements $\mathbf{A}_{ij}^{(k)}$ with $i,j \geq k$). Let S^k be the graph associated with the active submatrix. S^k is called *an elimination graph*; see Parter [171]. The nodes of this graph are $N - k + 1$ last-numbered nodes of S. S^k contains all members connecting those nodes that were present in S, and additional members corresponding to fill-ins produced during the $k - 1$ initial elimination steps. The sequence $S = S^1, S^2, S^3, \ldots$ can be obtained using the following rule:

To obtain S^{k+1} from S^k, delete node k and add all possible members between nodes that are adjacent to node k in S^k.

For example, consider a graph S and the corresponding adjacency matrix, as shown in Figure 5.40. Two steps of the Gaussian elimination and the corresponding elimination graphs are also illustrated.

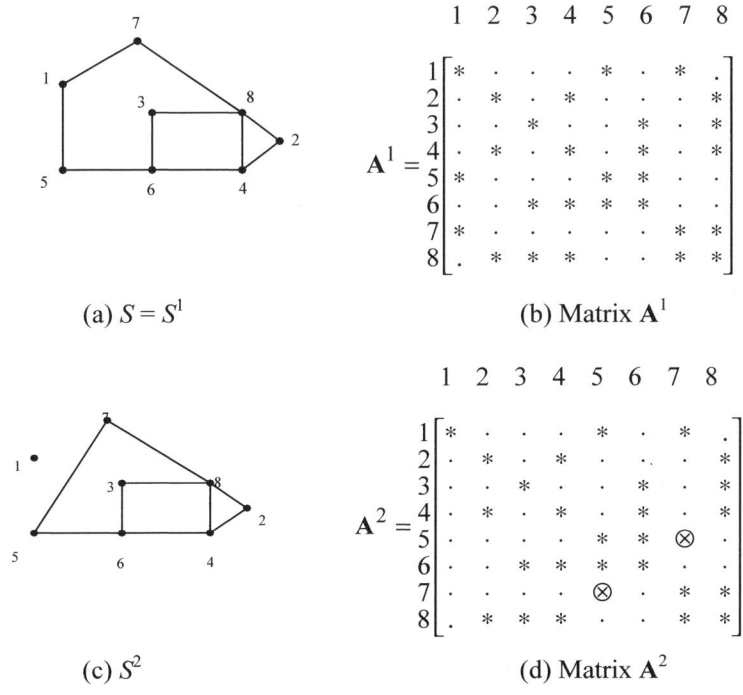

(a) $S = S^1$ (b) Matrix $\mathbf{A}^1$

(c) S^2 (d) Matrix $\mathbf{A}^2$

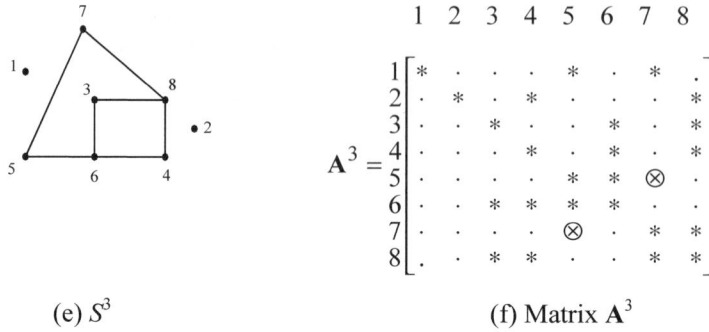

(e) S^3 (f) Matrix $\mathbf{A}^3$

Fig. 5.40 Illustration of two steps of the Gaussian elimination.

Eliminating the rest of the nodes, and considering a clique (a complete graph) between the nodes adjacent to each eliminated node (when such members are not present), matrix $\mathbf{U}$ is obtained. The structure of $\mathbf{U} + \mathbf{U}^t$ and the corresponding filled graph are shown in Figure 5.41.

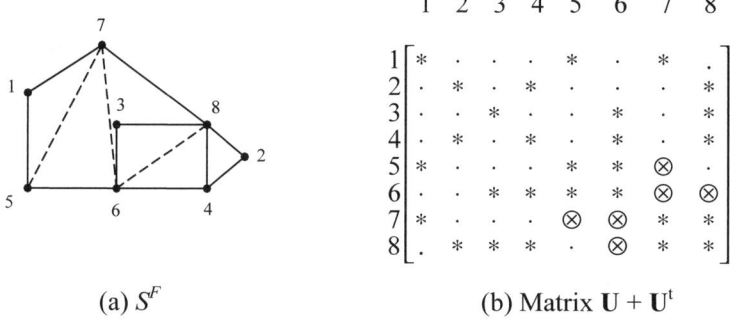

(a) S^F (b) Matrix $\mathbf{U} + \mathbf{U}^t$

Fig. 5.41 The structure of $\mathbf{U} + \mathbf{U}^t$ and the corresponding graph.

There are algorithms that try to reduce the number of fill-ins caused by elimination. The minimum degree algorithm of Tinney [218] is perhaps the best method for such a reduction. For brevity, this is not discussed here; the interested reader may refer to Tinney's original paper.

EXERCISES

5.1 Find a good starting node for nodal numbering of the following structural models, using graph-theory approaches:

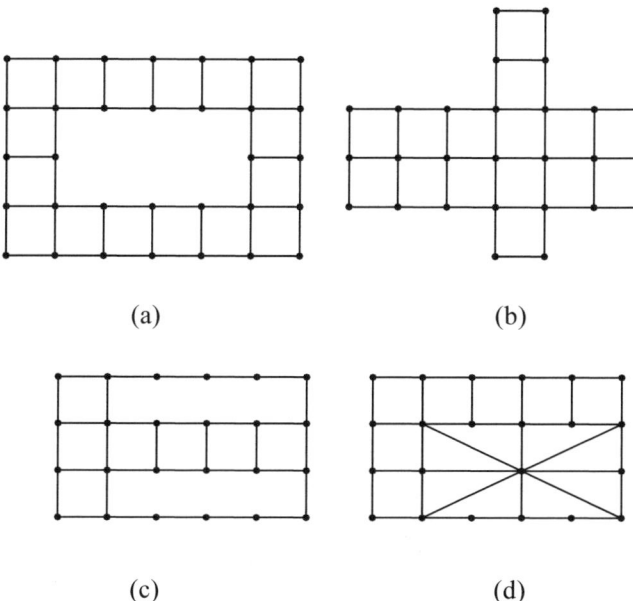

5.2 Find a good starting node for the following models, using an algebraic graph-theoretical method, that is, calculate the dominant eigenvector of the corresponding adjacency matrices:

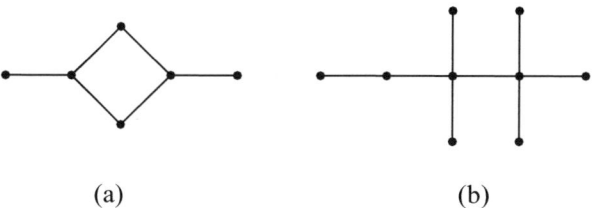

5.3 For the models of Exercise 5.1, find a suboptimal transversal and perform the ordering. Calculate the bandwidth of the corresponding stiffness matrices when the models are viewed as planar trusses.

5.4 Find the nodal ordering of the system for models (a) and (b) in Exercise 5.1 using the Fiedler vector of the Laplacian matrix of the model.

270   OPTIMAL STRUCTURAL ANALYSIS

5.5   For the following graph S, consider a mesh basis and order the cycles, using different distance numbers, to optimise the bandwidth of the corresponding cycle adjacency matrix:

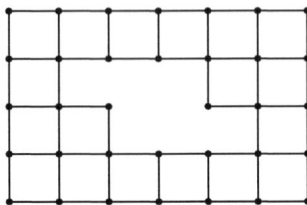

5.6   Order the nodes of the following FEM using the natural associate graph, incidence graph and skeleton graph of the model for bandwidth reduction:

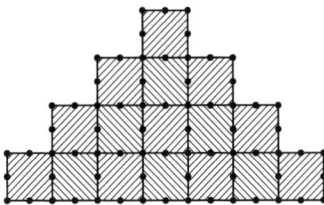

5.7   For the following FEM, construct the element clique graph, the skeleton graph, the representative graph and the element clique representative graph, with respect to the starting node n_0:

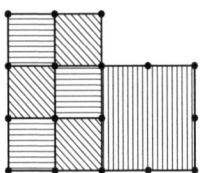

5.8 Order the nodes of the following FEM using the algorithm presented for frontwidth reduction:

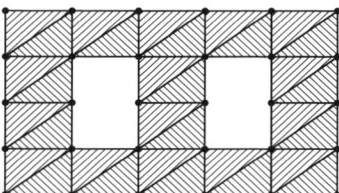

5.9 For nine graphs of an $m \times n$ mesh consisting of rectangular elements introduced in this chapter, compare the number of nodes, number of members and diameters of the graphs.

5.10 Order the members and elements of a fundamental cycle basis of the following graph to reduce the bandwidth of its cycle basis incidence matrix. Repeat the process to optimise the bandwidth of its cocycle basis incidence matrix.

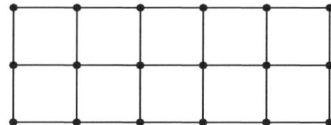

CHAPTER 6
Ordering for Optimal Patterns of Structural Matrices: Algebraic Graph Theory Methods

6.1 INTRODUCTION

There are different matrices associated with a graph, such as the incidence matrix, the adjacency matrix and the Laplacian matrix. One of the aims of algebraic graph theory is to determine how properties of graphs are reflected in algebraic properties of these matrices. The eigenvalues and eigenvectors of these matrices provide valuable tools for combinatorial optimisation and, in particular, for the ordering of sparse symmetric matrices, such as the stiffness and flexibility matrices of the structures.

In this chapter, algebraic graph–theoretical methods are discussed for nodal ordering for bandwidth, profile and frontwidth optimisation. Hybrid methods are also applied to nodal ordering, using graph theory and algebraic graph theory.

6.2 ADJACENCY MATRIX OF A GRAPH FOR NODAL ORDERING

6.2.1 BASIC CONCEPTS AND DEFINITIONS

There are several geographical papers dealing with the question of whether important places or well-connected sets of towns in a traffic network can be identified by an inspection of certain eigenvalues and the corresponding eigenvectors of the adjacency matrix **A** of the underlying graph model. Gould [65] appears to be the first important publication on this subject. Other ideas can be found in Straffing [213] and Maas [157].

In structural analysis, Kaveh [113] used the first eigenvalue and eigenvector of [**A** + **I**] for nodal ordering for bandwidth reduction. Grimes et al. [66] employed this concept for finding pseudo-peripheral nodes of a graph. This algebraic graph–theoretical method is studied in the following text.

A node n_i of S is called *peripheral* if its eccentricity is the same as the diameter of S, that is, $\delta(S) = e(n_i)$. If the eccentricity is close to the diameter, then n_i is called a *pseudo-peripheral* node or a *good starting node*.

Reordering the nodes of the graph model of a structure does not change the properties of the stiffness matrix. This fact stays true for the properties of the graph itself. Therefore, a natural question is, what can the theory of matrices and, in particular, the eigenvalues of the matrices associated with graphs tell us about the structure of the graph itself? In the following text, we shall endeavour to find out to what extent the eigenvalues of the adjacency matrix of a given graph reflect the properties of that graph.

Let **A** be the adjacency matrix of the graph S, which is a real symmetric (0, 1) matrix, and the sum of entries of any row or column is equal to the valency of the corresponding node. Denote the characteristic polynomial of **A** by $\phi(S;x)$. Since $\phi(S;x)$ is uniquely determined by the graph S, it is referred to as the *characteristic polynomial of S* and expressed as

$$\phi(S;x) = \det(x\mathbf{I} - \mathbf{A}) = \sum_{i=0}^{N} a_i x^{N-i}. \tag{6-1}$$

Since **A** is a real symmetric matrix, its eigenvalues (the roots of this polynomial) must be real, and can be ordered as $\lambda_1 \geq \lambda_2 \geq \lambda_3 \geq \ldots \geq \lambda_N$. These eigenvalues are called the *eigenvalues* of S, and the sequence of N eigenvalues is called the *spectrum* of G.

The important results are stated below; however, the reader may refer to Schwenk and Wilson [194] for further details and proofs.

1. The sum of the eigenvalues of a graph is equal to the trace of **A**, and is therefore zero.

2. If S is connected with N nodes, then $2\cos(\dfrac{\pi}{N+1}) \leq \lambda_1 \leq N-1$. The lower bound occurs only when S is a path graph, and the upper bound occurs when S is a complete graph.

3. If S is a connected graph with m distinct eigenvalues and with diameter d, then $m > d$.

By no means the spectrum specifies its graph uniquely; however, it does provide a wealth of information about the graph and hence about the structure. Some applications of such information will be given in this chapter and in Chapter 8. However, the writer strongly believes that in future many other applications in structural mechanics will be found.

Table 6.1 shows some simple examples to verify the results stated.

Table 6.1 Simple examples.

Graph	Adjacency matrix	Characteristic polynomial	Eigenvalues
K_2	$\begin{bmatrix} 0 & 1 \\ 1 & 0 \end{bmatrix}$	$x^2 - 1$	$1, -1$
P_3	$\begin{bmatrix} 0 & 0 & 1 \\ 0 & 0 & 1 \\ 1 & 1 & 0 \end{bmatrix}$	$x^3 - 2x$	$\sqrt{2}, -\sqrt{2}, 0$
C_4	$\begin{bmatrix} 0 & 1 & 0 & 1 \\ 1 & 0 & 1 & 0 \\ 0 & 1 & 0 & 1 \\ 1 & 0 & 1 & 0 \end{bmatrix}$	$x^4 - 4x^2$	$2, -2, 0, 0$

Perron–Frobenius Theorem: If S is a connected graph with at least two nodes, then:

(i) its largest eigenvalue λ_1 is a simple root of $\phi(S; x)$;

(ii) corresponding to the eigenvalue λ_1, there is an eigenvector $\mathbf{w}_1$, all of whose entries are positive;

(iii) if λ is any other eigenvalue of S, then $-\lambda_1 \leq \lambda \leq \lambda_1$;

(iv) the deletion of any member of S decreases the largest eigenvalue.

The largest eigenvalue λ_1 is often known as the *spectral radius* of S. Since the eigenvectors corresponding to any eigenvalue other than λ_1 must be orthogonal to

$\mathbf{w}_1$, we observe that the multiples of $\mathbf{w}_1$ are the only eigenvectors all of whose entries are positive.

Consider the node adjacency matrix $\mathbf{A}$ of S. Let,

$$\mathbf{Q} = \mathbf{A} + \mathbf{I}, \qquad (6\text{-}2)$$

where $\mathbf{I}$ is an $N(S) \times N(S)$ identity matrix. The eigenvalues of $\mathbf{Q}$ are one unit bigger than those of $\mathbf{A}$, and the eigenvectors of $\mathbf{Q}$ are exactly the same as those of $\mathbf{A}$. Matrix $\mathbf{Q}$ is real and symmetric, and it can easily be shown that all the entries of $\mathbf{Q}^k$ are positive; thus it is primitive and, according to the Perron–Frobenius theorem, λ_1 is real and positive and a simple root of the characteristic equation, $\lambda_1 > |\lambda|$ for any eigenvalue $\lambda \neq \lambda_1$, and λ_1 has a unique corresponding eigenvector $\mathbf{w}_1$ with all entries positive.

As $\mathbf{w}_i$ is the eigenvector corresponding to λ_i, $\mathbf{Q}\mathbf{w}_i = \lambda_i \mathbf{w}_i$ for $i = 1, \ldots, N(S)$. Multiplying the two sides by $\mathbf{Q}$, one obtains $\mathbf{Q}\mathbf{Q}\mathbf{w}_i = \lambda_i \mathbf{Q}\mathbf{w}_i = \lambda_i^2 \mathbf{w}_i$. Repeating this process results in $\mathbf{Q}^k \mathbf{w}_i = \lambda_i^k \mathbf{w}_i$. Now consider any vector $\mathbf{x}$ not orthogonal to $\mathbf{w}_1$ as follows:

$$\mathbf{x} = \alpha_1 \mathbf{w}_1 + \alpha_2 \mathbf{w}_2 + \ldots + \alpha_{N(S)} \mathbf{w}_{N(S)} \qquad \alpha_1 \neq 0. \qquad (6\text{-}3)$$

Multiplying the two sides by $\mathbf{Q}^k$ and using $\mathbf{Q}^k \mathbf{w}_i = \lambda_i^k \mathbf{w}_i$ for $i = 1, \ldots, N(S)$, we have,

$$\mathbf{Q}^k \mathbf{x} = \lambda_1^k \alpha_1 \mathbf{w}_1 + \lambda_2^k \alpha_2 \mathbf{w}_2 + \ldots + \lambda_{N(S)}^k \alpha_{N(S)} \mathbf{w}_{N(S)}, \qquad (6\text{-}4)$$

and, as $k \to \infty$,

$$\mathbf{Q}^k \mathbf{x} / \lambda_1^k = \alpha_1 \mathbf{w}_1 + (\lambda_2/\lambda_1)^k \alpha_2 \mathbf{w}_2 + \ldots + (\lambda_{N(S)}/\lambda_1)^k \alpha_{N(S)} \mathbf{w}_{N(S)} \to \alpha_1 \mathbf{w}_1, \qquad (6\text{-}5)$$

since λ_1 is the eigenvalue of the strictly largest modulus and (λ_i/λ_1) is less than unity and approaches zero when $k \to \infty$. In other words, the ratios of the components of $\mathbf{Q}^k \mathbf{x}$ approach the ratios of the components of $\mathbf{w}_1$ as k increases.

Let $\mathbf{v} = \{1, 1, \ldots, 1\}^t$. Then the ith component of $\mathbf{Q}^k \mathbf{v}$ is the number of walks of length k beginning at an arbitrary node of S and ending at n_i. If n_i is a good starting node (peripheral node), this number will be smaller. Thus, for $k \to \infty$, one should obtain some average number, defined as the *accessibility index* by Gould [65]. This number indicates how many walks go through a node on average. With a suitable normalisation, $\mathbf{Q}^k\mathbf{v}$ converges to the largest eigenvector $\mathbf{w}_1$ of $\mathbf{Q}$; see Straffing [213].

6.2.2 A GOOD STARTING NODE

Algorithm A

Step 1: Calculate the dominant eigenvector $\mathbf{w}_1 = \{w_1, w_2, ..., w_{N(S)}\}^t$ of matrix $\mathbf{Q}$.

Step 2: Find Min w_i in $\mathbf{w}_1$. The node corresponding to this entry is taken as a good starting node of S.

For calculating the dominant eigenvector $\mathbf{w}_1$ of $\mathbf{Q}$, an iterative method is used, which assumes $\mathbf{v} = \{1, 1, ..., 1\}^t$ and calculates $\mathbf{Qv}$. This vector is then normalised and multiplied by $\mathbf{Q}$. This process is repeated until the difference between two consecutive eigenvalues, obtained from $\mathbf{Qv} = \lambda\mathbf{v}$, is reduced to a small value, which, for example, can be taken as 10^{-3}.

6.2.3 PRIMARY NODAL DECOMPOSITION

Once a good starting node is selected, an SRT is constructed and its contours $\{C_1, C_2, ..., C_m\}$ are obtained. These subsets are then ordered according to their distances from the selected starting node. Obviously, many SRTs can be constructed on a node. Although all of them lead to the same nodal decompositions, different transversals will be obtained for different SRTs. Thus, in the generation process, the nodes of each contour C_i are considered in ascending order of their entries in eigenvector W_1 for selecting the nodes in C_{i+1}, in order to provide the conditions for the possibility of generating a minimal (or optimal) transversal as defined in the next section.

6.2.4 TRANSVERSAL P OF AN SRT

For selection of an optimal transversal, the weight of a node is defined as its value w_i in $\mathbf{w}_1$, when an algebraic graph–theoretical method is employed.

Algorithm B

Let $C_1, C_2, ..., C_m$ be the selected contours of the SRT, and correspondingly put these subsets in $\mathbf{w}_1$ into a similar order, that is,

$$\mathbf{w}_1 = \{W(C_1), W(C_2), ..., W(C_m)\}, \tag{6-6}$$

where $W(C_i)$ contains the entries of $\mathbf{w}_1$ corresponding to the nodes of C_i. Now the algorithm can be described as follows:

278 OPTIMAL STRUCTURAL ANALYSIS

Step 1: Label the root as N_1 and assign w_i of this node as its new weight, denoted by $\bar{w}_1$.

Step 2: Calculate the new weight $\bar{w}_i$ of each node of C_2 by adding the w_i's from $W(C_2)$ to $\bar{w}_1$.

Step 3: Repeat the process of Step 2, calculating $\bar{w}_i s$ for each node of $C_3, C_4, ..., C_m$.

Step 4: Take a node N_m of minimal weight from the last contour C_m of the selected SRT.

Step 5: Find N_{m-1} from C_{m-1}, which is connected to N_m by a branch of the SRT.

Step 6: Repeat the process of Step 5, selecting $N_{m-2}, N_{m-3}, ..., N_1$ as the representative nodes of the contours $C_{m-2}, C_{m-3}, ..., C_1$.

The set $P = \{N_1, N_2, ..., N_m\}$ forms a suboptimal transversal of the selected SRT.

6.2.5 NODAL ORDERING

Step 1: Number N_1 as "1".

Step 2: N_2 is given number "2" and an SR subtree is generated from N_2, numbering the nodes of C_2 in the order of their occurrence in this SR subtree.

Step 3: The process of Step 2 is repeated for numbering the nodes of $C_3, C_4, ..., C_m$, sequentially using $N_3, N_4, ..., N_m$ as the starting nodes of SR subtrees, until all the nodes of S are numbered.

Now the numbering can be reversed, in a way similar to that of the Reverse Cuthill–McKee algorithm, for possible reduction of fill-ins in the process of Gaussian elimination.

6.2.6 EXAMPLE

S is the model of a grid with uniform valency distribution, as shown in Figure 6.1(a). Using algorithm A, the following dominant eigenvector is obtained for the matrix $\mathbf{Q}$ of S, in which for simplicity only four digits are provided:

$\mathbf{w}_1$ = {0.3344, 0.5298, 0.6161, 0.5951, 0.4791, 0.3011, 0.1180, 0.3972, 0.7432, 0.9540, 1.0000, 0.8786, 0.6183, 0.2875, 0.2875, 0.6183, 0.8786, 1.000, 0.9540, 0.7432, 0.3972, 0.1180, 0.3011, 0.4791, 0.5951, 0.6160, 0.5298, 0.3344$\}^t$.

ORDERING: ALGEBRAIC GRAPH THEORY METHODS

Thus, node "7" is selected as a good starting node. An SRT is generated from this node and, using algorithm B, a transversal $P = \{7,14,21,28,27,26, 25,24, 23,22\}$ is selected, which is shown in bold lines in Figure 6.1(a). Final nodal numbering is illustrated in Figure 6.1(b).

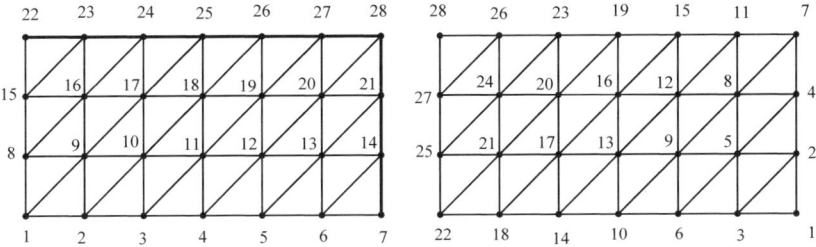

(a) Initial numbering and the selected transversal. (b) Final numbering.

Fig. 6.1 The graph model S and its nodal numbering.

6.3 LAPLACIAN MATRIX OF A GRAPH FOR NODAL ORDERING

6.3.1 BASIC CONCEPTS AND DEFINITIONS

Another interesting matrix associated with a graph is the Laplacian matrix of S, denoted by $\mathbf{L}(S)$.

Consider a directed graph S with arbitrary nodal numbering and member orientations. The adjacency matrix $\mathbf{A}(S)$, degree matrix $\mathbf{D}(S)$, node-member incidence matrix $C(S)$, and Laplacian matrix $\mathbf{L}(S)$ are defined as follows:

The *adjacency matrix* $\mathbf{A}(S) = [a_{ij}]_{N \times N}$ of the labelled graph S is defined as

$$a_{ij} = \begin{cases} 1 & \text{if node } n_i \text{ is adjacent to } n_j, \\ 0 & \text{otherwise.} \end{cases}$$

The *degree matrix* $\mathbf{D}(S) = [d_{ij}]_{N \times N}$ is the diagonal matrix of node degrees

$$d_{ij} = \begin{cases} \deg(n_i) & \text{if } i = j, \\ 0 & \text{otherwise.} \end{cases}$$

The *Laplacian matrix* $\mathbf{L}(S) = [l_{ij}]_{N \times N}$ is defined as

$$\mathbf{L}(S) = \mathbf{D}(S) - \mathbf{A}(S); \tag{6-7}$$

therefore, the components of **L**(S) are given as follows:

$$l_{ij} = \begin{cases} -1 & \text{if } n_i \text{ is adjacent to } n_j, \\ \deg(n_j) & \text{if } i = j, \\ 0 & \text{otherwise.} \end{cases}$$

The node-member incidence matrix $\mathbf{C}(G) = [c_{ij}]_{N \times M}$ for the arbitrarily oriented graph is defined as

$$c_{ij} = \begin{cases} +1 & \text{if } m_j \text{ points towards } n_i, \\ -1 & \text{if } m_j \text{ points away from } n_i, \\ 0 & \text{otherwise.} \end{cases}$$

Two distinct rows of **C**(S) have non-zero entries in the same column if and only if a member joins the corresponding nodes. These entries are 1 and −1. It can be shown that

$$\mathbf{L} = \mathbf{CC}^t. \tag{6-8}$$

It can also be shown that **L** is independent of the orientation of the members of the graph.

Hall [72] considered the problem of finding the minimum of the weighted sum,

$$Z = \frac{1}{2} \sum_{i,j} (x_i - x_j)^2 a_{ij}, \tag{6-9}$$

where a_{ij} are the elements of the adjacency matrix **A**. The sum is the over all pairs of squared distances between nodes that are connected, and so the solution should result in nodes with large numbers of inter-connection being clustered together.

The above equation can be rewritten as

$$\frac{1}{2} \sum_{i,j} (x_i^2 - 2x_i x_j + x_j^2) a_{ij} = \frac{1}{2} \sum_{i,j} x_i^2 a_{ij} - \frac{1}{2} \sum_{i,j} 2 x_i x_j a_{ij} + \frac{1}{2} \sum_{i,j} x_j^2 a_{ij}$$
$$= \sum_{i,j} x_i^2 a_{ij} + \sum_{i,j} x_i x_j a_{ij} = \mathbf{x}^t \mathbf{L} \mathbf{x}, \tag{6-10}$$

where **L** is the Laplacian. Hall also supplied the condition that $\mathbf{x}^t \mathbf{x} = 1$, that is, the distances are normalised. Using the Lagrange multiplier, we have

$$Z = \mathbf{x}^t \mathbf{L} \mathbf{x} - \lambda \mathbf{x}^t \mathbf{x}, \tag{6-11}$$

and, to minimise this expression, the derivative with respect to **x** is taken as

$$\mathbf{Lx} - \lambda \mathbf{x} = 0 \tag{6-12}$$

or
$$\mathbf{Lx} = \lambda \mathbf{x}, \tag{6-13}$$

which is the eigenvalue equation. The smallest eigenvalue of **L** is $\lambda_1 = 0$, and the corresponding eigenvector y_1 has all its normalised components equal to 1. The second eigenvalue λ_2 and the associated eigenvector y_2 have many interesting properties, which will be used for nodal numbering in this chapter and for domain decomposition in Chapter 8.

In order to get a feel of the magnitude of $\lambda_2 = \alpha(S)$, also known as the *algebraic connectivity* of a graph, some simple theorems are restated from the results of Fiedler [51] in the following text:

1. For a complete graph K_N with N nodes, $\alpha(K_N) = N$.

2. If $S_1 \subseteq S_2$ (S_1 and S_2, have the same nodes), then $\alpha(S_1) \leq \alpha(S_2)$.

3. Let S be a graph. Let S_1 be formed from S by removing k nodes and all adjacent members. Then

$$\alpha(S_1) \geq \alpha(S) - k. \tag{6-14}$$

4. For a non-complete graph S,

$$\alpha(S) \leq v(S) \leq e(S), \tag{6-15}$$

where $v(S)$ and $e(S)$ are the node connectivity and edge connectivity of S, respectively. The *node connectivity* of a graph S is the smallest number of nodes whose removal from S, along with members incident with at least one of the removed nodes, leaves either a disconnected graph or a graph with a single node. The *edge connectivity* of S is the smallest number of edges whose removal from S leaves a disconnected graph or a graph with one node. As an example, the node and edge connectivity of a complete graph K_N is equal to $N - 1$.

5. For a graph with $N(S)$ nodes,

$$\alpha(G) = \lambda_2 \leq \frac{N}{N-1} \min\{\deg(n); n \in N(G)\} \tag{6-16}$$

and the largest eigenvalue has the following bound:

$$\lambda_N \geq \frac{N}{N-1}\max\{\deg(n); n \in N(S)\}. \tag{6-17}$$

6. Let U be the set of all real N-tuple $\mathbf{x}$ such that $\mathbf{X}^t\mathbf{X} = 1$ and $\mathbf{x}^t\mathbf{e}_N = 0$. From the theory of symmetric matrices, the following characterisation for $\alpha(S)$ is obtained:

$$\alpha(S) = \min\{\mathbf{x}^t \mathbf{L} \mathbf{x} | \mathbf{x} \in U\}, \tag{6-18}$$

where $\quad\quad\quad\quad \mathbf{e}_N = \{1,1,\ldots,1\}^t. \tag{6-19}$

7. The following theorem is interesting since it relates the properties of the adjacency matrix $\mathbf{A}$ of a graph to those of its Laplacian matrix $\mathbf{L}$. Such theorems may establish firm relationships between the application of the largest eigenvalue and eigenvector of $\mathbf{A}$ for ordering to the second-smallest eigenvector and eigenvalue of the Laplacian matrix $\mathbf{L}$ of the graph for ordering and partitioning.

Theorem: Let S be a graph with adjacency matrix $\mathbf{A}$ and Laplacian matrix $\mathbf{L}$. Let D and d be the maximum and minimum node degrees of S, respectively. The second-largest eigenvalue μ_2 of $\mathbf{A}$ and the second-smallest eigenvalue λ_2 of $\mathbf{L}$ are then related as follows:

$$\delta - \lambda_2 \leq \mu_2 \leq \Delta - \lambda_2. \tag{6-20}$$

Proof: μ_2 is the second-largest eigenvalue of $\mathbf{A}$, and $\delta - \lambda_2$ is the second-largest eigenvalue of $\delta \mathbf{I} - \mathbf{L} \leq \mathbf{A} - (\text{diag}(\deg(v)) - \delta \mathbf{I})$, which differs from $\mathbf{A}$ only on the diagonal, where the non-negative values $\deg(v) - \delta$ are subtracted. Consequently, $\delta - \lambda_2 \leq \mu_2$. In a similar way, the other inequality is also obtained.

Lemma: If S is not a complete graph, then $\mu_2 \geq 0$ and $\lambda_2 \leq \Delta$.

6.3.2 NODAL NUMBERING ALGORITHM

On the basis of the concepts presented in the previous section, the method can be described as follows:

Step 1: Construct the Laplacian matrix $\mathbf{L}(S)$ of the given graph S.

Step 2: Compute the second eigenvalue λ_2 of L and its corresponding eigenvector $\mathbf{y}_2$. Different methods are available for such a calculation. Paulino et al. [174] used a special version of the subspace iteration method. However, the algorithm of Lanczos described in the next chapter can also be efficiently applied; $\mathbf{y}_2$ is also known as the *Fiedler* vector.

Step 3: Reorder the nodes of S in ascending order of the vector components in $\mathbf{y}_2$.

Similar to the previous algebraic method, this algorithm has the advantage of using global information of the graph model. However, although it does not use the pseudo-peripheral nodes and SRT and its transversal, its efficiency is very sensitive to the initial ordering of the nodes of the model. Preconditioning by pre-ordering can be used for improving the running time of the method, resulting in some kind of dependency on graph-theoretical properties.

6.3.3 EXAMPLE

Consider a FE mesh with 12 nodes, as shown in Figure 6.2(a), with an arbitrary nodal numbering. The element clique graph of the model is illustrated in Figure 6.2(b).

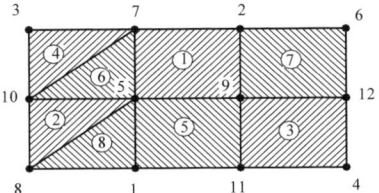

(a) A simple finite element model.

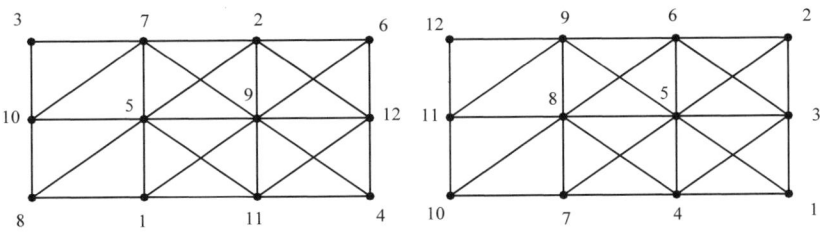

(b) Numbering before ordering.   (c) Numbering after ordering.

Fig. 6.2 A graph G.

The Laplacian matrix $\mathbf{L}(S)$ is constructed and its eigenvalue λ_2 and eigenvector $\mathbf{y}_2$ are calculated as follows:

$$\lambda_2 = 1.1071,$$

$$\mathbf{y}_2 = \{-0.0608, -0.2023, 1.0000, -0.5303, 0.0658, -0.4721, 0.3099, 0.3106, -0.2399, 0.5829, -0.3125, -0.4514\}^t.$$

Using $\mathbf{y}_2$, the new labelling is obtained, as illustrated in Figure 6.2(c).

284 OPTIMAL STRUCTURAL ANALYSIS

This method can also be applied to finite element nodal numbering, using any of the 10 graphs defined in Chapter 5.

6.4 A HYBRID METHOD FOR ORDERING

In this method, the advantages of both graph and algebraic graph methods are incorporated into an algorithm for ordering. In the algebraic graph method, general approaches are used to calculate the eigenvalues and eigenvectors, and the information available from the connectivity of their graph models is ignored. This is why the computational time and complexity of these algorithms are not low enough to compete with pure graph theory methods. In this section, graph parameters are used to increase the efficiency of the algebraic graph theory approaches. Typical graph parameters can be taken as the degrees of the nodes, the 1-weighted degrees of the nodes, the distances of the nodes from two pseudo-peripheral nodes, and the 2-weighted degrees of the nodes of the graph.

The algebraic graph theory method employed here is not the same as that employed in a general eigenproblem but rather a specific method is used in which the valuable features of graph parameters are incorporated.

6.4.1 DEVELOPMENT OF THE METHOD

Here, the graph parameters are considered as Ritz vectors, and the first eigenvector of the complementary Laplacian matrix $\mathbf{L}_c$ (Fiedler vector) is considered as a linear combination of Ritz vectors. The coefficients for these vectors are in fact the weights of the graph parameters, which are usually determined either by heuristic approaches or by experience.

Consider the following vector:

$$\bar{\phi} = \sum_{i=1}^{p} w_i \mathbf{v}_i, \qquad (6\text{-}21)$$

where $\bar{\phi}$ is an approximation to the Fiedler vector, $\mathbf{v}_i$ ($i = 1, ..., p$) are the normalised Ritz vectors representing the graph parameters, and w_i ($i = 1, ..., p$) are the coefficients of the Ritz vectors (Ritz coordinates), which are unknowns, and p is the number of parameters being employed. Equation (6-21) can be written as

$$\bar{\phi} = \mathbf{v}\mathbf{w}, \qquad (6\text{-}22)$$

where $\mathbf{w}$ is a $p \times 1$ vector and $\mathbf{v}$ is an $N \times p$ matrix containing the Ritz vectors.

Consider the eigenproblem of the complementary Laplacian:

$$\mathbf{L}_c \phi = \rho \phi. \tag{6-23}$$

Approximating ϕ by $\overline{\phi}$ and multiplying by $\mathbf{v}^t$, we have

$$\mathbf{v}^t \mathbf{L}_c \mathbf{v} \mathbf{w} = \rho \mathbf{v}^t \mathbf{v} \mathbf{w} \tag{6-24}$$

or $\quad\quad\quad\quad\quad \mathbf{A}\mathbf{w} = \rho \mathbf{B}\mathbf{w}, \tag{6-25}$

where $\mathbf{A} = \mathbf{v}^t \mathbf{L}_c \mathbf{v}$ and $\mathbf{B} = \mathbf{v}^t \mathbf{v}$. Both $\mathbf{A}$ and $\mathbf{B}$ are $p \times p$ matrices and therefore Eq. (6-21) has a much smaller dimension compared to Eq. (6-23); ρ is the approximate eigenvalue of the original problem.

The solution of the reduced problem, with dimensions far less than the original one, results in the first eigenvector $\mathbf{w}_1$ and hence $\overline{\phi}$. Nodal ordering is then performed considering the relative entries of $\overline{\phi}$ in an ascending order.

The present methods not only lead to a set of suitable coefficients for graph parameters but also provide efficient means for measuring the relative significance of each considered graph parameter. These coefficients may also be incorporated in the design of other specific graph-theoretical algorithms for ordering.

6.4.2 NUMERICAL RESULTS

Many examples are studied and the results of three models are presented in this section. In the tables presented, column 2 contains the results of the Pure Algebraic Graph Method (PAGM) of [174].

For the first case, four vectors, representing Ritz vectors, are considered. For these vectors, $\mathbf{v}_1$ contains the degrees of the nodes, $\mathbf{v}_2$ comprises the 1-weighted degrees of the nodes, and $\mathbf{v}_3$ and $\mathbf{v}_4$ are distances of the nodes from two pseudo-peripheral nodes. These nodes can be obtained using different algorithms; see Kaveh [113]. The results are provided in column 3 of the tables, denoted by $\mathbf{v}^4$.

For the second case, five Ritz vectors are employed. The first four vectors are the same as those of the previous case, and the fifth vector $\mathbf{v}_5$ contains the 2-weighted degrees of the nodes of the graph. The results are provided in column 4 of the tables, labelled as $\mathbf{v}^5$.

It should be noted that other vectors containing graph properties that influence the ordering may be considered in addition to the above five vectors. However, the formation of such additional vectors may require some extra computational time, reducing the efficiency of the algorithm.

Example 1: An FE mesh with one opening comprising 1248 nodes and 1152 rectangular elements is considered, as shown in Figure 6.3. The results for different methods and their computational time are illustrated in Table 6.2 for a comparison of their efficiency.

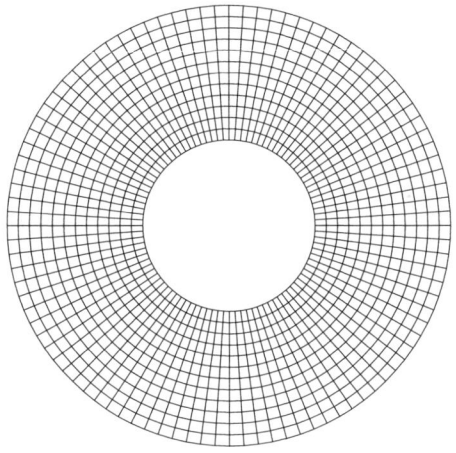

Fig. 6.3 An FE mesh with one opening.

Table 6.2 Results of Example 1.

	PAGM	$\mathbf{v}^4$	$\mathbf{v}^5$
B	46	43	45
P	34,848	36,243	36,189
$\tilde{F}$	28.07	29.44	29.25
F_{max}	35	39	39
Time (s)	1400.3	2.8	2.9

Example 2: An H-shaped FE mesh comprising 2096 nodes and 3900 triangular elements is considered, as shown in Figure 6.4. The results for different methods and their computational time are illustrated in Table 6.3 for a comparison of their efficiency.

ORDERING: ALGEBRAIC GRAPH THEORY METHODS   287

Fig. 6.4 An H-shaped FE mesh.

Table 6.3 Results of Example 2.

	PAGM	$\mathbf{v}^4$	$\mathbf{v}^5$
B	74	77	77
P	47,741	49,400	48,936
$\tilde{F}$	23.97	25.63	25.32
F_{max}	37	42	42
Time (s)	Large	2.63	2.89

Example 3: A two-dimensional finite element model (FEM) of a tunnel comprising 6888 nodes and 6720 rectangular elements is considered, as shown in Figure 6.5. The results of using different methods and their computational time are presented in Table 6.4.

288 OPTIMAL STRUCTURAL ANALYSIS

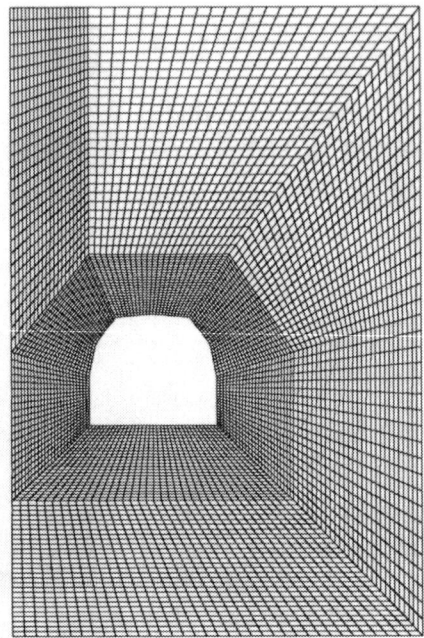

Fig. 6.5 A two-dimensional FEM of a tunnel.

Table 6.4 Results of Example 3.

	PAGM	$\mathbf{v}^4$	$\mathbf{v}^5$
B	455	331	332
P	731,694	733,738	733,738
$\tilde{F}$	112.99	112.93	112.93
F_{max}	164	175	175
Time (s)	10.6	27.6	28.9

Example 4: An FE mesh with four openings comprising 748 nodes and 1236 triangular elements is considered, as shown in Figure 6.6. The results for different methods and their computational time are illustrated in Table 6.5 for a comparison of their efficiency.

ORDERING: ALGEBRAIC GRAPH THEORY METHODS

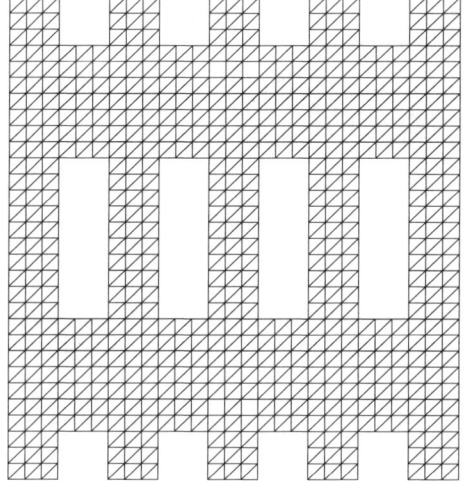

Fig. 6.6 An FE mesh with four openings.

Table 6.5 Results of Example 4.

	PAGM	v^4	v^5
B	39	49	47
P	13,118	13,162	13,126
$\tilde{F}$	18.42	18.61	18.56
F_{max}	29	29	29
Time (s)	1677	1.2	1.3

Example 5: A three-dimensional FEM of a nozzle is considered, as shown in Figure 6.7. This model contains 4000 rectangular shell elements. The results for different methods and their computational time are illustrated in Table 6.6 in order to compare their efficiency.

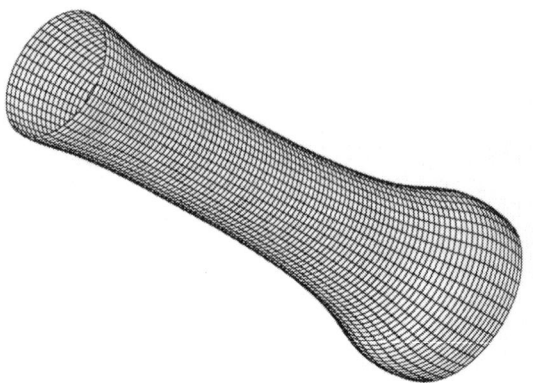

Fig. 6.7 A three-dimensional FEM of a nozzle.

Table 6.6 Results of Example 5.

	PAGM	$\mathbf{v}^4$	$\mathbf{v}^5$
B	39	49	47
P	13,118	13,162	13,126
$\tilde{F}$	18.42	18.61	18.56
F_{max}	29	29	29
Time (s)	1677	1.2	1.3

6.4.3 DISCUSSIONS

The performance of the hybrid method compares well with that of a PAGM, with a substantial reduction in the computational time. Naturally, addition of extra graph parameters will increase the computational time required. Relative values of the coefficients of the Ritz vectors show the importance of the corresponding parameters in the ordering algorithm. For the examples presented in the previous section, the coefficients corresponding to $\mathbf{v}_3$ and $\mathbf{v}_4$ (the distances from the pseudo-peripheral nodes) seem to be more important, since most of the examples have a more or less uniform distribution of nodal degrees. Naturally, for models with non-uniform degree distributions, the significance of the other graph parameters will also become apparent.

Though only nodal ordering is addressed here, the application of the present method can easily be extended to element ordering. For this purpose, the natural associate graph or the incidence graph of an FE mesh should be used in place of the element clique graph.

ORDERING: ALGEBRAIC GRAPH THEORY METHODS 291

EXERCISES

6.1 Find a good starting node for the following models using an algebraic graph–theoretical method, that is, calculate the dominant eigenvector of the corresponding adjacency matrices:

(a)   (b)

6.2 Find the nodal ordering of the system for models (a) and (b) in Exercise 6.1 using the Fiedler vector of the Laplacian matrix of the model.

6.3 Find a formula for calculating the eigenvalues of a path graph P_n.

6.4 Find a formula for calculating the eigenvalues of a cycle graph C_n.

CHAPTER 7
Decomposition for Parallel Computing: Graph Theory Methods

7.1 INTRODUCTION

In the last decade, parallel processing has come to be widely used in the analysis of large-scale structures [1,158,169,220]. This chapter is devoted to the optimal decomposition of structural models using graph theory approaches. First, efficient graph theory methods for the optimal decomposition of space structures are presented. Next, the subdomaining approaches for partitioning of finite element models (FEMs) are analysed. A substructuring technique for the force method of structural analysis is then discussed. Lastly, an efficient substructuring method for dynamic analysis of large-scale structures is explained.

Several partitioning algorithms are developed for solving multi-member systems, which can be categorised as graph theory methods and algebraic graph theory approaches.

For the graph theory method, Farhat [44] proposed an automatic finite element domain decomposer, which is based on a Greedy type algorithm and seeks to decompose an FEM into balanced domains, sharing a minimum number of common nodal points. To avoid domain splitting, Al-Nasra and Nguyen [4] incorporated geometrical information of the FEM into an automatic decomposition algorithm similar to the one proposed by Farhat [44]. The Sparspak uses nested dissection method of George and Liu [57], which uses a level tree for dissecting a model. Kaveh and Roosta [130,132] employed different expansion processes for decomposing space structures and finite element meshes.

Applications of the methods of this chapter are by no means confined to structural systems but can be equally applied to other large-scale problems like the analysis of hydraulic systems and electrical networks.

7.2 EARLIER WORKS ON PARTITIONING

7.2.1 NESTED DISSECTION

The term "nested dissection" was introduced by George [55], following a suggestion of Birkhoff. Its roots lie in finite element substructuring, and it is closely related to the tearing and interconnecting method of Kron [141].

The central concept for nested dissection is the removal of a set of nodes from the graph (separator) of a symmetric matrix (or the model of a structure) that leaves the remaining graph in two or more disconnected parts. In nested dissection, these parts are themselves further divided by the removal of sets of nodes, with the dissection nested to any depth.

If the variables of each subgraph are grouped together, by ordering the nodes of their nodes contiguously followed by numbering the nodes, in the separator, then the following block form will be obtained:

$$\begin{bmatrix} \mathbf{A}_{11} & \mathbf{0} & \mathbf{A}_{13} \\ \mathbf{0} & \mathbf{A}_{22} & \mathbf{A}_{23} \\ \mathbf{A}_{31} & \mathbf{A}_{32} & \mathbf{A}_{33} \end{bmatrix}. \quad (7\text{-}1)$$

The blocks $\mathbf{A}_{11}$ and $\mathbf{A}_{22}$ may themselves be ordered to such a form by using dissection sets. This way every level defines a nested dissection order.

The significance of the above partitioning of the matrix is twofold: first, the zero blocks are preserved in the factorisation, thereby limiting fill; second, factorisation of the matrices $\mathbf{A}_{11}$ and $\mathbf{A}_{22}$ can proceed independently, thereby enabling parallel execution on separate processors.

When a complicated design is assembled from simpler substructures, it makes sense to exploit these natural substructures. The resulting ordering is likely to be good, simply because, when each variable is eliminated, only the other variables of its substructures are involved.

7.2.2 A MODIFIED LEVEL-TREE SEPARATOR ALGORITHM

The separator routine in Sparspak, FNDSEP, finds a pseudo-peripheral node in the graph and generates a level structure from it. It then chooses the median level in the level structure as the node separator. However, this choice may separate the graph into widely disparate parts. In a modification made by Pothen et al. [182], the node separator is selected to the smallest level k, such that the first k levels together contain more than half of the nodes. A node separator is obtained by removing from the nodes in level k those nodes that are not adjacent to any node in level $k - 1$, and therefore these are added to the part containing the nodes in the first $k - 1$ levels. The other part has nodes in levels $k + 1$ and higher. Although such a method is simple, the spectral bisection method computes a smaller node separator compared to that by the Sparspak algorithm.

7.3 SUBSTRUCTURING FOR PARALLEL ANALYSIS OF SKELETAL STRUCTURES

7.3.1 INTRODUCTION

In many engineering applications, particularly in the analysis and design of large systems, it is convenient to allocate the design of certain components (substructures) to individual design groups. The study of each substructure is carried out more or less independently, and the dependencies between the substructures are resolved after the study of individual substructures is completed. The dependencies among the components may of course require the redesign of some of the substructures, so the above procedure may be iterated several times.

As an example, suppose for a structural model we choose a set of nodes I and their incident members, which, if removed, disconnect it into two substructures. If the variables associated with each substructure are numbered consecutively, followed by the variables associated with I, then the partitioning of the stiffness matrix $\mathbf{A}$ will be as in Eq. (7-1).

The Cholesky factor $\mathbf{L}$ of $\mathbf{A}$, correspondingly, will be partitioned as

$$\mathbf{L} = \begin{bmatrix} \mathbf{L}_{11} & \mathbf{0} & \mathbf{0} \\ \mathbf{0} & \mathbf{L}_{22} & \mathbf{0} \\ \mathbf{W}_{13}^t & \mathbf{W}_{23}^t & \mathbf{L}_{33} \end{bmatrix}, \tag{7-2}$$

where

$$\mathbf{A}_{11} = \mathbf{L}_{11}\mathbf{L}_{11}^t, \quad \mathbf{A}_{22} = \mathbf{L}_{22}\mathbf{L}_{22}^t, \quad \mathbf{W}_{13} = \mathbf{L}_{11}^t\mathbf{A}_{13}, \quad \mathbf{W}_{23} = \mathbf{L}_{22}\mathbf{L}_{23}^t$$

and

$$\mathbf{L}_{33}^t\mathbf{L}_{33} = \mathbf{A}_{33} - \mathbf{A}_{13}^t\mathbf{A}_{11}^{-1}\mathbf{A}_{23} - \mathbf{A}_{23}^t\mathbf{A}_{22}^{-1}\mathbf{A}_{23}. \tag{7-3}$$

Therefore, $\mathbf{A}_{11}$ and $\mathbf{A}_{22}$ correspond to each substructure, and the matrices $\mathbf{A}_{13}$ and $\mathbf{A}_{23}$ represent the "glue" that relates the substructures through the nodes of I.

Since the factors of $\mathbf{A}_{11}$ and $\mathbf{A}_{22}$ are independent, they can be computed in either order, or in parallel if two processors are available. Finally, in some design applications, several substructures may be identical, for example, have the same configuration and properties, and each substructure may be regarded as a super element, which is constructed once and used repeatedly in the design of several structures. In the above example, $\mathbf{A}_{11}$ and $\mathbf{A}_{22}$ could be identical.

7.3.2 SUBSTRUCTURING DISPLACEMENT METHOD

For the analysis of skeletal structures and for the finite element method, using the displacement approach, an appropriate formulation such as the Galerkian method reduces to solving the following matrix equation:

$$\mathbf{K}\mathbf{v} = \mathbf{p}, \tag{7-4}$$

where $\mathbf{K}$ is the global stiffness matrix and $\mathbf{v}$ and $\mathbf{p}$ are the nodal displacement and nodal force vectors, respectively. To distribute the computation after decomposing the model into q subdomains, each subdomain can be treated as a super element and mapped onto the processors. Various methods for decomposition will be presented in this chapter. The global stiffness matrix and nodal force vector are equivalent to the assembly of its components for q subdomains:

$$\mathbf{K} = \sum_{j=1}^{q} \mathbf{k}_j \quad \text{and} \quad \mathbf{p} = \sum_{j=1}^{q} \mathbf{p}_j. \tag{7-5}$$

Equation (7-4) can be written in the following partitioned form:

$$\begin{bmatrix} \mathbf{K}_{ii} & \mathbf{K}_{ib} \\ \mathbf{K}_{bi} & \mathbf{K}_{bb} \end{bmatrix} \begin{bmatrix} \mathbf{v}_i \\ \mathbf{v}_b \end{bmatrix} = \begin{bmatrix} \mathbf{p}_i \\ \mathbf{p}_b \end{bmatrix}. \tag{7-6}$$

In the above equation, a boundary node is defined as a node that is part of more than one subdomain. The degrees of freedom (DOF) at the boundary nodes are treated as boundary DOF. The vectors $\mathbf{v}_i$ and $\mathbf{v}_b$ are displacements, and $\mathbf{p}_i$ and $\mathbf{p}_b$ are forces, corresponding to internal and boundary nodes, respectively.

Each subdomain requires solution of an equation, similar to Eq. (7-4):

$$[\mathbf{k}]_j [\mathbf{d}]_j = [\mathbf{p}]_j. \tag{7-7}$$

For the full domain, Eq. (7-7) can be written in partitioned form as

$$\begin{bmatrix} \mathbf{k}_{ii} & \mathbf{k}_{ib} \\ \mathbf{k}_{bi} & \mathbf{k}_{bb} \end{bmatrix} \begin{bmatrix} \mathbf{v}_i \\ \mathbf{v}_b \end{bmatrix} = \begin{bmatrix} \mathbf{p}_i \\ \mathbf{p}_b \end{bmatrix}. \tag{7-8}$$

Using static condensation for eliminating the interior DOF of each subdomain, the effective stiffnesses and load vectors on the interface boundaries are obtained.

For internal nodes, we have

$$[\mathbf{k}_{ii}][\mathbf{v}_i] + [\mathbf{k}_{ib}][\mathbf{v}_b] = [\mathbf{p}_i], \tag{7-9}$$

or

$$[\mathbf{v}_i] = [\mathbf{k}_{ii}]^{-1}\{[\mathbf{p}_i] - [\mathbf{k}_{ib}][\mathbf{v}_b]\}. \tag{7-10}$$

Substituting in Eq. (7-8), we have

$$[\mathbf{k}_{bi}][\mathbf{k}_{ii}]^{-1}\{[\mathbf{p}_i] - [\mathbf{k}_{ib}][\mathbf{v}_b]\} + [\mathbf{k}_{bb}][\mathbf{v}_b] = [\mathbf{p}_b], \tag{7-11}$$

or

$$[\mathbf{k}^*][\mathbf{v}_b] = [\mathbf{p}_b] - [\mathbf{k}_{ib}]^{-1}[\mathbf{k}_{ii}][\mathbf{p}_i], \tag{7-12}$$

where

$$[\mathbf{k}^*] = [\mathbf{k}_{bb}] - \{[\mathbf{k}_{bi}][\mathbf{k}_{ii}]^{-1}[\mathbf{k}_{ib}]\} \tag{7-13}$$

is the condensed super-element stiffness matrix and

$$[\mathbf{p}^*] = [\mathbf{p}_b] - [\mathbf{k}_{bi}][\mathbf{k}_{ii}]^{-1}[\mathbf{p}_i], \tag{7-14}$$

is the modified load vector. A summation of the interface conditions for the subdomains leads to the formation of the global interface stiffness matrix $\mathbf{K}^*$ and the global interface load vector $\mathbf{p}^*$ as follows:

$$\mathbf{K}^* = \sum_{j=1}^{q} \mathbf{k}_j^* \quad \text{and} \quad \mathbf{p}^* = \sum_{j=1}^{q} \mathbf{p}_j^*. \tag{7-15}$$

K is symmetric and positive definite, and **K*** has the same properties. The following interface system can now be solved:

$$[\mathbf{K}^*][\mathbf{v}_b] = [\mathbf{p}^*]. \qquad (7\text{-}16)$$

Once $\mathbf{v}_b$ is found, the internal DOF for a subdomain can be evaluated using Eq. (7-10).

A natural route to parallelism now is to provide it through domain decomposition by distributing the substructures onto the processors available. Several approaches can be used to solve Eq. (7-4). In the following section, three broad classifications are briefly discussed.

7.3.3 METHODS OF SUBSTRUCTURING

Direct Methods

A substructuring method can be used to obtain the condensed stiffness matrix on each subdomain in parallel on the different processors. To create matrix **K***, it is necessary to condense the stiffness matrix of each substructure (subdomain), that is, from Eq. (7-13) the product $[\mathbf{k}_{bi}][\mathbf{k}_{ii}]^{-1}[\mathbf{k}_{ib}]$ should be calculated. The explicit formation of $[\mathbf{k}_{ii}]^{-1}[\mathbf{k}_{ib}]$ requires NB_{DOF} triangular system resolutions, where Mb_{DOF} is the number of subdomain boundary DOF. This step can be considered to be as follows:

Each internal DOF makes its contribution to the stiffness of each boundary DOF such that the behaviour of the condensed boundary is equivalent to the behaviour of the entire domain. This step can be executed step by step so that only the internal DOF connected to the boundary DOF updates the boundary stiffness matrix. This requires the internal DOF to appear at the bottom of the internal stiffness matrix $\mathbf{k}_{ii}$, so that they are modified by the elimination of all other internal DOF.

A frontal method can be used, which has the advantage of allowing very flexible strategies concerning the sequence of elimination of equations. When this method is applied to subdomain condensation, it is necessary to assemble the boundary DOF in the frontal matrix and to retain them until all the internal DOF have been eliminated. At the end of the frontal elimination process, the frontal matrix is exactly the condensed matrix $[\mathbf{k}_{bi}][\mathbf{k}_{ii}]^{-1}[\mathbf{k}_{ib}]$.

The interface system of equations is then solved employing a direct approach (e.g. skyline method) on a single machine. Although the direct methods are simple and terminate in a fixed number of steps, the interface solution dominates the overall computational cost when the interface system is large, thus limiting the overall efficiency. In such a case, however, a distributed algorithm can be used for factorisation of the direct method to overcome this difficulty.

Iterative Methods

A different method of avoiding the explicit inverse of k_{ii} in Eq. (7-13) is the use of an iterative approach. Among the iterative solutions, the conjugate gradient method is a promising candidate because of its inherent parallelism and its rate of convergence. The theory of the conjugate gradient method is well known [90]. One iteration of this method for solving a system of equations $\mathbf{Kv} = \mathbf{p}$ is given as follows:

$$\{\mathbf{u}\} = [\mathbf{K}]\{\mathbf{f}\}, \qquad (7\text{-}17a)$$

$$\alpha = \{\mathbf{r}\}^t \{\mathbf{r}\} / \{\mathbf{f}\}^t \{\mathbf{u}\}, \qquad (7\text{-}17b)$$

$$\{\mathbf{v}_{new}\} = \{\mathbf{v}\} + \alpha\{\mathbf{f}\}, \qquad (7\text{-}17c)$$

$$\{\mathbf{r}_{new}\} = \{\mathbf{r}\} + \alpha\{\mathbf{u}\}, \qquad (7\text{-}17d)$$

$$\lambda = \{\mathbf{r}_{new}\}^t \{\mathbf{r}_{new}\} / \{\mathbf{r}\}^t \{\mathbf{r}\}, \qquad (7\text{-}17e)$$

$$\{\mathbf{f}_{new}\} = \{\mathbf{r}_{new}\} + \lambda\{\mathbf{f}\}. \qquad (7\text{-}17f)$$

Before each iteration, the vectors $\{\mathbf{v}\}$, $\{\mathbf{f}\}$ and $\{\mathbf{r}\}$ are set to $\{\mathbf{v}_{new}\}$, $\{\mathbf{f}_{new}\}$ and $\{\mathbf{r}_{new}\}$, respectively.

The vectors are initialised as

$$\{\mathbf{r}\} = \{\mathbf{p}\} - [\mathbf{K}]\{\mathbf{v}_0\} \qquad (7\text{-}18a)$$

and

$$\{\mathbf{f}\} = \{\mathbf{r}\}, \qquad (7\text{-}18b)$$

where $\{\mathbf{v}_0\}$ is usually taken as null, unless some approximation to the solution is known. Iteration is terminated when the residual is small. One criterion for handling the iteration is

$$\|\mathbf{r}\| / \|\mathbf{p}\| < \varepsilon, \qquad (7\text{-}19)$$

where ε is the tolerance specified for the problem.

In structural analysis, the vector **r** is the potential gradient and is identical to the residual force vector (**p** − **Kv**) in the linear case. The vector **f** is the gradient direction to generate the displacement vector **v**. For discussion and further details, the reader may refer to Law [147].

Preconditioned Conjugate Gradient (PCG) methods form a large class of the many iterative methods that have been suggested to reduce the cost of forming condensed stiffness matrices. A saving in total time may be achieved since the predominant matrix-vector product at each iteration is computed in parallel. For further detailed discussion, the interested reader may refer to Keyes and Gropp [134].

Hybrid Methods

These methods use a combination of the direct and iterative methods. For instance, the components of the condensed matrix **k*** may be obtained for the substructures using the direct method, and the resulting interface can be solved using an iterative approach.

A comparative study of direct, iterative and hybrid methods is made by Chadha and Baugh [25].

In the following sections, algorithms are presented for partitioning of the nodes of structural graph models, which can be incorporated in any program available for the analysis of skeletal structures. Domain decomposition algorithms are presented in Chapter 8.

7.3.4 MAIN ALGORITHM FOR SUBSTRUCTURING

Let S be the graph model of a structure. The following algorithm decomposes S into q subgraphs with equal or near-equal number of nodes (support nodes are not counted) having the least number of interface nodes:

Step 1: Delete all the support nodes with their incident members, and denote the remaining subgraphs by S_r.

Step 2: Determine the distance between each pair of nodes of S_r, and evaluate the eccentricities of its nodes.

Step 3: Sort the remaining nodes (RN) in ascending order of their eccentricities.

Step 4: Select the first node of RN as the representative node of the subgraph S_1 to be determined and find a second node as the representative node of subgraph S_2 with the maximum distance from S_1.

Step 5: Find the third representative node with the maximum least distance from S_1 and S_2, and denote it with S_3.

Step 6: Subsequently, select a representative node of subgraph S_k for which the least distance from S_1, S_2, ..., S_{k-1} is maximum. Repeat this process until q representative nodes of the subgraphs to be selected are found.

Step 7: For each subgraph S_j ($j = 1, ..., q$), add an unselected node n_i of RN if it is adjacent only to S_j and its least distance from all nodes of other subgraphs is maximum.

Step 8: Continue the process of Step 7, without the restriction of transforming one node to each subgraph S_j, until no further node can be transferred. The remaining nodes in RN are interface nodes.

Step 9: Transfer the support nodes to the nearest subgraph.

Once the nodes for each subgraph S_j are found, the incidence members can easily be specified.

The algorithm is recursively applied to the selected substructures, decomposing each substructure into smaller ones, resulting in further refinement.

7.3.5 EXAMPLES

Example 1: A double-layer grid supported at four corner nodes is considered and partitioned into $q = 2$, 4 substructures; see Figure 7.1. The corresponding node adjacency matrices (pattern of their stiffness matrices) are illustrated in Figure 7.2(a and b). For the case $q = 2$, the selected substructures are further refined with $q' = 2$ and 3, and the corresponding matrices are shown in Figure 7.3(a and b).

302 OPTIMAL STRUCTURAL ANALYSIS

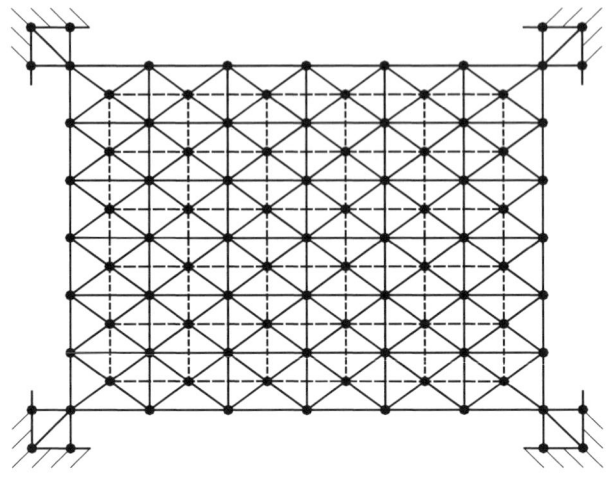

Fig. 7.1 A double-layer grid S.

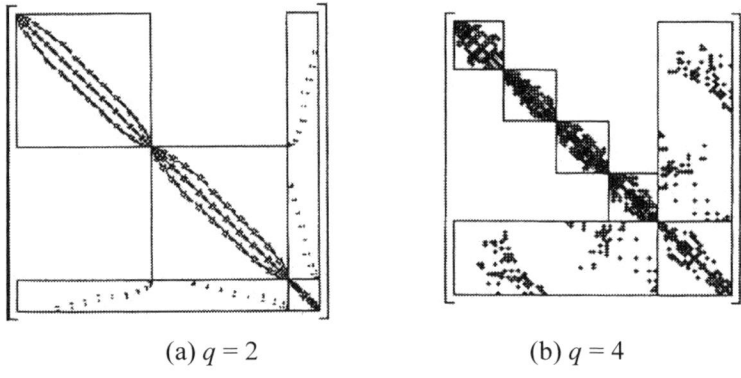

(a) $q = 2$ (b) $q = 4$

Fig. 7.2 Patterns of the adjacency matrices for different values of q.

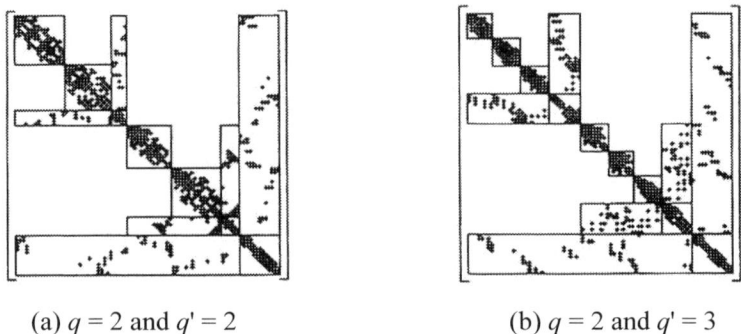

(a) $q = 2$ and $q' = 2$ (b) $q = 2$ and $q' = 3$

Fig. 7.3 Patterns of the adjacency matrices for $q = 2$ and $q' = 2$ and 3.

Example 2: A dome-type space structure supported at six nodes is considered and partitioned into $q = 2, 3, 4$ and 5 substructures; see Figure 7.4.

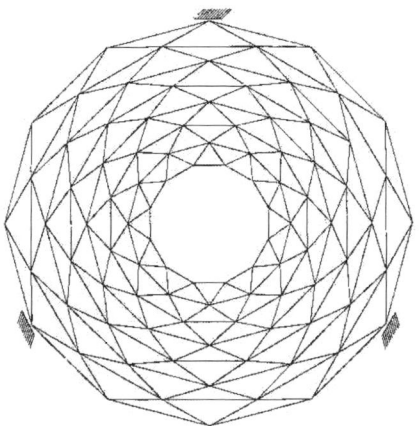

Fig. 7.4 A dome-type space structure.

The corresponding node adjacency matrices are illustrated in Figure 7.5(a–d). For the case $q = 2$, the selected substructures are further refined with $q' = 2$ and 3, and the corresponding matrices are shown in Figure 7.6(a and b).

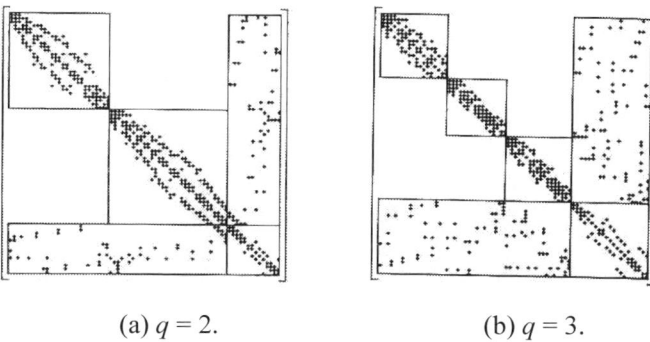

(a) $q = 2$.   (b) $q = 3$.

304   OPTIMAL STRUCTURAL ANALYSIS

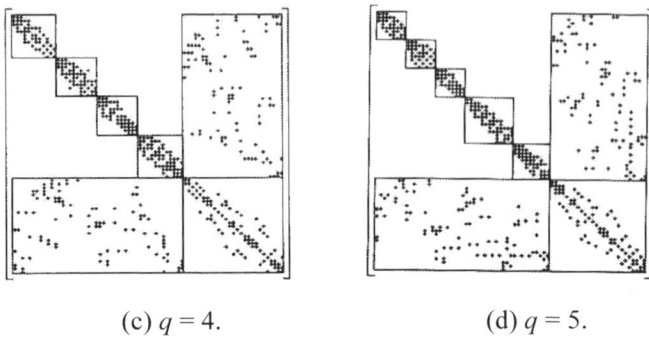

(c) $q = 4$.   (d) $q = 5$.

Fig. 7.5 Patterns of the adjacency matrices for different values of q.

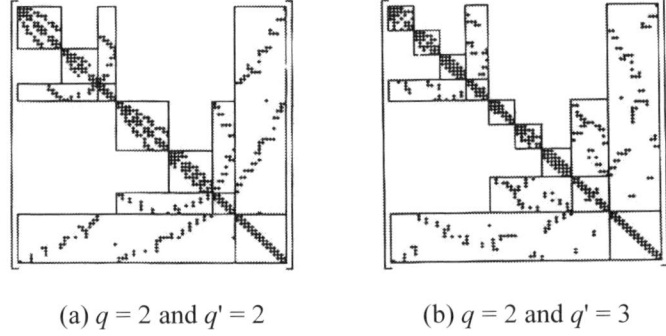

(a) $q = 2$ and $q' = 2$ (b) $q = 2$ and $q' = 3$

Fig. 7.6 Patterns of the adjacency matrices for $q = 2$ and different values of q'.

Once the subgraphs and the interface nodes are specified, ordering the nodes of each subgraph reduces the bandwidth of each block, and appropriate numbering of the interface nodes results in banded border for the entire matrix.

7.3.6 SIMPLIFIED ALGORITHM FOR SUBSTRUCTURING

In the following text, a simplified algorithm is presented that requires less storage and computer time than the main algorithm, at the expense of selecting subgraphs with a slightly higher number of interface nodes for some structural models. In this approach, the number of distances to be considered and compared for finding the nodes of substructures is far less than when the main algorithm is used, where the distances between each pair of nodes of S are required. This simplified algorithm consists of the following steps:

Step 1: Form a shortest route tree (SRT) rooted from an arbitrary node to find a representative node of S_1 with the maximum distance from the root. The selected node is also denoted by S_1.

Step 2: Form an SRT rooted from S_1 to calculate the distance between each node of S and S_1 and find the representative node S_2 with the maximum distance from S_1.

Step 3: Form an SRT rooted from S_2 to calculate the distance between each node of S and S_2 and find the representative node S_3 with the maximum least distance from the selected nodes. Repeat this process until q representative nodes $S_1, S_2, \ldots, S_q$, forming a transversal, are selected.

Step 4: For each subgraph S_i, find a node adjacent to the previously formed S_i only, with the maximum least distance from other representative nodes, in turn.

Step 5: Continue the process of Step 4, without the restriction of transforming one node to each subgraph S_i, until no further node can be transferred.

7.3.7 GREEDY TYPE ALGORITHM

In this algorithm, the *weight* of a node is taken as the number of elements incident with that node. The *interior boundary* of a subdomain D_i is defined as the subset of its boundary that will interface with another subdomain D_j. The total number of elements in a given mesh is denoted by $M(\text{FEM})$.

Step 1: Start with a node and add incident elements having the least current weight one by one. The *current weight* is taken as the number of unselected elements at that stage incident with that node. Continue this process until $M(\text{FEM})/q$ elements are selected as D_1.

Step 2: Select an interior node of D_1, and repeat Step 1 to form D_2.

Step k: Repeat Step 2 for $k = 3, 4, \ldots, q$ with an interior node of D_{k-1} and form the subdomain D_k.

This process is a Greedy type algorithm, which selects one element of minimal current weight at a time and completes a domain when $N(\text{FEM})/q$ (+1 if remainder $\neq 0$) elements are selected for the formation of that subdomain. The current weight of an element is updated when an incident element is joined to the expanding subdomain.

7.4 DOMAIN DECOMPOSITION FOR FINITE ELEMENT ANALYSIS

In this section, efficient algorithms are developed for automatic partitioning of unstructured meshes for the parallel solution of problems in the finite element method. These algorithms partition a domain into subdomains with approximately equal loads and good aspect ratios, while the interface nodes are confined to the smallest possible. Examples are included to illustrate the performance and efficiency of the presented algorithms.

7.4.1 INTRODUCTION

Domain decomposition is attractive in finite element computations on parallel architectures, because it allows individual subdomain operations to be performed concurrently on separate processors and serial solutions to be obtained on a sequential computer to overcome the limitation of core storage capacity. Given a number of available processors q, an arbitrary FEM is decomposed into q subdomains, where formation of element matrices, assembly of global matrices, partial factorisation of the stiffness matrix and state determination or evaluation of generalised stresses can be carried out independent of similar computations for the other subdomains, and hence can be performed in parallel.

In parallel processing of subdomains, the time to complete a task will be the time to compute the longest subtask. An algorithm for domain decomposition will be efficient if it yields subdomains that require an equal amount of execution time. In other words, the algorithm has to achieve a load balance among the processors. In general, this will be particularly ensured if each subdomain contains an equal number of elements or an equal total number of DOF. However, for some numerical techniques based on domain decomposition, a balanced number of elements or total DOF among the subdomains does not imply balancing of the subdomain calculations themselves. The use of a frontal subdomain solver provides a relevant example. In this case, the computing load within a domain is not only a function of the number of elements within the subdomain but also of the element numbering. Thus, the optimal number of elements is a priori unknown and can vary significantly from one subdomain to another.

To reduce the cost of synchronisation and message passing between the processors in a parallel architecture, the amount of interface nodes should be minimised because the parallel solution for the generalised displacements usually requires explicit synchronisation on a shared-memory multi-processor and message passing on local-memory ones. In a domain decomposition method, another significant mesh partitioning factor that should be considered is the subdomain aspect ratio. This ratio has a vital impact on the convergence rate of the iterative approaches for the finite element tearing and interconnecting method.

The above features suggest that an automatic finite element domain decomposer should meet the following four basic requirements to be efficient:

1. It should be able to handle irregular geometry and arbitrary discretisation to be a general purpose one.

2. It must yield a set of balanced subdomains to ensure that the overall computational load be as evenly distributed as possible among the processors.

3. It should minimise the amount of interface nodes to reduce the cost of synchronisation and/or message passing between the processors.

4. It must result in subdomains with proper aspect ratios to improve the convergence rate of the domain decomposition–based iterative method.

Methods of subdomaining are well documented in the literature; see, for example, Farhat and Wilson [48], Farhat [44], Dorr [37] and Malone [160], Farhat and Roux [47], Al-Nasra and Nguyen [4] Farhat and Lesoinne [45], Khan and Topping [135,136] Topping and Khan [220], Topping and Sziveri [221], Vanderstraeten and Keunings [224], and Kaveh and Roosta [130,132]. Several automatic domain decomposition methods that address the load balance and minimum interprocessor computation problems have already been reported in the literature. In general, these algorithms can be grouped into two categories: engineering-based methods and graph theory–based methods. For engineering-based approaches, one can refer to the algorithms in [44,49], and for graph theory–based methods the algorithms in [113,130] can be referred to.

In this section, two efficient algorithms are presented for the decomposition of one- to three-dimensional FEMs of arbitrary shapes. The first method is a *graph-based* method and uses a general expansion process. The second is an *engineering-based* approach. In these algorithms, the resulted subdomains generally have good aspect ratios, especially when originally the elements have this property.

7.4.2 A GRAPH-BASED METHOD FOR SUBDOMAINING

In this algorithm, first the associate graph model or incidence graph model G of the FEM is generated. Then, a good starting node R_1 of G is selected. R_1 is taken as the first node of the first subgraph G_1. Next, G_1 is expanded from R_1. The process of expansion is continued such that the equality of the total DOFs of subdomains is provided. G_2 is formed similar to G_1 but it is expanded from R_2, which is an unselected node with the maximum distance from R_1. R_2 should contain no node of G_1. The process of expansion is executed in a manner that provides the connectedness of the subgraph being formed (if it is possible). A similar approach is employed and G_3, ..., G_q are generated, and the subdomains of the FEM corresponding to the selected subgraphs of G are identified. The steps of the algorithm are as follows:

Step 1: Use the associate or incidence graph G of the considered FEM and form an SRT rooted from an arbitrary node of G, to find a node R_1 with the maximum distance from the root.

Step 2: Generate subgraph G_i ($i = 1$ to q) as follows:

(a) Form an SRT rooted from R_i to calculate the distance between each node of G and R_i (R_i is taken as the first selected node of G_i), and find an unselected node R_{i+1} with the maximum distance from R_i.

(b) Find all the unselected boundary nodes of G_i, and denote them by UBN.

(c) Associate an integer with each node n_i of UBN that is the same as its distance from R_i plus the number of unselected nodes adjacent to n_i minus the number of selected nodes adjacent to n_j. Then, detect the node with the minimum integer and add it to G_i.

(d) If the total DOF of the corresponding subdomain is less than $[TDOF + W_0 (q - 1)]/q$, then repeat the above steps from Step (b); otherwise, execute Step 2 to generate subgraph G_{i+1}. *TDOF* is the total DOF of the FEM and W_0 is the total DOF for the nodes of the corresponding subdomain, which are also contained in unselected elements.

In the above algorithm, only the connectivity of the nodes of G is considered, and no labels for edges of G, list or matrices of edges are needed. Therefore, the formation of SRTs of G and data keeping will be more simple and efficient. Since valencies of the nodes of an associate or incidence graph of an FEM are not generally very different, the adjacency list is an efficient means of keeping the connectivity data of G. The adjacency list of a graph G is a matrix containing $N(G)$ rows and Δ columns, where Δ is the maximum degree of the nodes of G. The *i*th row contains the labels of the nodes adjacent to the node *i*.

Step 1 is carried out to select a good starting node in the generated associate or incidence graph G. Using the adjacency list of G, Step (a) can be performed as explained below; however, any other type of list may also be used.

1. Select all the nodes of the R_ith row of the adjacency list of G. The distance between these nodes and the root is equal to unity.

2. Select all the unselected nodes of the row *j* (*j* is an element of the set of the selected nodes of the previous step). The distance of these nodes from the root is one unit more.

3. Repeat Step 2 until all the nodes are selected.

The last instruction of Step (a) is carried out to select the first node of the next subgraph. This node should not be included in the previously generated subgraphs (i.e. it should be an unselected node). In Step (b), UBN contains unselected nodes that are adjacent to selected nodes of G_i. To extend G_i, a node of UBN will be added to G_i in every execution of Step (c). In this step, an integer will be associated with each node of UBN that defines the best possible node, having the following properties:

1. It is near the root.

2. It does not make the next UBN very large.

3. It is connected to G_i with more nodes, which leads to a desirable configuration for G_j.

This integer is equal to the distance from R_i plus the valency of the node minus the number of selected adjacent nodes multiplied by two. The value of $[TDOF + W_0(q - 1)]/q$ is not needed to be calculated in every execution of Step (d). Since every subdomain should have at least $TDOF/q$ degrees of freedom, W_0 can be calculated when the DOF of a subdomain becomes more than $TDOF/q$. Additional value, $W_0(q - 1)/q$, is considered, since the DOF of the interface nodes of subdomains are calculated in two or more subdomains and the DOF of the subdomains should be equal or nearly equal.

In this algorithm, a disconnected subdomain may be generated. This happens when no node can be found in Step (b). In such a case, an unselected node with minimum distance should be added to the considered subgraph. To avoid such situations, one should avoid decomposing a small FEM into many subdomains. However, the following modifications can always be used:

1. Formation of a single SRT from an arbitrary node to find a good starting node may not lead to the best node; however, the existing good starting node algorithms can be used to select a better node.

2. If a subgraph G_i contains two components G_i' and G_i'', one can exchange nodes of G_i' or G_i'' with the adjacent subgraphs to provide connectedness for G_i.

3. Use a non-deterministic heuristic of combinatorial optimisation such as Simulated Annealing to improve the initial partitioning to avoid the formation of multi-connected subdomains.

7.4.3 RENUMBERING OF DECOMPOSED FINITE ELEMENT MODELS

Once the subdomains and interface nodes are specified, the nodes and/or elements of each subdomain and the interface nodes can be renumbered for bandwidth, profile or frontwidth reduction, depending on whether a band, profile or frontal solver is exploited, respectively. The process of renumbering includes the following steps:

1. Renumber the internal nodes/elements of the subdomains $M_1, ..., M_q$ using an available algorithm.

2. Select an interface node connected to M_1 that is contained in a minimum number of elements as the starting node, and number the interface nodes using a nodal ordering algorithm. In the process of renumbering, when possible, priority is given to the nodes connected to lower-numbered subdomains.

It should be noted that, for a specified solver such as a frontal solver, the resulting subdomains and interface nodes should also satisfy additional conditions. For example in a frontal solver, a necessary condition for the applied domain decomposition approach to be feasible is that the number of DOF lying on the interface of any subdomain be smaller than the frontwidth associated with the direct (one domain) approach. However, such conditions cannot always be satisfied using the existing decomposition heuristics, because they generally depend on the shape and the connectivity of FEMs; see Lesoinne et al. [149].

7.4.4 COMPLEXITY ANALYSIS OF THE GRAPH-BASED METHOD

The direct formation of an associate graph or an incidence graph of a FEM with l elements has time complexity $O(\theta^2 \lambda^2)$, where θ is the maximum number of nodes of an element. For checking l elements to have a common node or boundary, the following loops should be executed:

    for i from 2 to n
        for j from 1 to $i-1$
            for k from 1 to θ
                for l from 1 to θ
                -
                end for
            end for
        -
        end for
    end for

The first two loops are designed to control elements i and j that have common nodes. However, if the maximum difference δ between the labels of two elements with a common boundary for the associate graph or with a common node for the incidence graph is given, then the second loop will be replaced by the following:

    for j from $i - \delta$ to $i-1$.

This modification becomes very efficient in large-scale problems and reduces the time complexity of the process to $O(\theta^2 \delta \lambda)$. Hence, ordering of the elements should be performed in the process of mesh generation. However, this method for generating an associate graph or an incidence graph is dependent on the order of the data. In the following text, a different method that does not depend on the data order and requires a fewer number of operations is presented. This method consists of the following steps:

1. Generate an adjacency list with α rows and ε columns. In this list, the ith row contains labels of the elements that contain node i, and α and ε are the numbers of the nodes of the FEM and the maximum number of elements containing a

specified node, respectively. Each pair of elements has at least one common node in any row of this list.

2. If the associate graph is needed, then control each pair of elements of every row to have more common nodes.

3. When two elements of equal or different dimensions have interface nodes equal to or more than the smallest dimension of the elements, then, in the associate graph, the corresponding nodes are connected to each other by a member. In the case of the incidence graph, the nodes corresponding to each pair of elements of every row are connected to each other by a member.

This method has the time complexity $O(\theta^2\varepsilon^2\alpha)$ for the formation of the associate graph, and $O(\varepsilon^2\delta\alpha)$ for the construction of the incidence graph of an FEM.

Formation of an SRT, using a proper list for keeping connectivity of nodes of the graph, has time complexity $O(\Delta\lambda)$, where Δ is the maximum valency of the nodes of the graph.

Every execution of Step 2 includes the following steps with the time complexities:

Step (a) has the time complexity $O(\Delta\lambda)$.

In every execution of Step 2, Step (b) is carried out m times, where *m* is the maximum number of elements of each subdomain, and its bound is less than λ and its average value is equal to n/q. Since each execution of Step (b) has time complexity $O(\Delta m)$, the time complexity of this step is $O(\Delta\lambda^2)$.

Step (c) is carried out *m* times in every execution of Step 2. Each execution of this step has time complexity $O(\Delta\kappa)$, where κ is the maximum cardinality of the UBN, and its bound is less than λ. Hence this step also has the time complexity $O(\Delta\lambda^2)$.

Step (d) includes the following two processes:

1. Calculate the DOF for the corresponding subdomain. This process is also carried out *m* times in every execution of the Step 2 and has the time complexity $O(\theta\lambda)$.

2. Calculate $[TDOF + W_0(q - 1)]/q$. This process may be carried out at most *k* times in every execution of Step 2 and has the time complexity $O(\theta\lambda^2)$. Step 2 should be carried out *q* times; hence the time complexity of this step is $O(\delta\lambda^2 q)$.

312   OPTIMAL STRUCTURAL ANALYSIS

7.4.5 COMPUTATIONAL RESULTS OF THE GRAPH-BASED METHOD

Example 1: An FEM with $\lambda = 606$, $\alpha = 1961$ is considered; each element has four corner nodes and four mid-side nodes and each node has 2 DOF and is decomposed into 2, ..., 6 subdomains, as shown in Figures 7.7(a–d) for $q = 4$ and 6, where λ and α denote the numbers of elements and nodes, respectively. The DOF of the selected subdomains and interface nodes for $q = 2$, ..., 6 are illustrated in Table 7.1, when the associate graph and the incidence graph are used.

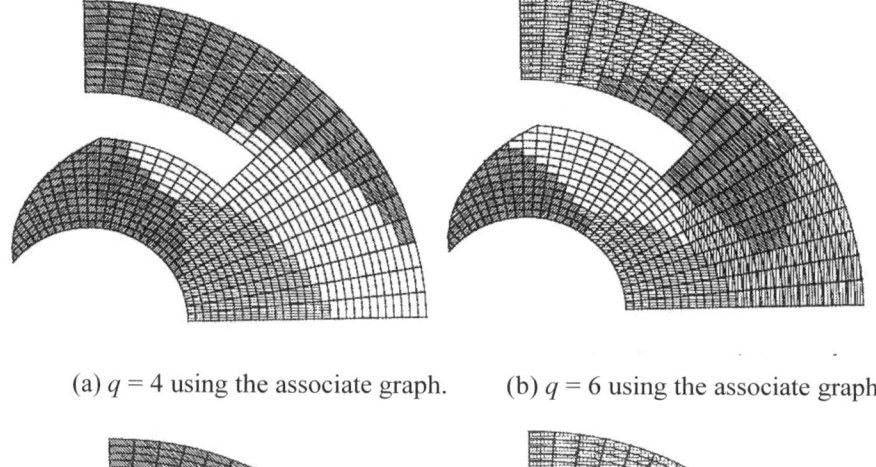

(a) $q = 4$ using the associate graph.   (b) $q = 6$ using the associate graph.

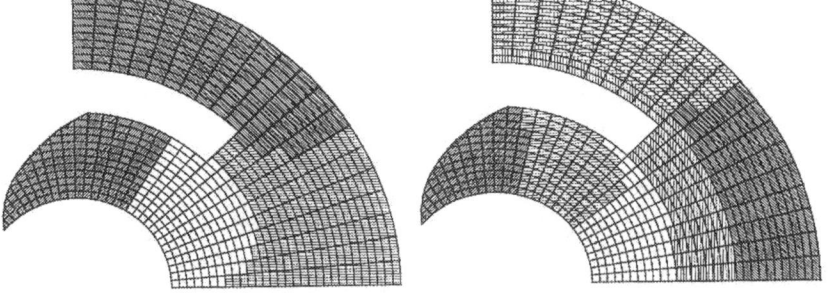

(c) $q = 4$ using the incidence graph.   (d) $q = 6$ using the incidence graph.

Fig. 7.7 A finite element model and its decompositions.

Table 7.1 Results of Example 1.

q	Type of graph	DOFs of subdomains; interface nodes
2	Associate	2002, 1994; 74
	Incidence	2016, 2008; 102
3	Associate	1370, 1360, 1352; 160
	Incidence	1352, 1370, 1352; 150
4	Associate	1048, 1052, 1060, 1044; 280
	Incidence	1022, 1030, 1030, 1022; 182
5	Associate	856, 860, 868, 852, 842; 352
	Incidence	828, 844, 848, 826, 816; 240
6	Associate	724, 730, 744, 748, 672, 706; 394
	Incidence	700, 728, 714, 692, 692, 694; 296

Example 2: An L-shaped FEM with $\lambda = 2400$, $\alpha = 1281$ and each node having DOF equal to 2 is considered. The model is decomposed into 6 and 12 subdomains, as shown in Figure 7.8(a–d). The DOF of the subdomains and interface nodes using the associate graph and the incidence graph are illustrated in Table 7.2.

(a) $q = 6$ using the associate graph.   (b) $q = 12$ using the associate graph.

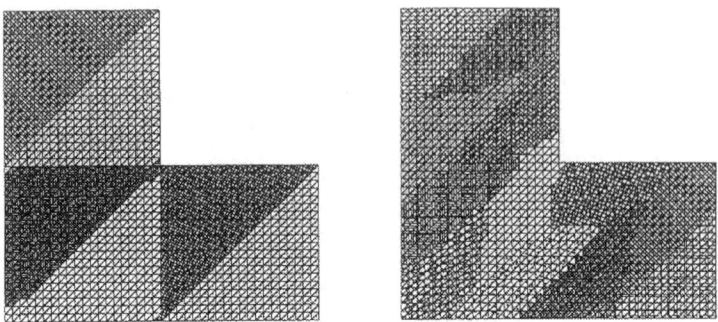

(c) $q = 6$ using the incidence graph. (d) $q = 12$ using the incidence graph.

Fig. 7.8 An L-shaped finite element model and its decompositions.

Table 7.2 Results of Example 2.

q	Type of graph	DOFs of subdomains; interface nodes
8	Associate	462, 462, 462, 462, 462, 462; 210
	Incidence	462, 462, 462, 462, 462, 462; 216
12	Associate	244, 244, 248, 248, 246, 246, 242, 252, 246, 232, 236, 232; 342,
	Incidence	252, 252, 250, 250, 268, 268, 246, 268, 260, 232, 234, 242; 440

Example 3: An FEM with $\lambda = 528$, $\alpha = 307$ with each node having two DOF is considered. The model is decomposed into 2, 3 and 4 subdomains, and the decomposed models for $q = 4$ are shown in Figure 7.9(a and b). The DOF of the subdomains and interface nodes using an associate graph and an incidence graph are illustrated in Table 7.3. The patterns of the node adjacency matrices employing the associate graph for the model, after ordering, are shown in Figure 7.10(a–c).

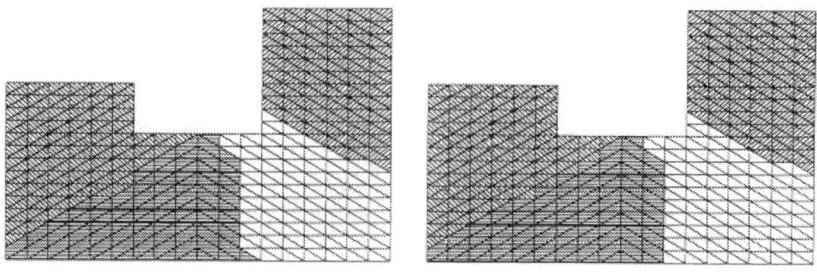

(a) $q = 4$ using the associate graph.   (b) $q = 4$ using the incidence graph.

Fig. 7.9 A finite element model and its decompositions.

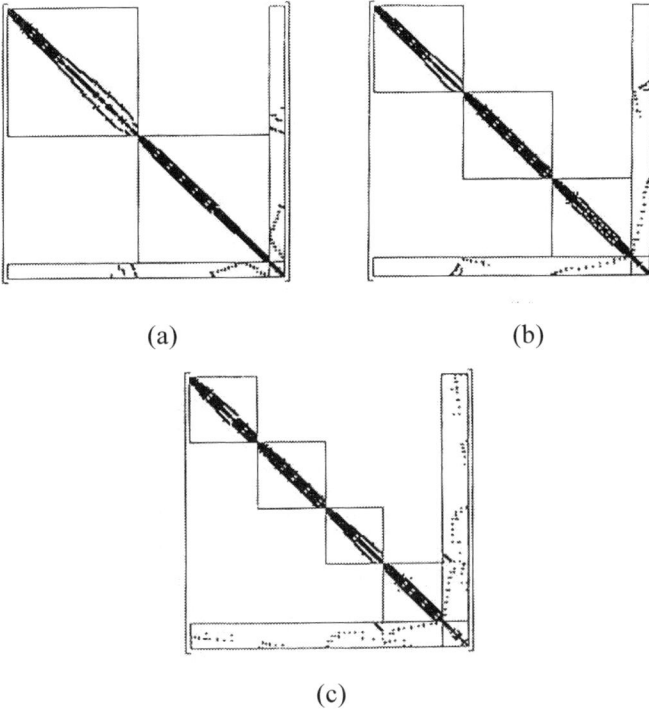

Fig. 7.10 The patterns of the ordered node adjacency matrices.

Table 7.3 Results of Example 3.

q	Type of graph	DOFs of subdomains; interface nodes
2	Associate	324, 324; 34
	Incidence	322, 322; 30
3	Associate	224, 216, 218; 42
	Incidence	222, 222, 220; 50
4	Associate	170, 164, 170, 168; 58
	Incidence	176, 164, 170, 170; 66

7.4.6 DISCUSSIONS ON THE GRAPH-BASED METHOD

This algorithm has low time complexity, is simple to program and leads to efficient partitioning of an FEM into subdomains with the required properties; therefore, it can also be considered as a good educational approach. The FEM that should be partitioned can contain meshes with different dimensions, types and sizes. Although the problem of aspect ratios of the subdomains is not dealt with explicitly in this section, the algorithm has the feature of expansion in all directions, leading to good aspect ratios.

7.4.7 ENGINEERING-BASED METHOD FOR SUBDOMAINING

Definitions

A *level structure* $L(r)$ of an FEM rooted from an element r (as the root) is defined as a partitioning of the set of elements into levels $l_1(r), l_2(r), \ldots, l_d(r)$ such that the following are true:

1. $l_1(r) = \{r\}$.

2. All elements adjacent to elements in level $l_i(r)$ ($1 < i < d$) are in levels $l_{i-1}(r), l_i(r)$ and $l_{i+1}(r)$.

3. All elements adjacent to elements in level $l_{di}(r)$ are in levels $l_{d-1}(r)$ and $l_d(r)$.

The overall level structure may be expressed as the set $L(r) = \{l_1(r), l_2(r), \ldots, l_d(r)\}$, where d is the *depth* of the level structure and is simply the total number of levels, and two elements are adjacent if they share a common node.

The *element adjacency list* of a finite element mesh contains the lists of elements adjacent to each element. The element-node list of an FEM contains the lists of nodes of each element and is generally employed as an input for data connectivity of FEMs. Following Webb and Froncioni [225], the *node-element list* contains the lists of elements containing each node of the finite element mesh.

A *genre structure* is a level structure in which each level is divided into one or more genres, and the index of each genre, as defined below, simply shows the pseudo-distance between the root and its elements. The overall genre structure rooted from an element r may be expressed as the set $G(r) = \{g_0(r), g_1(r), g_2(r), \ldots, g_s(r)\}$, in which the pseudo-distance between r and the elements of genre $g_i(r)$ is equal to i. The *index vector* $IV_r(i)$ of a genre structure rooted from an element r is a vector of $(n + 1)$ dimensions whose ith array ($i = 0, \ldots, n$) defines the total number of elements of g_i ($j = 0, \ldots, i$), that is,

$$IV_r(i) = \sum_{j=0}^{i} |g_i(r)|. \tag{7-20}$$

Thus, the cardinality of genre i ($0 < i \leq n$) is simply equal to $IV_r(i) - IV_r(i-1)$, and the cardinality of $g_0(r)$ is equal to 1. The following scheme (in pseudo code) should be used to form a genre structure from an arbitrary starting element r to generate its index vector and to find the pseudo-distances $pd(r, e_i)$ between the root r and all elements $e_i(i = 1, ..., \lambda$, where λ denotes the number of elements) of the considered FEM. In this scheme, $D \in \{1, 2, 3\}$ denotes the highest dimension of the elements in the model, and $CCN(g_i(r), e)$ denotes the set of common corner nodes between the elements of genre $g_i(r)$, and the element e.

1. Set $g_0(r) = \{r\}$, $IV_r(0) = 1$, $pd(r, r) = 0$ and mask r.
2. Set $i = 1$, $a = 0$ and $b = 0$.
3. for $j = D$ to 1 Step 1
   for $k = a$ to b
   (I) put each unmasked element e with $|CCN(g_k(r), e)| \geq j$ into $g_i(r)$.
   (II) if $|g_i(r)| \neq 0$ then set $IV_r(i) = IV_r(i-1) + |g_i(r)|$, $pd(r, e) = i$
   ($e \in g_i(r)$), $i = i + 1$ and mask the elements of $g_i(r)$.
   end for
   end for
4. If $IV_r(i-1) < \lambda$ then set $a = b + 1$, $b = i$ and repeat Step 3.

7.4.8 GENRE STRUCTURE ALGORITHM

Step 1: Form a genre structure rooted from an arbitrary element and select an element e_s^1 from its last genre.

Step 2: Calculate the pseudo-distance between e_s^1 and each element and select an element e_s^2 with the maximum pseudo-distance from e_s^1.

Step 3: Calculate the pseudo-distance between e_s^2 and each element. If $q = 2$, then go to Step 5.

Step 4: Find an unselected element e_s^i ($i = 3, 4, ..., q$) contained in genres $g_{j1}(e_s^1)$, $g_{j2}(e_s^2)$, $g_{j3}(e_s^3)$, ..., $g_{ji-1}(e_s^{i-1})$ such that the least value of $IV_r(j_k - 1)$ is the maximum, where $j_k > 0$, $k = 1, ..., i - 1$, and then calculate the pseudo-distances between e_s^i and the elements.

Step 5: For each selected element e_s^j ($j = 1, ..., q$) and each element e_k ($k = 1, ..., \lambda$), assign an integer in(e_s^j, e_k) as follows:

$$\text{in}(e_s^j, e_k) = \lambda + \text{mpd}(e_s^j, e_k) - pd(e_s^j, e_k), \tag{7-21}$$

where

$$\text{mpd}(e_s^j, e_k) = \min\{pd(e_s^l, e_k) \mid 1 \leq l \leq q, l \neq j\}$$

Step 6: Let e_s^j be the first element of the subdomain M_i. Calculate the weight of M_i and mask e_s^i, where $i = 1, ..., q$.

Step 7: Find an expandable subdomain M_i with minimum weight, add an unmasked element e_k with maximum non-zero priority number $P_i = CN \times \text{in}(e_s^i, e_k)$ to M_i, update the weight M_i and mask e_k, where if $|CCN(M_j, e_k)| \leq 3$ then $CN = |CCN(M_i, e_k)|$, else $CN = 3$. If there is no element to be added to M_i, this subdomain is not expandable and should be masked. If there are several elements with the maximum priority number P_i, then select the one with the minimum sum of integers corresponding to e_s^i ($i = 1, ..., q$). Repeat this step until all the elements are masked.

In this algorithm, the weight of a subdomain M_i can be taken as an arbitrary single number such as the number of the elements of M_i, the total DOF of the nodes of M_j, a function of the number and labels of the elements of M_i, and so on. However, here the total DOF of the nodes of a subdomain is considered as the *weight* of the subdomain.

Obviously, in this method only the corner nodes of a finite element mesh should be provided, that is, mid-side nodes and interior nodes are not needed. This increases the efficiency of the algorithm and results in saving computer storage space for FEMs with high-order elements.

An important problem that should be contemplated in a domain decomposition method is the connectedness of the elements of a single subdomain. In this algorithm, multi-component subdomains are avoided. Since the integers that are calculated in Step 5 are more than zero, the priority number P_i of an element e_i corresponding to the subdomain M_i will be zero if $|CCN(M_i, e_i)| = 0$, that is, the element is not connected to M_i with a corner node. As stated in Step 7, an element with priority number $P_i = 0$ cannot be added to M_i. This provides the connectedness of M_i ($i = 1, ..., q$); however, it leads to differences between the weights of the subdomains because when a subdomain cannot be expanded and is masked, the other unmasked subdomains are still expanded. This problem has been nearly remedied in the present algorithm by Steps 2 and 4. In these steps, the first elements of the subdomains are selected in such a manner that there are enough elements to be added to them for further expansion of the subdomains. For complete balanced loads for subdomains, one can let elements with zero priority numbers to be also added to a subdomain, in which case multi-component subdomains will be generated. However, there are several non-deterministic heuristics used in combinatorial optimisation such as Simulated Annealing, Stochastic Evolution and Tabu Search, which can be used for better load balancing of subdomains and reduction in the number of interface nodes; see, for example [13].

These combinatorial optimisation methods are normally included in an FEM decomposition algorithm as follows:

Step I: Invoke a direct partitioning scheme to produce an initial decomposition of reasonable quality.

Step II: Use an optimisation procedure to improve the initial partitioning.

The second step generally needs high computer time. Hence this algorithm is designed for careful partitioning of the finite element meshes to avoid (as far as possible) the use of optimisation procedures for general cases. However, this method can be applied as a direct method in Step I. This will be efficient, since the more the load balancing of subdomains and the less the number of interface nodes produced by a direct partitioning scheme, the less the cost for the applied optimisation method.

Step 1 of the algorithm presented in this section is carried out to find a good starting element e_s^1 for the first subdomains M_1. Step 2 is executed to calculate the pseudo-distance between e_s^1 and each element and to find an element e_s^2 as the good starting element of the second subdomain M_2. It should be noted that, when $q > 2$, the index of genres containing a specified element should be known because it is needed for the selection of the starting elements of subdomains M_i ($i = 3, ..., q$). Step 3 should be carried out to calculate the pseudo-distance between e_s^2 and each element of the considered finite element mesh. Also in this step, the index of genres containing a specified element should be defined for $q > 2$. Step 4 is executed to find good starting elements for subdomains M_i ($i = 3, ..., q$). The condition imposed in step 4 is included in order to provide the starting elements of subdomains to be unobtrusive when the process of expansion is performed in Step 7. This condition increases the probability that a subdomain will remain expandable while the other subdomains are being expanded. Step 5 is carried out to calculate an integer for each selected (starting) element and each element of the finite element mesh. This integer is always more than zero since λ is always more than or equal to a pseudo-distance between two elements, and a pseudo-distance is always equal to or more than zero. The integers calculated in this step affect the priority number of elements in two ways when the process of expansion is performed: (1) the elements that are added to a subdomain have lower priority numbers for other subdomains and (2) the elements of a subdomain do not have flange positions in relation to the region of the subdomain (loosely speaking). These effects make the number of boundary interior nodes of a subdomain low and its aspect ratio a desired value. The less the differences between the geometrical dimensions of a subdomain with a given area/volume, the smaller the boundary and the better the aspect ratio of the subdomain. However, this remark is true when the elements have good aspect ratios originally. For more details about the aspect ratio of a

320   OPTIMAL STRUCTURAL ANALYSIS

subdomain, see the recent paper of Farhat et al. [46], in which their final choice has been to compute the aspect ratio AR of a subdomain M_i as follows:

$$AR(M_i) = c_2 \times \frac{\text{Surface }(M_i)}{\text{Surface of circumscribed circle}} \text{ (two-dimensional problems)}$$

$$AR(M_i) = c_3 \times \frac{\text{Volume }(M_i)}{\text{Volume of circumscribed sphere}} \text{ (three-dimensional problems)}$$

(7-22)

where c_2 and c_3 are scaling constants designed such that $0 < AR \leq 1$.

Step 6 is executed to initialise the subdomains M_i ($i = 1, ..., q$) and their weights and to mask their first (starting) elements. The elements of a subdomain are masked only to forbid their repeated selection. Step 7 contains the expansion process of the algorithm. In every execution of this step, an element with maximum priority number corresponding to a subdomain M_i is added to M_i, where M_i is the subdomain with the current minimum weight. This method of expansion leads to equal loads for subdomains such that the subdomains remain expandable, and this condition is provided in the process of selecting e_s^i ($i = 1, ..., q$) and giving a priority number to an element corresponding to the subdomain being formed. The priority number defined in this step is simply designed to give more priority to an element connected to a subdomain M_k with more corner nodes in comparison with an element connected to M_k with less corner nodes having the same integers.

7.4.9 EXAMPLE

Consider the simple finite element mesh, as shown in Figure 7.11(a), with each node having two DOF and assume that it is decomposed into three subdomains. The steps of the present algorithm are performed as follows:

Step 1: A genre structure is rooted from an arbitrary element such as the element 15. The elements of each genre are recognised with the index of the genre as illustrated in Figure 7.11(b). The last genre, $g_8(15)$, contains the element 6; hence $e_s^1 = 6$.

Step 2: $G(6)$ is formed to calculate the pseudo-distance between the element 6 and other elements. The elements of each genre are assigned with the index of the genre; this index is same as the pseudo-distances between the root and the elements of the genre. In Figure 7.11(c) the pseudo-distance between the root (element 6) and other elements are depicted; the element 19 belongs to the last genre of $G(6)$, having the highest pseudo-distance from the root, and thus $e_s^2 = 19$.

DECOMPOSITION VIA GRAPH THEORY METHODS   321

Step 3: $G(19)$ is generated, and the pseudo-distances between the root and the elements are shown in Figure 7.11(d). Since $q > 2$, Step 4 should be executed.

Step 4: Two elements 2 and 23 satisfy the condition of this step, since

$$2 \in g_{10}(6) \text{ and } g_7(19)$$

$$IV_6(9) = 16 \text{ , } IV_{19}(6) = 11$$

$$23 \in g_7(6) \text{ and } g_{10}(19)$$

$$IV_6(6) = 11 \text{ , } IV_{19}(9) = 16,$$

and

$$\min\{IV_6(i), IV_{19}(j)\} < 11,$$

where

$$0 \le i, j \le 16 \text{ and } (i, j) \ne (9, 6) \text{ and } (6, 9).$$

Element 2 or 23 can be selected for e_s^3 arbitrarily; suppose $e_s^3 = 23$. Figure 7.11(e) shows the pseudo-distances between e_s^3 and the other elements.

Step 5: For each element, three integers are assigned corresponding to e_s^3, e_s^2 and e_s^3. These integers are illustrated in Figure 7.11(f) for each element.

Step 6: Execution of this step leads to $M_1 = \{6\}$, $M_2 = \{19\}$ and $M_3 = \{23\}$. The weights of M_1, M_2 and M_3 are the same and equal to 8, and their elements are masked.

Step 7: This step is carried out $\lambda - q = 21$ times, and in each execution one element with maximum priority number is added to a subdomain with the current minimum weight as follows:

All subdomains have the same weight; hence the subdomain M_1 is selected arbitrarily for expansion. The elements with non-zero priority numbers that are connected to M_1 are 5, 11 and 12, and their priority numbers are 2×29, 1×25 and 2×27, respectively. Thus, element 5 is added to M_1 and is masked. The weight of M_1 is now equal to 12. The subdomains M_2 and M_3 have the minimum current weight. The subdomain M_2 is selected arbitrarily for expansion. The elements 13, 14 and 20 are connected to M_2, and their priority numbers are 2×34, 1×29 and

322 OPTIMAL STRUCTURAL ANALYSIS

2 × 29, respectively. Hence the element 13 is added to M_2 and is masked. The current weight of M_2 is now equal to 12. The subdomain M_3 has the least current weight. The elements 16, 17, 18, 22 and 24 are connected to M_3, and their priority numbers are 1 × 27, 2 × 27, 1 × 25, 2 × 29 and 2 × 29, respectively. The priority numbers of the elements 22 and 24 are maximum; however, element 24 is added to M_3 because the sum of its integers is less than that of the element 22. The element 24 is masked. The weight of the subdomain M_3 is now equal to 12. The repetitions of this step lead to the decomposition as illustrated in Figure 7.11(g).

1	2	3	4	5	6
7	8	9	10	11	12
13	14	15	16	17	18
19	20	21	22	23	24

(a) A simple two-dimensional FEM.

5	4	3	4	5	8
4	2	1	2	4	7
3	1	0	1	3	6
4	2	1	2	4	7

(b) Genres of $G(15)$.

14	10	6	3	1	0
15	11	7	4	2	1
16	12	8	5	4	3
17	13	9	8	7	6

(c) Genres of $G(6)$.

6	7	8	9	13	17
3	4	5	8	12	16
1	2	4	7	11	15
0	1	3	6	10	14

(d) Genres of $G(19)$.

13	9	8	7	6	7
12	8	5	4	3	4
11	7	4	2	1	2
10	6	3	1	0	1

(e) Genres of $G(23)$.

16.31 17	33.26 22	26.21 22	28.22 20	29.12 19	31.7 17
12.33 15	17.28 20	22.24 24	26.20 24	25.14 23	27.9 21
9.34 14	14.29 19	20.24 24	21.19 37	21.14 27	23.11 28
7.34 14	12.28 19	18.21 21	17.19 29	17.14 33	19.11 29

(f) Integers of the elements.

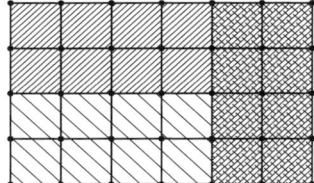

(g) Decomposition of the FEM for $q = 3$.

Fig. 7.11 Illustration of the steps for the example.

7.4.10 COMPLEXITY ANALYSIS OF THE ENGINEERING-BASED METHOD

In the following text, an efficient data structure is searched and the domain decomposition heuristic of this section is analysed for the worst case.

The element-node list of an FEM is usually used as input for data connectivity of elements, and here such a list is used. However, for efficient programming of algorithms such as the present one, this list should not be used exclusively, because finding elements connected to a specified element takes $O(\lambda\theta^2)$ operations, where θ denotes the maximum number of nodes of an element. Hence, another data structure (list) together with the element-node list should be used. This list should (1) be addressed simply, (2) make high reduction in the time complexity of the algorithm, (3) require low computer space and (4) be generated with a low number of operations.

Two lists commonly employed are the node-element list and the element adjacency list. These are compared with respect to the above four conditions (for the present algorithm) as follows:

1. They can both be simply addressed.

2. The element adjacency list leads to further reduction in the time complexity. For example, it needs $O(\Delta)$ operations for finding elements connected to a specified element, and the node-element list takes $O(\theta\varepsilon)$ operations, where Δ and ε denote the maximum number of elements connected to an element and to a node, respectively.

3. In most FEMs for which the number of nodes α is higher than λ, the element adjacency list takes less computer memory than the node-element list.

4. Using the element-node list, generation of the node-element and element adjacency lists takes $O(\lambda\theta)$ and $O(\lambda^2\theta^2)$ operations, respectively. The element

adjacency list is inefficient only from this point of view. Hence a method with low time complexity is needed for generating the element adjacency list. In the following text, a method is proposed for the formation of the element adjacency list, which has time complexity $O(\lambda\omega\theta^2)$, where ω denotes the maximum number of elements contained in a level of an arbitrary level structure.

Step I: Generate the node-element list of the FEM.

Step II: Using the element-node and node-element lists, form a level structure rooted from an arbitrary element r.

Step III: Erase the node-element list.

Step IV. For each element ($i = 1, ..., \lambda$) contained in level l_j, control levels $l_a, ..., l_b$ to find elements connected to i, where, if $j = 0$, then $a = 0$ and $b = 1$; else, if $0 < j < d$, then $a = j - 1$ and $b = j + 1$; else $a = j - 1$ and $b = j$.

In Step III, the node-element list is erased from the working memory to provide enough memory for the formation of the element adjacency list. However, one can generate the node-element list and then form the element adjacency list directly. This scheme takes $O(\alpha\Delta\varepsilon^2)$ operations. This method is efficient from the point of time requirement; however, it needs the element-node, node-element and element adjacency lists to be in the working memory simultaneously. Hence, it is proposed that the element adjacency list be generated by this method when the considered FEM is not a very large-scale model; otherwise Steps I–IV should be performed.

Formation of the element adjacency list, using the element-node list only (which has time complexity $O(\lambda^2\theta^2)$), can also be improved such that the maximum difference δ between the labels of the elements with a common node is given. In this case, the time complexity reduces to $O(\lambda\delta\theta^2)$; hence ordering of elements should be performed in the process of mesh generation.

Generation of a genre structure takes $O(D\lambda\Delta\theta^2)$ operations; however, one can reduce it to $O(D\lambda\Delta)$ by saving the number of common corner nodes of each pair of elements when the element adjacency list is formed. Therefore, an additional vector that has the same size as the element adjacency list is needed. Thus, Steps 1, 2 and 3 have time complexity $O(D\lambda\Delta)$, where D denotes the dimension of the considered FEM and varies from 1 to 3.

In each execution of Step 4, one should find an element e_s^i in which the least number of selected elements of each generated genre structure contained in the previous genres (before e_s^i is selected as a member of the next genre of each genre structure) is maximum; hence the time complexity of this step is $O(\lambda q^2)$. Step 5 also has the time complexity $O(\lambda q^2)$. The time complexity of Step 6 is $O(q\theta)$. Step 7 has time complexity $O(\lambda^2 \theta)$, and is the critical step of the algorithm.

The time complexity of the critical step of the algorithm shows that this algorithm is efficient for FEMs with high-order elements as well as for FEMs with low-order elements. It depends only on the number of elements of the FEM (λ) and the maximum number of nodes of an element (θ).

7.4.11 COMPUTATIONAL RESULTS OF THE ENGINEERING-BASED METHOD

Two examples are studied in this section, using the direct method for the formation of their element adjacency list.

Example 1: A multi-connected finite element mesh is shown in Figure 7.12(a), and decomposed into 2, 3, 4, 8 and 16 subdomains as illustrated in Figures 7.12 (b–f). In this example, each node has two DOF. The computational time is provided in Table 7.4.

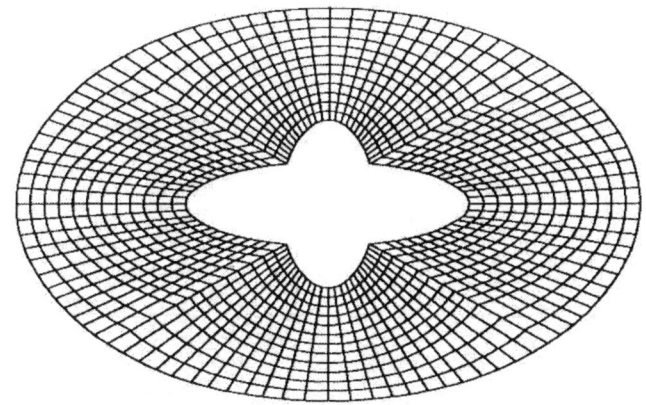

(a) A multi-connected FEM with 1152 elements and 1248 nodes.

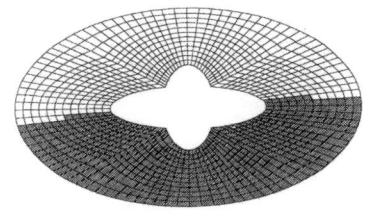

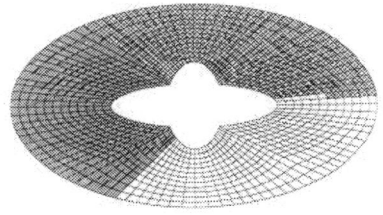

(b) $q = 2$ (c) $q = 3$

326 OPTIMAL STRUCTURAL ANALYSIS

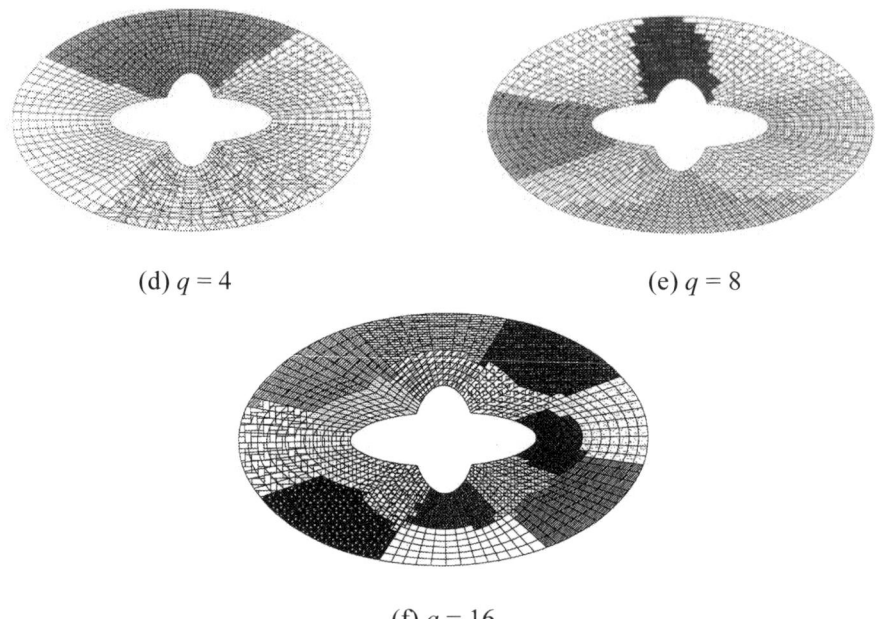

(d) $q = 4$ (e) $q = 8$

(f) $q = 16$

Fig. 7.12 Decompositions of the multi-connected finite element mesh.

Table 7.4 Computational time.

q	2	3	4	8	16
Time (s)	29.00	29.28	29.44	30.59	33.39

Example 2: A multi-connected H-shaped finite element mesh with each node having two DOF is shown in Figure 7.13(a), and decomposed into 2, 4, 5, 8, 16 and 32 subdomains as illustrated in Figures 7.13(b–g). The computational time is provided in Table 7.5.

(a) A multi-connected H-shaped FEM with 1340 elements and 1042 nodes.

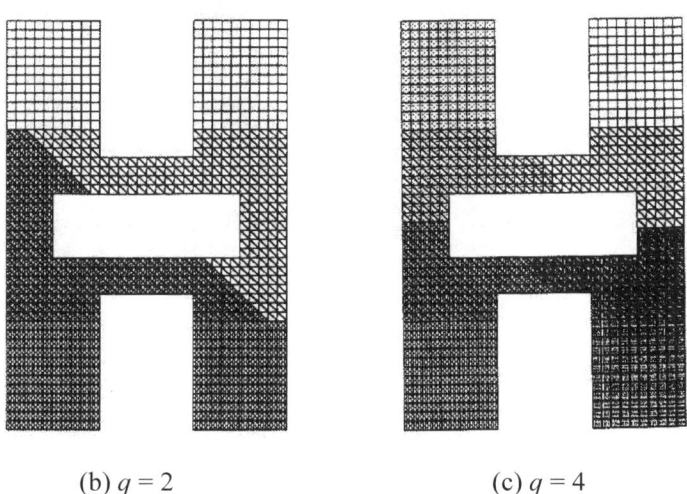

(b) $q = 2$ (c) $q = 4$

328 OPTIMAL STRUCTURAL ANALYSIS

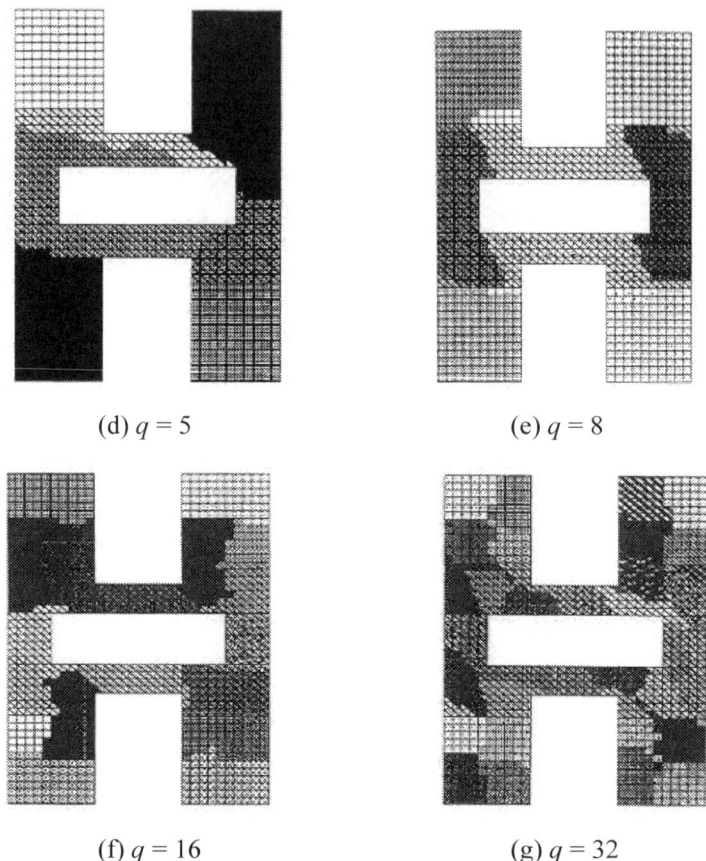

(d) $q = 5$ (e) $q = 8$

(f) $q = 16$ (g) $q = 32$

Fig. 7.13 Decompositions of a multi-connected H-shaped finite element mesh.

Table 7.5 Computational time.

q	2	4	5	8	16	32
Time (s)	33.95	31.75	32.90	37.14	40.70	52.79

7.4.12 DISCUSSIONS

The algorithm developed in this section is designed as a pre-processor for concurrent finite element computations. It may also serve as an automatic decomposer for serial solutions on a sequential computer to overcome limited core storage capacity. This algorithm has low time complexity and leads to efficient partitioning of a finite element mesh into subdomains with required properties. A finite element mesh to be partitioned may contain various meshes with different dimensions, types and sizes. The algorithm uses a simultaneous expansion process, which is an improved version of the algorithm presented in the previous section for substructuring. In this algorithm, the method for selecting the first (representative) element for each subdomain is improved, and the better priority numbers for elements to be added to the expanding subdomains are defined to form subdomains with more appropriate properties.

This algorithm is designed to have properties required for an efficient decomposition and leads to subdomains with the following properties:

1. Low computer space and time requirements. In the present algorithm, only the corner nodes need to be given, and this leads to a large space saving in FEMs with high-order elements. The time complexity of the algorithm is independent of the number of nodes for the considered FEM, and the critical step of the algorithm takes $O(\lambda^2\theta)$ operations in the worst case.

2. General in use. The algorithm can be employed to decompose unstructured FEMs without any restriction, and an arbitrary parameter can be considered as the loads of the subdomains.

3. Balance loads for subdomains. Selection of the starting elements of subdomains and the expansion process are performed in a manner that leads to an efficient balancing of loads. However, to decrease the differences between the loads of subdomains, the following steps are included, which should be executed in place of Steps 1–4 of the original algorithm:

(a) Find the pseudo-distance between each element and all the elements of the finite element mesh.

(b) Find q elements $e_s^1, e_s^2, ..., e_s^q$, provided e_s^i ($i = 1, ..., q$), which is contained in genres $g_{j1}(e_s^1)$, $g_{j1}(e_s^2)$, ..., $g_{j1}(e_s^q)$, is selected in such a way that the least value of $IV(j_k - 1)$ is maximum, where $j_k \neq 0$ ($k = 1, ..., q$).

However, this takes more operations than those of Steps 1-4.

4. Close to minimum number of interface nodes. In this algorithm, the number of interface nodes is kept to the least possible by selecting the elements to be added to a subdomain which have not high priority numbers for the other subdomains, and have a proper position in relation to the previously selected elements of the subdomains.

5. Good aspect ratios for subdomains. When the elements of the considered finite element mesh have aspect ratios with proper values, the algorithm leads to a

330   OPTIMAL STRUCTURAL ANALYSIS

decomposition with subdomains having reasonable aspect ratios. This is because the subdomains are expanded in all directions, which makes the denominators of the equations introduced by Farhat et al. [46] to be increased.

7.5 SUBSTRUCTURING: FORCE METHOD

The force method can be employed in parallel analysis of structures. In this section, the formulation of substructuring is provided, and an algorithm is presented for such analysis. The computational process is illustrated using simple examples.

In this section, the notations and formulations presented in Chapter 3 will be used.

7.5.1 ALGORITHM FOR THE FORCE METHOD SUBSTRUCTURING

Once a structural model has been decomposed using any of the methods presented in the previous sections, the following approach can be used for the analysis employing the force method:

To support a substructure in a statically determinate fashion, cuts are introduced at members incident with the interface nodes contained in the corresponding substructure, except at one arbitrary node where the substructure is connected to the previous one.

For a given substructure S_i, let the external forces be denoted by $\mathbf{p}_i$ and redundant forces by $\mathbf{q}_i$. Then the substructure S_i can be analysed for the internal forces in the substructure (not coupling redundants) $\mathbf{q}_i$ in the aforementioned manner, that is,

$$\begin{bmatrix} \mathbf{v}_{0i} \\ \mathbf{v}_{1i} \end{bmatrix} = \begin{bmatrix} \mathbf{D}_{00} & \mathbf{D}_{01} \\ \mathbf{D}_{10} & \mathbf{D}_{11} \end{bmatrix}_i \begin{bmatrix} \mathbf{p}_i \\ \mathbf{q}_i \end{bmatrix}. \tag{7-23}$$

For continuity within the substructure,

$$\mathbf{q}_i = -(\mathbf{D}_{11}^{-1} \mathbf{D}_{10})_i \mathbf{p}_i. \tag{7-24}$$

Deflections corresponding to the nodal force are

$$\mathbf{v}_{0i} = (\mathbf{D}_{00} - \mathbf{D}_{10}^t \mathbf{D}_{11}^{-1} \mathbf{D}_{10})_i \mathbf{p}_i, \tag{7-25}$$

that is,

$$\mathbf{v}_{0i} = \mathbf{F}_i \mathbf{p}_i, \tag{7-26}$$

where $\mathbf{F}_i$ is the flexibility transformation matrix for the ith substructure. Internal forces are obtained as

$$\mathbf{r}_i = (\mathbf{B}_0 - \mathbf{B}_1 \mathbf{D}_{11}^{-1} \mathbf{D}_{10})_i \mathbf{p}_i, \qquad (7\text{-}27)$$

or

$$\mathbf{r}_i = \mathbf{B}_i \mathbf{p}_i \qquad (7\text{-}28)$$

and

$$\mathbf{B}_i = (\mathbf{B}_0 - \mathbf{B}_1 \mathbf{D}_{11}^{-1} \mathbf{D}_{10})_i, \qquad (7\text{-}29)$$

where $\mathbf{B}_i$ is the force transformation matrix in the redundant substructure. The matrices $\mathbf{F}_i$ and $\mathbf{B}_i$ are formed for each substructure, in turn.

For the complete structure S composed of q substructures (S_1, S_2, ..., S_q), the force vector $\mathbf{p}_i$ acting on a substructure "s" is given by

$$\mathbf{p}_{si} = [\mathbf{a}_e \quad \mathbf{b}_e] \begin{bmatrix} \mathbf{p}_e \\ \mathbf{q}_c \end{bmatrix}, \qquad (7\text{-}30)$$

where $\mathbf{q}_c$ are the coupling redundants. On a particular substructure, there will be three different types of forces: $\mathbf{p}_{ee}$ is the external force vector, $\mathbf{p}_{ec}$ is the coupling redundant forces vector and $\mathbf{p}_{eb}$ contain the statically determinate connection forces.

For the entire structure, the following matrices $\mathbf{A}_{ee}$ and $\mathbf{B}_{ec}$ are defined:

$$\mathbf{A}^t = \{a_{e(1)}, a_{e(2)}, ..., a_{e(q)}\}, \qquad (7\text{-}31)$$

and

$$\mathbf{B}^t = \{b_{e(1)}, b_{e(2)}, ..., b_{e(q)}\}. \qquad (7\text{-}32)$$

Then,

$$\mathbf{p}_s = \begin{bmatrix} \mathbf{p}_{s(1)} \\ \mathbf{p}_{s(2)} \\ \vdots \\ \vdots \\ \mathbf{p}_{s(q)} \end{bmatrix} = \begin{bmatrix} \mathbf{A}_{ee} & \mathbf{B}_{ec} \end{bmatrix} \begin{bmatrix} \mathbf{p}_e \\ \mathbf{q}_c \end{bmatrix}. \qquad (7\text{-}33)$$

The forces $\mathbf{p}_s$ can be partitioned according to three types of forces $\mathbf{p}_{ei}$, $\mathbf{p}_{eb}$, and $\mathbf{p}_{ec}$ as mentioned earlier. Then,

$$\begin{bmatrix} \mathbf{p}_{ei} \\ \mathbf{p}_{eb} \\ \mathbf{p}_{ec} \end{bmatrix} = \begin{bmatrix} \mathbf{a}_{ei} & \mathbf{0} \\ \mathbf{a}_{eb} & \mathbf{b}_{eb} \\ \mathbf{a}_{ec} & \mathbf{b}_{ec} \end{bmatrix} \begin{bmatrix} \mathbf{p}_e \\ \mathbf{q}_c \end{bmatrix}. \qquad (7\text{-}34)$$

It is obvious that, whereas $\mathbf{q}_c$ may produce $\mathbf{p}_{eb}$ and $\mathbf{p}_{ec}$ forces, it does not produce $\mathbf{p}_{ei}$ forces.

The flexibility matrix of the entire structure corresponding to $\mathbf{p}_e$ and $\mathbf{q}_c$ can be formed using Eq. (7-25) and Eq. (7-33) as follows:

$$\begin{bmatrix} \mathbf{f}_{ee} & \mathbf{f}_{ec} \\ \mathbf{f}_{ce} & \mathbf{f}_{cc} \end{bmatrix} = \begin{bmatrix} \mathbf{A}_{ee}^t \\ \mathbf{A}_{ec}^t \end{bmatrix} \begin{bmatrix} \mathbf{F}_{e(1)} & & \\ & \ddots & \\ & & \mathbf{F}_{e(q)} \end{bmatrix} \begin{bmatrix} \mathbf{A}_{ee} & \mathbf{B}_{ec} \end{bmatrix}, \qquad (7\text{-}35)$$

and

$$\begin{bmatrix} \mathbf{v}_e \\ \mathbf{v}_c \end{bmatrix} = \begin{bmatrix} \mathbf{f}_{ee} & \mathbf{f}_{ec} \\ \mathbf{f}_{ce} & \mathbf{f}_{cc} \end{bmatrix} \begin{bmatrix} \mathbf{p}_e \\ \mathbf{p}_c \end{bmatrix}. \qquad (7\text{-}36)$$

For continuity across the cut sections of the structure,

$$\mathbf{v}_c = 0. \qquad (7\text{-}37)$$

Hence,

$$\mathbf{q}_c = -\mathbf{f}_{cc}^{-1} \mathbf{f}_{ce} \mathbf{p}_e \qquad (7\text{-}38)$$

DECOMPOSITION VIA GRAPH THEORY METHODS   333

Deflections of the structure are then given as,

$$\mathbf{v}_e = (\mathbf{f}_{ee} - \mathbf{f}_{ec}\mathbf{f}_{cc}^{-1}\mathbf{f}_{ce})\mathbf{p}_e, \qquad (7\text{-}39)$$

making the complete analysis of the structure feasible.

7.5.2 EXAMPLES

Example 1: A single-bay four-storey frame is considered, as shown in Figure 7.14. The forces are depicted in Figure 7.14(a) and the nodal and element orderings are given in Figure 7.14(b). For this frame, $I = 41{,}623.14$ cm^4 (for all members) and $E = 2.1 \times 10^5$ N/m^2.

The model is decomposed into two substructures as illustrated in Figure 7.15. The analysis is performed and the bending moments are obtained as provided in Table 7.6.

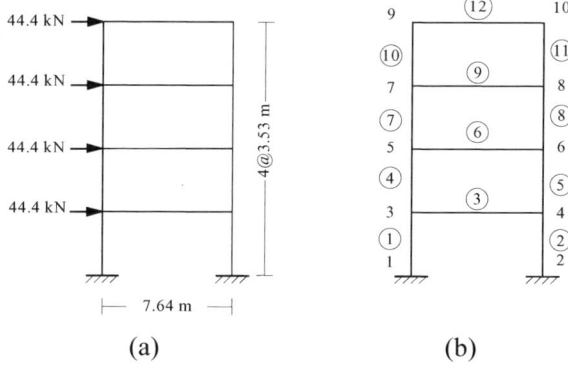

(a)   (b)

Fig. 7.14 A single-bay four-storey frame with geometric and connectivity properties.

Table 7.6 Bending moments of Example 1.

Nodes	End nodes of members	Bending moments(kN.m)
1	1-3	−219.17
	3-1	78.93
3	3-4	185.07
	3-5	−106.14
	5-3	0.34
5	5-6	2.94
	5-7	7.34

334 OPTIMAL STRUCTURAL ANALYSIS

	7	7-5	96.2
		7-8	112.7
		7-9	−16.48
	9	9-7	58.04
		9-10	58.04
	10	10-9	58.04
		10-8	−58.04
	8	8-10	−16.48
		8-7	112.7
		8.6	−96.2
	6	6-8	−52.84
		6-5	170.28
		6-4	−117.43
	4	4-6	−78.93
		4-3	185.07
		4-2	−106.14
	2	2-4	−219.17

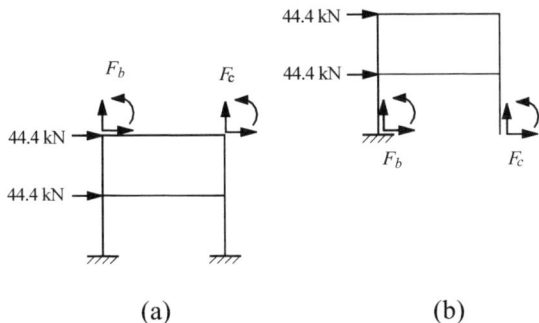

(a)   (b)

Fig. 7.15 Decomposition of the structural model.

Example 2: A three-bay pitched-roof frame together with material properties and dimensions is shown in Figure 7.16.

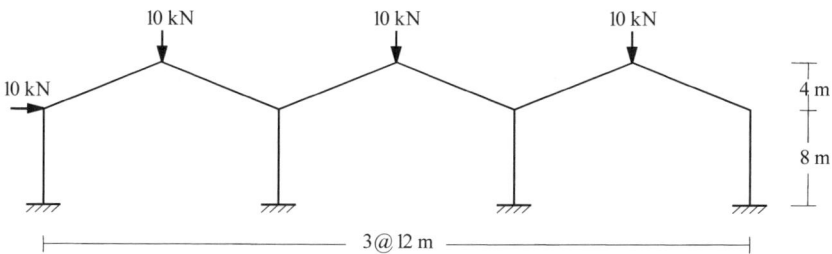

Fig. 7.16 A three-bay pitched-roof frame.

This model is partitioned into two substructures, as illustrated in Figure 7.17, where different groups of loads on each substructure are shown. For all the members, $I = 0.2 \text{ m}^4$ and $E = 2.1 \times 10^5 \text{ N/m}^2$. The bending moments for members of this frame are presented in Table 7.7.

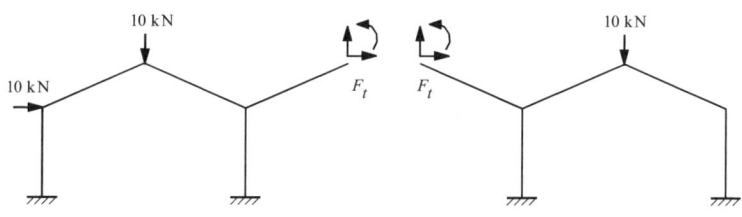

Substructure I   Substructure II

Fig. 7.17 Decomposition of the structural model.

Table 7.7 Bending moments of Example 2.

Nodes	End nodes of members	Bending moments(kN m)
1	1-5	−72
5	5-1	−30
	5-6	30
6	6-5	3
	6-7	3
	7-6	26
7	7-8	65
	7-2	−91
2	2-7	−116
8	8-7	45
	8-8	45

	9-8	28
9	9-10	3.03
	9-3	60
3	3-9	40.02
10	10-9	52
	10-11	52
11	11-10	79

The substructuring analysis, using the force method for frame structures, can be generalised to the analysis of other types of structures when the algebraic force method is employed; see Plemmons and White [179]. In this method, appropriate partitioning of the incidence matrices of the structural graph models is performed, leading to well-structured equilibrium equations. It is proved that sparse null bases can then be constructed in parallel, using the proposed decomposition. The performance of the method is illustrated by some examples from skeletal structures.

7.6 SUBSTRUCTURING FOR DYNAMIC ANALYSIS

Many substructuring methods have been developed, such as those of Hurty [85], Hale and Meirovitch [70], and Craig and Chang [30]. In these methods, the substructure modes, geometric compatibility and force equilibrium equations are used to assemble the system matrix. For analysis with many substructures, the system matrix of the entire structure may be large, and therefore the number of calculations for the solution increases. Benfield and Hruda [11] and Rubin [191] and Hintz [81] studied component modal synthesis by transformations or approximations. In the latter, the order of system matrix is also reduced, but accuracy is decreased. The projection matrix can be used for reducing the order of the system matrix to the relevant DOF. This method requires many matrix inversions. A substructure modal analysis for determining the exact modal parameters of the synthesised structure, based on the method developed by Yee and Tsuei [233], will now be presented, and applied to the analysis of some simple examples.

7.6.1 MODAL ANALYSIS OF A SUBSTRUCTURE

Once a system has been divided into q substructures, each substructure will have its own equation of motion and can be connected to other substructures through nodes.

The equation of motion for a substructure can be written as

$$\mathbf{M}\ddot{\mathbf{x}}(t) + \mathbf{K}\mathbf{x}(t) = \mathbf{f}(t), \qquad (7\text{-}40)$$

where $\mathbf{M}$ is the mass matrix, $\mathbf{x}(t)$ is the response vector in the time domain and $\mathbf{f}(t)$ is the force vector in the time domain.

The natural frequencies and mode shapes of the substructure are then obtained by the solution of the following differential equation:

$$\mathbf{M}\ddot{\mathbf{x}}(t) + \mathbf{K}\mathbf{x}(t) = 0. \tag{7-41}$$

Let λ_r and ϕ_r be the rth mode eigenvalue and eigenvector, with the following orthogonality properties:

$$\mathbf{\Phi}^t \mathbf{M} \mathbf{\Phi} = \mathbf{I}, \text{ and } \mathbf{\Phi}^t \mathbf{K} \mathbf{\Phi} = \mathbf{\Lambda}, \tag{7-42}$$

where $\mathbf{\Lambda}$ is a diagonal matrix of eigenvalues and

$$\mathbf{\Phi} = [\phi_1 \; \phi_2 \; \ldots \; \phi_n] \tag{7-43}$$

is the modal matrix obtained by solving Eq. (7-40); the displacement vector $\mathbf{x}(t)$ can be written as

$$\mathbf{x}(t) = \mathbf{\Phi}\mathbf{p}(t), \tag{7-44}$$

where $\mathbf{p}(t)$ is the modal coordinate vector of the substructure. By this transformation, Eq. (7-40) becomes

$$\mathbf{M}\mathbf{\Phi}\ddot{\mathbf{p}}(t) + \mathbf{K}\mathbf{\Phi}\,\mathbf{p}(t) = \mathbf{f}(t). \tag{7-45}$$

For harmonic excitation at a frequency ω, which is different from the natural frequency λ, the response function can be expressed as

$$\mathbf{x}(t) = \mathbf{\Phi}[\mathbf{\Lambda} - \omega^2 \mathbf{I}]^{-1} \mathbf{\Phi}^t \mathbf{f}(t), \tag{7-46}$$

or

$$\mathbf{x}(t) = \mathbf{H}(\omega)\mathbf{f}(t), \tag{7-47}$$

where $f(t) = fe^{i\omega t}$ and $x(t) = xe^{i\omega t}$ is used. It should be noted that vector $\mathbf{x}$ is a function of frequency. Since $[\mathbf{\Lambda} - \omega^2 \mathbf{I}]$ is a diagonal matrix, the transfer matrix $\mathbf{H}(\omega) = \mathbf{\Phi}[\mathbf{\Lambda} - \omega^2 \mathbf{I}]^{-1} \mathbf{\Phi}^t$ can be obtained by simple multiplication when the eigensolution $\mathbf{\Phi}$ of the substructure is provided.

7.6.2 PARTITIONING OF THE TRANSFER MATRIX H(w)

Owing to the nature of Eq. (7-47), the complete substructure model set is not needed to analyse the system dynamics at a given frequency range. As an example, if the **H**(w) of a substructure at a frequency range $w_b < w < w_c$ is desired, the substructure mode below w_a and above w_d can be excluded, where $w_a \ll w_b$ and $w_d \gg w_c$. The contribution of the modes above w_d and below w_a have no considerable effect on the transfer matrix **H**(w) at the given frequency range. Hence, only the substructure modes within the frequency range $w_a < w < w_d$ need to be considered. For a complicated substructure, these modes are usually obtained using a finite element analysis or one can also use the test results on the models.

There is no universal rule for selecting the lower and upper frequency bounds of the modes to be included in the analysis. The frequency bounds selected depend on the degree of accuracy required for the results of the analysis.

Equation (7-47) for substructure i can be partitioned as

$$\begin{bmatrix} \mathbf{x}_{ii} \\ \mathbf{x}_{ij} \\ \cdots \\ \cdots \\ \cdots \\ \mathbf{x}_{ik} \end{bmatrix} = \begin{bmatrix} \mathbf{H}^i_{ii}(\omega) & \mathbf{H}^i_{ij}(\omega) & \cdots & \cdots & \cdots & \mathbf{H}^i_{ii}k(\omega) \\ \mathbf{H}^i_{ji}(\omega) & \mathbf{H}^i_{jj}(\omega) & \cdots & \cdots & \cdots & \mathbf{H}^i_{jk}(\omega) \\ \cdots & \cdots & & & & \cdots \\ \cdots & \cdots & & & & \cdots \\ \cdots & \cdots & & & & \cdots \\ \mathbf{H}^i_{ki}(\omega) & \mathbf{H}^i_{kj}(\omega) & \cdots & \cdots & \cdots & \mathbf{H}^i_{kk}(\omega) \end{bmatrix} \begin{bmatrix} \mathbf{f}_{ii} \\ \mathbf{f}_{ij} \\ \cdots \\ \cdots \\ \cdots \\ \mathbf{f}_{ik} \end{bmatrix}, \qquad (7\text{-}48)$$

where $\mathbf{x}_{ij}$ is the response (displacement) vector and $\mathbf{f}_{ij}$ is the force vector, common to substructures i and j, and superscript i refers to substructure i.

7.6.3 DYNAMIC EQUATION OF THE ENTIRE STRUCTURE

Substructure modes, geometric compatibility and force equilibrium provide sufficient tools to set up the dynamic equation of the entire structure.

For the structure, the dynamic equation can be written as

$$\tilde{\mathbf{M}}\ddot{\mathbf{q}} + \tilde{\mathbf{K}}\mathbf{q} = 0 . \qquad (7\text{-}49)$$

If the structure is composed of q substructures and substructure i has N_i substructure modes, the total number of substructures modes N_t will be as follows:

$$N_t = \sum_{i=1}^{q} N_i. \qquad (7\text{-}50)$$

For a structure with N_t substructure modes and r constraint equations, the final dynamic equation of the entire structure is of the order $N_t - r$. For a complex structure with many substructure modes, the eigen solution for Eq. (7-49) may require a considerable amount of numerical computation. Therefore, a different approach can be used as follows:

Consider a structure with four substructures as shown in Figure 7.18. Substructures 1 and 3 are connected to substructure 2, but not to each other, and substructure 3 is connected to substructure 2 and 4, which are connected, but not to each other. For natural vibration, the external forces acting on the system are zero.

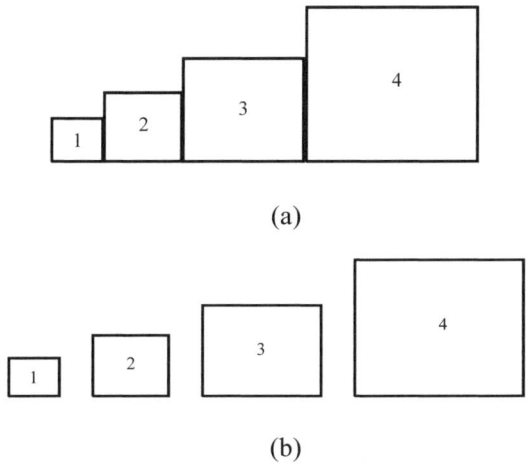

Fig. 7.18 Synthesised system and its four substructures.

The response equations for substructures 1-4 can be written, using Eq. (7-48), as follows:

$$\mathbf{x}^1 = \begin{bmatrix} \mathbf{x}_{11} \\ \mathbf{x}_{12} \end{bmatrix} = \begin{bmatrix} \mathbf{H}_{11}^1(\omega) & \mathbf{H}_{12}^1(\omega) \\ \mathbf{H}_{21}^1(\omega) & \mathbf{H}_{22}^1(\omega) \end{bmatrix} \begin{bmatrix} \mathbf{0} \\ \mathbf{f}_{12} \end{bmatrix} \qquad (7\text{-}51)$$

$$\mathbf{x}^2 = \begin{bmatrix} \mathbf{x}_{22} \\ \mathbf{x}_{21} \\ \mathbf{x}_{23} \end{bmatrix} = \begin{bmatrix} \mathbf{H}_{11}^2(\omega) & \mathbf{H}_{12}^2(\omega) & \mathbf{H}_{13}^2(\omega) \\ \mathbf{H}_{21}^2(\omega) & \mathbf{H}_{22}^2(\omega) & \mathbf{H}_{23}^2(\omega) \\ \mathbf{H}_{31}^2(\omega) & \mathbf{H}_{32}^2(\omega) & \mathbf{H}_{33}^2(\omega) \end{bmatrix} \begin{bmatrix} \mathbf{0} \\ \mathbf{f}_{21} \\ \mathbf{f}_{23} \end{bmatrix} \qquad (7\text{-}52)$$

340 OPTIMAL STRUCTURAL ANALYSIS

$$\mathbf{x}^3 = \begin{bmatrix} \mathbf{x}_{33} \\ \mathbf{x}_{32} \\ \mathbf{x}_{34} \end{bmatrix} = \begin{bmatrix} \mathbf{H}^3_{11}(\omega) & \mathbf{H}^3_{12}(\omega) & \mathbf{H}^3_{13}3(\omega) \\ \mathbf{H}^3_{21}(\omega) & \mathbf{H}^3_{22}(\omega) & \mathbf{H}^3_{23}(\omega) \\ \mathbf{H}^3_{31}(\omega) & \mathbf{H}^2_{32}(\omega) & \mathbf{H}^3_{33}(\omega) \end{bmatrix} \begin{bmatrix} 0 \\ \mathbf{f}_{32} \\ \mathbf{f}_{34} \end{bmatrix} \quad (7\text{-}53)$$

$$\mathbf{x}^4 = \begin{bmatrix} \mathbf{x}_{44} \\ \mathbf{x}_{43} \end{bmatrix} = \begin{bmatrix} \mathbf{H}^4_{11}(\omega) & \mathbf{H}^4_{12}(\omega) \\ \mathbf{H}^4_{21}(\omega) & \mathbf{H}^4_{22}(\omega) \end{bmatrix} \begin{bmatrix} 0 \\ \mathbf{f}_{43} \end{bmatrix}. \quad (7\text{-}54)$$

The conditions of geometric compatibility and force equilibrium can be expressed as follows:

$$\mathbf{x}_{ij} = \mathbf{x}_{ji} \text{ and } \mathbf{f}_{ij} = -\mathbf{f}_{ji}. \quad (7\text{-}55)$$

System synthesis and the modal force matrix can be obtained by retaining only the nodal coordinates:

$$\mathbf{x}_{12} = \mathbf{H}^1_{22}(\omega)\mathbf{f}_{12} \quad (7\text{-}56a)$$

$$\mathbf{x}_{21} = \mathbf{H}^2_{22}(\omega)\mathbf{f}_{21} + \mathbf{H}^2_{23}(\omega)\mathbf{f}_{23} \quad (7\text{-}56b)$$

$$\mathbf{x}_{23} = \mathbf{H}^2_{32}(\omega)\mathbf{f}_{21} + \mathbf{H}^2_{33}(\omega)\mathbf{f}_{23} \quad (7\text{-}56c)$$

$$\mathbf{x}_{32} = \mathbf{H}^3_{22}(\omega)\mathbf{f}_{32} + \mathbf{H}^3_{23}(\omega)\mathbf{f}_{34} \quad (7\text{-}56d)$$

$$\mathbf{x}_{34} = \mathbf{H}^3_{32}(\omega)\mathbf{f}_{32} + \mathbf{H}^3_{33}(\omega)\mathbf{f}_{34} \quad (7\text{-}56e)$$

$$\mathbf{x}_{43} = \mathbf{H}^4_{22}(\omega)\mathbf{f}_{43} \quad (7\text{-}56f)$$

Applying the conditions of compatibility and equilibrium at the interface nodes of the substructures, we have

$$\begin{bmatrix} \mathbf{H}^1_{22}(\omega) + \mathbf{H}^2_{22}(\omega) & -\mathbf{H}^2_{23}(\omega) & 0 \\ -\mathbf{H}^2_{32}(\omega) & \mathbf{H}^2_{33}(\omega) + \mathbf{H}^3_{22}(\omega) & -\mathbf{H}^3_{23}(\omega) \\ 0 & -\mathbf{H}^3_{32}(\omega) & \mathbf{H}^3_{33}(\omega) + \mathbf{H}4^2_{22}(\omega) \end{bmatrix} \begin{bmatrix} \mathbf{f}_{12} \\ \mathbf{f}_{23} \\ \mathbf{f}_{34} \end{bmatrix} = 0, \quad (7\text{-}57)$$

or

$$\overline{\mathbf{H}}\mathbf{f} = 0, \quad (7\text{-}58)$$

where $\bar{\mathbf{H}}$ is defined as the modal force matrix and $\bar{\mathbf{f}}$ is the modal force vector.

The next set of Equations (7-59) is called the *set of modal force* equations. These are the most important equations in this method. They satisfy the compatibility, equilibrium and substructure response equations of the entire structure. The order of the modal force equation depends on the number of interface nodes of the structure. The transfer matrix $\mathbf{H}(w)$ is a function of the frequency of the entire system and is obtained by simple multiplication.

Once the modal force matrix has been formed, the natural frequencies of the entire structure are evaluated by equating its determinant to zero as follows:

$$\det \begin{bmatrix} \mathbf{H}_{22}^1(\omega)+\mathbf{H}_{22}^2(\omega) & -\mathbf{H}_{23}^2(\omega) & 0 \\ -\mathbf{H}_{32}^2(\omega) & \mathbf{H}_{33}^2(\omega)+\mathbf{H}_{22}^3(\omega) & -\mathbf{H}_{23}^3(\omega) \\ 0 & -\mathbf{H}_{32}^3(\omega) & \mathbf{H}_{33}^3(\omega)+\mathbf{H}4_{22}^2(\omega) \end{bmatrix} = 0. \quad (7\text{-}59)$$

Multiplying Eq. (7-59) by

$$\prod_{i=1}^{q} \prod_{j=1}^{n_i}(\lambda_j^i - \omega^2), \quad (7\text{-}60)$$

the following polynomial equation is obtained:

$$\sum_{j=0}^{n} a_j \omega^{2j} = 0, \quad (7\text{-}61)$$

where q is the number of substructures and the jth eigenvalue of the ith substructure is the total number of normal modes (total number of DOF) for the entire structure and a_j are the coefficients of the polynomial.

Once the value of ω_r is calculated from the polynomial equation (7-61), the eigenvector $\mathbf{f}$ for the rth mode can be obtained from the modal force equations (7-56) as

$$\tilde{\mathbf{f}} = \begin{bmatrix} \mathbf{f}_{12} \\ \mathbf{f}_{23} \\ \mathbf{f}_{34} \end{bmatrix}. \quad (7\text{-}62)$$

The rth mode eigenvector $\mathbf{x}$ is recovered by back substitution of the corresponding modal force vector $\mathbf{f}$ back in Eq. (7-51). If a natural frequency of the system is identical to that of the substructure, Eq. (7-48) cannot be employed, and a modified approach should be considered; see Yee and Tsuei [233].

7.6.4 EXAMPLES

Example 1: An industrial pitched-roof frame is considered, as shown in Figure 7.19. The structure is decomposed into two substructures. The information obtained from decomposition is provided in Table 7.8. The analysis is performed, and the frequencies are calculated for the first three modes and compared with the analysis without substructuring in Table 7.9. All the elements are selected as 2IPB40, having $A = 396$ cm^2, $I = 115{,}360$ cm^4, and the weight per unit length $G = 310$ kg/m.

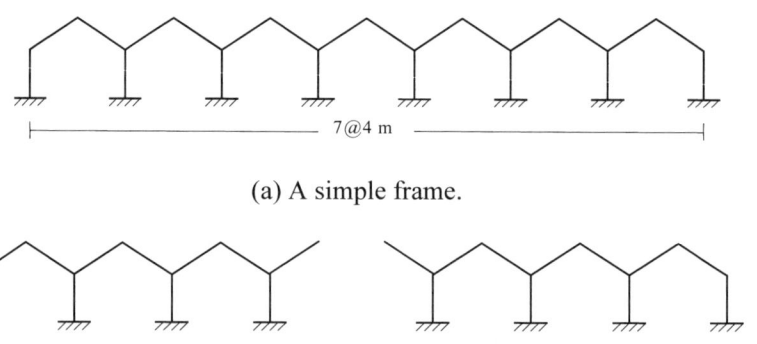

(a) A simple frame.

(b) Substructures of the frame.

Fig. 7.19 A simple industrial frame and its decomposition.

Table 7.8 Data for Example 1.

	DOF	Number of nodes	Number of elements	Number of supports
Structure	45	23	22	8
Substructure 1	24	12	11	4
Substructure 2	24	12	11	4

Table 7.9 Results of Example 1.

Frequency	With substructuring	Without substructuring
First mode	50.92	50.92
Second mode	68.42	68.42
Third mode	98.66	98.66

Example 2: A simple frame consisting of three types of elements is considered as shown in Figure 7.20(a).

For type 1 $A = 396$ cm^2, $I = 115,360$ cm^4, and $G = 310$ kg/m;

for type 2 $A = 198$ cm^2, $I = 5,768,015,360$ cm^4, and $G = 155$ kg/m;

and for type 3 $A = 33.4$ cm^2, $I = 2770$ cm^4, and $G = 26.6$ kg/m.

All the columns up to the level of 18 m are Type 1, the rest of the columns are Type 2, and beams are Type 3 elements. The structure is decomposed into two and four substructures, as illustrated in Figure 7.20(b–c). The results are depicted in Tables 7.10 to 7.12.

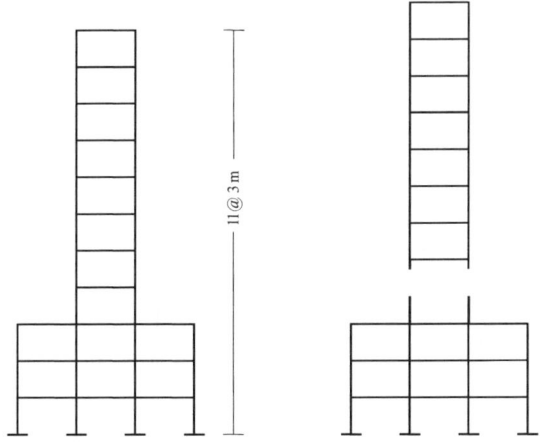

(a) A simple frame. (b) Frame decomposed into two substructures.

344   OPTIMAL STRUCTURAL ANALYSIS

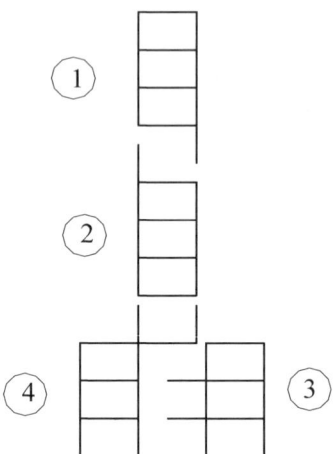

(c) Frame decomposed into four substructures.

Fig. 7.20 A simple frame and its decompositions.

Table 7.10 Data for Example 2.

	DOF	Number of nodes	Number of elements	Number of supports
Structure	84	32	45	8
Substructure 1	48	16	22	0
Substructure 2	42	18	23	4

Table 7.11 Data for substructuring of Example 2.

	DOF	Number of nodes	Number of elements	Number of supports
Structure	84	32	45	4
Substructure 1	27	9	11	0
Substructure 2	27	9	11	0
Substructure 3	24	10	11	2
Substructure 4	27	11	12	2

Table 7.12 Results of Example 2.

Frequency	Entire structure	Two substructures	Four substructures
First mode	9.32	9.35	9.35
Second mode	27.97	27.95	27.95
Third mode	56,60	56.55	56.55

DECOMPOSITION VIA GRAPH THEORY METHODS 345

Example 3: The model of a tall building is considered as shown in Figure 7.21(a) and decomposed into two and four substructures as illustrated in Figures 7.21 (b–c). The results are provided in Tables 7.13 to 7.15.

The properties of the elements of different types are the same as in Example 2. All columns up to the level of 18 m are of Type 1, and the rest of the columns are of Type 2. Beams are of Type 3.

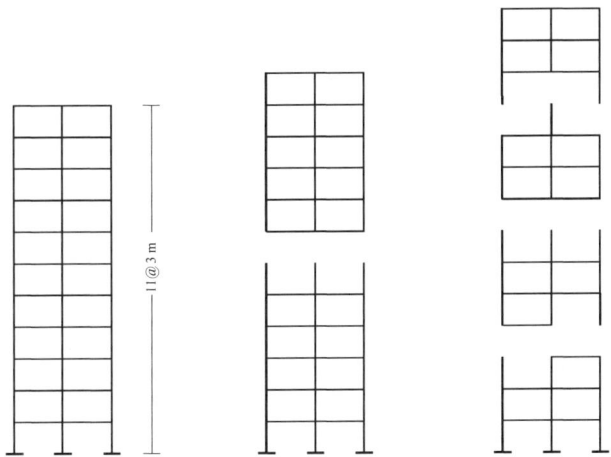

(a) A tall building frame.

(b) Frame decomposed into two substructures.

(c) Frame decomposed into four substructures.

Fig. 7.21 A tall building frame and its decompositions.

Table 7.13 Data for Example 3.

	DOF	Number of nodes	Number of elements	Number of supports
Structure	99	36	55	3
Substructure 1	54	18	27	0
Substructure 2	54	21	28	3

Table 7.14 Data for substructuring of Example 3.

	DOF	Number of nodes	Number of elements	Number of supports
Structure	99	36	55	3
Substructure 1	33	11	14	0
Substructure 2	30	10	13	0
Substructure 3	36	12	14	0
Substructure 4	27	12	14	3

346   OPTIMAL STRUCTURAL ANALYSIS

Table 7.15 Results of Example 3.

Frequency	Entire structure	Two substructures	Four substructures
First mode	8.53	8.55	8.55
Second mode	27.34	27.35	27.35
Third mode	59.15	59.15	59.155

Remark: The dimension of the modal force matrix in the method presented is smaller than the dimension of the system matrix, resulting in considerable reduction in the numerical computations. The method is general and applicable to skeletal structures as well as FEMs.

EXERCISES

7.1   Construct the associate graph G of the following finite element model and bisect G. Then form the corresponding subdomains.

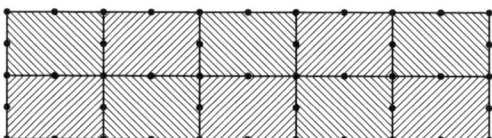

7.2   Decompose the following structure into $q = 2$ substructures and make a complete displacement analysis of the model.

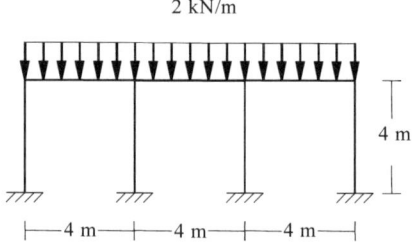

7.3 Perform a complete force method analysis for the following skeletal structure after decomposing its model into $q = 2$ substructures.

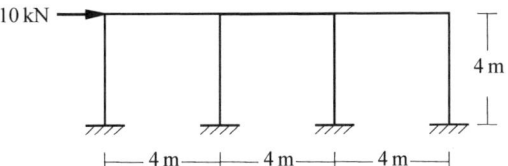

7.4 Decompose the following skeletal structure into $q = 3$ substructures and perform a complete dynamic analysis.

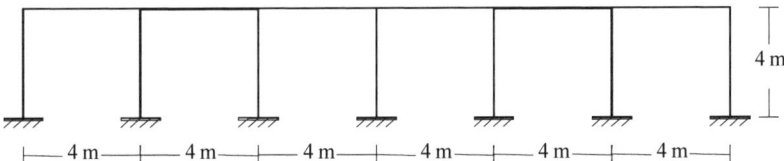

CHAPTER 8

Decomposition for Parallel Computing: Algebraic Graph Theory Methods

8.1 INTRODUCTION

Mathematicians have done a great deal of research on algebraic graph-theory methods; see Mohar [165,166] and Cevtković et al. [34]. In the field of structural engineering, Simon [202] suggested the spectral bisection (SB) method. Pothen et al. [182] proposed the recursive spectral bisection (RSB) algorithm for hypercube architectures. Hendrickson and Leland [79] extended SB through the use of multiple eigenvectors to allow the partitioning of a domain into four or eight subdomains at each stage of recursive decomposition. Shang–Hsieh et al. [199] generalised the RSB to an arbitrary number of processors developing the recursive spectral sequential-cut (RSS) and the recursive spectral two-way (RST) algorithms.

A mixed graph theory and algebraic graph-theory approach, which uses the selected features of both types of methods, is due to Kaveh and Davaran [116]. The RSB method is generalised to a weighted graph for use in adaptive meshes by Kaveh and Davaran [117].

An efficient method is developed for finite element domain decomposition; see Kaveh and Rahimi Bondarabady [124]. A weighted incidence graph is first constructed for the finite element model (FEM). A spectral partitioning heuristic is then applied to the graph using the second and the third eigenvalues of the Laplacian matrix of the graph, to partition it to three subgraphs and correspondingly trisect the FEM.

Optimal Structural Analysis   A. Kaveh
© 2006 Research Studies Press Limited

A fast algorithm is developed for the decomposition of large-scale FEMs; see Kaveh and Rahimi Bondarabady [127]. A weighted incidence graph G_0 is constructed and then reduced to a graph G_n of the desired size by a sequence of contractions $G_0 \rightarrow G_1 \rightarrow G_2 \rightarrow \ldots \rightarrow G_n$. Two Ritz vectors are constructed for G_0. A similar process is repeated for G_i ($i = 1,2, \ldots, n$), and the sizes of the vectors obtained are then extended to N, a Ritz matrix consisting of $2(n + 1)$ normalised Ritz vectors. The first eigenvector of this matrix is used as an approximate Fiedler vector to bisect G_0.

8.2 ALGEBRAIC GRAPH THEORY FOR SUBDOMAINING

8.2.1 BASIC DEFINITIONS AND CONCEPTS

The adjacency and Laplacian matrices of a graph are essential in the study of this chapter, which are already defined in Section 6.3.1.

Consider a directed graph G with N nodes, an arbitrary nodal numbering, and arbitrary member orientations. Then,

$$\mathbf{L}(G) = \mathbf{D}(G) - \mathbf{A}(G), \qquad (8\text{-}1)$$

It can also be shown that $\mathbf{L}$ is independent of the orientation of the members of the graph. The quadratic form associated with the Laplacian matrix $\mathbf{L}$ can be constructed as follows:

$$\mathbf{x}^t \mathbf{L} \mathbf{x} = \mathbf{x}^t \mathbf{C} \mathbf{C}^t \mathbf{x} = (\mathbf{C}^t \mathbf{x})^t (\mathbf{C}^t \mathbf{x}) = \sum_{\substack{\{n_i,n_j\} \in M \\ i,j \leq N}} (x_i - x_j)^2, \qquad (8\text{-}2)$$

where x_i is the ith component of $\mathbf{x}$. From this equation it follows that $\mathbf{L}$ is a positive semi-definite matrix.

The smallest eigenvalue of $\mathbf{L}$ is $\lambda_1 = 0$, and the corresponding eigenvector $\mathbf{y}_1$ has all its normalised components equal to 1. The second eigenvalue λ_2 and the associated eigenvector $\mathbf{y}_2$ have many interesting properties, which are used in this chapter, for domain decomposition.

A finite element domain D is to be divided into two subdomains D_1 and D_2; with their interfaces shown by S-S, Figure 8.1(a). Consider the natural associate graph of D, and denote it by G, as illustrated in Figure 8.1(b). The second eigenvector of the Laplacian matrix or weighted Laplacian (to be defined in the next section) of G can be used for partitioning of the nodes of G. Let x_m be the median value of the components of the eigenvector. Let A be the set of nodes whose node components

DECOMPOSITION VIA ALGEBRAIC GRAPH THEORY METHODS 351

are less than or equal to x_m, and B be the remaining set of nodes. If there is a single node with the components corresponding to x_m, then A and B differ in size by at most one. If there are several such nodes, the nodes can arbitrarily be assigned to A or B to make these sets differ in size by at most unity.

The initial partition of G results in an edge separator of the graph. Let A_1 denote the nodes of A, which are adjacent to some nodes in B, and similarly B_1 be the set of nodes in B that are adjacent to some nodes in A. Consider the set of edges E of G with one end at A and the other end at B. Then E is an edge separator of G, and we have a bipartite graph $H = (A_1, B_1, E)$.

The node separator of G can be obtained from the edge separator E using different methods. One can simply choose the smaller of the two end point sets A_1 and B_1. In order to choose the smallest node separator, one should select a set S consisting of some nodes from both sets of end points A_1 and B_1, such that every edge in E is incident on the least one of the nodes of S. The set S is then a node separator, since the removal of these nodes causes the deletion of the edges incident to them. S is called a *node cover* of the bipartite graph H. A minimum cover can then easily be found by a maximal matching in H.

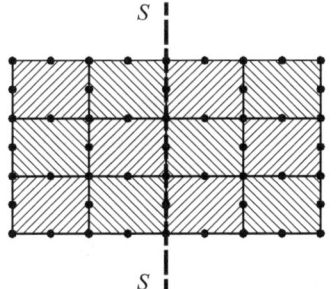 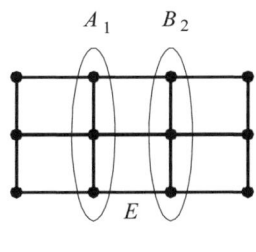

(a) A finite element domain.   (b) The natural associate graph of D.

Fig. 8.1 A domain decomposed into two subdomains.

The dividing vector and Laplacian matrix of the mesh are denoted by $\mathbf{x}$ and $\mathbf{L}$, respectively. The first and the second eigenvectors $\mathbf{y}_1$ and $\mathbf{y}_2$ of the Laplacian matrix have the following properties:

$$\mathbf{x}^t \mathbf{y}_1 = 0,$$

$$|C(\mathbf{x})| = \tfrac{1}{4} \sum_{(i,j) \in I} (x_i - x_j)^2 = \tfrac{1}{4} \mathbf{x}^t \mathbf{L} \mathbf{x}. \qquad (8\text{-}3)$$

352 OPTIMAL STRUCTURAL ANALYSIS

Matrix **L** is positive semi-definite and

$$0 = \lambda_1 \leq \lambda_2 \leq \lambda_3 \leq \ldots \leq \lambda_n,$$

$$\mathbf{x}'\mathbf{L}\mathbf{x} \geq \tfrac{1}{4}\mathbf{y}'_2\mathbf{L}\mathbf{y}_2, \tag{8-4}$$

and

$$\operatorname{Min}|C(\mathbf{x})| = |C(\mathbf{y}_2)| = \tfrac{1}{4}\mathbf{y}'_2\mathbf{L}\mathbf{y}_2 \tag{8-5}$$

The second-smallest eigenvector $\mathbf{y}_2$ is the Fiedler vector. Equation (8-5) shows that of all the partitioning vectors **x**, which bisect a mesh and are orthogonal to $\mathbf{y}_1$, the Fiedler vector minimises the number of cut interfaces; see Seale and Topping [195].

Unfortunately, the components of $\mathbf{y}_2$ are not discrete and the elements of the mesh cannot be partitioned on a (+ 1, − 1) basis. Then the $N/2$ elements corresponding to $N/2$ lowest components of $\mathbf{y}_2$ are assigned to one subdomain and the remainder to the other subdomain. If N is odd, then one subdomain is given the additional element. This is where the heuristic nature of the algorithm appears, and theoretical optimality changes to suboptimality.

When using such a heuristic, it is worthwhile to have a lower bound on what can be achieved. Such bounds have been established by many researchers; a few results are included in the following theorems.

Theorem 1: Let A and B be disjoint subsets of nodes of graph G that are at a distance $r = 2$ from each other. Let S denote the set of nodes not belonging to A that are at a distance less than 2 from A. Then,

$$|S|^2 + \beta|S| - 4|A|(1-|A|) \geq 0,$$
$$\text{with } \beta = (\Delta/\lambda_2) + 4|A| - 1. \tag{8-6}$$

Theorem 2: Let S be a node separator that divides a graph G into two parts A and B, with $|A| \geq |B| \geq |S|$. Then,

$$|S| \geq \frac{(1-|A|)\lambda_2}{2\Delta - (\lambda_3 - \lambda_2)}. \tag{8-7}$$

For calculating the second eigenvalue λ_2 of **L**, first it is tri-diagonalised using the Lanczos method, and then the spectral properties of Sturm polynomials are applied; see Jennings and McKeown [90]. Because of the importance of the Lanczos algorithm, a brief description of it is given in the next section.

Before the end of this section, it is convenient to describe another important relevant number, known as the *isoperimetric number*. This is defined as

$$i(G) = \min_X \frac{|\partial X|}{|X|}, \qquad (8\text{-}8)$$

where the minimum is taken over all non-empty subsets $X \subseteq N(G)$ satisfying

$$|X| \leq \tfrac{1}{2}|N(G)|. \qquad (8\text{-}9)$$

∂X denotes the set of edges of G having one end in X and the other end at $N(G) \setminus X$.

For example, for a path P_7, shown in Figure 8.2, $i(G) = 1/3$, and for the complete graph K_5, shown in Figure 8.3, the value of $i(G)$ is equal to 3.

Fig. 8.2 A path graph with $i(G) = 1/3$.

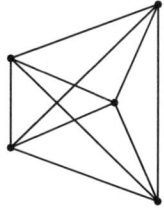

Fig. 8.3 A complete graph K_5 with $i(G) = 3$.

This problem is closely related to the minimum-cut problem, known as the *bisection width* problem; see Goldberg and Gardner [62]. Some basic properties of $i(G)$ are explored by Mohar [165] as follows:

(a) $i(G) = 0$ if and only if G is disconnected.

(b) If δ is the minimal degsree of nodes of G, then $i(G) \leq \delta$.

(c) If Δ is the maximal degree of nodes of G, then $i(G) \leq 2/(\Delta-2) + 2/|\langle N(G)/2\rangle|$.

(d) For a complete graph K_N, $i(K_N) = [N/2]$.

(e) For the path graph P_N on N nodes, $i(P_N) = 1/|\langle N/2\rangle|$.

(f) For a cycle graph C_N on N nodes, . $i(C_N) = 2/|\langle N/2\rangle|$.

Here $|\langle x\rangle|$ denotes the integer part of the real number, that is, the largest integer, which is smaller than x, and $[x]$ shows the smallest integer that is greater than or equal to x.

In general, the calculation of $i(G)$ is NP-hard. However, the following theorems provide bound on $i(G)$ of a general graph:

Theorem 1: Let G be a graph of order N with M edges, then

$$i(G) \leq \begin{cases} M/(N-1) \text{ if } N \text{ is even,} \\ M(N+1)/N(N-1) \text{ if } N \text{ is odd.} \end{cases} \quad (8\text{-}10)$$

Theorem 2: Let G be a graph with maximal node degree Δ, and let λ_2 be the second-smallest eigenvalue of its Laplacian matrix. If G is not identical to K_1, K_2, or K_3, then:

$$i(G) \leq \sqrt{\lambda_2(2\Delta - \lambda_2)}. \quad (8\text{-}11)$$

For proofs of these theorems and further information, the reader may refer to Mohar [165,166].

8.2.2 LANCZOS METHOD

One of the well-known algorithms for computing a few eigenvalues and eigenvectors of large, sparse symmetric matrices is the Lanczos algorithm. This algorithm can be used to approximate the largest and smallest eigenvalues of a large symmetric matrix, whose order is so large that similarity transformations are not feasible. The algorithm builds up a tri-diagonal matrix row by row.

Consider the eigensolution of a symmetric matrix **A**. Start with a single trial vector. The algorithm generates a sequence of mutually orthogonal vectors by means of a process, which includes pre-multiplications by the $N \times N$ matrix **A**. Theoretically, the sequence of vectors should terminate after N vectors have been generated. The orthogonal vectors combine to produce a transformation matrix, which has the effect of transforming **A** to a tri-diagonal form. The elements of the tri-diagonal matrix are generated as the orthogonal vectors are formed, and when the formation of the tri-diagonal matrix is complete, its eigenvalues can be computed by any suitable technique, such as the Sturm sequence or LR or QR methods.

The pre-multiplication process in this method amplifies the components of the eigenvectors corresponding to the largest modulus. Therefore, for a large **A**, the process of vector generation is terminated after p steps, where p is far less than N. Eigensolution of the resulting tri-diagonal matrix of order p will yield an approximation to the set of p dominant eigenvalues of **A**. The largest eigenvalues of this tri-diagonal matrix give good approximations to the dominant eigenvalues of **A**. Therefore, the Lanczos algorithm may be used for either full or partial eigensolution of a matrix.

The standard Lanczos algorithm transforms a symmetric matrix into a non-symmetric tri-diagonal form. The method described below modifies the procedure so that a symmetric tri-diagonal matrix is formed. If $\mathbf{Y} = \{y_1\ y_2\ y_3,\ \ldots,\ y_N\}$ is the compounded set of mutually orthogonal vectors, the transformation to tri-diagonal form may be described by

$$[\mathbf{A}][\mathbf{y}_1 \ldots \mathbf{y}_N] = [\mathbf{y}_1 \ldots \mathbf{y}_N]\begin{bmatrix} \alpha_1 & \beta_1 & & & & \\ \beta_1 & \alpha_2 & \beta_2 & & & \\ . & . & . & . & . & \\ & & \beta_{N-2} & \alpha_N & \beta_{N+1} \\ & & & \beta_{N-1} & \alpha_N \end{bmatrix}. \qquad (8\text{-}12)$$

The expanded form of this matrix leads to N vector equations of the form

$$\mathbf{Ay}_1 = \alpha_1 \mathbf{y}_1 + \beta_1 \mathbf{y}_2$$

$$\mathbf{Ay}_2 = \beta_1 \mathbf{y}_1 + \alpha_2 \mathbf{y}_2 + \beta_2 \mathbf{y}_3$$

$$\ldots\ldots\ldots\ldots \qquad (8\text{-}13)$$

$$\mathbf{Ay}_j = \beta_{j-1} \mathbf{y}_{j-1} + \alpha_j \mathbf{y}_j + \beta_j \mathbf{y}_{j+1}$$

$$\ldots\ldots\ldots\ldots$$

$$\mathbf{Ay}_N = \beta_{N-11} \mathbf{y}_{N-1} + \alpha_N \mathbf{y}_N.$$

356 OPTIMAL STRUCTURAL ANALYSIS

The first vector $\mathbf{y}_1$, is chosen to be an arbitrary non-null vector, normalised such that $\mathbf{y}_1^t \mathbf{y}_1 = 1$. If the first equation of the above equations is pre-multiplied by $\mathbf{y}_1^t$, it can be established that, when $\alpha_1 = \mathbf{y}_1^t \mathbf{A}\mathbf{y}_1$, then $\mathbf{y}_2^t \mathbf{y}_2 = 0$. Thus, choosing α_1 in this way, ensures that $\mathbf{y}_2$ is orthogonal to $\mathbf{y}_1$. Substitution of this value of α_1 into the first equation yields $\beta_1 \mathbf{y}_2$. Then, since $\mathbf{y}_2$ can be obtained from $\beta_1 \mathbf{y}_2$ by Euclidean normalisation, $1/\beta_1$ being the normalising factor, the remaining vectors of the set may be generated in a similar way, using the second, third, ... of the equations. The process is described by the following relationships:

$$\mathbf{v}_j = \mathbf{A}\mathbf{y}_j - \beta_j \mathbf{y}_{j-1} \quad (\beta_0 = 0)$$

$$\alpha_j = \mathbf{y}_j^t \mathbf{v}_j$$

$$\mathbf{z}_j = \mathbf{v}_j - \alpha_j \mathbf{y}_j \qquad (8\text{-}14)$$

$$\beta_j = (\mathbf{z}_j^t \mathbf{z}_j)^{1/2}.$$

It can be shown that the vectors are all mutually orthogonal and also that the last equation is implicitly satisfied.

Example: For the following matrix, calculate $\mathbf{Y}$ and $\mathbf{T}$ matrices, and find their eigenvalues and eigenvectors.

$$\mathbf{A} = \begin{bmatrix} 1 & -3 & -2 & 1 \\ -3 & 10 & -3 & 6 \\ -2 & -3 & 3 & -2 \\ 1 & 6 & -2 & 1 \end{bmatrix}.$$

Symmetric Lanczos tri-diagonalisation of this matrix is formed as follows:

Using an initial vector as $\mathbf{y}_1 = \begin{bmatrix} 1 & 0 & 0 & 0 \end{bmatrix}^t$, the magnitude of α_1 is obtained as

$$\alpha_1 = \mathbf{y}_1^t \mathbf{A}\mathbf{y}_1 = \begin{bmatrix} 1 & 0 & 0 & 0 \end{bmatrix} \begin{bmatrix} 1 & -3 & -2 & 1 \\ -3 & 10 & -3 & 6 \\ -2 & -3 & 3 & -2 \\ 1 & 6 & -2 & 1 \end{bmatrix} \begin{bmatrix} 1 \\ 0 \\ 0 \\ 0 \end{bmatrix} = 1.0.$$

Substituting in the first equation of (8-13) result in:

$$\boldsymbol{\beta}_1\mathbf{y}_2 = \mathbf{A}\mathbf{y}_1 - \alpha_1\mathbf{y}_1 = \begin{bmatrix} 1 & -3 & -2 & 1 \\ -3 & 10 & -3 & 6 \\ -2 & -3 & 3 & -2 \\ 1 & 6 & -2 & 1 \end{bmatrix}\begin{bmatrix} 1 \\ 0 \\ 0 \\ 0 \end{bmatrix} - 1 \times \begin{bmatrix} 1 \\ 0 \\ 0 \\ 0 \end{bmatrix} = \begin{bmatrix} 0 \\ -3 \\ -2 \\ 1 \end{bmatrix}.$$

Since

$$\beta_1^2 = (0)^2 + (-3)^2 + (-2)^2 + (1)^2 = 14,$$

hence $\beta_1 = 3.7417$, and

$$\mathbf{y}_2 = \begin{bmatrix} 0.0000 \\ -0.8018 \\ -0.5345 \\ 0.8041 \end{bmatrix}.$$

Now α_2 can be obtained as

$$\alpha_2 = \mathbf{y}_1^t \mathbf{A}\mathbf{y}_1 = 2.7857,$$

and, from the second relation of Eq. (8-13), $\beta_2 = 5.2465$ and

$$\mathbf{y}_3 = \begin{bmatrix} 0.0000 \\ -3.3348 \\ -0.8041 \\ 0.8041 \end{bmatrix}.$$

Repeating a similar process results in $\alpha_3 = 10.1993$ and $\beta_3 = 4.4796$, leading to

$$\mathbf{y}_4 = \begin{bmatrix} 0.000 \\ -3.404 \\ 0.776 \\ 0.5310 \end{bmatrix},$$

and $\alpha_4 = 1.0150$. Therefore,

$$\mathbf{Y} = [\mathbf{y}_1 \ \mathbf{y}_2 \ \mathbf{y}_3 \ \mathbf{y}_4] = \begin{bmatrix} 1.0000 & 0 & 0 & 0 \\ 0 & -0.8018 & -0.4912 & -0.3404 \\ 0 & 0.5345 & 0.3348 & 0.7760 \\ 0 & 0.2673 & -0.80412 & 0.5310 \end{bmatrix},$$

and

$$\mathbf{T} = \begin{bmatrix} 1.0000 & 3.7417 & 0 & 0 \\ 3.7417 & 2.7857 & 5.2465 & 0 \\ 0 & 5.2465 & 10.1993 & 4.4796 \\ 0 & 0 & 4.4796 & 1.0150 \end{bmatrix}.$$

Now the eigenvalues and eigenvectors for $\mathbf{T}$ and $\mathbf{A}$ can be calculated by any method preferred. $\mathbf{u}_i$ are the eigenvectors of $\mathbf{T}$, and $\mathbf{v}_i$ are the normalised eigenvectors of $\mathbf{A}$.

$$\lambda_1 = -4.5694, \ \mathbf{u}_1 = \begin{bmatrix} -3.4719 \\ 2.6554 \\ -1.2466 \\ 1.0000 \end{bmatrix} \text{ and } \mathbf{v}_1 = \begin{bmatrix} -0.7460 \\ -0.3990 \\ -0.2279 \\ 0.4820 \end{bmatrix},$$

$$\lambda_2 = -1.3035, \ \mathbf{u}_2 = \begin{bmatrix} 0.4187 \\ 0.2809 \\ -0.5176 \\ 1.0000 \end{bmatrix} \text{ and } \mathbf{v}_2 = \begin{bmatrix} 0.3394 \\ -0.2524 \\ 0.3669 \\ 0.4286 \end{bmatrix},$$

$$\lambda_3 = 4.4568, \ \mathbf{u}_3 = \begin{bmatrix} -1.8343 \\ -1.6948 \\ 0.7683 \\ 1.0000 \end{bmatrix} \text{ and } \mathbf{v}_3 = \begin{bmatrix} -0.6556 \\ 0.2291 \\ 0.6931 \\ -0.1930 \end{bmatrix},$$

$$\lambda_4 = 14.3299, \ \mathbf{u}_4 = \begin{bmatrix} 0.4181 \\ 1.4863 \\ 2.9723 \\ 1.0000 \end{bmatrix} \text{ and } \mathbf{v}_4 = \begin{bmatrix} 0.1196 \\ -0.8560 \\ 0.2794 \\ -0.4182 \end{bmatrix}.$$

For this small matrix, the eigenvalues of the tri-diagonalised matrix are similar to the eigenvalues of the original matrix. However, for large matrices, the orthogonality condition $\mathbf{y}_i^t \mathbf{y}_j = 0$ for $j > i+1$ will not be satisfied. This is because each step of the process magnifies the rounding errors present. Therefore, the eigenvalue estimates obtained by using the basic algorithm cannot always be guaranteed. This difficulty may be overcome by improving the algorithm, so that $\mathbf{y}_{i+1}$ is orthogonalised with respect to vectors $\mathbf{y}_1, \mathbf{y}_2, \mathbf{y}_3, \ldots, \mathbf{y}_{j-1}$.

The additional re-orthogonalisation naturally increases the computational cost. However, this is necessary for large matrices. For large N, the number of iterations k should be restricted to a small number $k \ll N$. Therefore, after the kth iteration, Eq. (8-12) is replaced by

$$\mathbf{A}\mathbf{Y}_k = \mathbf{Y}_k \mathbf{T}_k + \mathbf{E}_k, \qquad (8\text{-}15)$$

where

$$\mathbf{E}_k = [0 \quad 0 \quad \ldots \quad \beta_k \mathbf{y}_{k+1 \, N \times k}]. \qquad (8\text{-}16)$$

Let the eigen-pair of $\mathbf{T}_k$ be $(\mu_i, \mathbf{P}_i)$, then,

$$\mathbf{T}_k \mathbf{P}_i = \mu_i \mathbf{P}_i, \qquad (8\text{-}17)$$

and post-multiplication of Eq. (8-15) by $\mathbf{P}_i$ yields

$$\mathbf{A}(\mathbf{Y}_k \mathbf{P}_i) = \mu_i (\mathbf{Y}_k \mathbf{P}_i) + \mathbf{E}_k \mathbf{P}_i. \qquad (8\text{-}18)$$

Since the second term of the above equation is reduced, by increasing k, the eigenvalues of $\mathbf{A}$ can be approximated by those of $\mathbf{T}_k$.

8.2.3 RECURSIVE SPECTRAL BISECTION PARTITIONING ALGORITHM

Step 1: Construct the natural associate graph, incidence graph or skeleton graph of the given finite element mesh.

Step 2: Compute the second eigenvector (Fiedler vector) $\mathbf{y}_2$ of the Laplacian of the graph using the Lanczos algorithm.

Step 3: Order the nodes of the graph according to their associate components in the vector $\mathbf{y}_2$.

Step 4: Assign half of the nodes for each subdomain.

360 OPTIMAL STRUCTURAL ANALYSIS

Step 5: Repeat recursively for each subdomain.

Natural associate graph and integer graph are used to transform the topological properties of the FEMs into those of the graphs; however, any of the 10 graphs introduced in Chapter 5, or their combinations, can also be used for this purpose. Naturally, the efficiency will be different for each graph.

Example: Consider a simple finite element mesh as shown in Figure 8.4(a). The model is decomposed into two subdomains using the associate graph of the model, as shown in Figure 8.4(b).

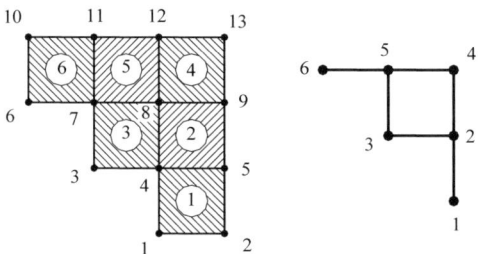

(a) A finite element mesh. (b) Associate graph of the mesh.

Fig. 8.4 A finite element mesh and its associate graph.

The Laplacian matrix of the associate graph is formed as a 6 × 6 matrix:

$$\mathbf{L}(G) = \begin{bmatrix} 1 & -1 & 0 & 0 & 0 & 0 \\ -1 & 3 & -1 & -1 & 0 & 0 \\ 0 & -1 & 2 & 0 & -1 & 0 \\ 0 & -1 & 0 & 2 & -1 & 0 \\ 0 & 0 & -1 & -1 & 3 & -1 \\ 0 & 0 & 0 & 0 & -1 & 1 \end{bmatrix}$$

Using the Lanczos method, **Y** is obtained as

$$\mathbf{Y} = \begin{bmatrix} 1.00 & 0.00 & 0.00 & 0.00 & 0.00 \\ 0.00 & -1.00 & 0.00 & 0.00 & 0.00 \\ 0.00 & 0.00 & 0.7071 & 5.09 \times 10^{-32} & 4.8 \times 10^{-32} \\ 0.00 & 0.00 & 0.7071 & 5.09 \times 10^{-32} & 4.8 \times 10^{-32} \\ 0.00 & 0.00 & 0.00 & -1.00 & 1.80 \times 10^{-47} \\ 0.00 & 0.00 & 0.00 & 0.00 & 1.00 \end{bmatrix}.$$

DECOMPOSITION VIA ALGEBRAIC GRAPH THEORY METHODS 361

Five orthogonal vectors are obtained, since the rank of $\mathbf{L}(G) = 5 < 6$. Therefore, $\mathbf{T}$ is a 5×5 matrix whose entries are as follows:

$$\alpha_1 = 1.000 \text{ and } \beta_1 = 1.000,$$

$$\alpha_2 = 0.000 \text{ and } \beta_2 = 1.414,$$

$$\alpha_3 = 2.000 \text{ and } \beta_3 = 1.414,$$

$$\alpha_4 = 3.000 \text{ and } \beta_4 = 1.000,$$

$$\alpha_5 = 3.000.$$

Using the Sturm polynomial method, the second eigenvalue and the corresponding normalised eigenvector of $\mathbf{L}(G)$ are obtained as

$$\lambda_2 = 0.585798 \text{ and } \mathbf{v}_2 = \begin{bmatrix} -6.5328 \times 10^{-1} \\ -2.7061 \times 10^{-1} \\ -1.0359 \times 10^{-5} \\ -1.0359 \times 10^{-5} \\ 2.7061 \times 10^{-1} \\ 6.5328 \times 10^{-1} \end{bmatrix}.$$

Using this vector, the model can be decomposed into two different forms as $\{(1,2,4),(3,5,6)\}$ and $\{(1,2,3),(4,5,6)\}$. The decomposed submodels are illustrated in Figure 8.5(a) and (b), respectively.

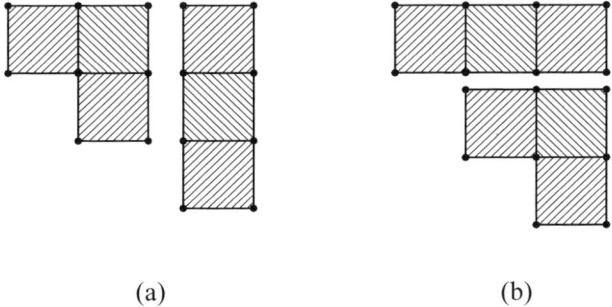

(a)   (b)

Fig. 8.5 Two different decompositions for the given mesh.

8.2.4 RECURSIVE SPECTRAL SEQUENTIAL-CUT PARTITIONING ALGORITHM

The RSS algorithm partitions the graph in such a way that subgraphs are cut out of the original graph one by one (or sequentially) in a recursive fashion as follows:

Step 1: Construct the natural associate graph, incidence graph or skeleton graph of the given finite element mesh.

Step 2: Compute the second eigenvector (Fiedler vector) of the Laplacian of the graph using the Lanczos algorithm.

Step 3: Order the nodes of the graph according to their associate components in the Fiedler vector.

Step 4: Assign N/q of the nodes (discard the remainder) to one subdomain and the remaining to the other, in which n_p is initially equal to the number of partitions (or processors) desired.

Step 5: Repeat recursively for the larger subdomain in the previous step with $q = q - 1$ until all subdomains are defined (i.e. $q = 1$).

8.2.5 RECURSIVE SPECTRAL TWO-WAY PARTITIONING ALGORITHM

In this algorithm, instead of using a bisection approach, the RST algorithm uses a two-way partitioning approach, which partitions the graph into two parts not necessarily of equal size. M_i denotes the number of nodes in each subdomain D_i, when the partitioning task is completed. This is computed in advance by sequentially employing the following equation for $i = 1, ..., q$

$$M_i = \frac{N - \sum_{j=1}^{i-1} M_j}{q - (i-1)} (+1 \text{ if remainder} \neq 0), \qquad (8\text{-}19)$$

where q is the number of processors, and N is the total number of nodes of the whole domain. The number of partitions desired in the intermediate subdomain in each two-way partitioning step is denoted by n_p (initially $n_p = q$ for the whole domain). Each intermediate subdomain maintains a list l of subdomain members associated with D_i, which has been assigned to it (initially the list is (1,2, ..., q) for the whole domain). The kth component of this list is denoted by $l(k)$.

Obviously, RSB is a particular case of the RST algorithm when the number of processors q is equal to an integer power of 2. The algorithm consists of the following steps:

Step 1: Construct the natural associate graph, incidence graph or the skeleton graph of the given finite element mesh.

Step 2: Compute the second eigenvector (Fiedler vector) of the Laplacian of the graph, using the Lanczos algorithm.

Step 3: Order the nodes of the graph, according to their associate components in the Fiedler vector.

Step 4: Compute the following integers:

$$p_1 = n_p/2 \text{ (discard the remainder)}, \; p_2 = n_p - p_1 \text{ and } N = \sum_{k=1}^{p1} M_{l(k)}.$$

Assign N nodes and the list of the first p_1 components in $l(k)$ to one subdomain, and set $n_p = p_1$ for this subroutine.

Assign the remaining nodes and the remaining components in $l(k)$ to the other subdomain, and set $n_p = p_2$ for this subdomain.

Step 5: Repeat recursively for each subdomain with $n_p > 1$.

8.3 MIXED METHOD FOR SUBDOMAINING

In this section, a mixed method is presented for domain decomposition employing combinatorial and algebraic graph–theoretical algorithms. The method uses combinatorial graph theory for partial decomposition, followed by a SB approach based on concepts from the algebraic graph theory. Examples are presented to illustrate the efficiency of the mixed method. The effects of nodal ordering on the performance of the SB and the mixed methods are also investigated.

8.3.1 INTRODUCTION

Parallel processing is often used for the analysis of large-scale structures and FEMs. As discussed earlier, pure algebraic graph–theoretical methods require the calculation of the Fiedler vector for very large matrices, which can experience some difficulties when the Lanczos method is employed. Therefore, a mixed method that uses the advantages of both graph–theoretical and algebraic graph–theoretical approaches is beneficial. This section is devoted to developing such an algorithm.

8.3.2 MIXED METHOD FOR GRAPH BISECTION

A finite element domain D is divided into two subdomains D_1 and D_2; with their interfaces shown by S-S, Figure 8.6(a). The dividing vector and Laplacian matrix of the mesh are denoted by $\mathbf{x}$ and $\mathbf{L}$, respectively. As studied earlier, the first and the second eigenvectors $\mathbf{y}_1$ and $\mathbf{y}_2$ of the Laplacian matrix have the following properties:

$$\mathbf{x}^t \mathbf{y}_1 = 0$$

and
$$\text{Min}\left| C(\mathbf{x}) = |C(\mathbf{y}_2)| \right| = \tfrac{1}{4} \mathbf{y}_2^t \mathbf{L} \mathbf{y}_2. \tag{8-20}$$

Therefore, $\mathbf{y}_2$ will be used as a heuristic to partition the nodes of the FEMs.

Now let us consider the problem of partitioning a finite element mesh by a mixed method. Consider a given domain D. Two subdomains R_2 and R_3 with an equal number of elements ($|R_2| = |R_3|$) are separated from subdomains D_1 and D_2 respectively, as illustrated in Figure 8.6(b). Let n be the total number of elements of the model, and i be the total number of elements in each subdomain R_2 and R_3, respectively. Then the number of elements in subdomain R_1 will be $m = n - 2i$. By reordering the elements, the partitioning vector $\mathbf{x}$ can be written as follows:

$$\mathbf{x}^t = \begin{bmatrix} \mathbf{X}_1^t & \mathbf{X}_2^t & \mathbf{X}_3^t \end{bmatrix}, \tag{8-21}$$

$$\mathbf{x}^t = [x_1\, x_2,\, \ldots,\, x_m\, |x_{m+1}\, x_{m+2},\, \ldots,\, x_{m+i}\, |x_{m+i+1}\, x_{m+i+2},\, \ldots,\, x_{m+2i}], \tag{8-22}$$

where $\mathbf{X}_1$, $\mathbf{X}_2$ and $\mathbf{X}_3$ are partitions of vector $\mathbf{x}$, corresponding to subdomains R_1, R_2 and R_3, respectively.

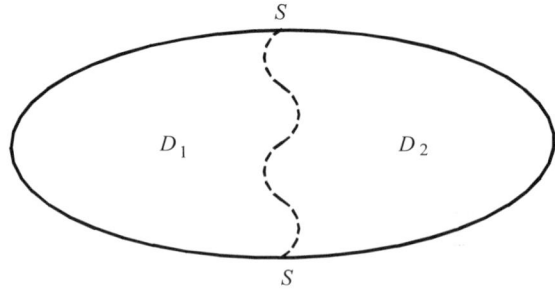

(a) A domain divided into two subdomains

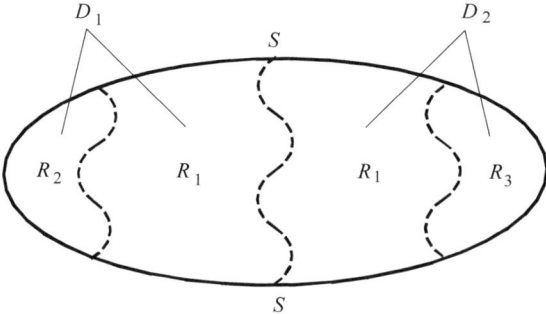

(b) A domain divided into three subdomains

Fig. 8.6 A domain and its subdivisions.

Since it is assumed that R_2 and R_3 are completely contained in different subdomains after final bisection,

$$x_{m+1} = \ldots = x_{m+i} = x_{m+i+1} = \ldots x_{m+2i} = t, \tag{8-23}$$

$$t = \begin{cases} +1 & \text{if } R_1 \text{ is in subdomain } D_1 \\ -1 & \text{if } R_2 \text{ is in subdomain } D_1 \end{cases} \tag{8-24}$$

$$\mathbf{x}^t \mathbf{L} \mathbf{x} = \begin{bmatrix} \mathbf{X}_1^t & \mathbf{X}_2^t & \mathbf{X}_3^t \end{bmatrix} \begin{bmatrix} \mathbf{L}_{11} & \mathbf{L}_{12} & \mathbf{L}_{13} \\ \mathbf{L}_{21} & \mathbf{L}_{22} & \mathbf{L}_{23} \\ \mathbf{L}_{31} & \mathbf{L}_{32} & \mathbf{L}_{33} \end{bmatrix} \begin{bmatrix} \mathbf{X}_1 \\ \mathbf{X}_2 \\ \mathbf{X}_3 \end{bmatrix}. \tag{8-25}$$

Since R_2 and R_3 are disjoint, it can easily be deduced that

$$\mathbf{L}_{23} = \mathbf{L}_{32}^t = 0. \tag{8-26}$$

Then:

$$\mathbf{x}^t \mathbf{L} \mathbf{x} = \mathbf{X}_1^t \mathbf{L}_{11} \mathbf{X}_1 + 2(\mathbf{X}_1^t \mathbf{L}_{12} \mathbf{X}_2 + \mathbf{X}_1^t \mathbf{L}_{13} \mathbf{X}_3) + (\mathbf{X}_2^t \mathbf{L}_{22} \mathbf{X}_2 + \mathbf{X}_3^t \mathbf{L}_{33} \mathbf{X}_3). \tag{8-27}$$

Considering Eq. (8-25), it is known that, the vectors $\mathbf{X}_2$ and $\mathbf{X}_3$ can be replaced by two scalar unknowns $\bar{x}_2$ and $\bar{x}_3$, and $\bar{x}_2 = -\bar{x}_3$. Therefore, Eq. (8-27) can be rewritten as

$$\mathbf{x}^t \mathbf{L} \mathbf{x} = \mathbf{X}_1^t \mathbf{L}_{11} \mathbf{X}_1 + 2(\mathbf{X}_1^t \bar{\mathbf{L}}_{12} \bar{x}_2 + \mathbf{X}_1^t \bar{\mathbf{L}}_{13} \bar{x}_3) + (\mathbf{X}_2^t \bar{\mathbf{L}}_{22} \bar{x}_2 + \bar{\mathbf{X}}_3^t \bar{\mathbf{L}}_{33} \bar{x}_3), \tag{8-28}$$

where

$$\bar{x}_2 = -\bar{x}_3 = \begin{cases} +1 \text{ if } R_1 \text{ is in subdomain } D_1 \\ -1 \text{ if } R_2 \text{ is in subdomain } D_1 \end{cases} \quad (8\text{-}29)$$

In the above relations, $\bar{\mathbf{L}}_{22}$ is the number of interfaces between R_1 and R_2, and $\bar{\mathbf{L}}_{33}$ is the number of interfaces between R_1 and R_3. $\bar{\mathbf{L}}_{12}$ and $\bar{\mathbf{L}}_{13}$ are two vectors with the same dimensions as $\mathbf{X}_1$, and their components are equal to the number of interfaces between the elements of R_1 and subdomains R_1 and R_3, respectively, such that

$$\bar{x}_2 \bar{\mathbf{L}}_{12} = -\mathbf{L}_{12}\mathbf{X}_2, \quad (8\text{-}30)$$

$$\bar{x}_3 \bar{\mathbf{L}}_{13} = -\mathbf{L}_{13}\mathbf{X}_3, \quad (8\text{-}31)$$

By these assumptions, Eq. (8-25) can be reduced as follows:

$$\mathbf{x}'\mathbf{L}\mathbf{x} = \begin{bmatrix} \mathbf{X}_1^t & \bar{x}_2 & \bar{x}_3 \end{bmatrix} \begin{bmatrix} \mathbf{L}_{11} & \bar{\mathbf{L}}_{12} & \bar{\mathbf{L}}_{13} \\ \bar{\mathbf{L}}_{12}^t & \bar{\mathbf{L}}_{22} & 0 \\ \bar{\mathbf{L}}_{13}^t & 0 & \bar{\mathbf{L}}_{33} \end{bmatrix} \begin{bmatrix} \mathbf{X}_1 \\ \bar{x}_2 \\ \bar{x}_3 \end{bmatrix}. \quad (8\text{-}32)$$

Hence

$$(\mathbf{x}'\mathbf{L}\mathbf{x}) = (\mathbf{y}'\mathbf{L}^*\mathbf{y}), \quad (8\text{-}33)$$

$$\text{Min }(\mathbf{x}'\mathbf{L}\mathbf{x}) = \text{Min }(\mathbf{y}'\mathbf{L}^*\mathbf{y}). \quad (8\text{-}34)$$

Relations (8-32), (8-33) and (8-34) mean that the problem of finding the second eigenvector of the Laplacian matrix $\mathbf{L}$ can be converted into that of finding the second eigenvector of the reduced Laplacian matrix $\mathbf{L}^*$. The dimension of $\mathbf{L}^*$ is $r = n - 2i + 2$. The matrix $\mathbf{L}^*$ is tri-diagonalised using the Lanczos method. The second-smallest eigenvector $\mathbf{L}^*$ can be found using the special properties of Sturm sequence polynomials, and the second eigenvector of $\mathbf{L}^*$ can be obtained from back substitution into $\mathbf{L}^*\mathbf{y}_2 = \lambda_2^* \mathbf{y}_2$.

The graph-theory interpretation of the above method is illustrated in Figure 8.7.

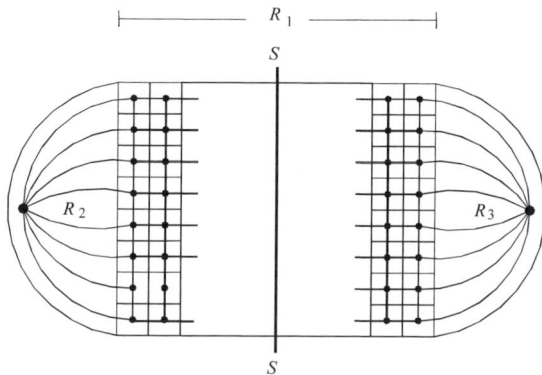

Fig. 8.7 Graph-theory interpretation of the mixed method.

In the first phase, two subdomains R_2 and R_3 with equal number of elements are separated from the main domain, using a suitable graph-theory method, as shown in Figure 8.5. These subdomains are replaced by two super elements, and their interfaces with the remaining subdomain, R_1, are obtained. The reduced associate graph and its Laplacian matrix $\mathbf{L}^*$ are constructed. The degree of the super elements is the same as the number of their interfaces with the subdomain R_1. In the second phase, the SB method is applied to this graph and the domain is bisected with R_2 and R_3 contained in each half.

For initial partitioning in the first phase, an expansion process of Sections 7.3.4 or 7.3.5 can be used. In this algorithm, the first two pseudo-peripheral nodes r_2 and r_3 of the associate graph are constructed. The shortest-route (SR) subtrees rooted from these nodes are named R_2 and R_3, and the nodes are added to these subdomains contour by contour. The expansion process is halted as soon as a suitable number of elements is assigned to R_2 and R_3.

It should be noted that, if the above expansion process is continued for initial partitioning of the domain, then the unrestricted growth of the SR subtrees yields a suboptimal solution, unless some kind of priority is employed in the process of expansion; see Figure 8.8.

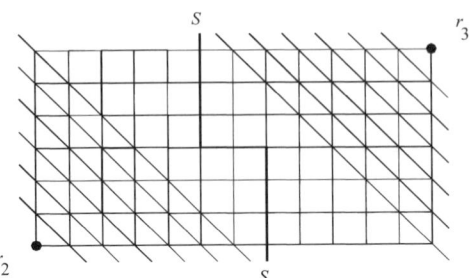

Fig. 8.8 Initial decomposition of a rectangular domain.

In order to obtain a subdomain R_1 with suitable aspect ratio for decomposition in the second phase, the following condition is imposed in the process of expansion,

$$d(\text{Srsubtree}_{r2}) + d(\text{Srsubtree}_{r3}) \leq (d(\text{SRT}) - w(\text{SRT})), \qquad (8\text{-}35)$$

where $d(\text{Srsubtree}_{r2})$ and $d(\text{Srsubtree}_{r3})$ are the numbers of contours of the shortest-route trees (SRTs) generated from r_2 and r_3, and $d(\text{SRT})$ and $w(\text{SRT})$ are the depth and width of the SRT for the main domain.

The algorithm for the mixed method can be summarised as follows:

Step 1: Construct the associate graph of the FEM.

Step 2: Select two pseudo-peripheral nodes of the associate graph.

Step 3: Generate two SR subtrees rooted from the selected nodes and form subdomains R_2 and R_3 by an expansion process using these SRsubtrees. Terminate the expansion process as soon as the above condition is fulfilled.

Step 4: Replace each subdomain R_2 and R_3 with a single node, and form the new reduced associate graph and construct its Laplacian.

Step 5: Calculate the second eigenvalue and eigenvector of the reduced Laplacian.

Step 6: Bisect the graph and the corresponding FEM.

The above algorithm can be applied recursively, for further bisection of the constructed subdomains.

8.3.3 EXAMPLES

Three examples are proposed for illustrating the performance of the mixed method compared to the SB method. The effect of element ordering (or the choice of the initial trial Lanczos vector) in the convergence of the Lanczos and SB methods is illustrated in the third example.

Example 1: A rectangular finite element domain with 4-node quadrilateral elements is considered, as shown in Figure 8.9. This is a mesh with 341 nodes and 300 elements.

The number of elements assigned to each subdomain R_2 and R_3 by the graph–theoretical method is denoted by i, and the number of the remaining elements is $m = n - 2i$. The results are presented in Tables 8.1 and 8.2.

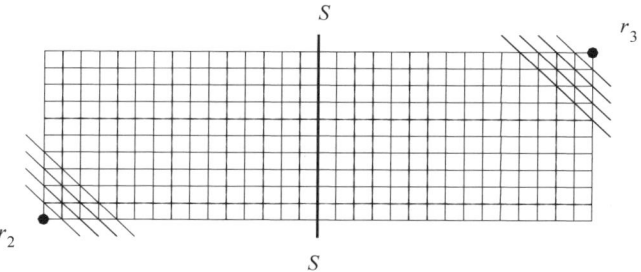

Fig. 8.9 A rectangular FEM.

Table 8.1 Computational time for the mixed and SB methods for Example 1.

Method	i	m	t/t_0	λ_2
SB	0	300	1.0	0.011
Mixed	1	298	1.07	0.011
	15	270	0.99	0.0133
	110	190	0.76	0.0269
	150	150	0.64	0.043
	170	130	0.58	0.057
	190	110	0.56	0.079

Table 8.2 The number of interface edges for different numbers of iterations.

Number of iterations	59	50	45	40	35	30	25
SB method	10	12	>12	>12	>12	>12	>12
Mixed method ($i = 55$)	10	10	10	10	10	10	>10

370 OPTIMAL STRUCTURAL ANALYSIS

Example 2: An I-shaped finite element domain with 4-node quadrilateral elements is considered as illustrated in Figure 8.10. Here, r_2 and r_3 are selected as pseudo-peripheral nodes of the associate graph of the model. The length and width of the SRT for the main domain are 34 and 19, respectively. The results are shown in Tables 8.3 and 8.4, where i is the number of elements in subdomains R_2 and R_3 obtained from the expansion of the SRTs from the selected peripheral nodes.

Table 8.3 Computational time for the mixed and SB methods for Example 2.

Method	i	m	t/t_0	λ_2
SB	0	250	1.0	0.0202
Mixed	1	248	1.10	0.0202
	15	220	0.98	0.0219
	35	180	0.85	0.028

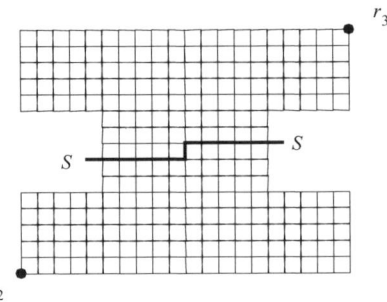

Fig. 8.10 An I-shaped FEM with $n = 250$ quadrilateral elements.

Table 8.4 The number of interface edges for different methods.

Number of iteration	59	55	50	45	40
SB method	11	11	>13	>13	>13
Mixed method ($i = 35$)	11	11	13	13	>13

Example 3: An unsymmetric domain with $n = 300$ elements is considered, as shown in Figure 8.11.

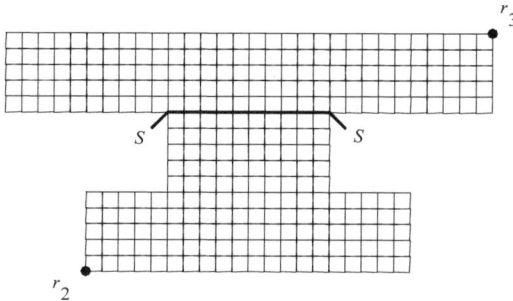

Fig. 8.11 An unsymmetric FEM.

In this example, ordering of the elements is first started from the elements incident to r_2, and then ordering is performed from the elements connected to r_3. For 59 iterations, the result obtained from the first ordering leads to $\lambda_2 = 0.01406$, while the second ordering results in $\lambda_2 = 0.01310$. The first is optimal, and the second is suboptimal. In both ordering, the initial vector was taken as $\mathbf{y}_0 = [\,1\ 0\ 0\ \ldots\ 0]^t$. If in the second ordering the initial vector is changed to $\mathbf{y}_0 = [0\ 0\ 0\ \ldots\ 1]^t$, then the second ordering will also lead to optimal decomposition. The results of the other examples being studied reveal that, for certain numbers of Lanczos iterations, different decompositions will be obtained for different orderings. Further studies are necessary to explore the significance of initial element ordering in the process of decomposition by the SB or the mixed method.

8.3.4 DISCUSSIONS

The mixed method is simple, and incorporates features from both the graph–theoretical and the SB methods. Reduction of the size of the problem in the first phase of the method makes the algebraic process of the second phase more efficient. Evaluation of the eigenvalues and eigenvectors becomes simpler than in the pure algebraic method. In this method, much larger problems can be dealt with because the first phase requires far less computational effort, while using the efficiency of the second phase for better bisection. The mixed method is less sensitive to initial element ordering than the pure algebraic approach.

8.4 SPECTRAL BISECTION FOR ADAPTIVE FEM; WEIGHTED GRAPHS

In this section, the SB method is improved to solve weighted graph-partitioning problems. Then the partitioning of the adaptive FEM, which is mapped into the partitioning of the associate weighted graph, is performed using the improved spectral bisection (ISB) method. The ISB method, similar to the subdomain generation method (SGM), operates on a coarse background mesh. First, the modified (weighted) Laplacian matrix of the coarse mesh is combined with the nodal weight matrix. Then, the eigenvectors corresponding to some of the smallest eigenvalues (3 or 4) of the combined matrix are derived. Employing some heuristic, the mesh is bisected according to the derived eigenvectors and gives a mode, which results in better partitioning. It is shown that the first smallest mode of the combined matrix often yields the best partitioning.

8.4.1 BASIC CONCEPTS

Consider a graph with weights assigned to its nodes and edges. The nodal weight vector is

$$\mathbf{NW} = [nw_i]; \; i = 1, 2, \ldots, N, \tag{8-36}$$

and the edge weight matrix is defined as

$$\mathbf{EW} = [Ew_{ij}]; \; i, j = 1, \ldots, N, \tag{8-37}$$

$$EW_{ij} = \begin{cases} ew_{ij} & \text{if } i \text{ and } j \text{ are adjacent,} \\ 0 & \text{otherwise.} \end{cases} \tag{8-38}$$

The entries of the weighted Laplacian matrix $\mathbf{WL}$ of a weighted graph are defined similarly to those of $\mathbf{L}$ as follows:

$$wl_{ij} = \begin{cases} \sum_{j=1}^{D_i} ew_{ij}, & \text{if } i = j \\ -ew_{ij} & \text{if nodes } i \text{ and } j \text{ are adjacent,} \\ 0 & \text{otherwise.} \end{cases} \tag{8-39}$$

The vector $\mathbf{y}_1 = \{1 \; 1 \; \ldots 1\}^t$ is the eigenvector corresponding to the first smallest eigenvalue $\lambda_1 = 0$ of $\mathbf{WL}$. The matrix $\mathbf{WL}$ is positive semi-definite.

The aim is to partition a given graph into two subgraphs, such that the subgraphs have equal or near equal weights, and the weight of the cutset is minimum. The cutset (edge separator) is a set of edges whose removal results in two disjoint subgraphs. The weight of a cutset is the sum of the weights of its edges. The weight of a subgraph is the sum of the weights of its nodes.

The partitioning vector **x** is defined by Eq. (8-5), but the number of $+1$'s and -1's is not equal in this case. Hence it is not orthogonal to y_1. Instead it is required that $|\mathbf{NW}^t\mathbf{x}|$ be minimised.

The weight of a cutset produced by the partitioning vector **x** is

$$|C(\mathbf{x})| = \tfrac{1}{4} \sum_{(i,j) \in I} (x_i - x_j)^2 ew_{ij}, \qquad (8\text{-}40)$$

which should be minimised. It can be shown that

$$|C(\mathbf{x})| = \tfrac{1}{4} \mathbf{x}^t \mathbf{L} \mathbf{x}. \qquad (8\text{-}41)$$

For a partitioning to be optimal, $|C(\mathbf{x})|$ should be minimised. It is seen that the optimal **x** should minimise $|C(\mathbf{x})|$ and $|\mathbf{NW}^t\mathbf{x}|$, simultaneously. The minimisation of these two functions can be cast in the form of the minimisation of the following objective function:

$$Z = \tfrac{1}{4} \mathbf{x}^t \mathbf{W}\mathbf{L}\mathbf{x} + r|\mathbf{NW}^t\mathbf{x}|. \qquad (8\text{-}42)$$

In the above equation, r is a penalty parameter that balances the weight of the second term against the first one. The lower value of r, forces the minimisation of the first term.

The second term in Eq. (8-42) is now replaced by $r(\mathbf{x}^t \mathbf{NW}^t \mathbf{x})^2$. Therefore

$$Z = \tfrac{1}{4} \mathbf{x}^t \mathbf{W}\mathbf{L}\mathbf{x} + r(\mathbf{x}^t \mathbf{NW})(\mathbf{NW}^t \mathbf{x}), \qquad (8\text{-}43)$$

or

$$Z = \mathbf{x}^t (\tfrac{1}{4} \mathbf{WL} + r\mathbf{NWNW}^t)\mathbf{x}, \qquad (8\text{-}44)$$

or

$$Z = \mathbf{x}^t \mathbf{WL}^* \mathbf{x}. \qquad (8\text{-}45)$$

Since $Z > 0$ for all **x**, the matrix $\mathbf{L}^* = \tfrac{1}{4} \mathbf{WL} + r\mathbf{NWNW}^t$ is positive definite, and its eigenvalues have the following properties:

$$\lambda_1 \leq \lambda_2 \leq \ldots \leq \lambda_n. \qquad (8\text{-}46)$$

The eigenvalue $\mathbf{y}_1$ corresponding to the smallest eigenvalue of $\mathbf{WL}^*$, minimises the quadratic form in Eq. (8-43). Hence the smallest eigenvalue is found, and the optimal partitioning is created from the first eigenvector of $\mathbf{WL}^*$.

It is at this point that the heuristic should be applied. For this purpose, the components of $\mathbf{y}_1$ are stored in ascending order. The first N_1 elements corresponding to the N_1 lowest components of $\mathbf{y}_1$, with the sum of weights equal to $\frac{1}{2}\sum_{i=1}^{N} m_i$, are assigned to one subdomain. The remaining elements are assigned to the second subdomain. It should be noted that N_1 is not necessarily equal to $N/2$ in this case.

In some cases, it is observed that the second or third eigenvector corresponds to an optimal solution. Therefore, in this method the first four eigenvectors are obtained and tested applying the heuristic.

8.4.2 PARTITIONING OF ADAPTIVE FE MESHES

For parallel implementation of adaptive FE meshes, a suitable domain decomposition algorithm should be selected. In this method, the initial FE analysis is performed on a coarse mesh, and element re-meshing parameters are obtained. The nodal mesh parameters are then calculated by nodal averaging.

Employing nodal mesh parameters, the re-meshing is performed and a refined mesh is obtained. If, before re-meshing, the number of elements and the number of edges in each element of the coarse mesh is specified, then the decomposition process can be performed on the coarse mesh. Using a predictor such as that of a neural network, to estimate the number of elements generated after adaptive re-meshing, the initial mesh can be divided into two subdomains by employing the ISB method of the following section.

The initial coarse mesh is partitioned into the desired number of subdomains, such that each subdomain of the refined mesh contains approximately equal numbers of elements, and the number of interfacing boundary edges is minimum or near minimum.

For example, consider an initial coarse mesh and its re-meshing in Figure 8.12.

DECOMPOSITION VIA ALGEBRAIC GRAPH THEORY METHODS 375

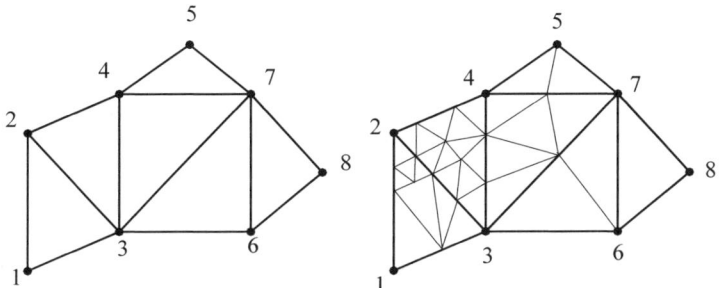

(a) Initial coarse mesh. (b) Fine mesh after re-meshing.

Fig. 8.12 A simple finite element model.

The information about the fine mesh in Figure 8.12(b) is transformed into that of an associate graph as follows:

1. Construct the natural associate graph of the coarse mesh. For this purpose, assign a node to each element and join two nodes by an edge if their corresponding elements have a common edge.

2. Assign an integer to each node of the associate graph, equal to the number of fine elements to be included in the coarse element corresponding to that node.

3. Assign an integer to each edge of the associate graph, equal to the number of interface edges that will be generated on the corresponding common edge, after re-meshing. The weighted associate graph of Figure 8.12 will be obtained as illustrated in Figure 8.13.

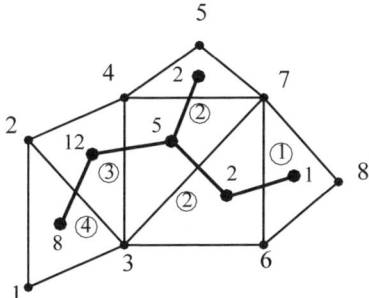

Fig. 8.13 The weighted associate graph.

Now the ISB method can be applied and the partitions of the coarse mesh can be obtained, leading to the optimality criteria being satisfied on the refined mesh.

8.4.3 COMPUTATIONAL RESULTS

Three examples are presented in this section. In the first two examples, the models are taken from [91] and compared to the SGM. In the latter example, the model is bisected by the ISB method. For all the models, the penalty parameter, r, is taken as unity.

Example 1: A square finite element domain is bisected, as shown in Figure 8.14. The results of the ISB and SGMs are illustrated by bold and dashed lines, respectively. The results are presented in Table 8.5, with '*' indicating the best result obtained. Comparison of the two methods is illustrated in Table 8.6.

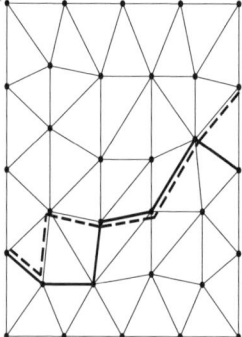

Fig. 8.14 A square finite element domain.

Table 8.5 Bisection obtained from the first four modes, Example 1.

Eig. no.	λ	Interface (1)	Weight imbalance (2)	(1) + (2)
1	0.0404	15	24	39
2*	0.0472	10	6	16
3	0.1214	16	14	30
4	0.1414	20	20	40

* indicating the best result obtained.

Table 8.6 Comparison of ISB and SGM.

Method	Weight imbalance (1)	Cut interface (2)	(1) + (2)
SGM	18	11	29
ISB	6	10	16

Example 2: An L-shaped finite element domain is bisected, as shown in Figure 8.15. The results of the ISB and the SGMs, are illustrated by bold and dashed lines, respectively. In this example, the first mode is selected, since its resulting subdomains are connected. The results are presented in Table 8.7, with '*' indicating the best result obtained. Comparison of the two methods is illustrated in Table 8.8.

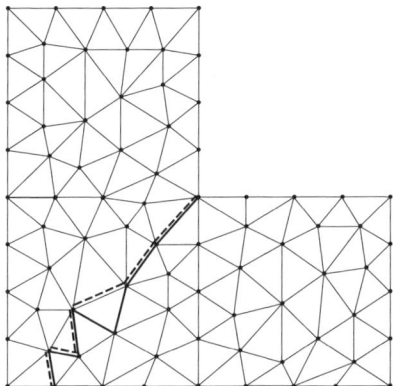

Fig. 8.15 An L-shaped finite element domain.

Table 8.7 Bisection obtained from the first four modes, Example 2.

Eig. no.	λ	Interface (1)	Weight imbalance (2)	(1) + (2)
1	0.0088	17	8	25
2*	0.0181	17	6	23
3	0.0373	22	4	26
4	0.0485	23	2	25

* indicating the best result obtained.

Table 8.8 Comparison of ISB and SGM.

Method	Weight imbalance (1)	Cut interface (2)	(1) + (2)
SGM	16	15	31
ISB	8	17	25

Example 3: An I-shaped finite element domain is considered, as shown in Figure 8.16, and the cut interface is distinguished by bold lines. The results are presented in Table 8.9.

378   OPTIMAL STRUCTURAL ANALYSIS

Table 8.9 Bisection obtained from the first four modes, Example 3.

Eig. no.	λ	Interface (1)	Weight imbalance (2)	(1) + (2)
1*	0.0094	24	8	32
2	0.0219	36	8	44
3	0.0323	46	8	54
4	0.0698	55	0	55

* indicating the best result obtained.

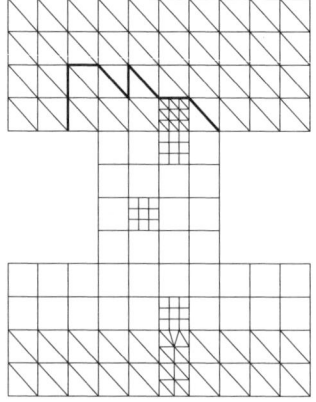

Fig. 8.16 An I-shaped domain bisected by ISB.

8.5 SPECTRAL TRISECTION OF FINITE ELEMENT MODELS

In this section, a new efficient method is presented for finite element domain decomposition. A weighted incidence graph is first constructed for the FEM. A spectral partitioning heuristic is then applied to the graph using the second and the third eigenvalues of the Laplacian matrix of the graph, to partition it to three subgraphs and correspondingly trisect the FEM.

8.5.1 CRITERIA FOR PARTITIONING

In the process of analysis, it is necessary to have information exchange between adjacent subdomains. For example, for the finite element mesh shown in Figure 8.17, the two elements A and B are contained in two different adjacent subdomains. The interrelation between these elements in the overall stiffness matrix is illustrated in Figure 8.18. It can be easily observed that the number of informations exchanged between these two elements is equal to 3, considering the symmetry of the stiffness matrix. This number is given, considering only one unknown per node, and it should be multiplied by the degrees of freedom of each node. This number is called the *communication number* of the two elements. It can easily be proved that, if the number of common nodes of two elements is denoted by n_c, then the communication number will be given as

$$\frac{n_c(n_c+1)}{2}.$$

Obviously, for an optimal partitioning it is necessary to minimise the sum of the communication numbers. Such a minimisation reduces the interaction of the elements, decreasing the time for computation.

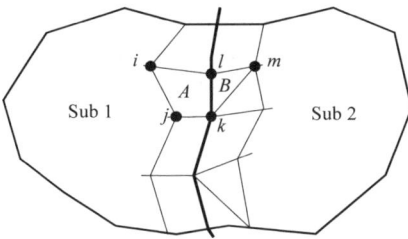

Fig. 8.17 A finite element mesh, and the two elements A and B contained in two different adjacent subdomains.

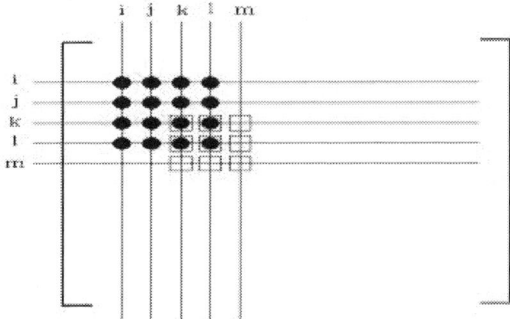

Fig. 8.18 The inter-relation between the elements in the overall stiffness matrix.

On the other hand, in the parallel processing, if the computational time is proportional to the maximum time required for processing each subdomain, the partitioning should be performed in such a way that all subdomains require nearly the same computational time. If the processing time for each subdomain is proportional to the total number of elements in each subdomain, the model should be partitioned into subdomains with equal number of elements. Thus, for an optimal subdomaining, the following considerations are vital:

1. equal number of elements in each subdomain;

2. minimum communication between the elements of each pair of adjacent subdomains.

8.5.2 WEIGHTED INCIDENCE GRAPHS FOR FINITE ELEMENT MODELS

An incidence graph G of an FEM has its vertices in a one-to-one correspondence with the elements of the considered FEM, and two vertices of G are connected by an edge if the corresponding elements have at least one common node. Weights are assigned to the edges of G as the communication numbers for each pair of elements in the FEM. Such a graph is called the *weighted incidence graph* of the considered FEM.

For example, an FEM and its corresponding weighted incidence graph are illustrated in Figure 8.19.

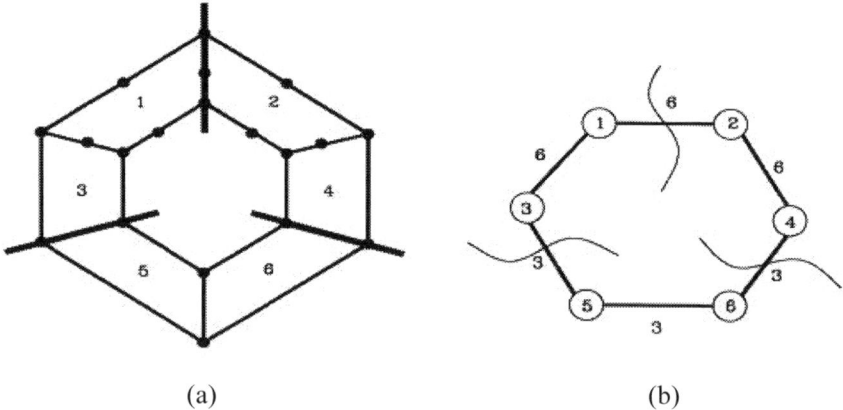

Fig. 8.19 A finite element model and the corresponding weighted incidence graph.

Using this graph, the problem of efficient decomposition of an FEM is transferred into an optimal partitioning of the corresponding graph, with the following properties:

1. The number of nodes for the subgraphs is nearly equal.

2. The sum of the weights of the edges for which the two ends are in two different subgraphs, is minimum.

A graph partitioning into three parts and the corresponding trisection of the FEM, are illustrated in Figure 8.19.

8.5.3 GRAPH TRISECTION ALGORITHM

The vector **x** will partition the graph into three subgraphs containing equal numbers of vertices ($N/3$), if the number of $z_i s$ is $N/3$, and it follows that

$$\sum_{k=1}^{N} x_k = 0, \tag{8-47}$$

where N is assumed to be a multiple of three.

If, on the other hand, $\mathbf{x}^t.e = \mathbf{x}^t.\phi_1 = 0$, this means that the partitioning vector **x** is orthogonal to ϕ_1. It is necessary to assign a measure of how well this vector minimises the communication between the vertices of each pair of subgraphs.

Consider two arbitrary adjacent vertices i and j of the graph. If these two vertices belong to the same subgraph, then,

$$\|x_i - x_j\|^2 = 0, \tag{8-48}$$

where for each complex number $C = A + B_i$, $\|C\| = A^2 + B^2$, and if they belong to different subgraphs, then:

$$\|x_i - x_j\|^2 = 3. \tag{8-49}$$

Thus, the resulting communication between them is equal to $\frac{1}{3}\|x_i - x_j\|^2 w_{ij}$.

So the total communication between adjacent subgraphs is

$$t_c(x) = \frac{1}{3} \sum_{(i,j) \in E} \|x_i - x_j\|^2 \times w_{ij} \tag{8-50}$$

382   OPTIMAL STRUCTURAL ANALYSIS

On the other hand, from the fact that $w_{ij} = 0$ for non-adjacent vertices i and j, it can be written as,

$$t_c(x) = \frac{1}{2} \times \frac{1}{3} \sum_{i=1}^{N} \sum_{j=1}^{N} \|x_i - x_j\|^2 \times w_{ij}, \qquad (8\text{-}51)$$

where the coefficient 1/2 is due to twice consideration of edge (i, j).

According to the definition of the Laplacian matrix of the weighted graph, it can be shown that,

$$t_c(x) = \frac{1}{6} x' L^* x, \qquad (8\text{-}52)$$

and the problem of efficient partitioning is converted to the following optimisation problem:

$$\text{Minimise} \quad t_c(x) = \frac{1}{6} x' L^* x$$

$$\text{s.t.} \quad x'.e = 0 \qquad (8\text{-}53)$$

$$\text{and} \quad x_k \in \left\{+1, \; -\frac{1}{2} + \frac{\sqrt{3}}{2}i, \; -\frac{1}{2} - \frac{\sqrt{3}}{2}i\right\}$$

Separating real and imaginary parts of vector **x**, this vector can be written as

$$\mathbf{x} = u + v_i, \qquad (8\text{-}54)$$

where $u, v \in R^N$. Hence $\sum u_i = \sum v_i = 0$ and $\|x_i - x_j\|^2 = (u_i - u_j)^2 + (v_i - v_j)^2$. This implies that, the problem defined by Eq. (8-53) is equivalent to the following problem:

$$\left| \begin{array}{l} \text{Minimise} \quad t_c(\mathbf{x}) = \dfrac{1}{6}\left(u' L^* u + v' L^* v\right) \\[4pt] \text{s.t.} \quad \mathbf{u}'.\mathbf{e} = \mathbf{v}'.\mathbf{e} = 0 \\[4pt] \text{and} \quad (u_i, v_i) \in \left\{(+1, 0), \left(\dfrac{-1}{2}, \dfrac{\sqrt{3}}{2}\right), \left(\dfrac{-1}{2}, \dfrac{-\sqrt{3}}{2}\right)\right\} \end{array} \right. \qquad (8\text{-}55)$$

On the other hand, since $\sum_{i=1}^{N} u_i = \sum_{i=1}^{N} v_i = 0,$ one can write

$$\mathbf{u}^t.\mathbf{v} = \frac{N}{3}\left(1\times 0 + \frac{-1}{2}\times\frac{\sqrt{3}}{2} + \frac{-1}{2}\times\frac{-\sqrt{3}}{2}\right) = 0$$

$$\mathbf{u}^t.\mathbf{u} = \frac{N}{3}\left[(0)^2 + \left(\frac{-1}{2}\right)^2 + \left(\frac{-1}{2}\right)^2\right] = \frac{N}{2} \qquad (8\text{-}56)$$

$$\mathbf{v}^t.\mathbf{v} = \frac{N}{3}\left[(0)^2 + \left(\frac{\sqrt{3}}{2}\right)^2 + \left(\frac{-\sqrt{3}}{2}\right)^2\right] = \frac{N}{2}$$

A property of real, symmetric matrices such as $\mathbf{L}^*$ is that their eigenvectors are orthogonal and form a basis in R^N. A set of vectors is said to form a *basis* in R^N, if all vectors of R^N can be written as a linear combination of these vectors.

Consequently, for an appropriately chosen set of coefficients a_i and b_i, i = 1,2, ..., N, the vectors $\mathbf{u}$ and $\mathbf{v}$ can be written as

$$\mathbf{u} = \sum_{i=1}^{N} a_i \phi_i \;,$$

and
$$\mathbf{v} = \sum_{i=1}^{N} b_i \phi_i. \qquad (8\text{-}57)$$

On the other hand, $\phi_1 = \mathbf{e}$ and $\mathbf{u}^t.\mathbf{e} = \mathbf{v}^t.\mathbf{e} = 0$; consequently, $a_1 = b_1 = 0$, and

$$\mathbf{u} = \sum_{i=2}^{N} a_i \phi_i,$$

and
$$\mathbf{v} = \sum_{i=2}^{N} b_i \phi_i \;. \qquad (8\text{-}58)$$

Normalising the vectors ϕ_i so that $\phi_i^t.\phi_i = N/2$, $i = 2, ..., N$, we have,

$$\mathbf{u}^t.\mathbf{u} = \frac{N}{2}\left(\sum_{i=2}^{N} a_i \phi_i^t\right)\left(\sum_{i=2}^{N} a_i \phi_i\right) = \frac{N}{2}\left(\sum_{i=2}^{N} a_i^2\right) \qquad (8\text{-}59)$$

from which,

$$\sum_{i=2}^{N} a_i^2 = 1, \qquad (8\text{-}60)$$

and similarly,

$$\sum_{i=2}^{N} b_i^2 = 1. \qquad (8\text{-}61)$$

On the other hand

$$\mathbf{u}.\mathbf{v} = 0 = \left(\sum_{i=2}^{N} a_i \phi_i^t\right)\left(\sum_{i=2}^{N} b_i \phi_i\right) = \frac{N}{2}\left(\sum_{i=2}^{N} a_i b_i\right) \qquad (8\text{-}62)$$

from which,

$$\sum_{i=2}^{N} a_i b_i = 1. \qquad (8\text{-}63)$$

If $\mathbf{L}*\phi_i = \lambda_i \phi_i$, then

$$\mathbf{u}'\mathbf{L}*\mathbf{u} = \left(\sum_{i=2}^{N} a_i \phi_i^t\right)\left(\sum_{i=2}^{N} a_i L*\phi_i\right) = \left(\sum_{i=2}^{N} a_i \phi_i^t\right)\left(\sum_{i=2}^{N} a_i \lambda_i \phi_i\right) = \frac{N}{2}\sum_{i=2}^{N} a_i^2 \lambda_i \qquad (8\text{-}64)$$

and similarly,

$$\mathbf{v}'\mathbf{L}*\mathbf{v} = \frac{N}{2}\sum_{i=2}^{N} b_i^2 \lambda_i. \qquad (8\text{-}65)$$

In this manner, the optimisation problem defined by Eq. (8-55) can be written as

$$\left|\begin{array}{ll} \text{Minimise} & t_c(x) = \dfrac{N}{12}\sum_{i=2}^{N}\left(a_i^2 + b_i^2\right)\lambda_i \\[2mm] \text{s.t.} & \sum_{i=2}^{N} a_i^2 = 1 \; , \; \sum_{i=2}^{N} b_i^2 = 1 \; , \; \sum_{i=2}^{N} a_i b_i = 0 \\[2mm] & u_i = \sum_{i=2}^{N} a_i \phi_i \; , \; v_i = \sum_{i=2}^{N} b_i \phi_i \\[2mm] & (u_i, v_i) \in \{(\;), (\;), (\;)\} \end{array}\right. \qquad (8\text{-}66)$$

It is at this point that the heuristic must be applied. If $N = 3$ is considered, then the conditions (8-60), (8-61) and (8-63) are as follows:

$$a_2^2 + a_3^2 = 1, \quad b_2^2 + b_3^2 = 1, \quad a_2 b_2 + a_3 b_3 = 0. \tag{8-67}$$

and the other conditions are not applied.

The solution of these equations can be considered as follows:

$$\begin{vmatrix} a_2 = \cos\theta, & a_3 = \sin\theta \\ b_2 = -\sin\theta, & b_3 = \cos\theta \end{vmatrix} \tag{8-68}$$

This implies that

$$\begin{bmatrix} u \\ v \end{bmatrix} = \begin{bmatrix} \cos\theta & \sin\theta \\ -\sin\theta & \cos\theta \end{bmatrix} \begin{bmatrix} \phi_2 \\ \phi_3 \end{bmatrix}. \tag{8-69}$$

With this approximation, the components of (u_i, v_i) are not generally the prescribed discrete values and the vertices of the graph cannot be clustered on the $z_k, k = 1, 2, 3$ basis. Instead, the $N/3$ vertices corresponding to the $N/3$ closest numbers to $z_k s$ are assigned to each subgraph.

After computation of ϕ_2 and ϕ_3 by changing θ, values of (u_i, v_i) for each vertex can be computed. According to the above criteria, the optimum value of θ is the value that results in minimum communication between the subgraphs.

For applying the above criteria, for finding closest (u_i, v_i) vectors to $z_k s$, it is sufficient to compute and compare angles between this vector and the z_k vectors.

Figure 8.20(a) shows a typical vector (u_i, v_i) that is close to z_1, implying that the vertex i belongs to the subgraph 1.

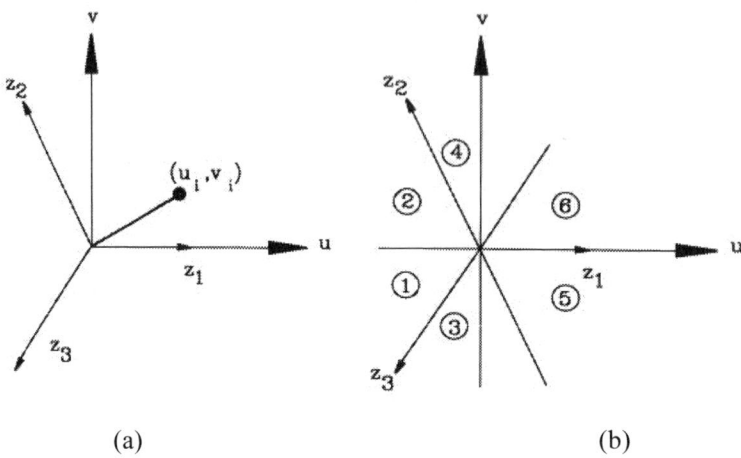

Fig. 8.20 A typical vector.

For example, for the graph shown in Figure 8.19(b), and with

$$\mathbf{L}^* = \begin{bmatrix} 12 & -6 & -6 & & & \\ -6 & 12 & 0 & -6 & & \\ -6 & 0 & 9 & 0 & -3 & \\ & -6 & 0 & 9 & 0 & -3 \\ & & -3 & 0 & 6 & -3 \\ & & & -3 & -3 & 6 \end{bmatrix},$$

the solution of the eigenproblem $\mathbf{L}^*\phi_i = \lambda_i \phi_i$, results in

$\lambda_2 = 3.80$ and $\phi_2 = [-0.41, -0.41, -0.15, -0.15, 0.56, 0.56]^t$
$\lambda_3 = 4.40$ and $\phi_3 = [-0.25, 0.25, -0.56, 0.56, -0.36, 0.36]^t$

For $\theta = 0$, we have $u = \phi_2$ and $v = \phi_3$. (u_i, v_i) for $i = 1,\ldots,6$ in a u-v coordinate system are depicted in Figure 8.20(b). From this Figure, it can be seen that vertices 5 and 6 belong to subgraph 1, vertices 2 and 4 belong to subgraph 2, and vertices 1 and 3 belong to subgraph 3. The corresponding partitioned graph and FEM are illustrated in Figure 8.19.

8.5.4 NUMERICAL RESULTS

Many examples are trisected, and the results for a few of the considered models are presented in the following text.

Example 1: The model of a shell structure is considered, as shown in Figure 8.21(a). This model has 2016 triangular elements. The model is trisected as illustrated in Figure 8.21(b).

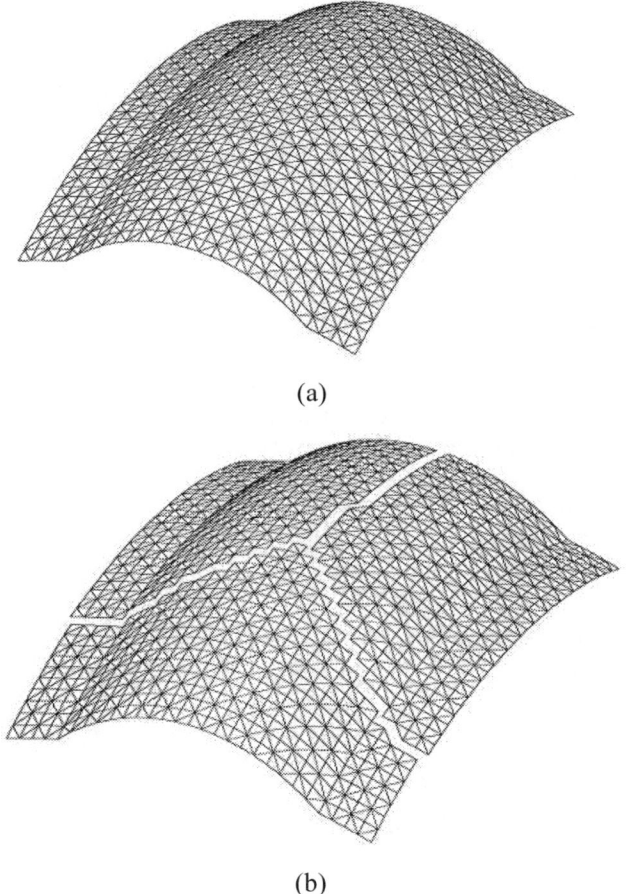

(a)

(b)

Fig. 8.21 A shell-type FEM.

Example 2: The FEM of a nozzle is considered, as shown in Figure 8.22(a). This model contains 1485 rectangular elements. The trisected model is illustrated in Figure 8.22(b).

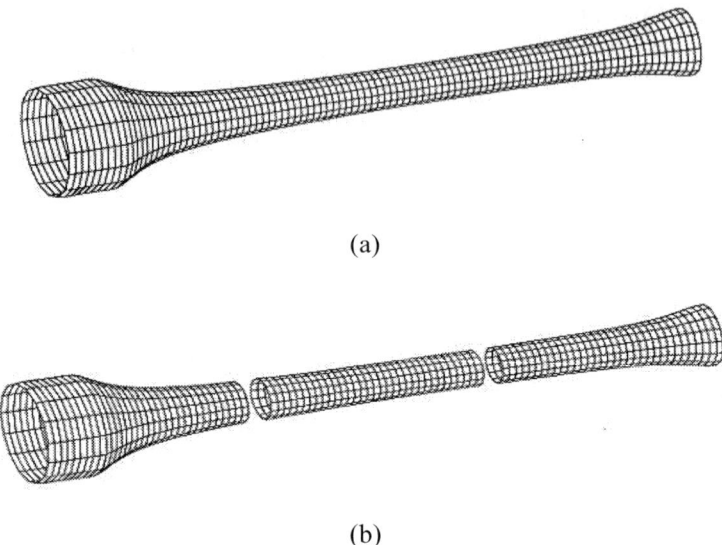

(a)

(b)

Fig. 8.22 The FEM of a nozzle.

Example 3: The FEM of a fan with 1 opening is considered, as shown in Figure 8.23(a). This model contains 1350 rectangular elements. The trisected model is illustrated in Figure 8.23(b) with absolutely balanced nodes.

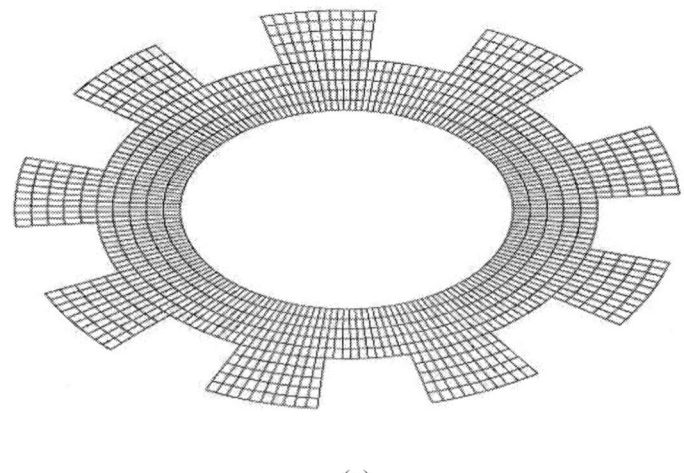

(a)

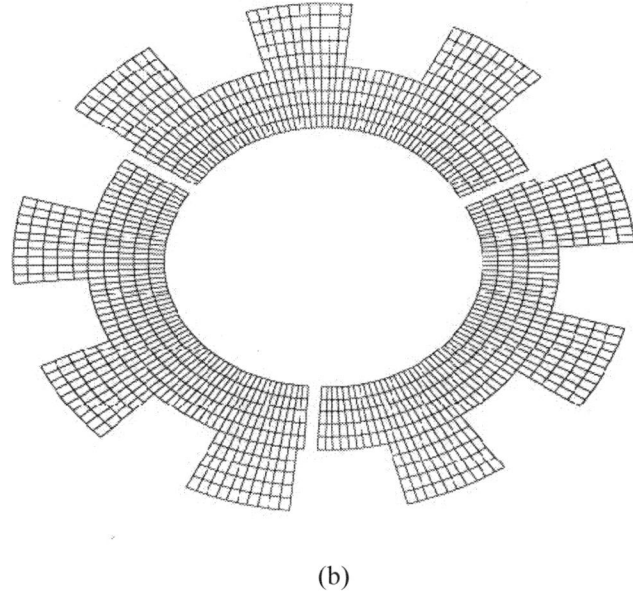

(b)

Fig. 8.23 The FEM of a fan.

8.5.5 DISCUSSIONS

The proposed algorithm for finite element domain decomposition is simple and efficient. A weighted incidence graph is first constructed, and a spectral partitioning heuristic is then applied to the graph, using the second and the third eigenvalues of the Laplacian matrix of the graph, to partition it into three subgraphs, and trisect the corresponding FEM. This method can recursively be applied for further decomposition of the models.

8.6 BISECTION OF FINITE ELEMENT MESHES USING RITZ AND FIEDLER VECTORS

In this section, an efficient algorithm is developed for the decomposition of large-scale FEMs. A weighted incidence graph is used to transform the connectivity properties of finite element meshes into those of graphs. A graph G_0 constructed in this manner is then reduced to a graph G_n of desired size by a sequence of contractions $G_0 \rightarrow G_1 \rightarrow G_2 \rightarrow \ldots \rightarrow G_n$. For G_0, two pseudo-peripheral nodes s_0 and t_0 are selected, and two SRTs are expanded from these nodes. For each starting node, a vector is constructed with N entries, each entry being the shortest distance of a node n_i of G_0 from the corresponding starting node. Hence, two vectors v_1 and v_2 are formed as Ritz vectors for G_0. A similar process is repeated for G_i ($i = 1, 2, \ldots, n$), and the size of the vectors obtained is then extended to N. A Ritz matrix consisting of $2(n + 1)$ normalised Ritz vectors each having N entries is constructed. This matrix is then used in the formation of an eigenvalue problem. The first eigenvector is calculated, and an approximate Fiedler vector is constructed for the bisection of G_0. The performance of the method is illustrated by some practical examples.

8.6.1 DEFINITIONS AND ALGORITHMS

The weighted Laplacian matrix $\mathbf{L}^* = [l^*_{ij}]_{N \times N}$ of a weighted incidence graph is defined as follows:

$$l^*_{ij} = \begin{cases} \sum_k w_{ik} & \text{if } i = j, \\ -w_{ij} & \text{if nodes } i \text{ and } j \text{ are adjacent}, \\ 0 & \text{otherwise}. \end{cases} \quad (8\text{-}70)$$

where N is the number of nodes of the graph corresponding to the number of the elements of the FEM.

An incidence graph G of an FEM has its nodes in a one-to-one correspondence with the elements of the considered FEM, and two nodes of G are connected by an edge if the corresponding elements have at least one common node. Weights are assigned to the edges of G as the communication number for each pair of elements in the FEM. Such a graph is called the *weighted incidence graph* of the considered FEM.

Using this graph, the problem of efficient decomposition of an FEM is transferred into an optimal partitioning of the corresponding graph, with the following properties.

The number of nodes for subgraphs is nearly equal, and the sum of the weights of the edges for which the two ends are in two different subgraphs is minimum.

8.6.2 GRAPH PARTITIONING

The optimal graph-partitioning problem can be formulated as follows:

Find $\mathbf{x} = [x_i]_{N \times 1}$

to minimise communication $= \mathbf{x}^t \mathbf{L}_* \mathbf{x}$ \quad (8-71)

s.t. $\mathbf{x}^t \mathbf{e} = 0$, $\mathbf{e}^t = [1,1,\ldots,1]$

$x_i = +1$ or -1.

The condition $\mathbf{x}^t\mathbf{e} = 0$ means that the number of +1s is the same as the number of −1s (i.e. the equilibrium of the nodes is satisfied). This condition can be fulfilled for graphs with an even number of nodes; however, for those with an odd number of nodes this cannot be fulfilled. Therefore, this condition should be replaced by

$$\text{Minimise } \mathbf{x}^t\mathbf{e}. \tag{8-72}$$

8.6.3 DETERMINATION OF PSEUDO-PERIPHERAL NODES

A pair of pseudo-peripheral nodes (starting nodes) is determined by the following process.

Algorithm

Step 1: Select an arbitrary node v of minimum degree.

Step 2: Generate an $\text{SRT}_v = \{C_1^v, C_2^v, ..., C_d^v\}$ rooted at v. Decompose the subgraph containing the nodes of C_d^v into its components (subcontours).

Step 3: Select one node of minimum degree from each component. Generate an SRT from each of such nodes, and choose the one corresponding to the smallest width.

Step 4: Repeat Steps 2 and 3 as far as reduction in width of the SRT can be observed.

The root "s" and the selected end node "e" are the pseudo-peripheral nodes required.

8.6.4 FORMATION OF AN APPROXIMATE FIEDLER VECTOR

Here, the graph parameters are considered as Ritz vectors, and the second eigenvector of the Laplacian matrix **L** (Fiedler vector) is considered as a linear combination of Ritz vectors [11]. The coefficients for these vectors are in fact the weights of the graph parameters.

Consider the following vector,

$$\overline{\phi} = \sum_{i=1}^{p} w_i \mathbf{v}_i, \tag{8-73}$$

where $\overline{\phi}$ is an approximation to the Fiedler vector, $\mathbf{v}_i$ ($i = 1, ..., p$) are the normalised Ritz vectors representing the graph parameters, and w_i ($i = 1, ..., p$) are the coefficients of the Ritz vectors (Ritz coordinates) which are unknowns, and p is the number of parameters being employed. Equation (8-73) can be written as

$$\bar{\phi} = \mathbf{vw}, \tag{8-74}$$

where $\mathbf{w}$ is a $p \times 1$ vector, and $\mathbf{v}$ is an $N \times p$ matrix containing the Ritz vectors.

Consider the eigenproblem of the Laplacian as

$$\mathbf{L}\phi = \lambda\phi, \tag{8-75}$$

Approximating ϕ by $\bar{\phi}$ and multiplying by $\mathbf{v}^t$ results in,

$$\mathbf{v}^t \mathbf{L} \mathbf{v} \mathbf{w} = \lambda \mathbf{v}^t \mathbf{v} \mathbf{w}, \tag{8-76}$$

or

$$\mathbf{A}\mathbf{w} = \lambda \mathbf{B}\mathbf{w}, \tag{8-77}$$

where $\mathbf{A} = \mathbf{v}^t \mathbf{L} \mathbf{v}$ and $\mathbf{B} = \mathbf{v}^t \mathbf{v}$. Both $\mathbf{A}$ and $\mathbf{B}$ are p × p matrices, and therefore Eq. (8-77) has a much smaller dimension compared to Eq. (8-75); λ is the approximate eigenvalue of the original problem.

Solution of the reduced problem, with dimensions far less than the original one, results in λ_2, and the corresponding eigenvector $\mathbf{w}$ can easily be calculated. Substituting $\mathbf{w}$ in Eq. (8-74) leads to the approximate Fiedler vector.

8.6.5 GRAPH COARSENING

During the coarsening phase, a sequence of $G_0 \rightarrow G_1 \rightarrow G_2 \rightarrow \ldots \rightarrow G_n$ smaller graphs $G_i = (V_i, E_i)$ is constructed from the original graph $G_0 = (V_0, E_0)$, such that $V_i > V_{i+1}$. The graph is constructed from G_i by finding a maximal matching $M_i \subseteq E_i$ of G_i and identifying the nodes that are incident on each edge of the matching. In this process, no more than two nodes are collapsed together, since a matching of a graph is a set of edges, no two of which are incident on the same node. Nodes are not incident on any edge of the matching, but are simply copied over to G_{i+1}. When $u,v \in V_i$ are contracted to form node $w \in V_{i+1}$, the weight of node w is set to be equal to the sum of the weights of nodes u and v, while the edges incident on w are set to be equal to the weight of the edges incident on u and v minus the weight of the edge (u,v). If there is an edge that is incident on both u and v, then the weight of this edge is set to be equal to the sum of the weights of these edges. There are different matchings [79]; here, heavy-edge matching is used. In this method, a matching M_i is computed such that the weight of the edges of M_i is high.

DECOMPOSITION VIA ALGEBRAIC GRAPH THEORY METHODS 393

8.6.6 DOMAIN DECOMPOSITION USING RITZ AND FIEDLER VECTORS

Algorithm

Step 1: Construct the weighted incidence graph G_0 of the FE mesh, and find two pseudo-peripheral nodes s_0 and t_0 using the algorithm of Section 5.5.1.

Step 2: Construct two Ritz vectors $\mathbf{v}_{0s}$ and $\mathbf{v}_{0t}$ for G_0, by forming two SRTs from the selected nodes s_0 and t_0.

Step 3: Condense the graph G_0 by heavy weight matching to obtain a graph of desired size G_n by a sequence of contractions, that is, $G_0 \rightarrow G_1 \rightarrow G_2 \rightarrow \ldots \rightarrow G_n$.

Step 4: Repeat Steps 1 and 2 for G_1, forming two vectors $\mathbf{v}'_{1s}$ and $\mathbf{v}'_{1t}$.

Step 5: Expand $\mathbf{v}'_{1s}$ and $\mathbf{v}'_{1t}$ to $\mathbf{v}_{1s}$ and $\mathbf{v}_{1t}$, as described in Section 8.6.7.

Step 6: Construct two Ritz vectors $\mathbf{v}''_{2s}$ and $\mathbf{v}''_{2t}$ for G_2 in a similar manner, and expand $\sum_{k=1}^{N} = 0$ and $\mathbf{v}''_{2t}$ to $\mathbf{v}'_{2s}$ and $\mathbf{v}'_{2t}$, followed by expansion to $\mathbf{v}_{2s}$ and $\mathbf{v}_{2t}$.

Step 7: Repeat steps similar to Step 5 to G_3, G_4, ..., G_n, and form a pair of Ritz vectors for each $G_i(i = 3, 4, \ldots, n)$.

Step 8: Normalise each vector $\mathbf{v}_{is}$ and $\mathbf{v}_{it}$ ($i = 0, 1, \ldots, n$) and form a Ritz matrix as $\mathbf{v} = [\mathbf{v}_{0s}\ \mathbf{v}_{0t}\ \mathbf{v}_{1s}\ \mathbf{v}_{1t}\ \mathbf{v}_{2s}\ \mathbf{v}_{2t}\ \ldots \mathbf{v}_{ns}\ \mathbf{v}_{nt}]$.

Step 9: Construct $\mathbf{A} = \mathbf{v}^t \mathbf{L} \mathbf{v}$ and $\mathbf{B} = \mathbf{v}^t \mathbf{v}$, obtaining an eigenvalue problem of the form $\mathbf{Aw} = \lambda \mathbf{Bw}$.

Step 10: Calculate the first eigenvector $\mathbf{w}_1$ and form the approximate Fiedler vector as $\mathbf{vw}_1$.

Step 11: Order the entries and construct the status vector $\mathbf{Sv}$ for bisection.

The algorithm is further described by the following example.

8.6.7 ILLUSTRATIVE EXAMPLE

Consider the model G_0 shown in Figure 8.24(a). For the corresponding graph illustrated in Figure 8.24(b), the pseudo-peripheral nodes and the Ritz vectors are as follows:

Graph G_0: start node = s_0 = 18, end node = e_0 = 7

$$\mathbf{v}_{0s} = [d(s_0, i)] = [5,4,3,2,2,2,7,6,5,4,3,2,1,1,7,6,5,0],$$

$$\mathbf{v}_{0t} = [d(e_0, i)] = [2,3,4,5,6,7,0,1,2,3,4,5,6,7,1,1,2,7].$$

The entries of $\mathbf{v}_{0s}$ and $\mathbf{v}_{0t}$ are the distances of each node of G_0 from the starting nodes s_0 and e_0, respectively.

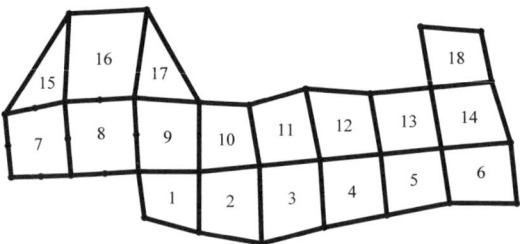

(a) The finite element model.

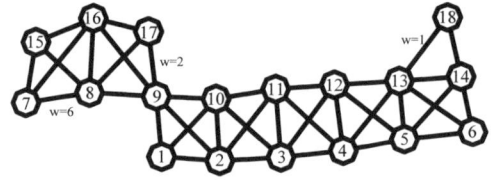

(b) Incidence graph of the model.

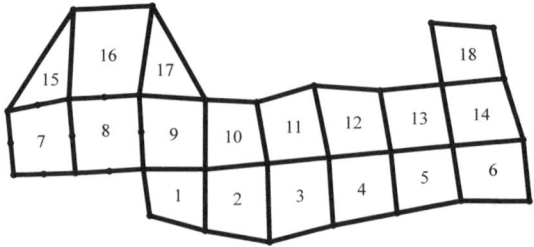

(c) The model of G_1.

DECOMPOSITION VIA ALGEBRAIC GRAPH THEORY METHODS 395

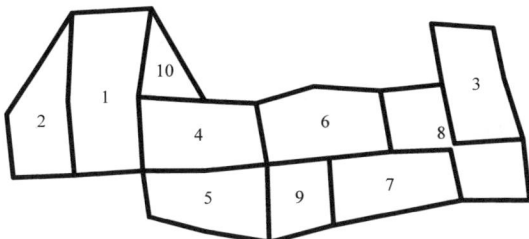

(d) The model of G_2.

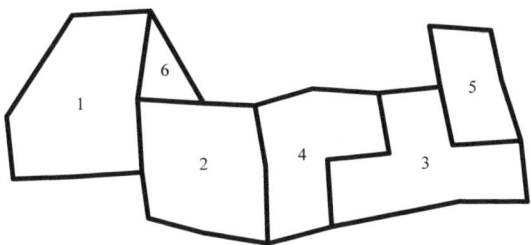

(e) The bisected model.

Fig. 8.24 The process of contraction of the example.

For the contracted graph, the model G_1 shown in Figure 8.24(c), the starting nodes and the corresponding Ritz vectors are as follows:

Graph G_1: start node = $s_1 = 2$, end node = $e_1 = 3$

$$\mathbf{v}''_{2t} = [d(s_1,i)] = [1,0,5,2,\underline{2},3,4,4,3,2],$$

$$\mathbf{v}'_{1t} = [d(e_1,i)] = [4,5,0,3,3,2,1,1,2,4].$$

These vectors are now expanded to the following vectors:

Vector expansion: the nodes 1,2 of G_0 are the same as node 5 of G_1 and therefore

$$\mathbf{v}_{1s} = [\underline{2},\underline{2},3,4,4,4,0,1,2,2,3,3,4,5,0,1,2,5].$$

396   OPTIMAL STRUCTURAL ANALYSIS

Similarly

$$\mathbf{v}_{1t} = [3,3,2,1,1,1,5,4,3,3,2,2,1,0,5,4,4,0].$$

For the contracted graph, the model G_2 shown in Figure 8.24(d), the starting nodes and the corresponding Ritz vectors are as follows:

Graph G_2: start node = s_2 = 5 , end node = e_2 = 1:

$$\mathbf{v}''_{2s} = [d(s_2,i)] = [\underline{4},3,1,2,0,4]$$
$$\mathbf{v}''_{2t} = [d(e_2,i)] = [0,1,3,2,4,1]$$

These vectors are now expanded to the following vectors:

Vector expansion: the nodes 1,2 of G_1 are the same as node 1 of G_2.

Vector expansion: the nodes 8, 16 of G_0 are the same as node 1 of G_1, and the nodes 7, 15 of G_0 are the same as node 2 of G_1, Therefore,

$$\mathbf{v}'_{2s} = [\underline{4},\underline{4},0,3,3,2,1,1,2,4],$$
$$\mathbf{v}'_{2t} = [0,0,4,1,1,2,3,3,2,1].$$

Further expansions of these vectors result in

$$\mathbf{v}_{2s} = [3,3,2,1,1,1,\underline{4},\underline{4},3,3,2,2,1,0,\underline{4},\underline{4},4,0],$$
$$\mathbf{v}_{2t} = [1,1,2,3,3,3,0,0,1,1,2,2,3,4,0,0,1,4]$$

Now orthogonalising the vectors $\mathbf{v}_i$ (i = 1,2,3) leads to the Ritz matrix as:

$$\mathbf{V} = [\mathbf{v}_{1s},\mathbf{v}_{1t},\mathbf{v}_{2s},\mathbf{v}_{2t},\mathbf{v}_{3s},\mathbf{v}_{3t}].$$

For the reduced eigenvalue problem $\mathbf{Aw} = \lambda\mathbf{Bw}$, the first eigenvector is

$$\mathbf{w}_1 = [0.757, -0.889, 1.000, 0.216, 0.701, -0.129].$$

Now the approximate Fiedler vector is calculated as

$$\mathbf{vw}_1 = [-0.45, -0.17, 0.13, 0.42, 0.58, 0.73, -0.89, -0.74, -0.45, -0.17, 0.13, 0.41,$$
$$0.71, 0.87, -0.74, -0.75, -0.61, 1.00],$$

and the status vector results as

$$\mathbf{Sv} = [- - + + + + - - - - + + + + - - - +]$$

DECOMPOSITION VIA ALGEBRAIC GRAPH THEORY METHODS   397

The model is now bisected, as illustrated in Figure 8.24(e), into two subdomains as subdomain 1 containing the set of elements 1,2,7,8,9,10,15,16,17, and subdomain 2 having the set of elements 3,4,5,6,11,12,13,14,18.

8.6.8 NUMERICAL RESULTS

Example 1: An H-shaped FE mesh with one opening comprising of 13,380 nodes and 13,000 rectangular elements is considered, as shown in Figure 8.25(a).

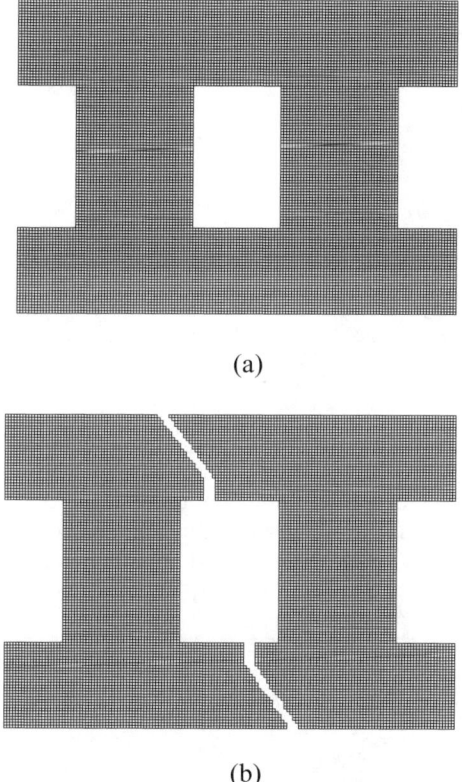

(a)

(b)

Fig. 8.25 An FE mesh with one opening and its bisection.

After nine phases of contraction, G_9 with 200 nodes is obtained. The number of Ritz vectors is 20. The bisected model is shown in Figure 8.25(b). Note that, for this problem, the spectral method solves a 13,380 × 13,380 matrix, while the present method uses a 200 × 200 matrix.

398 OPTIMAL STRUCTURAL ANALYSIS

Example 2: The FE mesh of a tunnel comprising of 6888 nodes and 6720 rectangular elements is considered, as shown in Figure 8.26(a). After nine phases of contraction, G_9 with 100 nodes is obtained. The number of Ritz vectors is 20. The bisected model is shown in Figure 8.26(b). Here, we have an eigensolution of a 100×100 matrix in place of a $6,888 \times 6,888$ matrix.

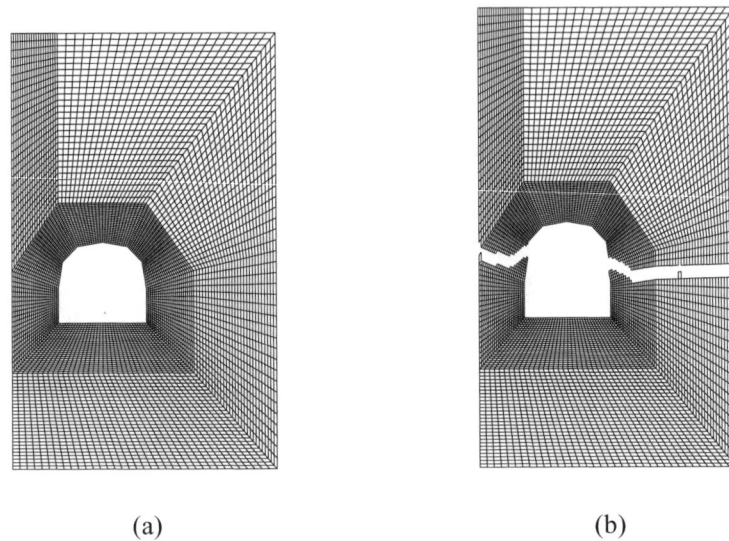

(a) (b)

Fig. 8.26 An FE mesh for a tunnel and its bisection.

Example 3: The FE mesh of a twin tunnel comprising of 13,699 nodes and 13,440 rectangular elements is considered, as shown in Figure 8.27(a). After eight phases of contraction, G_8 with 200 nodes is obtained. The number of Ritz vectors is 18. The bisected model is shown in Figure 8.27(b).

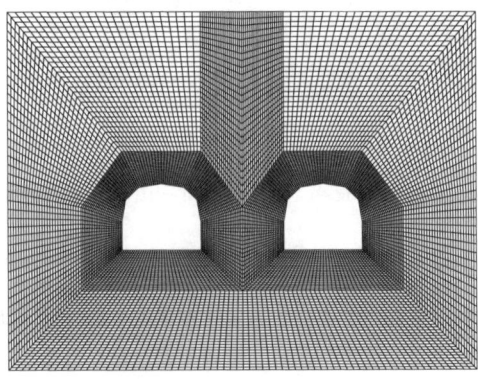

(a)

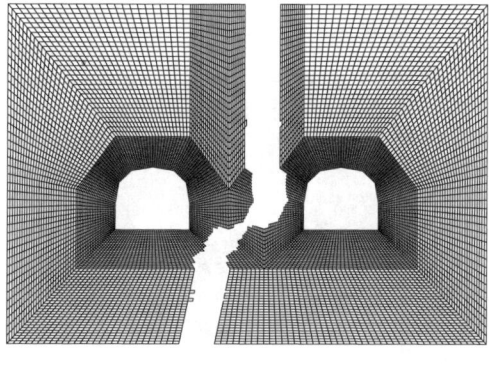

(b)

Fig. 8.27 The FE mesh of a twin tunnel and its bisection.

Example 4: The FE mesh of a castellated beam comprising of 13,464 nodes and 13,056 rectangular elements is considered, as shown in Figure 8.28(a).

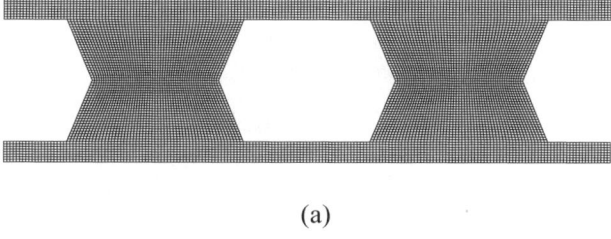

(a)

After five phases of contraction, G_5 with 1000 nodes is obtained. The number of Ritz vectors is 12. The bisected model is shown in Figure 8.28(b).

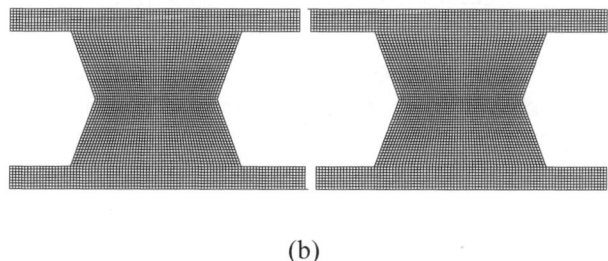

(b)

Fig. 8.28 The FE mesh of a castellated beam and its bisection.

Example 5: The FE mesh of a torus-shaped model comprising of 12,000 nodes and 12,000 rectangular elements is considered, as shown in Figure 8.29(a). After

400   OPTIMAL STRUCTURAL ANALYSIS

five phases of contraction, G_5 with 1000 nodes is obtained. The number of Ritz vectors is 12. The bisected model is shown in Figure 8.29(b).

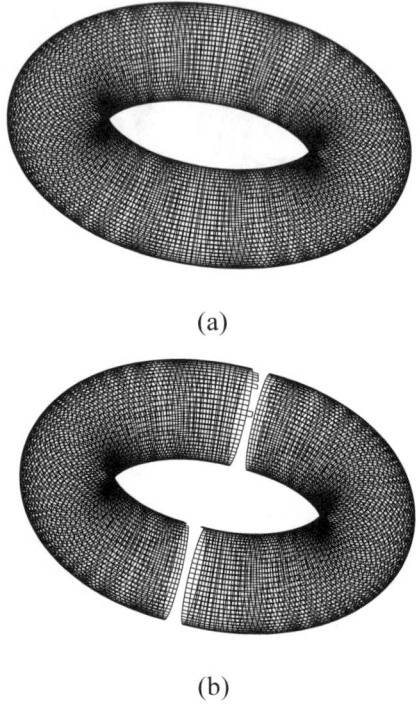

(a)

(b)

Fig. 8.29 A torus-shaped FE mesh and its bisection.

Details of the examples are collectively presented in Table 8.10

Table 8.10 The results of bisection.

No. of Ritz vectors	No. of phases	Final no. of nodes	No. of elements	No. of nodes	Example
20	9	200	13,000	13,380	1
20	9	100	6720	6888	2
18	8	200	13,440	13,699	3
12	5	1000	13,056	13,464	4
12	5	1000	12,000	12,000	5

8.6.9 DISCUSSIONS

The performance of the present method, illustrated in Section 8.6.8, compares well with that of pure algebraic graph methods, with a substantial reduction in the size of the eigensolution involved. In order to notice the difference, columns 2 and 4 of Table 8.10 can be compared for the dimensions of the matrices involved in the original spectral method and the present approach. This, in turn, increases the accuracy and reduces the computational time. Large-scale problems can be handled much more easily in the condensed form.

EXERCISES

8.1 Form the Laplacian matrix of the following graph, and calculate its second eigenvalue.

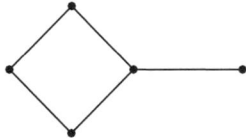

8.2 Construct the Fiedler vector for the following graph.

8.3 Decompose the nodes of the following graph using its Fiedler vector.

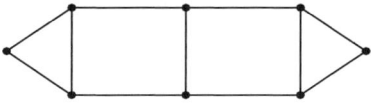

8.4 Construct the associate graph G of the following finite element model and bisect G. Then form the corresponding subdomains.

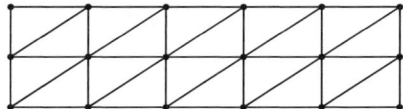

402   OPTIMAL STRUCTURAL ANALYSIS

8.5   Decompose the following FEM into $q = 3$ subdomains using the incidence graph to represent its connectivity property:

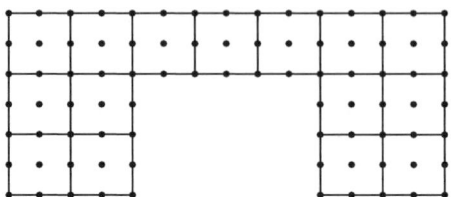

8.6   Use the spectral bisection method to decompose the following graph, and control its lower bound.

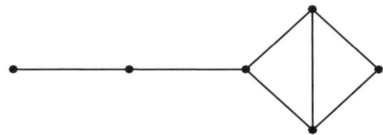

CHAPTER 9

Decomposition and Nodal Ordering of Regular Structures Using Graph Products

9.1 INTRODUCTION

Algebraic graph theory can be considered as a branch of graph theory in which eigenvalues and eigenvectors of certain matrices are employed to deduce the principal properties of a graph. In fact, eigenvalues are closely related to most of the invariants of a graph, linking one extremal property to another. Eigenvalues play a central role in the fundamental understanding of graphs. There are a number of interesting books on algebraic graph theory: Biggs [13], Cvetković et al. [34,35], Seidel [196], Chung [27], and Godsil and Royle [61].

One of the major contributions to algebraic graph theory is due to Fiedler [51], who introduced the properties of the second eigenvalue and eigenvector of the Laplacian of a graph. The latter, known as the *Fiedler vector*, is used in graph nodal ordering and bipartition; see Kaveh and Rahimi Bondarabady [123–127].

General methods are available in the literature for calculating the eigenvalues of matrices; however, for matrices corresponding to regular models, it is beneficial to make use of their regularity.

In this chapter, efficient methods are presented for calculating the eigenvalues of regular structural models. The eigenvalues of the adjacency and Laplacian matrices for a regular graph model are easily obtained by evaluation of the eigenvalues of its generators. The second eigenvalue of the Laplacian of a graph is also obtained using a much faster and simpler approach than existing methods. Once the second eigenvector $\mathbf{v}_2$ of the Laplacian matrix is calculated, the bisection of the model can be performed. This is achieved by arranging the entries of $\mathbf{v}_2$ in an ascending order and partitioning the nodes into two subsets according to their occurrence in $\mathbf{v}_2$.

It should be noted that in the present chapter the term "generator" is used in its literal sense and the term "regular" is employed in both literal and mathematical senses.

9.2 DEFINITIONS OF DIFFERENT GRAPH PRODUCTS

Many structures have regular patterns and can be viewed as the Cartesian product, strong Cartesian product, or direct product of a number of simple graphs. These subgraphs, used in the formation of the entire model, are called the *generators* of that model. Graph products were developed in the past 50 years (see e.g. Berge [12], Sabidussi [193], Harary and Wilcox [74], and Imrich and Klavzar [86]), and are employed in nodal ordering and domain decomposition by Kaveh and Rahami [121–122]. In this chapter, very efficient methods are presented for graph partitioning, domain decomposition and nodal ordering of regular structures.

9.2.1 BOOLEAN OPERATION ON GRAPHS

To explain the products of graphs, let us consider a graph S as a subset of all unordered pairs of its nodes. The node set and member set of S are denoted by $N(S)$ and $M(S)$, respectively. The nodes of S are labelled as $v_1, v_2, \ldots, v_M$, and the resulting graph is a *labelled graph*. Two distinct adjacent nodes, v_i and v_j, form a member, denoted by $v_i v_j \in M(S)$.

A Boolean operation on an ordered pair of disjoint labelled graphs K and H results in a labelled graph S, which has $N(K) \times N(H)$ as its nodes. The set $M(S)$ of members of S is expressed in terms of the members in $M(K)$ and $M(H)$, differently for each Boolean operations. Three different operations are discussed in this chapter, corresponding to Cartesian product, strong Cartesian product and direct product of two graphs.

9.2.2 CARTESIAN PRODUCT OF TWO GRAPHS

Many structures have regular patterns and can be viewed as the Cartesian product of a number of simple graphs. These subgraphs, which are used in the formation of a model, are called the *generators* of that model.

DECOMPOSITION AND NODAL ORDERING

The simplest Boolean operation on a graph is the Cartesian product $K \times H$ introduced by Sabidussi [193]. The Cartesian product is a Boolean operation $S = K \times H$, in which, for any two nodes $u = (u_1, u_2)$ and $v = (v_1, v_2)$ in $N(K) \times N(H)$, the member uv is in $M(S)$ whenever

$$u_1 = v_1 \text{ and } u_2 v_2 \in M(H) \qquad (9\text{-}1a)$$

or

$$u_2 = v_2 \text{ and } u_1 v_1 \in M(K). \qquad (9\text{-}1b)$$

As an example, the Cartesian product of $K = P_2$ and $H = P_3$ is shown in Figure 9.1.

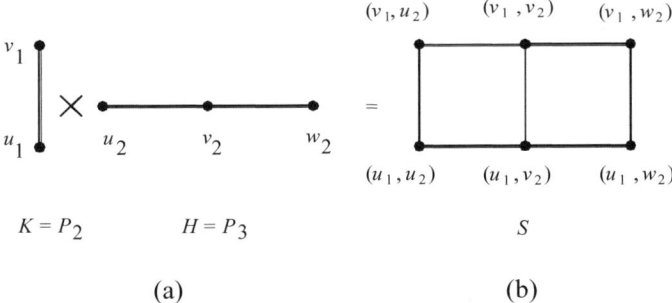

Fig. 9.1 The Cartesian product of two simple graphs.

In this product, the two nodes (u_1, v_2) and (v_1, v_2) are joined by a member, since the condition (9-1b) is satisfied.

The Cartesian product of two graphs K and H can be constructed by taking one copy of H for each node of K and joining copies of H corresponding to adjacent nodes of K by matching the size $N(H)$.

The graphs K and H will be referred to as the *generators* of S. The Cartesian product operation is symmetric, that is, $K \times H \cong H \times K$. For other useful graph operations, the reader may refer to the work by Gross and Yellen [69].

Examples: In this example, the Cartesian product $C_7 \times P_5$ of the path graph with 5 nodes denoted by P_5 and a cycle graph shown by C_7 is illustrated in Figure 9.2.

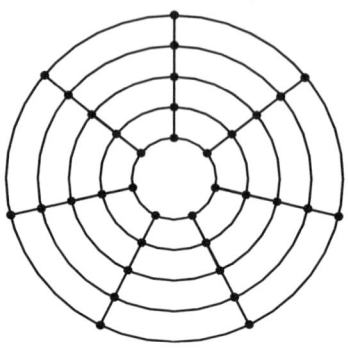

Fig. 9.2 Representation of $C_7 \times P_5$.

Two representations of the Cartesian product $C_3 \times P_4$ are illustrated in Figure 9.3.

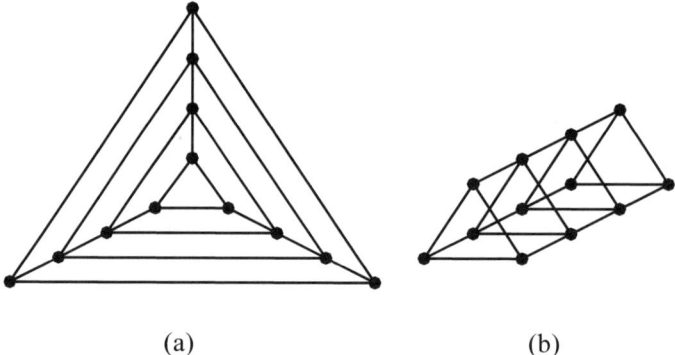

(a)   (b)

Fig. 9.3 Two different representations of $C_3 \times P_4$.

The Cartesian product $P_{m1} \times P_{m2} \times P_{m3}$ of three paths forms a three-dimensional mesh. As an example, the Cartesian product of $P_6 \times P_4 \times P_5$, resulting in a $5 \times 3 \times 4$ mesh, is shown in Figure 9.4.

A graph can be the product of more than two specific graphs, such as paths and cycles. As an example, the product of three graphs, $P_2 \times K_3 \times P_4$, is shown in Figure 9.5. The product of a general graph and a path, $S \times P_4$, is illustrated in Figure 9.6.

DECOMPOSITION AND NODAL ORDERING

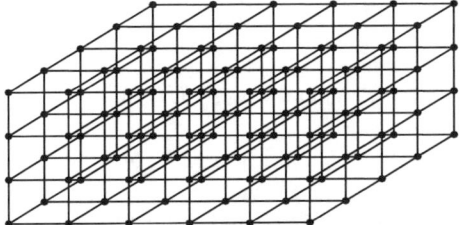

Fig. 9.4 Representation of a 5 × 3 × 4 mesh.

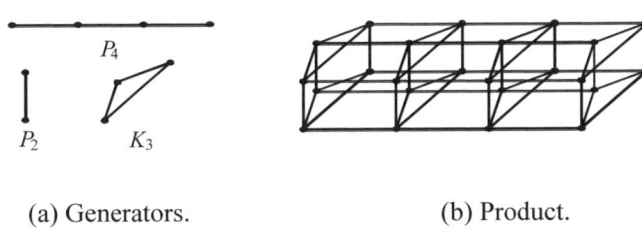

(a) Generators.   (b) Product.

Fig. 9.5 The Cartesian product of three graphs $P_2 \times K_3 \times P_4$.

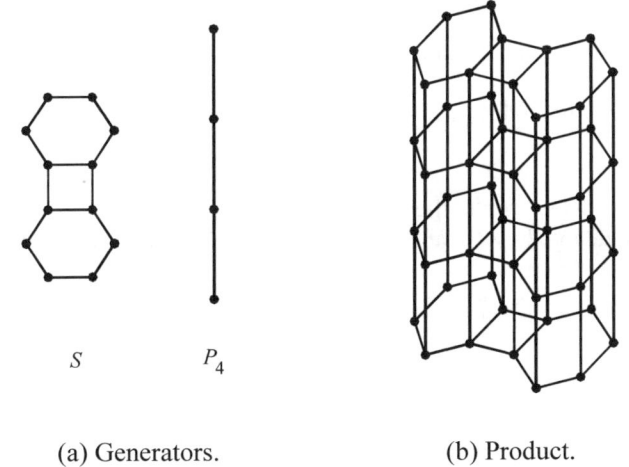

(a) Generators.   (b) Product.

Fig. 9.6 The Cartesian product of S by P_4.

9.2.3 STRONG CARTESIAN PRODUCT OF TWO GRAPHS

This is another Boolean operation, known as the *strong Cartesian product*. The strong Cartesian product is a Boolean operation $S = K \boxtimes H$ in which, for any two distinct nodes $u = (u_1, u_2)$ and $v = (v_1, v_2)$ in $N(K) \times N(H)$, the member uv is in $M(S)$ if

$$u_1 = v_1 \text{ and } u_2 v_2 \in M(H) \text{ or} \qquad (9\text{-}2\text{a})$$

$$u_2 = v_2 \text{ and } u_1 v_1 \in M(K) \text{ or} \qquad (9\text{-}2\text{b})$$

$$u_1 v_1 \in M(K) \text{ and } u_2 v_2 \in M(H). \qquad (9\text{-}2\text{c})$$

As an example, the strong Cartesian product of $K = P_2$ and $H = P_3$ is shown in Figure 9.7.

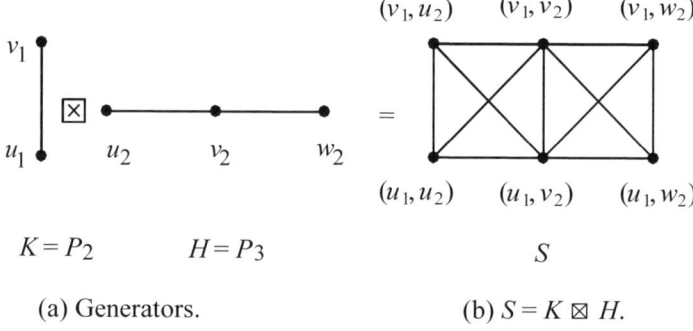

(a) Generators. (b) $S = K \boxtimes H$.

Fig. 9.7 The strong Cartesiasn product of two simple graphs.

In this example, the nodes (u_1, u_2) and (v_1, v_2) are joined, since the condition (9-2c) is satisfied.

Examples: In this example, the strong Cartesian product $P_7 \boxtimes P_5$ of a path graph with 7 nodes, denoted by P_7, and the path graph P_5 is illustrated in Figure 9.8.

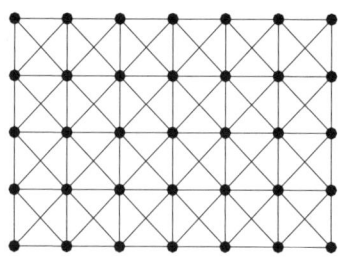

Fig. 9.8 Strong product representation of $P_7 \boxtimes P_5$.

As a second example, the strong Cartesian product $C_7 \boxtimes P_4$ is shown in Figure 9.9.

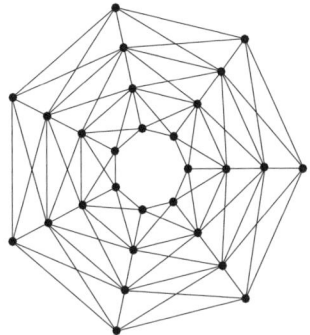

Fig. 9.9 Strong product representation of $C_7 \boxtimes P_4$.

9.2.4 DIRECT PRODUCT OF TWO GRAPHS

This is another Boolean operation, known as the *direct product*, introduced by Weichsel [226], who called it the *Kronecker Product*. The direct product is a Boolean operation $S = K * H$, in which, for any two nodes $u = (u_1, u_2)$ and $v = (v_1, v_2)$ in $N(K) \times N(H)$, the member uv is in $M(S)$ if

$$u_1 v_1 \in M(K) \text{ and } u_2 v_2 \in M(H). \tag{9-3}$$

As an example, the direct product of $K = P_2$ and $H = P_3$ is shown in Figure 9.10.

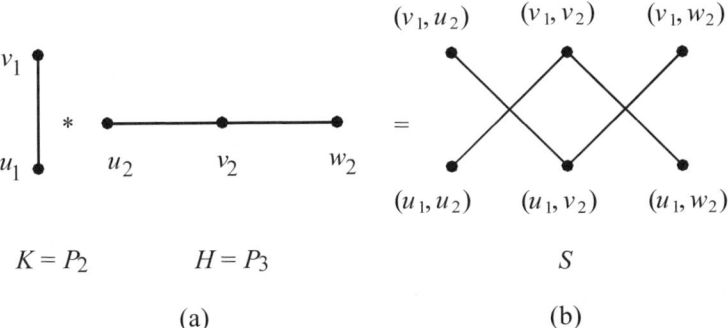

Fig. 9.10 The direct product of two simple graphs.

Here, the two nodes (u_1, u_2) and (v_1, v_2) are joined, since the condition (9-3) is satisfied.

Examples: The direct product $P_7 * P_5$ of the path graph P_7 and path graph P_5 is illustrated in Figure 9.11.

410   OPTIMAL STRUCTURAL ANALYSIS

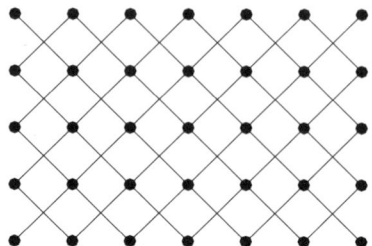

Fig. 9.11 Direct product representation of $P_7 * P_5$.

As a second example, the direct product $C_7 * P_4$ is shown in Figure 9.12.

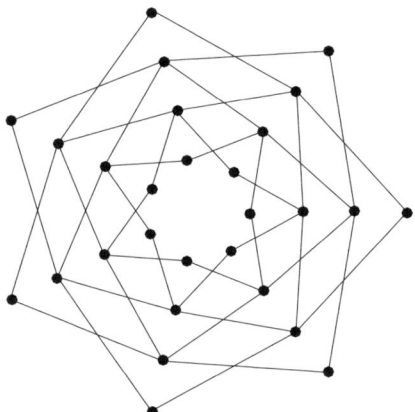

Fig. 9.12 Direct product representation of $C_7 * P_4$.

9.3 EIGENVALUES OF GRAPH MATRICES FOR DIFFERENT PRODUCTS

In this section, the basic definitions and theorems necessary for the study of eigenvalues and eigenvectors of regular graphs are presented. These products consist of Cartesian product, strong Cartesian product and direct product. The primary tool for this study is the Kronecker product as defined in the following section.

9.3.1 KRONECKER PRODUCT

The *Kronecker product* of two matrices **A** and **B** is the matrix we get by replacing the *ij*th entry of **A** by $a_{ij}\mathbf{B}$, for all i and j.

As an example,

$$\begin{bmatrix} 1 & 1 \\ 1 & 0 \end{bmatrix} \otimes \begin{bmatrix} a & b \\ c & d \end{bmatrix} = \begin{bmatrix} a & b & a & b \\ c & d & c & d \\ a & b & 0 & 0 \\ c & d & 0 & 0 \end{bmatrix}, \qquad (9\text{-}4)$$

where entry 1 in the first matrix has been replaced by a complete copy of the second matrix.

The Kronecker product has the property that, if **B**, **C**, **D** and **E** are four matrices such that **BD** and **CE** exist, then

$$(\mathbf{B} \otimes \mathbf{C})(\mathbf{D} \otimes \mathbf{E}) = \mathbf{BD} \otimes \mathbf{CE}. \qquad (9\text{-}5)$$

Thus, if **u** and **v** are vectors of the correct dimensions, then

$$(\mathbf{B} \otimes \mathbf{C})(\mathbf{u} \otimes \mathbf{v}) = \mathbf{Bu} \otimes \mathbf{Cv}. \qquad (9\text{-}6)$$

If **u** and **v** are eigenvectors of **B** and **C**, with eigenvalues λ and μ, respectively, then,

$$\mathbf{Bu} \otimes \mathbf{Cv} = \lambda\mu\, \mathbf{u} \otimes \mathbf{v}, \qquad (9\text{-}7)$$

where $\mathbf{u} \otimes \mathbf{v}$ is an eigenvector of $\mathbf{B} \otimes \mathbf{C}$ with eigenvalue $\lambda\mu$.

The associativity property of the Kronecker product will be used in the proof of the theorem in the following section.

9.3.2 CARTESIAN PRODUCT

For a Cartesian product $K \times H$, the adjacency matrix **A** can be written as

$$\mathbf{A}(K \times H) = \mathbf{A}(K) \otimes \mathbf{I}_{N(H)} + \mathbf{I}_{N(K)} \otimes \mathbf{A}(H). \qquad (9\text{-}8)$$

In this relation, $\mathbf{A}(K) \otimes \mathbf{I}_{N(H)}$ is the adjacency matrix of $N(H)$ node-disjoint copies of K, and $\mathbf{I}_{N(K)} \otimes \mathbf{A}(H)$ is the adjacency matrix of $N(K)$ node-disjoint copies of H.

412   OPTIMAL STRUCTURAL ANALYSIS

As an example, the adjacency matrix of the graph S in Figure 9.1 can be obtained as

$$\mathbf{A}(K) = \begin{bmatrix} 0 & 1 \\ 1 & 0 \end{bmatrix} \text{ and } \mathbf{A}(H) = \begin{bmatrix} 0 & 1 & 0 \\ 1 & 0 & 1 \\ 0 & 1 & 0 \end{bmatrix},$$

$$\mathbf{A}(S) = \begin{bmatrix} 0 & 0 & 0 & 1 & 0 & 0 \\ 0 & 0 & 0 & 0 & 1 & 0 \\ 0 & 0 & 0 & 0 & 0 & 1 \\ 1 & 0 & 0 & 0 & 0 & 0 \\ 0 & 1 & 0 & 0 & 0 & 0 \\ 0 & 0 & 1 & 0 & 0 & 0 \end{bmatrix} + \begin{bmatrix} 0 & 1 & 0 & 0 & 0 & 0 \\ 1 & 0 & 1 & 0 & 0 & 0 \\ 0 & 1 & 0 & 0 & 0 & 0 \\ 0 & 0 & 0 & 0 & 1 & 0 \\ 0 & 0 & 0 & 1 & 0 & 1 \\ 0 & 0 & 0 & 0 & 1 & 0 \end{bmatrix} = \begin{bmatrix} 0 & 1 & 0 & 1 & 0 & 0 \\ 1 & 0 & 1 & 0 & 1 & 0 \\ 0 & 1 & 0 & 0 & 0 & 1 \\ 1 & 0 & 0 & 0 & 1 & 0 \\ 0 & 1 & 0 & 1 & 0 & 1 \\ 0 & 0 & 1 & 0 & 1 & 0 \end{bmatrix}.$$

In the case of the Cartesian product, a similar relationship holds for the Laplacian matrices, that is,

$$\mathbf{L}(K \times H) = \mathbf{L}(K) \otimes \mathbf{I}_{N(H)} + \mathbf{I}_{N(K)} \otimes \mathbf{L}(H). \tag{9-9}$$

As an example, consider K and H as P_4 and P_3, respectively, for the nodal numbering shown in Figure 9.13.

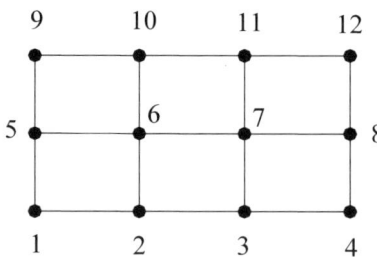

Fig. 9.13 The Cartesian product of P_4 and P_3.

The Laplacian of $P_4 \times P_3$ can be written as

$$\mathbf{L}(P_4 \times P_3) = \begin{bmatrix} \mathbf{A}_1 & -\mathbf{I}_4 & \mathbf{0} \\ -\mathbf{I}_4 & \mathbf{A}_2 & -\mathbf{I}_4 \\ \mathbf{0} & -\mathbf{I}_4 & \mathbf{A}_1 \end{bmatrix},$$

where the adjacency matrix of P_4 is written as

$$\mathbf{A}_1 = \begin{bmatrix} 0 & 1 & 0 & 1 \\ 1 & 0 & 1 & 0 \\ 0 & 1 & 0 & 1 \\ 1 & 0 & 1 & 0 \end{bmatrix}.$$

The above equation can be written as

$$\begin{bmatrix} \mathbf{A}_1 & -\mathbf{I}_4 & 0 \\ -\mathbf{I}_4 & \mathbf{A}_2 & -\mathbf{I}_4 \\ 0 & -\mathbf{I}_4 & \mathbf{A}_1 \end{bmatrix} = (\mathbf{A}_1 - \mathbf{I}_4) \otimes \mathbf{I}_3 + \begin{bmatrix} \mathbf{I}_4 & -\mathbf{I}_4 & 0 \\ -\mathbf{I}_4 & \mathbf{A}_2 - (\mathbf{A}_1 - \mathbf{I}_4) & -\mathbf{I}_4 \\ 0 & -\mathbf{I}_4 & \mathbf{I}_4 \end{bmatrix}$$

$$= (\mathbf{A}_1 - \mathbf{I}_4) \otimes \mathbf{I}_3 + \begin{bmatrix} 1 & -1 & 0 \\ -1 & 2 & -1 \\ 0 & -1 & 1 \end{bmatrix} \otimes \mathbf{I}_4,$$

$$= \mathbf{L}(P_4) \otimes \mathbf{I}_3 + \mathbf{L}(P_3) \otimes \mathbf{I}_4.$$

A similar argument can be repeated for arbitrary subgraphs K and H, and a proof for Eq. (9-9) can then be obtained.

Theorem 1: Let $\lambda_1, \lambda_2, \ldots, \lambda_m$ and $\mu_1, \mu_2, \ldots, \mu_n$ be the eigenvalues of the adjacency matrices of K and H, respectively. Then, the $m \times n$ eigenvalues of $S = K \times H$ are $\{\lambda_i + \mu_j\}$ for $i = 1, \ldots, m$ and $j = 1, \ldots, n$.

Proof: Using Eq. (9-9), we have

$$\mathbf{L}(K \times H)(\mathbf{u}_i \otimes \mathbf{v}_j) = (\mathbf{L}(K) \otimes \mathbf{I}_{N(H)})(\mathbf{u}_i \otimes \mathbf{v}_j) + (\mathbf{I}_{N(K)} \otimes \mathbf{L}(H))(\mathbf{u}_i \otimes \mathbf{v}_j). \quad (9\text{-}10)$$

The associativity property of the Kronecker product results in

$$(\mathbf{L}(K) \otimes \mathbf{I}_{N(H)})(\mathbf{u}_i \otimes \mathbf{v}_j) + (\mathbf{I}_{N(K)} \otimes \mathbf{L}(H))(\mathbf{u}_i \otimes \mathbf{v}_j) =$$

$$(\mathbf{L}(K))(\mathbf{u}_i) \otimes (\mathbf{I}_{N(H)})(\mathbf{v}_j) + (\mathbf{I}_{N(H)})(\mathbf{u}_j) \otimes (\mathbf{L}(K))(\mathbf{u}_i) = \lambda_i \mathbf{u}_i \otimes \mathbf{v}_j + \mathbf{u}_i \otimes \mu_j \mathbf{v}_j$$

leading to

$$\mathbf{L}(K \times H)(\mathbf{u}_i \otimes \mathbf{v}_j) = (\lambda_i + \mu_j)\mathbf{u}_i \otimes \mathbf{v}_j, \quad (9\text{-}11)$$

and the proof is complete.

414   OPTIMAL STRUCTURAL ANALYSIS

Corollary 1: Let $\lambda_1, \lambda_2, \ldots, \lambda_m$ and $\mu_1, \mu_2, \ldots, \mu_n$ be the eigenvalues of the Laplacian matrices of K and H, respectively. Then, the $m \times n$ eigenvalues of $S = K \times H$ are $\{\lambda_i + \mu_j\}$ for $i = 1, \ldots, m$ and $j = 1, \ldots, n$.

If S is a k-regular graph, then λ is an eigenvalue of the adjacency graph $\mathbf{A}(S)$, that is,

$$\lambda_i(\mathbf{L}(S)) = k - \lambda_{N+1-i}(\mathbf{A}(S)) \text{ for } i = 1, 2, \ldots, N, \qquad (9\text{-}12)$$

where N is the number of nodes of S.

This result enables us to use the known results of the eigenvalues of the adjacency matrix of a regular graph in the study of its Laplacian matrix.

9.3.3 STRONG CARTESIAN PRODUCT

Theorem 2: Let $\lambda_1, \lambda_2, \ldots, \lambda_m$ and $\mu_1, \mu_2, \ldots, \mu_n$ be the eigenvalues of the adjacency matrices of K and H, respectively. Then $m \times n$ eigenvalues of $S = K \boxtimes H$ are $\{\lambda_i + \mu_j + \lambda_i\mu_j\}$ for $i = 1, \ldots, m$ and $j = 1, \ldots, n$.

For the direct product $K \boxtimes H$, the adjacency matrix can be written as

$$\mathbf{A}(K \boxtimes H) = \mathbf{A}(K) \otimes \mathbf{I}_{N(H)} + \mathbf{I}_{N(K)} \otimes \mathbf{A}(H) + \mathbf{A}(K) \otimes \mathbf{A}(H). \qquad (9\text{-}13)$$

Consider the product $S = P_4 \boxtimes P_3$ as illustrated in Figure 9.14.

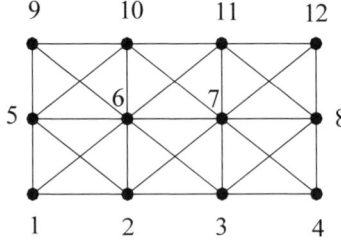

Fig. 9.14 The strong Cartesian product of P_4 and P_3.

The adjacency matrix of S can be written as

$$\mathbf{A}(S) = \begin{bmatrix} \mathbf{A}_1 & \mathbf{A}_2 & \mathbf{0} \\ \mathbf{A}_2 & \mathbf{A}_1 & \mathbf{A}_2 \\ \mathbf{0} & \mathbf{A}_2 & \mathbf{A}_1 \end{bmatrix}, \qquad (9\text{-}14)$$

DECOMPOSITION AND NODAL ORDERING   415

where

$$\mathbf{A}_1 = \begin{bmatrix} 0 & 1 & 0 & 0 \\ 1 & 0 & 1 & 0 \\ 0 & 1 & 0 & 1 \\ 0 & 0 & 1 & 0 \end{bmatrix} \text{ and } \mathbf{A}_2 = \begin{bmatrix} 1 & 1 & 0 & 0 \\ 1 & 1 & 1 & 0 \\ 0 & 1 & 1 & 1 \\ 0 & 0 & 1 & 1 \end{bmatrix}. \qquad (9\text{-}15)$$

Since $\mathbf{A}_1 + \mathbf{I}_4 = \mathbf{A}_2$, the matrix $\mathbf{A}(S)$ can be written as

$$\begin{aligned}
&\begin{bmatrix} \mathbf{A}_1 & 0 & 0 \\ 0 & \mathbf{A}_1 & 0 \\ 0 & 0 & \mathbf{A}_1 \end{bmatrix} + \begin{bmatrix} 0 & \mathbf{I}_4 & 0 \\ \mathbf{I}_4 & 0 & \mathbf{I}_4 \\ 0 & \mathbf{I}_4 & 0 \end{bmatrix} + \begin{bmatrix} 0 & \mathbf{A}_1 & 0 \\ \mathbf{A}_1 & 0 & \mathbf{A}_1 \\ 0 & \mathbf{A}_1 & 0 \end{bmatrix} \\
&= (\mathbf{A}_1 \otimes \mathbf{I}_3 + \begin{bmatrix} 0 & 1 & 0 \\ 1 & 0 & 1 \\ 0 & 1 & 0 \end{bmatrix} \otimes \mathbf{I}_4 + \mathbf{A}_1 \otimes \begin{bmatrix} 0 & 1 & 0 \\ 1 & 0 & 1 \\ 0 & 1 & 0 \end{bmatrix} \qquad (9\text{-}16) \\
&= \mathbf{A}_1 \otimes \mathbf{I}_3 + \mathbf{B}_1 \otimes \mathbf{I}_4 + \mathbf{A}_1 \otimes \mathbf{B}_1 \\
&= \mathbf{A}_1 \times \mathbf{B}_1 + \mathbf{A}_1 \otimes \mathbf{B}_1,
\end{aligned}$$

where $\mathbf{B}_1$ is obtained from $\mathbf{A}(S)$ by substituting $\mathbf{A}_1 = 0$ and $\mathbf{A}_2 = 1$. Therefore, the eigenvalues of $\mathbf{A}(S)$ can be obtained using $\{\lambda_i + \mu_j + \lambda_i\mu_j\}$ for $i = 1,\ldots, m$ and $j = 1, \ldots, n$.

Corollary 2: Let $\lambda_1, \lambda_2, \ldots, \lambda_m$ and $\mu_1, \mu_2, \ldots, \mu_n$ be the eigenvalues of the Laplacian matrices of K and H, respectively. Augment S by adding a sufficient number of edges, E_a, to the boundary nodes of S, to transform it into a regular graph S_r. A regular graph is a graph whose nodes all have equal degree, and it is k-regular with common degree k. Then, $m \times n$ eigenvalues of $S_r = K \boxtimes H \cup E_a$ are $[3(\lambda_i + \mu_j) - \lambda_i\mu_j]$ for $i = 1, \ldots, m$ and $j = 1, \ldots, n$.

Proof: Consider the graph $S = P_4 \boxtimes P_3$ in Figure 9.14. The Laplacian matrix of S has the following form:

$$\mathbf{L}(S) = \begin{bmatrix} \mathbf{A}_1 & \mathbf{A}_2 & 0 \\ \mathbf{A}_2 & \mathbf{A}_3 & \mathbf{A}_2 \\ 0 & \mathbf{A}_2 & \mathbf{A}_1 \end{bmatrix}, \qquad (9\text{-}17)$$

where

$$\mathbf{A}_2 = \begin{bmatrix} -2 & -1 & 0 & 0 \\ -1 & -1 & -1 & 0 \\ 0 & -1 & -1 & -1 \\ 0 & 0 & -1 & -2 \end{bmatrix}.$$

Adding boundary members such that $\mathbf{A}_1 = \mathbf{A}_2 + \mathbf{A}_3$, we have

$$\mathbf{L}(S) = (\mathbf{A}_1 + \mathbf{A}_2) \otimes \mathbf{I}_3 + \begin{bmatrix} 1 & -1 & 0 \\ -1 & 2 & -1 \\ 0 & -1 & 1 \end{bmatrix} \otimes \mathbf{A}_2 \quad (9\text{-}18)$$
$$= (\mathbf{A}_1 + \mathbf{A}_2) \otimes \mathbf{I}_3 + \mathbf{B}_2 \otimes \mathbf{A}_2,$$

where
$$\mathbf{A}_1 + \mathbf{A}_2 = 3 \begin{bmatrix} 1 & -1 & 0 & 0 \\ -1 & 2 & -1 & 0 \\ 0 & -1 & 2 & -1 \\ 0 & 0 & -1 & 1 \end{bmatrix} = 3\mathbf{B}_1. \quad (9\text{-}19)$$

On the other hand,

$$-\mathbf{A}_2 = \begin{bmatrix} 2 & 1 & 0 & 0 \\ 1 & 1 & 1 & 0 \\ 0 & 1 & 1 & 1 \\ 0 & 0 & 1 & 2 \end{bmatrix} = \begin{bmatrix} 3 & 0 & 0 & 0 \\ 0 & 3 & 0 & 0 \\ 0 & 0 & 3 & 0 \\ 0 & 0 & 0 & 3 \end{bmatrix} - \begin{bmatrix} 1 & -1 & 0 & 0 \\ -1 & 2 & -1 & 0 \\ 0 & -1 & 2 & -1 \\ 0 & 0 & -1 & 1 \end{bmatrix} = 3\mathbf{I}_4 - \mathbf{B}_1. \quad (9\text{-}20)$$

Therefore,
$$\begin{aligned}\mathbf{L}(S) &= 3\mathbf{B}_1 \otimes \mathbf{I}_3 + \mathbf{B}_2 \otimes (3\mathbf{I}_4 - \mathbf{B}_1) \\ &= 3(\mathbf{B}_1 \otimes \mathbf{I}_3 + \mathbf{B}_2 \otimes \mathbf{I}_4) - \mathbf{B}_2 \otimes \mathbf{B}_1 \\ &= 3(\mathbf{B}_1 \times \mathbf{B}_2) - \mathbf{B}_2 \otimes \mathbf{B}_1,\end{aligned} \quad (9\text{-}21)$$

and the eigenvalues of $\mathbf{L}(S)$ can be obtained using $[3(\lambda_i + \mu_j) - \lambda_i\mu_j]$; and the proof is complete.

9.3.4 DIRECT PRODUCT

Theorem 3: Let $\lambda_1, \lambda_2, \ldots, \lambda_m$ and $\mu_1, \mu_2, \ldots, \mu_n$ be the eigenvalues of the adjacency matrices of K and H, respectively. Then $m \times n$ eigenvalues of $S = K * H$ are $\{\lambda_i \mu_j\}$ for $i = 1, \ldots, m$ and $j = 1, \ldots, n$.

For the direct product $K * H$, the adjacency matrix can be written as

$$\mathbf{A}(K * H) = \mathbf{A}(K) \otimes \mathbf{A}(H). \tag{9-22}$$

Consider the product $S = P_4 * P_3$ as illustrated in Figure 9.15.

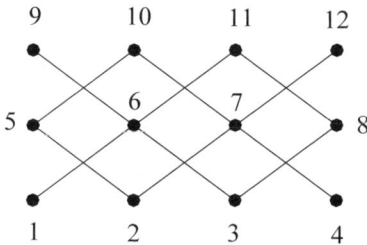

Fig. 9.15 The direct product of P_4 and P_3.

The adjacency matrix of S can be written as

$$\mathbf{A}(S) = \begin{bmatrix} \mathbf{A}_1 & \mathbf{A}_2 & \mathbf{0} \\ \mathbf{A}_2 & \mathbf{A}_1 & \mathbf{A}_2 \\ \mathbf{0} & \mathbf{A}_2 & \mathbf{A}_1 \end{bmatrix}, \tag{9-23}$$

where

$$\mathbf{A}_1 = \begin{bmatrix} 0 & 0 & 0 & 0 \\ 0 & 0 & 0 & 0 \\ 0 & 0 & 0 & 0 \\ 0 & 0 & 0 & 0 \end{bmatrix} \text{ and } \mathbf{A}_2 = \begin{bmatrix} 0 & 1 & 0 & 0 \\ 1 & 0 & 1 & 0 \\ 0 & 1 & 0 & 1 \\ 0 & 0 & 1 & 0 \end{bmatrix}. \tag{9-24}$$

The matrix $\mathbf{A}(S)$ can be written as

$$\mathbf{A}(S) = \begin{bmatrix} \mathbf{A}_1 & 0 & 0 \\ 0 & \mathbf{A}_1 & 0 \\ 0 & 0 & \mathbf{A}_1 \end{bmatrix} + \begin{bmatrix} 0 & \mathbf{A}_2 & 0 \\ \mathbf{A}_2 & 0 & \mathbf{A}_2 \\ 0 & \mathbf{A}_2 & 0 \end{bmatrix} = \begin{bmatrix} 0 & 1 & 0 \\ 1 & 0 & 1 \\ 0 & 1 & 0 \end{bmatrix} \otimes \mathbf{A}_2, \qquad (9\text{-}25)$$

or
$$\mathbf{A}(S) = \mathbf{B}_2 \otimes \mathbf{A}_2, \qquad (9\text{-}26)$$

where $\mathbf{B}_2$ is obtained from $\mathbf{A}(S)$ by substituting $\mathbf{A}_1 = 0$ and $\mathbf{A}_2 = 1$. The matrix $\mathbf{A}_2$ is the adjacency matrix of P_4, and $\mathbf{B}_2$ is the adjacency matrix of P_3. Therefore, the eigenvalues of $\mathbf{A}(S)$ can be obtained using $\{\lambda_i \mu_j\}$ for $i = 1, \ldots, m$ and $j = 1, \ldots, n$.

Corollary 3: Let $\lambda_1, \lambda_2, \ldots, \lambda_m$ and $\mu_1, \mu_2, \ldots, \mu_n$ be the eigenvalues of the Laplacian matrices of K and H, respectively. Augment S by adding a sufficient number of edges, E_a, to the boundary nodes of S, to transform it into a regular graph S_r. Then, the $m \times n$ eigenvalues of $S_r = K * H \cup E_a$ are $[2(\lambda_i + \mu_j) - \lambda_i \mu_j]$ for $i = 1, \ldots, m$ and $j = 1, \ldots, n$.

Proof: Consider the graph $S = P_4 * P_3$ in Figure 9.15. The Laplacian matrix of S has the following form:

$$\mathbf{L}(S) = \begin{bmatrix} \mathbf{A}_1 & \mathbf{A}_2 & 0 \\ \mathbf{A}_2 & \mathbf{A}_3 & \mathbf{A}_2 \\ 0 & \mathbf{A}_2 & \mathbf{A}_1 \end{bmatrix}. \qquad (9\text{-}27)$$

Adding boundary edges such that $\mathbf{A}_1 = \mathbf{A}_2 + \mathbf{A}_3$, we have

$$\mathbf{A}_2 = \begin{bmatrix} 1 & -1 & 0 & 0 \\ -1 & 2 & -1 & 0 \\ 0 & -1 & 2 & -1 \\ 0 & 0 & -1 & 1 \end{bmatrix}, \qquad (9\text{-}28)$$

or
$$\mathbf{L}(S) = (\mathbf{A}_1 + \mathbf{A}_2) \otimes \mathbf{I}_3 + \mathbf{B}_2 \otimes \mathbf{A}_2, \qquad (9\text{-}29)$$

where
$$\mathbf{A}_1 + \mathbf{A}_2 = 2 \begin{bmatrix} 1 & -1 & 0 & 0 \\ -1 & 2 & -1 & 0 \\ 0 & -1 & 2 & -1 \\ 0 & 0 & -1 & 1 \end{bmatrix} = 2\mathbf{B}_1. \qquad (9\text{-}30)$$

On the other hand,

$$-\mathbf{A}_2 = 2\mathbf{I}_4 - \mathbf{B}_1. \tag{9-31}$$

Therefore,

$$\begin{aligned}\mathbf{L}(S) &= 2\mathbf{B}_1 \otimes \mathbf{I}_3 + \mathbf{B}_2 \otimes (3\mathbf{I}_4 - \mathbf{B}_1) \\ &= 2(\mathbf{B}_1 \otimes \mathbf{I}_3 + \mathbf{B}_2 \otimes \mathbf{I}_4) - \mathbf{B}_2 \otimes \mathbf{B}_1 \\ &= 2(\mathbf{B}_1 \times \mathbf{B}_2) - \mathbf{B}_2 \otimes \mathbf{B}_1\end{aligned} \tag{9-32}$$

and the eigenvalues of $\mathbf{L}(S)$ can be obtained using $[2(\lambda_i + \mu_j) - \lambda_i\mu_j]$; and the proof is complete.

9.3.5 SECOND EIGENVALUES FOR DIFFERENT GRAPH PRODUCTS

In the above two cases, addition of boundary edges does not change the order of the nodes considerably, and therefore, S_r can be used for partitioning in place of S.

To find the smallest eigenvalue of S_r in the above corollaries, the smallest of λ_i and μ_j are both equal to zero, and therefore, $\lambda_1(S_r) = 0$. For the second eigenvalue, $[2(\lambda_i + \mu_j) - \lambda_i\mu_j]$ should be minimised. For this purpose, the second eigenvalues of the two subgraphs should be compared and the smallest one for the corresponding subgraph, together with zero for the other subgraph, should be used as a pair for calculating λ_2 for S_r. The term $\lambda_i\mu_j$ vanishes, and therefore,

$\lambda_2(S) = \text{Min}\{\lambda_2(K), \mu_2(H)\}$ for the Cartesian product;

$\lambda_2(S_r) = 3\text{Min}\{\lambda_2(K), \mu_2(H)\}$ for the strong Cartesian product; and

$\lambda_2(S_r) = 2\text{Min}\{\lambda_2(K), \mu_2(H)\}$ for the direct product.

As mentioned earlier, $\lambda_2(S_r)$ is a good approximation to $\lambda_2(S)$. This is especially true for large-scale graphs, that is, when the ratio of the number of boundary nodes to the total number of nodes for S is low. In the following, a simple method is presented for modifying $\lambda_2(S_r)$.

Consider λ and $\mathbf{x}$ as an approximate eigenvalue and eigenvector of a matrix $\mathbf{K}$, respectively. According to the Rayleigh quotient method, given an approximate eigenvector $\mathbf{x}$ for the real matrix $\mathbf{K}$, determination of the best estimate for the corresponding eigenvalue λ can be considered as a linear least squares approximation problem:

$$\mathbf{x}\lambda \cong \mathbf{K}\mathbf{x}. \tag{9-33}$$

From the normal equation $\mathbf{x}^t\mathbf{x}\lambda = \mathbf{x}^t\mathbf{K}\mathbf{x}$, the least squares solution is given by

$$\lambda = \frac{\mathbf{x}^t\mathbf{K}\mathbf{x}}{\mathbf{x}^t\mathbf{x}}. \tag{9-34}$$

This quantity, known as the *Rayleigh quotient*, can accelerate convergence of iterative methods such as power iteration, since it gives the better approximation to eigenvalue at iteration k than does the basic method alone. For normalised $\mathbf{x}$ with $\mathbf{x}^t\mathbf{x} = 1$, we have

$$\lambda = \mathbf{x}^t\mathbf{K}\mathbf{x}. \tag{9-35}$$

As an example, the above method is used for three graphs, and the approximate, improved and exact values of λ_2 are provided in Table 9.1 and Table 9.2 for the strong Cartesian product and the direct product, respectively.

Table 9.1 Approximate, modified and exact values of λ_2 for strong Cartesian product.

Graph	Approximate λ_2	Improved λ_2	Exact λ_2
$P_7 \boxtimes P_{10}$	0.1958	0.1678	0.1664
$P_{13} \boxtimes P_{23}$	0.0373	0.0344	0.0341
$P_{29} \boxtimes P_{29}$	0.0272	0.0266	0.0266
$P_{33} \boxtimes P_{33}$	0.0181	0.0176	0.0175
$P_{36} \boxtimes P_{19}$	0.0228	0.0220	0.0220

Table 9.2 Approximate, modified and exact values of λ_2 for direct product.

Graph	Approximate λ_2	Improved λ_2	Exact λ_2
$P_7 * P_{10}$	0.2937	0.2657	0.2649
$P_{13} * P_{23}$	0.0559	0.0530	0.0530
$P_{29} * P_{29}$	0.0234	0.0226	0.0226
$P_{33} * P_{33}$	0.0272	0.0266	0.0266
$P_{36} * P_{19}$	0.0152	0.0144	0.0144

Further improvement can be obtained by repeating the above process, though at the expense of additional computer time.

DECOMPOSITION AND NODAL ORDERING 421

9.4 EIGENVALUES OF A AND L MATRICES FOR CYCLES AND PATHS

Theorem 1: Consider the adjacency matrix **A** of a cycle graph C_n. When **A** is expanded with respect to the first row, the characteristic polynomial of C_n is obtained as $|\mathbf{A} - \lambda \mathbf{I}| = 0$. The eigenvalues corresponding to this relation can simply be expressed as

$$\lambda_r = 2\cos(2\pi r/n) \quad (r = 0, 1, \ldots, n-1). \tag{9-36}$$

Proof: Consider the jth eigenvector of a cycle as

$$\mathbf{x}_j = \sum_{i=1}^{n} \cos(\frac{2\pi i r}{p}) e_i, \quad \text{for } 0 \le r \le [\frac{n}{2}], \tag{9-37}$$

where e_i denotes the ith unit vector. Using a trigonometric identity,

$$\begin{aligned}\mathbf{A}\mathbf{x}_j &= \sum_{i=1}^{n} \{\cos(\frac{2\pi r(i-1)}{n}) \cos\frac{2\pi r(i+1)}{n}\} e_i \\ &= \sum_{i=1}^{n} 2\cos(\frac{2\pi r}{n} r) \cos(\frac{2\pi r i}{n}) e_i \\ &= 2\cos(\frac{2\pi r}{n})\mathbf{x}_j.\end{aligned} \tag{9-38}$$

For $j \ne 0$ and $j \ne n/2$, this eigenvector is non-symmetrical under a rotation of coordinates, so a one-step rotation through $2\pi/n$ radians provides a new linearly independent eigenvector. Whether p is even or odd, a complete list of n eigenvectors is found. Furthermore, since $\cos(2\pi r/n) = \cos(2\pi(n-r)/n)$, one can list the spectrum as

$$\{2\cos(\frac{2\pi r}{n}): \; r = 0, 1, \ldots, n-1\}. \tag{9-39}$$

This relationship shows that all the eigenvalues lie in the interval [2, –2], where the bound –2 corresponds to an even n. Since λ_r is a cosine function, except for +2 and –2, the other values are repeated twice [34].

Now λ_r is obtained for the Laplacian matrix **L** of a graph. For a cycle graph C_n, we have $\mathbf{D} = 2\mathbf{I}$. In fact, for a regular graph $\mathbf{D} = k\mathbf{I}$, and for a cycle, $k = 2$. Therefore,

$$\det(\mathbf{L} - \lambda\mathbf{I}) = 0 \Rightarrow \det[(\mathbf{D}-\mathbf{A}) - \lambda\mathbf{I}] = 0 \tag{9-40}$$

$$\Rightarrow \det[-\mathbf{A} - (\lambda - 2)\mathbf{I}] = 0$$

422 OPTIMAL STRUCTURAL ANALYSIS

The eigenvalues of $-\mathbf{A}$ are the same as those of $\mathbf{A}$ but with a reverse sign. Therefore, once the eigenvalues of $\mathbf{A}$ are obtained, the signs are reversed and two units are added.

For the Laplacian matrix $\mathbf{L}$, the equation $|\mathbf{L} - \lambda \mathbf{I}| = 0$ should be solved as follows:

$$\lambda_r = 2 - 2\cos(2\pi r/n) = (2\sin\pi r/n)^2 \quad (r = 0, \ldots, n-1) \tag{9-41}$$

$$\text{and for } r = 0 \Rightarrow \lambda_1 = 0.$$

Obviously, the maximum value of λ_r for matrix $\mathbf{A}$ is equal to 2, and the minimum value of λ_r for the Laplacian, ignoring 0, is equal to $(2\sin\pi/n)^2$. As an example, for $n = 10$, the second eigenvalue of $\mathbf{A}$ is obtained as $\lambda_2 = 0.3820$. Equation (9-41) reveals that $\lambda_r \in \{0,4\}$.

Theorem 2: For the adjacency matrix of a path graph P_n, we have

$$\lambda_r = 2\cos[\pi r/(n+1)] \quad (r = 1, \ldots, n). \tag{9-42}$$

Proof: Given any eigenvector x_j of the path graph P_n, one can immediately construct eigenvectors of the cycle graph C_{2n+2}, namely, $(x_j, 0, -x_j, 0)$ and $(0, x_j, 0, -x_j)$. Therefore, any eigenvector of P_n is a "double eigenvalue" of C_{2n+2}, so

$$\lambda_r = 2\cos[\pi r/(n+1)] \quad (r = 1, \ldots, n) \tag{9-43}$$

Repeating an argument similar to that of a cycle, for the Laplacian matrix of P_n,

$$\lambda_r = 2 - 2\cos(\pi r/n) = (2\sin\pi r/2n)^2 \quad (r = 0, \ldots, n-1) \tag{9-44}$$

$$\text{and } r = 0 \Rightarrow \lambda_1 = 0.$$

Again, here we have $\lambda_r \in \{0, 4\}$.

Obviously, the maximum value of λ_r for the adjacency matrix is $2\cos[\pi/(n+1)]$, and λ_2 for the Laplacian matrix is $(2\sin\pi/2n)^2$. As an example, for $n = 10$, we have $\lambda_2 = 0.0979$.

Example: Consider the Cartesian product of two paths P_6 and P_4, as shown in Figure 9.16. To calculate the eigenvalues of the Laplacian matrix $\mathbf{L}$ of the graph obtained by the Cartesian product of P_6 and P_4, first the eigenvalues for these paths should be obtained. Then, the first eigenvalue of P_6 is added to all the eigenvalues of P_4. Next, the second eigenvalue of P_6 is added to those of P_4. This process is continued until all eigenvalues of $S = P_6 \times P_4$ are obtained.

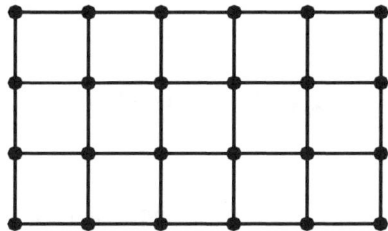

Fig. 9.16 Cartesian product $P_6 \times P_4$ of two path graphs.

As a numerical example, consider two path graphs P_6 and P_4. The eigenvalues for the Laplacian matrix **L** of S are calculated as follows:

$\lambda(P_6)$: {0; 0.2679; 1; 2; 3; 3.7321},

$\lambda(P_4)$: {0; 0.5858; 2; 3.4142},

$\lambda_S = \lambda(P_6) + \lambda(P_4) = $ {0; 0.5858; 2; 3.4142; 0.2679; … ; 7.1463}.

For the Laplacian matrix **L** of each path graph, the eigenvalues lie in the range [4]. Thus, the final results are contained in the range [8]. For this case, the first eigenvalue is zero, and therefore, for calculating the second eigenvalues, λ_2, one should choose the smallest eigenvalue from P_6 and P_4 so that, when added to zero, λ_2 for S is obtained. Obviously, as the number of nodes of a graph increases, the corresponding λ_2 will decrease. Hence, for evaluating λ_2 of S, it is sufficient to select the generator with the higher number of nodes (i.e. $\lambda_2(P_6)$ in this example), that is,

$$\lambda_2(S) = \lambda_2(P_6) \text{ since } n(P_6) > n(P_4).$$

As a second example, consider a product graph in polar coordinates, as shown in Figure 9.17.

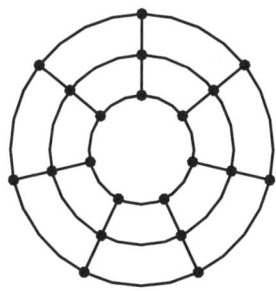

Fig. 9.17 Product of a path graph P_3 and a cycle graph C_7.

424 OPTIMAL STRUCTURAL ANALYSIS

For this case, the eigenvalues of the path P_3 and the cycle C_7 are calculated:

$\lambda(P_3)$: {0; 1; 3},

$\lambda(C_7)$: {0; 0.7530; 0.7530; 2.4450; 2.4450; 3.8019; 3.8019},

$\lambda_2(S) = \lambda_2(C_7) = 0.7530$.

and all the eigenvalues of **L** for S can be obtained, similar to the previous example. Further simplification can be achieved if the cycle has an even number of nodes more than 2. For such a case, two eigenvalues are 0 and 4. For evaluating the remaining eigenvalues, one can divide the number of nodes by 2 and consider an equivalent path graph with this number of nodes, and each eigenvalue obtained should be repeated once.

9.4.1 COMPUTING λ_2 FOR LAPLACIAN OF REGULAR MODELS

In the method of the previous section, if only the magnitude of λ_2 is required, then the method can further be simplified. As an example, in the above problem half of the nodes of the cycle should be compared to the number of nodes of the path generator, and the one with the largest number of nodes should be selected. For this graph, the second eigenvalue λ_2 has the same value as that of the main graph S.

The above idea can easily be generalised to three-dimensional models. Consider a grid as the product of three paths, $S = P_8 \times P_5 \times P_4$. This graph has 160 nodes, and P_8, P_5 and P_4 have 8, 5 and 4 nodes, respectively. Thus, $\lambda_2(S) = \lambda_2(P_8) = 0.1522$, since P_8 has more nodes than the other generators.

Consider a simple model as shown in Figure 9.18(a). After performing the geometric transformations,

$$z = [(x+iy)^6 + 1/(x+iy)^2]^{(1/4)} \text{ and } z = [(x+iy)^8 + 1]^{(1/4)}, \qquad (9\text{-}45)$$

the models shown in Figures 9.18 (b–c) are obtained. These models are equivalent to a circle with 25 and 72 sectors. Since $72/2 = 36 > 26$, $\lambda_2 = (2\sin\pi/72)^2 = 0.0076$.

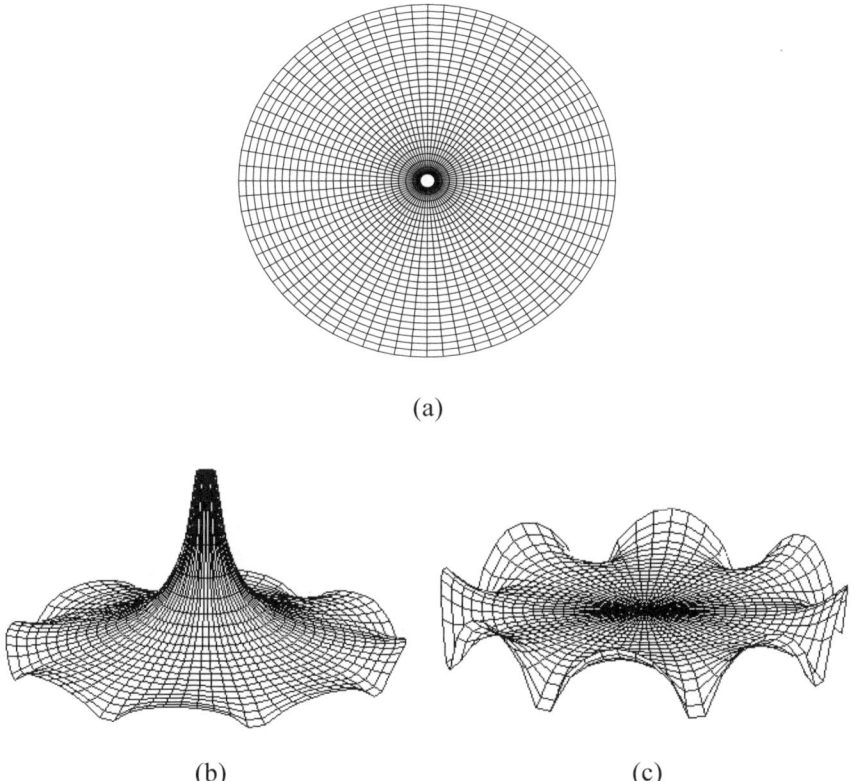

Fig. 9.18 A model as the Cartesian product of a circle and a path.

Once the eigenvalues are found, the corresponding eigenvectors can be calculated. However, this can be done much more simply considering that the eigenvectors of S are the Kronecker product of the eigenvectors of K and H, that is, $\mathbf{w}_k = \mathbf{u}_i \otimes \mathbf{v}_j$, where $\mathbf{w}_k$, $\mathbf{u}_i$ and $\mathbf{v}_j$ are the eigenvectors of S, K and H, respectively.

9.4.2 ALGORITHM

This method is simple and consists of the following steps:

Step 1: Calculate the second eigenvalue λ_2 of the Laplacian matrix $\mathbf{L}$ of the model.

Step 2: Construct the second eigenvector $\mathbf{v}_2$ corresponding to λ_2.

426 OPTIMAL STRUCTURAL ANALYSIS

Step 3: Order the entries of $\mathbf{v}_2$ in an ascending order.

Step 4: Bisect the graph and, correspondingly, the model.

The above algorithm can easily be used for graph nodal ordering and correspondingly for the nodal numbering of the FE meshes.

9.5 NUMERICAL EXAMPLES

Many examples are studied, and the results for three FE meshes are presented. The models are chosen from those encountered in practice, having different topologies. Here, no computational time is provided for the examples, since, unlike the known methods, for each example it takes a fraction of a second to calculate the eigenvalues and eigenvectors.

9.5.1 EXAMPLES FOR CARTESIAN PRODUCT

Example 1: A simply connected rectangular FE mesh with rectangular elements is considered, as shown in Figure 9.19(a). This model consists of 2168 nodes and 3045 elements. The skeleton graph is considered as $P_{88} \times P_{36}$ and is partitioned with $\lambda_2 = 0.00127$, corresponding to P_{88}, and the corresponding FEM is bisected. The process is repeated for further decomposition of the FEM into four, and eight subdomains, as illustrated in Figure 9.19(b).

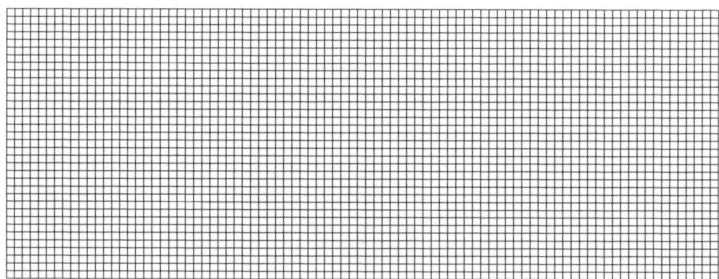

(a) A simple FE mesh.

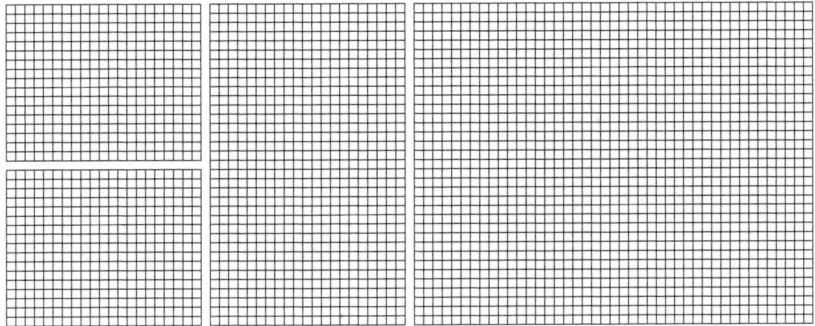

(b) decomposition of the model.

Fig. 9.19 A rectangular FE mesh and its decomposition.

Example 2: A circular FE mesh is considered, as shown in Figure 9.20. This model consists of 1872 nodes and 1800 rectangular elements. The skeleton graph is considered as $C_{72} \times P_{26}$ and is partitioned with $\lambda_2 = 0.007611$, corresponding to C_{72}, and the corresponding FEM is bisected. The process is repeated for further decomposition of the FEM into four, and eight subdomains. Typically such subdomains are shown in Figure 9.19(b).

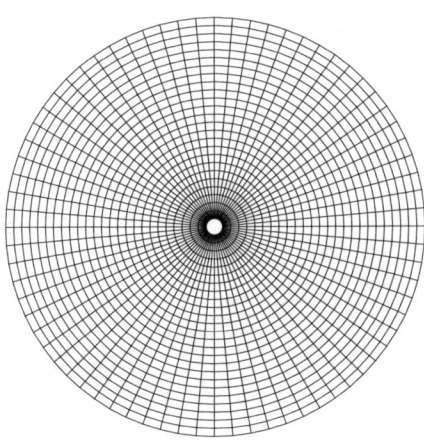

(a) A circular model S.

428   OPTIMAL STRUCTURAL ANALYSIS

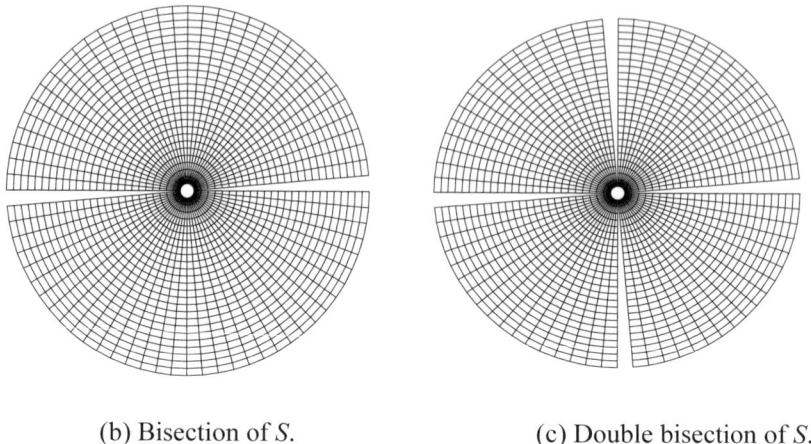

(b) Bisection of S.   (c) Double bisection of S.

Fig. 9.20 A circular FE mesh and its decompositions.

Example 3: The finite element of a nozzle is considered as shown in Figure 9.21(a). This model consists of 4000 nodes and 3960 rectangular shell elements. The skeleton graph is considered as $P_{100} \times C_{40}$ and is partitioned with $\lambda_2 = 0.000987$, corresponding to P_{100}, and the corresponding FEM is bisected; see Figure 9.21(b). The process is repeated for further decomposition of the FEM into four subdomains as illustrated in Figure 9.21(c).

In the above examples, the generators were chosen as paths and/or cycles. In the following examples, one of the generators is selected as an arbitrary graph. Again, λ_2 for the entire model can easily be obtained by a comparison of the λ_2 for the generators. In this case, however, the second eigenvalue of the generators should be calculated using classical methods.

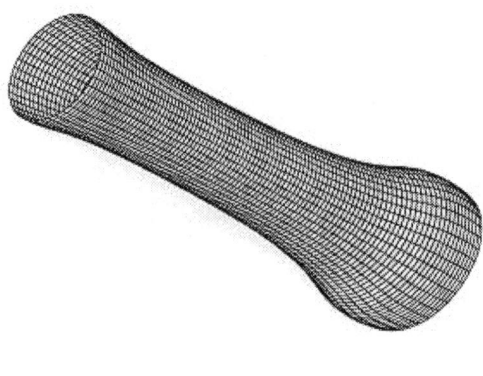

(a)

DECOMPOSITION AND NODAL ORDERING   429

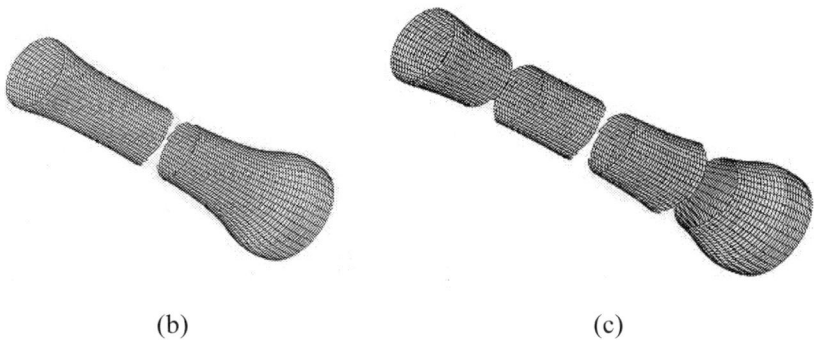

(b)   (c)

Fig. 9.21 A cylindrical shaped FE mesh and its decompositions.

Example 4: Consider a model with an arbitrary subgraph S_1 as its generator; see Figure 9.22. The Cartesian products of $S_1 \times P_4$ and $S_1 \times C_4$ are considered. The corresponding eigenvalues are as follows:

$\lambda_2(S_1) = 1.2679$ and $\lambda_2(P_4) = 0.5858$, and therefore $\lambda_2(S_1 \times P_4) = 0.5858$;
$\lambda_2(S_1) = 1.2679$ and $\lambda_2(C_4) = 2.0000$, and therefore $\lambda_2(S_1 \times C_4) = 1.2679$.

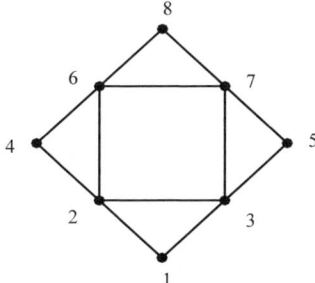

Fig. 9.22 The arbitrary generator S_1 of the considered model.

Example 5: Consider a model with an arbitrary subgraph S_2 as its generator; see Figure 9.23. The Cartesian products of $S_2 \times P_{10}$ and $S_2 \times C_{10}$ are considered. The corresponding eigenvalues are as follows:

$\lambda_2(S_2) = 0.2765$ and $\lambda_2(P_{10}) = 0.0979$, and therefore $\lambda_2(S_2 \times P_{10}) = 0.0979$;
$\lambda_2(S_2) = 0.2765$ and $\lambda_2(C_{10}) = 0.3820$, and therefore $\lambda_2(S_2 \times C_{10}) = 0.2765$.

430   OPTIMAL STRUCTURAL ANALYSIS

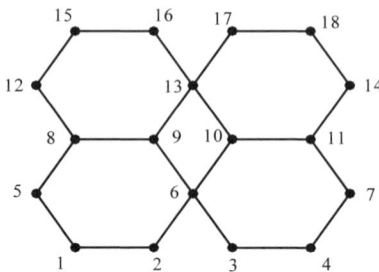

Fig. 9.23 The arbitrary generator S_2 of the considered model.

9.5.2 EXAMPLES FOR STRONG CARTESIAN PRODUCT

A dome S with beam elements is shown in Figure 9.24(a). This structure has two generators P_{19} and C_{35} and contains 665 nodes and 2520 members. Since $\lambda_2(P_{19}) = 0.0273$ and $\lambda_2(C_{35}) = 0.0321$, $\lambda_2(S) = 3 \times 0.0273 = 0.0818$ (see Figure 9.24(b)). Further partitioning results in four subgraphs, as illustrated in Figure 9.24(c).

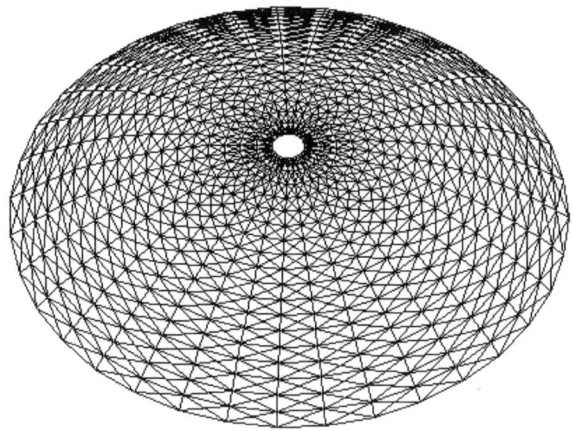

(a) The graph model S of a dome.

DECOMPOSITION AND NODAL ORDERING 431

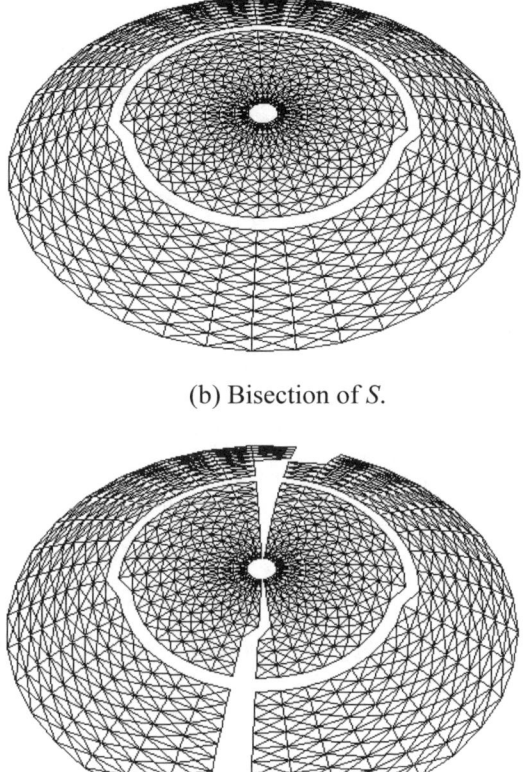

(b) Bisection of S.

(c) Further partitioning of S.

Fig. 9.24 A circular dome S and its partitions.

The evaluation of λ_2 for the entire graph using any standard program (e.g. Matlab) takes 33.4 s, while the above calculation needs only a fraction of second to be performed.

9.5.3 EXAMPLES FOR DIRECT PRODUCT

A barrel vault S with beam elements is shown in Figure 9.25(a). This structure has two generators P_{36} and P_{19}, with 684 nodes and 1260 members. Since $\lambda_2(P_{19}) = 0.0273$ and $\lambda_2(P_{36}) = 0.0076$, $\lambda_2(S) = 2 \times 0.0076 = 0.0152$ (see Figure 9.25(b)). The modified value and the exact value of $\lambda_2(S)$ are both equal to 0.0144. Further partitioning results in four subgraphs as shown in Figure 9.25(c).

432   OPTIMAL STRUCTURAL ANALYSIS

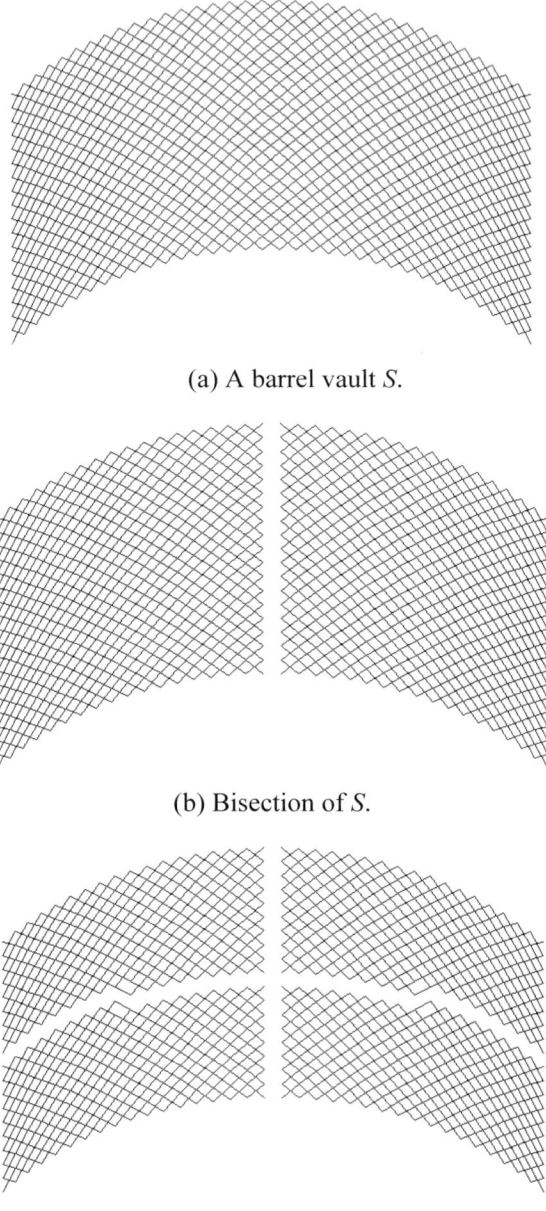

(a) A barrel vault S.

(b) Bisection of S.

(c) Further partitioning of S.

Fig. 9.25 A barrel vault S and its partitions.

Evaluating λ_2 for the entire graph using Matlab takes 29.5 s, while the above calculation needs only a fraction of second to be performed.

9.6 SPECTRAL METHOD FOR PROFILE REDUCTION

9.6.1 ALGORITHM

This algorithm is simple and consists of the following steps:

Step 1: Calculate the second eigenvalue λ_2 of the Laplacian matrix **L** of the model.

Step 2: Construct the second eigenvector $\mathbf{v}_2$ corresponding to λ_2.

Step 3: Order the entries of $\mathbf{v}_2$ in an ascending order.

Step 4: Renumber the nodes of the model according to their occurrence in $\mathbf{v}_2$.

The above algorithm leads to well-structured stiffness matrices with low profile.

9.6.2 EXAMPLES

Example 1: Figure 9.26 is a cylindrical grid S with two generators P_{16} and C_{15}. This model has 240 nodes and 465 members. Since $16 > 15/2$, for P_{16} we have $\lambda_2 = (2\sin\pi/2 \times 16)^2 = 0.038429$, corresponding to the second eigenvalue of the entire model. The results are depicted in Table 9.3.

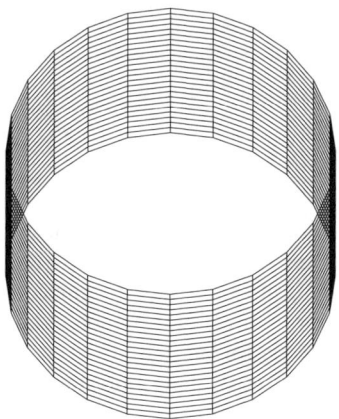

Fig. 9.26 A cylindrical grid S.

434   OPTIMAL STRUCTURAL ANALYSIS

Table 9.3 Results of Example 1.

	Profile	Bandwidth
Initial	16,258	239
New	3642	16

Example 2: A circular grid S with beam elements is shown in Figure 9.27. This grid has two generators P_{13} and C_{96}. The entire model contains 1248 nodes and 1152 members. Since 96/2 >; 13, for C_{96} we have $\lambda_2 = (2\sin\pi/96)^2 = 0.00428$, corresponding to the second eigenvalue of the entire model. The results are depicted in Table 9.4.

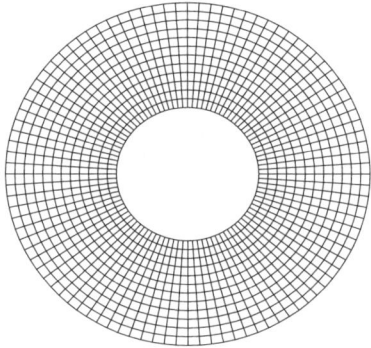

Fig. 9.27 A circular grid S.

Table 9.4 Results of Example 2.

	Profile	Bandwidth
Initial	49,972	239
New	34,848	46

Remarks: The present methods are highly efficient and make the calculation of eigenvalues for adjacency and Laplacian matrices of regular structural models very simple and fast, compared to the use of standard methods. The only restriction corresponds to irregular models, for which additional effort is needed to achieve regularisation. However, for models with small irregularity, the results of the present approach can always be used as a primary approximation.

9.7 NON-COMPACT EXTENDED P-SUM

In this chapter, three products are used for generating graph models. These products are special cases of a more general operation known as the *NEPS (non-compact extended p-sum)* of graphs. The notion of the NEPS was introduced independently by Cvetković and Lučić [35] and Shee [200]. The corresponding definitions and theorems are briefly introduced below:

Definition: Let $\boldsymbol{B}$ be a set of non-zero binary n-tuples, that is, $\boldsymbol{B} \subseteq \{0, 1\}^n \setminus \{0, ..., 0\}$. The NEPS of graph S_1, S_2, ..., S_n with basis $\boldsymbol{B}$ is the graph with node set $N(S_1) \times ... \times N(S_n)$ in which two nodes, say $(x_1, ..., x_n)$ and $(y_1, ..., y_n)$, are adjacent if and only if there exists an n-tuple $\{\beta_1, \beta_2, ..., \beta_n\} \in \boldsymbol{B}$ such that $x_i = y_i$ whenever $\beta_i = 0$ and x_i is adjacent to y_i (in S_i) whenever $\beta_i = 1$.

Consider the special cases in which a graph is the NEPS of only two graphs ($n = 2$):

1. the Cartesian product $S_1 \times S_2$ when $\boldsymbol{B} = \{(0, 1), (1, 0)\}$,

2. the strong Cartesian product $S_1 \boxtimes S_2$ when $\boldsymbol{B} = \{(0, 1), (1, 0), (1, 1)\}$, and

3. the sum $S_1 * S_2$ when $\boldsymbol{B} = \{(1, 1)\}$.

The following two theorems are proved by Cvetković and Lučić [35] and also presented in Cvetković et al.[34].

Theorem 1: Let A_1, ..., A_n be the adjacency matrices of graphs S_1, ..., S_n, respectively. The NEPS S with basis $\boldsymbol{B}$ of graphs S_1, ..., S_n has the adjacency matrix as

$$\mathbf{A} = \sum_{\beta \in B} \mathbf{A}_1^{\beta_1} \otimes ... \otimes \mathbf{A}_n^{\beta_n}. \qquad (9\text{-}46)$$

where $\mathbf{A}_i^0 = \mathbf{I}$ (the identity matrix of the same order as $\mathbf{A}_i$), $\mathbf{A}_i^1 = \mathbf{A}_i$ and $\otimes$ denotes the Kronecker product of the matrices.

Theorem 2: If, for $i = 1, ..., n$, $\{\lambda_{i1}, ..., \lambda_{in_i}\}$ is the spectrum of S_i (with n_i being the number of nodes of S_i), then the spectrum of S, which is the NEPS of S_1, S_2, ..., S_n with basis $\boldsymbol{B}$, consists of all possible values $\Lambda_{i_1,...,i_n}$ where

$$\Lambda_{i_1,...,i_n} = \sum_{\beta \in B} \lambda_{1i_1}^{\beta_1} ... \lambda_{ni_n}^{\beta_n} \quad (i_k = 1,...,n_k; k = 1,...,n). \qquad (9\text{-}47)$$

In particular, if $\lambda_1, \lambda_2, \ldots, \lambda_n$ and $\mu_1, \mu_2, \ldots, \mu_m$ are eigenvalues of K and H, respectively, then

$\lambda_i + \mu_j$ ($i = 1, \ldots, n$; $j = 1, \ldots, m$) are the eigenvalues of the Cartesian product $K \times H$,

$\lambda_i + \mu_j + \lambda_i \mu_j$ ($i = 1, \ldots, n$; $j = 1, \ldots, m$) are the eigenvalues of the strong Cartesian product $K \boxtimes H$, and

$\lambda_i \mu_j$ ($i = 1, \ldots, n$; $j = 1, \ldots, m$) are the eigenvalues of the direct product $K * H$.

EXERCISES

9.1 Calculate the eigenvalues and eigenvectors of $S = P_3 \times C_4$.

9.2 Find the second eigenvalue of $S = P_{26} \times C_{50}$.

9.3 Find the second eigenvalue of $S = P_{26} \times P_{50} \times P_{20}$.

9.4 Calculate the eigenvalues and eigenvectors of $S = P_4 \otimes C_3$.

9.5 Find the second eigenvalue of $S = C_{12} \otimes C_8$.

9.6 Calculate the eigenvalues and eigenvectors of $S = P_5 * C_2$.

9.7 Find the second eigenvalue of $S = P_{20} * C_{30}$.

APPENDIX A
Basic Concepts and Definitions of Graph Theory

A.1 INTRODUCTION

In this appendix, basic concepts and definitions of graph theory are presented. Since some of the readers may be unfamiliar with the theory of graphs, simple examples are included to make it easier to understand the main concepts.

Some of the uses of the theory of graphs in the context of civil engineering are as follows: A graph can be a model of a structure, a hydraulic network, a traffic network, a transportation system, a construction system, or a resource allocation system, for example. In this book, the theory of graphs is used as the model of a skeletal structure, and it is also employed as a method of transforming the connectivity properties of finite element meshes to those of graphs. Many such graphs are defined in this book and employed throughout the combinatorial optimisations performed for optimal analysis of skeletal structures and finite element models. This appendix will also enable the readers to develop their own ideas and methods in the light of the principles of graph theory. For further definitions and proofs, the reader may refer to Harary [73], Berge [12], Bondy and Murty [15], Wilson and Beineke [231], Brualdi and Ryser [16], Gondran and Minoux [63] and West [228].

A.2 BASIC DEFINITIONS

The performance of a structure depends not only on the characteristics of its components but also on their relative location. In a structure, if the properties of one member are altered, the overall behaviour may be changed. This indicates that the performance of a structure depends on the detailed characteristics of its members. If the location of a member is altered, the properties of the structure may again be different. Therefore, the connectivity (topology) of the structure influences the performance of the whole structure and is as important as the mechanical properties of its members. Hence, it is important to represent a structure so that its topology can be understood clearly. The graph model of a structure provides a powerful means for this purpose.

A.2.1 DEFINITION OF A GRAPH

A *graph S* consists of a non-empty set *N(S)* of elements called *nodes* (vertices or points) and a set *M(S)* of elements called *members* (edges or arcs) together with a relation of *incidence*, which associates each member with a pair of nodes, called its *ends*.

Two or more members joining the same pair of nodes are collectively known as a *multiple member,* and a member joining a node to itself is called a *loop*. A graph with no loops or multiple members is called a *simple graph*. If *N(S)* and *M(S)* are countable sets, then the corresponding graph *S* is *finite*. In this book, only simple finite graphs are needed, which are referred to as *graphs*.

The above definitions correspond to abstract graphs; however, a graph may be visualized as a set of points connected by line segments in Euclidean space; the nodes of a graph are identified with points, and its members are identified as line segments without their end points. Such a configuration is known as a *topological graph*. These definitions are illustrated in Figure A.1.

(a) A simple graph.   (b) A graph with loop and multiple members.

Fig. A.1 Simple and non-simple graphs.

A.2.2 ADJACENCY AND INCIDENCE

Two nodes of a graph are called *adjacent* if these nodes are the end nodes of a member. A member is called *incident with a node* if this node is an end node of the member. Two members are called *incident* if they have a common end node. The *degree* (valency) of a node n_i of a graph, denoted by $\deg(n_i)$, is the number of members incident with that node. Since each member has two end nodes, the sum

APPENDIX A–BASIC CONCEPTS AND DEFINITIONS   439

of node-degrees of a graph is twice the number of its members (the handshaking lemma).

A.2.3 GRAPH OPERATIONS

A *subgraph* S_i of S is a graph for which $N(S_i) \subseteq N(S)$ and $M(S_i) \subseteq M(S)$ and each member of S_i has the same ends as in S.

The *union* of subgraphs $S_1, S_2, \ldots, S_k$ of S, denoted by $S^k = \bigcup_{i=1}^{k} S_i = S_1 \cup S_2 \cup \ldots \cup S_k$, is a subgraph of S with $N(S^k) = \bigcup_{i=1}^{k} N(S_i)$ and $M(S^k) = \bigcup_{i=1}^{k} M(S_i)$. The *intersection* of two subgraphs S_i and S_j is similarly defined using intersections of node-sets and member-sets of the two subgraphs. The intersection of two subgraphs does not need to consist only of nodes, but it is usually considered to do so in the substructuring technique of structural analysis. The *ring sum* of two subgraphs $S_i \oplus S_j$ is a subgraph that contains the nodes and members of S_i and S_j except those elements common to S_i and S_j. These definitions are illustrated in Figure A.2.

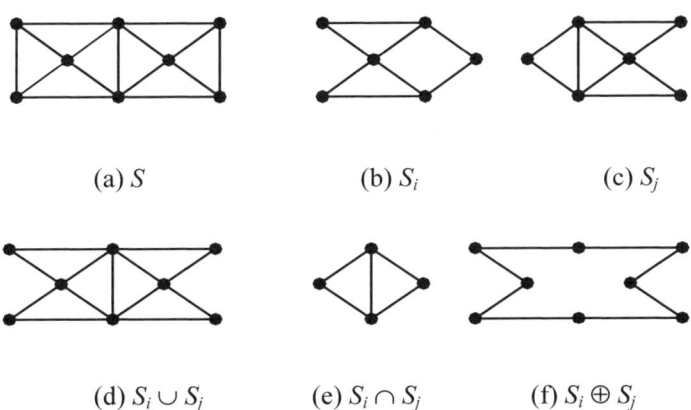

(a) S (b) S_i (c) S_j

(d) $S_i \cup S_j$ (e) $S_i \cap S_j$ (f) $S_i \oplus S_j$

Fig. A.2 A graph, two of its subgraphs, their union, intersection and ring sum.

Two graphs S and K are called *homeo-morphic* if one can obtain K from S by suppressing or inserting nodes of degree 2 in the members.

A.2.4 WALKS, TRAILS AND PATHS

A *walk* w of S is a finite sequence $w = \{n_0, m_1, n_1, \ldots, m_p, n_p\}$ whose terms are alternately nodes n_i and members m_i of S for $1 \leq i \leq p$, and n_{i-1} and n_i are the two ends of m_i. A *trail* t in S is a walk in which no member of S appears more than once. A *path* P is a trail in which no node appears more than once. The *length* of a path P_i, denoted by $L(P_i)$, is taken as the number of its members. P_i is called the *shortest path* between the two nodes n_0 and n_p if, for any other path P_j between these nodes, $L(P_i) \leq L(P_j)$. The *distance* between two nodes of a graph is defined as the number of the members of a shortest path between these nodes.

As an example in Figure A.3,

$$w = (n_1, m_3, \mathbf{n_4}, \mathbf{m_4}, n_5, m_9, n_2, m_2, n_3, m_7, \mathbf{n_4}, \mathbf{m_4}, \mathbf{n_5})$$

is a walk between n_1 and n_5 in which member m_4 and nodes n_4 and n_5 are repeated twice.

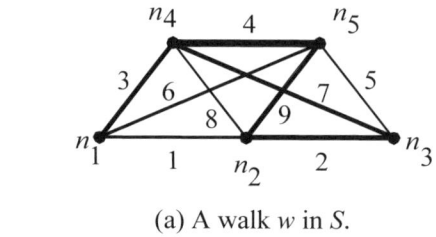

(a) A walk w in S.

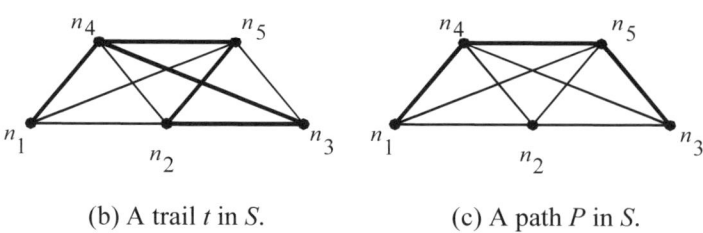

(b) A trail t in S.   (c) A path P in S.

Fig. A.3 A walk, a trail and a path in S.

$$t = (n_1, m_3, \mathbf{n_4}, m_4, n_5, m_9, n_2, m_2, n_3, m_7, \mathbf{n_4})$$

is a trail between n_1 and n_4 in which node n_5 is repeated twice.

$$P = (n_1, m_3, n_4, m_4, n_5, m_5, n_3)$$

is a path of length 3 in which no node and no member is repeated.

The path $(n_1, m_6, n_5, m_5, n_3)$ is a shortest path of length 2 between the two nodes n_1 and n_3, where the length of each member is taken as unity.

Two nodes n_i and n_j are said to be *connected* in S if there exists a path between these nodes. A graph S is called *connected* if all pairs of its nodes are connected. A

APPENDIX A–BASIC CONCEPTS AND DEFINITIONS 441

component of a graph S is a maximal connected subgraph, that is, it is not a subgraph of any other connected subgraph of S.

A.2.5 CYCLES AND CUTSETS

A *cycle* is a path $(n_0, m_1, n_1, \ldots, m_p, n_p)$ for which $n_0 = n_p$ and $p \geq 1$, that is, a cycle is a closed path. Similarly, a *closed trail (hinged cycle)* and a *closed walk* can be defined; see Figure A.4.

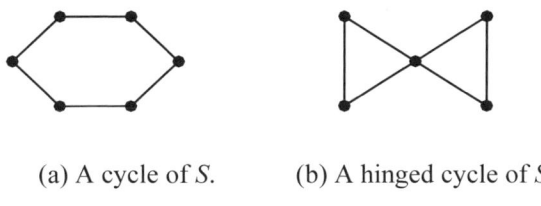

(a) A cycle of S.   (b) A hinged cycle of S.

Fig. A.4 Cycles of S.

A *cutset* is a collection of members whose removal from the graph increases the number of its components. If a cutset results in two disjoint subgraphs S_1 and S_2, then it is called a *prime cutset*. Notice that no proper subsets of a cutset have this property. A *link* is a member that has its ends in S_1 and S_2. Each S_1 and S_2 may or may not be connected. If both are connected, the cutset is called *prime*. If S_1 or S_2 consists of a single node, the cutset is called a *cocycle*. These definitions are illustrated in Figure A.5.

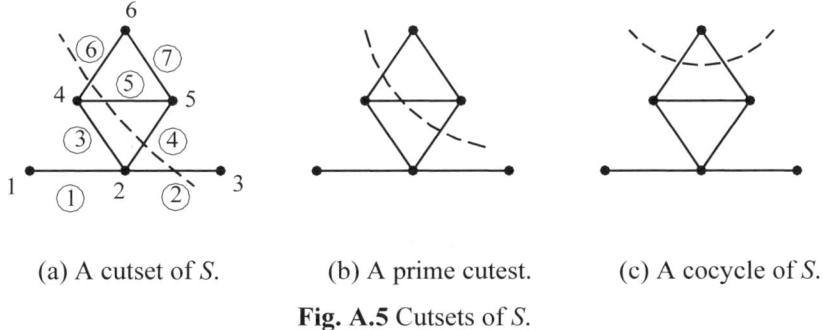

(a) A cutset of S.   (b) A prime cutest.   (c) A cocycle of S.

Fig. A.5 Cutsets of S.

A.2.6 TREES, SPANNING TREES AND SHORTEST ROUTE TREES

A *tree* T of S is a connected subgraph of S that contains no cycle. A set of trees of S forms a *forest*. Obviously, a forest with k trees contains $N(S) - k$ members. If a tree contains all the nodes of S, it is called a *spanning tree* of S. Henceforth, for simplicity it will be referred to as a *tree*.

A *shortest route tree* (SRT) rooted at a specified node n_0 of S is a tree for which the distance between every node n_j of T and n_0 is a minimum. An SRT of a graph can be generated by the following simple algorithm:

Label the selected root n_0 as "0" and the adjacent nodes as "1". Record the members incident to "0" as tree members. Repeat the process of labelling with "2" the unnumbered ends of all the members incident with nodes labelled as "1", again recording the tree members. This process terminates when each node of S is labelled and all the tree members are recorded. This algorithm has many applications in engineering and it is called a *breadth-first-search* algorithm.

A graph is called *acyclic* if it has no cycle. A tree is a connected acyclic graph. Any graph without cycles is a *forest*; thus the components of a forest are trees.

The above definitions are illustrated in Figure A.6.

It is easy to prove that, for a tree T,

$$M(T) = N(T) - 1, \qquad (A-1)$$

where $M(T)$ and $N(T)$ are the numbers of members and nodes of T, respectively.

The complement of T in S is called a *cotree*, denoted by T^*. The members of T are known as *branches* and those of T^* are called *chords*. For a connected graph S, the number of chords is given by

$$M(T^*) = M(S) - M(T). \qquad (A-2)$$

Since $N(T) = N(S)$,

$$M(T^*) = M(S) - N(S) + 1, \qquad (A-3)$$

where $M(S)$ and $N(S)$ are the numbers of members and nodes of S, respectively. Notice that the same notation is used for a set and its cardinality and the difference should be obvious from the context.

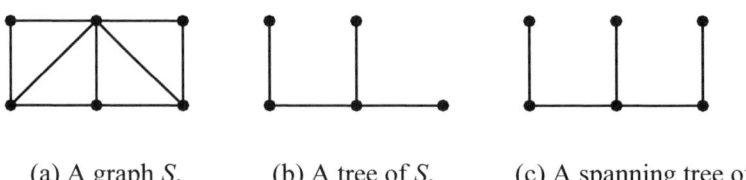

(a) A graph S. (b) A tree of S. (c) A spanning tree of S.

APPENDIX A–BASIC CONCEPTS AND DEFINITIONS   443

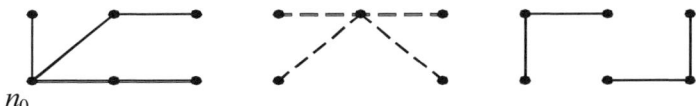

(d) An SRT rooted from n_0.   (e) The cotree of (c).   (f) A forest with 2 trees.

Fig. A.6 Different trees, cotree and a forest of S.

A.2.7 DIFFERENT TYPES OF GRAPHS

To simplify the study of properties of graphs, different types of graphs have been defined. Some important ones are as follows:

A *null graph* is a graph that contains no members. Thus, N_k is a graph containing k isolated nodes.

A *cycle graph* is a graph consisting of a single cycle. Therefore, C_k is a polygon with k members.

A *path graph* is a graph consisting of a single path. Hence, P_k is a path with k nodes and $(k-1)$ members.

A *wheel graph* W_k is defined as the union of a star graph with $(k-1)$ members and a cycle graph C_{k-1}, connected as shown in Figure A.7, for $k = 6$. Alternatively, a wheel graph W_k can be obtained from the cycle graph C_{k-1} by adding a node O and members (spokes) joining O to each node of C_{k-1}.

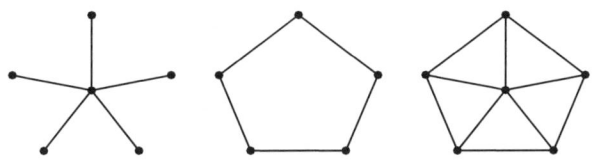

(a) Star graph S_6.   (b) Cycle graph C_5.   (c) Wheel graph W_6.

Fig. A.7 Wheel graph W_6.

A *complete graph* is a graph in which every pair of distinct nodes is connected by exactly one member; see Figure A.8. A complete graph with N nodes is denoted by K_N. It is easy to prove that a complete graph with N nodes has $N(N-1)/2$ members.

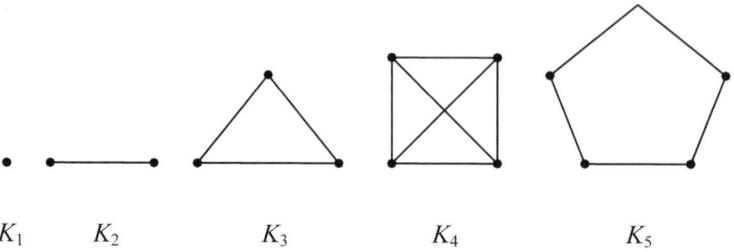

Fig. A.8 Five complete graphs.

A graph is called *bipartite* if the corresponding node set can be split into two sets N_1 and N_2 in such a way that each member of S joins a node of N_1 to a node of N_2. This graph is denoted by $B(S) = (N_1, M, N_2)$. A *complete bipartite* graph is a bipartite graph in which each node N_1 is joined to each node of N_2 by exactly one member. If the numbers of nodes in N_1 and N_2 are denoted by r and s, respectively, then a complete bipartite graph is denoted by $K_{r,s}$. Examples of bipartite and complete bipartite graphs are shown in Figure A.9.

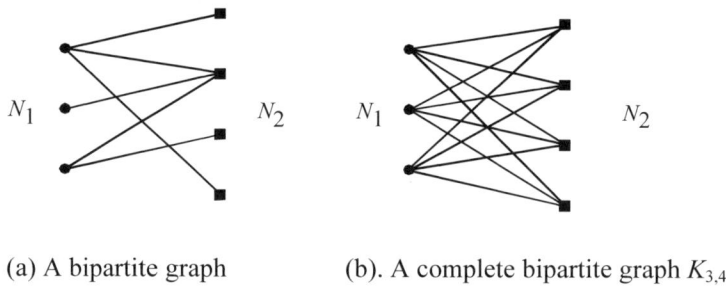

(a) A bipartite graph   (b). A complete bipartite graph $K_{3,4}$.

Fig. A.9 Two bipartite graphs.

A graph S is called *regular* if all of its nodes have the same degree. If this degree is k, then S is *k-regular graph*. For example, a triangle graph is 2-regular and a cubic graph is 3-regular.

Consider the set M of members of a graph S as a family of 2-node subsets of $N(S)$. The *line graph* $L(S)$ of S has its vertices in a one-to-one correspondence with members of S, and two vertices are connected by an edge if the corresponding members in S are incident. Thus, the vertices of $L(S)$ are the members of S, with two vertices of $L(S)$ being adjacent when the corresponding members of S are incident. As an example, the line graph of Figure A.10(a) is illustrated in Figure A.10(b).

APPENDIX A–BASIC CONCEPTS AND DEFINITIONS   445

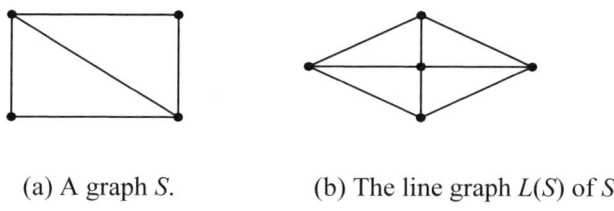

(a) A graph S.   (b) The line graph L(S) of S.

Fig. A.10 A simple graph and its line graph.

For the original graph S, the terms nodes and members are used, and for the line graph $L(S)$, the terms vertices and edges are used. In this book, many new graphs are defined and used for transforming the connectivity properties of the original models to those of the induced new graphs.

A.3 VECTOR SPACES ASSOCIATED WITH A GRAPH

A vector space can be associated with a graph by defining a vector, the field and the binary operations as follows:

Any subset of the $M(S)$ members of a graph S can be represented by a vector $\mathbf{x}$ whose $M(S)$ components are elements of the field of integer modulo 2, where component $x_i = 1$ when the ith member is an element of the subset, and $x_i = 0$ otherwise. The sum of two subset vectors $\mathbf{x}$ and $\mathbf{y}$ is a vector $\mathbf{z}$ with entries defined by $z_i = x_i + y_i$, representing the symmetric difference of the original subsets. The scalar product of $\mathbf{x}$ and $\mathbf{y}$ defined by $\Sigma x_i y_i$ is 0 or 1 according to whether the original subsets have an even or an odd number of members in common. Although this vector space can be constructed over an arbitrary field, for simplicity the field of integer modulo 2 is considered, in which $1 + 1 = 0$.

As an example, consider $\mathbf{x} = \{0, 0, 0, 1, 1, 1, 0\}^t$ and $\mathbf{y} = \{0, 0, 1, 1, 1, 0, 0\}^t$ representing two subgraphs of S. Then, their symmetric difference is obtained as $\mathbf{z} = \{0, 0, 1, 0, 0, 1, 0\}^t$, and the scalar product $\Sigma x_i y_i = 0 \pmod{2}$, since these subgraphs have two members in common.

Two important subspaces of the above vector space of a graph S are the cycle subspace and cutset subspace, known as the *cycle space* and the *cutset space* of S.

A.3.1 CYCLE SPACE

Let a cycle set of members of a graph be defined as a set of members that form a cycle or form several cycles having no common member, but perhaps common nodes. The null set is also defined as a cycle set. A vector representing a cycle set is called a *cycle set vector*. It can be shown that the sum of two cycle set vectors of a graph is also a cycle set vector. Thus, the cycle set vectors of a graph form a vector space over the field of integer modulo 2. The dimension of a cycle space is given by

$$\text{nullity } (S) = v(S) = b_1(S) = M(S) - N(S) + b_0(S), \qquad (A\text{-}4)$$

where $b_1(S)$ and $b_0(S)$ are the first and zero Betti numbers of S, respectively. As an example, the nullity of the graph S in Figure A.1(a) is $v(S) = 9 - 6 + 1 = 4$.

A.3.2 CUTSET SPACE

Consider a cutset vector similar to that of a cycle vector. Let the null set also be defined as a cutset. It can be shown that the sum of two cutset vectors of a graph is also a cutset vector. Therefore, the cutset vectors of a graph form a vector space, the dimension of which is given by

$$\text{rank } (S) = \rho(S) = N(S) - b_0(S). \qquad (A\text{-}5)$$

For example, the rank of S in Figure A.1(a) is $\rho(S) = 6 - 1 = 5$.

A.3.3 ORTHOGONALITY PROPERTY

Two vectors are called *orthogonal* if their scalar product is zero. It can be shown that a vector is a cycle set (cutset) vector if and only if it is orthogonal to every vector of a cutset (cycle set) basis. Since the cycle set and cutset spaces of a graph S containing $M(S)$ members are both subspaces of the $M(S)$-dimensional space of all vectors that represent subsets of the members, the cycle set and cutset spaces are *orthogonal components* of each other.

A.3.4 FUNDAMENTAL CYCLE BASES

A maximal set of independent cycles of a graph is known as its *cycle basis*. The cardinality of a cycle basis is the same as the first Betti number $b_1(S)$. A special basis known as a *fundamental cycle* basis can easily be constructed corresponding to a tree T of S, Kirchhoff [138]. In a connected S, a chord of T together with T contains a cycle known as a *fundamental cycle* of S. Moreover, the fundamental cycles obtained by adding the chords to T, one at a time, are independent, because each cycle has a member that is not in the others. Also, every cycle C_i depends on the set of fundamental cycles obtained by the above process, for C_i is the symmetric difference of the cycles determined by the chords of T that lie in C_i. Thus, the cycle rank (cyclomatic number, first Betti number, nullity) of graph S, which is the number of cycles in a basis of the cycle space of S, is given by

$$b_1(S) = M(S) - N(S) + 1, \qquad (A\text{-}6)$$

APPENDIX A–BASIC CONCEPTS AND DEFINITIONS   447

and if S contains $b_0(S)$ components, then,

$$b_1(S) = M(S) - N(S) + b_0(S). \tag{A-7}$$

As an example, the selected tree and three fundamental cycles of S are illustrated in Figure A.11.

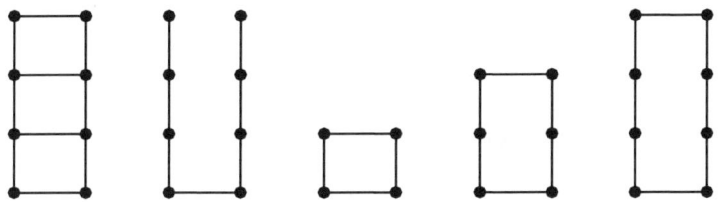

Fig. A.11 A graph S, a spanning tree, and the corresponding fundamental cycle basis.

A.3.5 FUNDAMENTAL CUTSET BASES

A basis can be constructed for the cutset space of a graph S. Consider the tree T and its cotree T^*. The subgraph of S consisting of T^* and any member of T (branch) contains exactly one cutset known as a *fundamental cutset*. The set of cutsets obtained by adding branches of T to T^*, one at a time, forms a basis for the cutset space of S, known as a *fundamental cutset basis* of S. The cutset rank (rank of S) is the number of cutsets in a basis for the cutset space of S, and it can be obtained by a reasoning similar to that of the cycle basis as

$$\rho(S) = N(S) - 1, \tag{A-8}$$

and for a graph with $b_0(S)$ components,

$$\rho(S) = N(S) - b_0(S) \tag{A-9}$$

A graph S and a fundamental cutset basis of S are shown in Figure A.12.

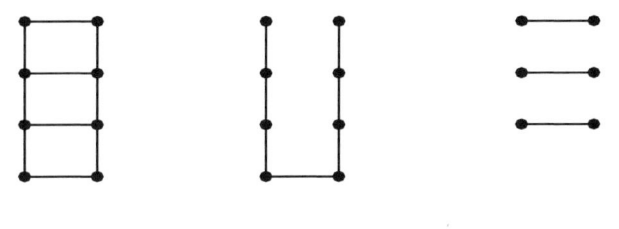

(a) A graph S.   (b) A tree T of S.   (c) Cotree T^* of T.

448   OPTIMAL STRUCTURAL ANALYSIS

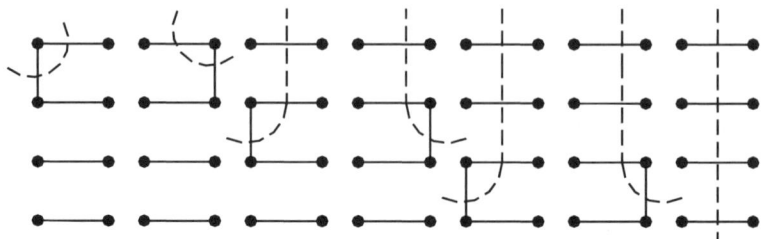

Fig. A.12 A graph S and a fundamental cutset basis of S.

A.4 MATRICES ASSOCIATED WITH A GRAPH

Matrices play a dominant role in the theory of graphs and especially in applications of structural analysis. Some of these matrices conveniently describe the connectivity properties of a graph, some provide useful information about the patterns of the structural matrices, and some others reveal additional information about transformations such as those of equilibrium and compatibility equations.

In this section, various matrices that reflect the properties of the corresponding graphs are studied. For simplicity, all graphs are assumed to be connected since the generalization to non-connected graphs is trivial and consists of considering the direct sum of the matrices for their components.

A.4.1 MATRIX REPRESENTATION OF A GRAPH

A graph can be represented in various forms. Some of these representations are of theoretical importance, and others are useful from the programming point of view when applied to realistic problems. In this section, five different representations of a graph are described.

Node Adjacency Matrix: Let S be a graph with N nodes. The *adjacency matrix* **A** is an $N \times N$ matrix in which the entry in row i and the entry in column j are 1 if node n_i is adjacent to n_j, and are 0 otherwise. This matrix is symmetric and the row sums of **A** are the degrees of the nodes of S.

The adjacency matrix of the graph S, shown in Figure A.13, is a 5 × 5 matrix as below:

$$\mathbf{A} = \begin{bmatrix} 0 & 1 & 1 & 1 & 0 \\ 1 & 0 & 1 & 1 & 0 \\ 1 & 1 & 0 & 0 & 1 \\ 1 & 1 & 0 & 0 & 1 \\ 0 & 0 & 1 & 1 & 0 \end{bmatrix}. \quad \text{(A-10)}$$

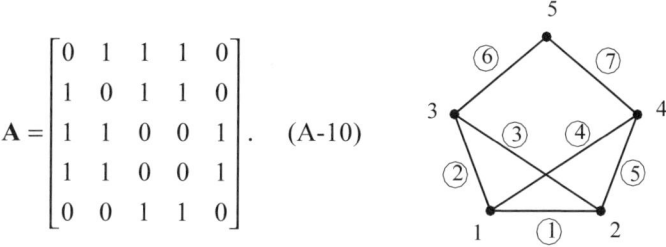

Fig. A.13 A graph S.

It can be noted that **A** is a symmetric matrix of trace zero. The (i, j)th entry of $\mathbf{A}^2$ shows the number of walks of length 2 with n_i and n_j as end nodes. Similarly, the entry in the (i, j) position of $\mathbf{A}^k$ is equal to the number of walks of length k with n_i and n_j as end nodes. The polynomial

$$f(\lambda) = \det(\mathbf{I}\lambda - \mathbf{A}), \quad \text{(A-11)}$$

is called the *characteristic polynomial* of S. The collection of $N(S)$ eigenvalues of **A** is known as the *spectrum* of S. Since **A** is symmetric, the spectrum of S consists of $N(S)$ real numbers. The sum of eigenvalues of **A** is equal to zero.

Node-Member Incidence Matrix: Let S be a graph with M members and N nodes. The *node-member incidence matrix* $\overline{\mathbf{B}}$ is an $N \times M$ matrix in which the entry in row i and the entry in column j are 1 if node n_i is incident with member m_j and are 0 otherwise. As an example, the node-member incidence matrix of the graph in Figure A.13 is a 5 × 7 matrix of the form

$$\overline{\mathbf{B}} = \begin{bmatrix} 1 & 1 & 0 & 1 & 0 & 0 & 0 \\ 1 & 0 & 1 & 0 & 1 & 0 & 0 \\ 0 & 1 & 1 & 0 & 0 & 1 & 0 \\ 0 & 0 & 0 & 1 & 1 & 0 & 1 \\ 0 & 0 & 0 & 0 & 0 & 1 & 1 \end{bmatrix}. \quad \text{(A-12)}$$

Obviously, the pattern of an incidence matrix depends on the particular way that its nodes and members are labelled. One incidence matrix can be obtained from another by simply interchanging rows (corresponding to re-labelling the nodes) and columns (corresponding to re-labelling the members).

450   OPTIMAL STRUCTURAL ANALYSIS

The incidence matrix $\overline{\mathbf{B}}$ and the adjacency matrix $\mathbf{A}$ of a graph S are related by

$$\overline{\mathbf{B}}\overline{\mathbf{B}}^t = \mathbf{A} + \mathbf{V}, \quad (A\text{-}13)$$

where $\mathbf{V}$ is a diagonal matrix of order $N(S)$ whose typical entry v_i is the valency of the node n_i of S for $i = 1, ..., N(S)$. For the example of Figure A.13, Eq. (A-13) becomes

$$\overline{\mathbf{B}}\overline{\mathbf{B}}^t = \begin{bmatrix} 0 & 1 & 1 & 1 & 0 \\ 1 & 0 & 1 & 1 & 0 \\ 1 & 1 & 0 & 0 & 1 \\ 1 & 1 & 0 & 0 & 1 \\ 0 & 0 & 1 & 1 & 0 \end{bmatrix} + \begin{bmatrix} 3 & & & & \\ & 3 & & & \\ & & 3 & & \\ & & & 3 & \\ & & & & 3 \end{bmatrix}. \quad (A\text{-}14)$$

The rows of $\overline{\mathbf{B}}$ are dependent and one row can arbitrarily be deleted to ensure the independence of the rest of the rows. The node corresponding to the deleted row is called a *datum (reference) node*. The matrix obtained after deleting a dependent row is called an *incidence matrix* of S and is denoted by $\mathbf{B}$.

Although $\mathbf{A}$ and $\mathbf{B}$ are of great theoretical value, the storage requirements for these matrices are high and proportional to $N \times N$ and $M \times N$, respectively. In fact, a large number of unnecessary zeros are stored in these matrices. In practice, one can use different approaches to reduce the storage required, some of which are described in the following text.

Member List: This type of representation is a common approach in structural mechanics. A member list consists of two rows (or columns) and M columns (or rows). Each column (or row) contains the labels of the two end nodes of each member, in which members are arranged sequentially. For example, the member list of S in Figure A.13 is as follows:

$$\mathbf{ML} = \begin{matrix} & m_1 & m_2 & m_3 & m_4 & m_5 & m_6 & m_7 \\ & \begin{bmatrix} 1 & 1 & 2 & 1 & 2 & 3 & 4 \\ 2 & 3 & 3 & 4 & 4 & 5 & 5 \end{bmatrix} \end{matrix}. \quad (A\text{-}15)$$

It should be noted that a member list can also represent orientations on members. The storage required for this representation is $2 \times M$. Some engineers prefer to add a third row containing the member's labels for easy addressing. In this case, the storage is increased to $3 \times M$.

A different way of preparing a member list is to use a vector containing the end nodes of members sequentially, for example, for the previous example this vector becomes

$$(1, 2\; ; 1, 3\; ; 2, 3\; ; 1, 4\; ; 2, 4\; ; 3, 5\; ; 4, 5\;). \tag{A-16}$$

This is a compact description of a graph; however, it is impractical because of the extra search required for its use in various algorithms.

Adjacency List: This list consists of N rows and D columns, where D is the maximum degree of the nodes of S. The ith row contains the labels of the nodes adjacent to node i of S. For the graph S shown in Figure A.13, the adjacency list is as follows:

$$\mathbf{AL} = \begin{array}{c} n_1 \\ n_2 \\ n_3 \\ n_4 \\ n_5 \end{array} \begin{bmatrix} 2 & 3 & 4 \\ 1 & 3 & 4 \\ 1 & 2 & 5 \\ 1 & 2 & 5 \\ 3 & 4 & \end{bmatrix}_{N \times D} \tag{A-17}$$

The storage needed for an adjacency list is $N \times D$.

Compact Adjacency List: In this list, the rows of **AL** are continually arranged in a row vector **R**, and an additional vector of pointers **P** is considered. For example, the compact adjacency list of Figure A.13 can be written as

$$\mathbf{R} = (2, 3, 4, 1, 3, 4, 1, 2, 5, 1, 2, 5, 3, 4),$$

$$\mathbf{P} = (1, 4, 7, 10, 13, 15) \tag{A-18}$$

P is a vector $(p_1, p_2, p_3, \ldots)$ that helps to list the nodes adjacent to each node. For node n_i, one should start reading **R** at entry p_i and finish at $p_{i+1} - 1$.

An additional restriction can be put on **R** by ordering the nodes adjacent to each node n_i in ascending order of their degrees. This ordering can be of some advantage; an example is nodal ordering for bandwidth optimisation. The storage required for this list is $2M + N + 1$.

A.4.2 CYCLE BASES MATRICES

The cycle-member incidence matrix $\overline{\mathbf{C}}$ of a graph S has a row for each cycle or hinged cycle and a column for each member. An entry c_{ij} of $\overline{\mathbf{C}}$ is 1 if cycle C_i

contains member m_j and it is 0 otherwise. In contrast to the node adjacency and node-member incidence matrix, the cycle-member incidence matrix does not determine a graph up to isomorphism, that is, two totally different graphs may have the same cycle-member incidence matrix.

For a graph S, there exist $2^{b_1(S)} - 1$ cycles or hinged cycles. Thus, $\overline{\mathbf{C}}$ is a $(2^{b_1(S)} - 1) \times M$ matrix. However, one does not need all the cycles of S, and the elements of a cycle basis are sufficient. For a cycle basis, a cycle-member incidence matrix becomes a $b_1(S) \times M$ matrix, denoted by $\mathbf{C}$, known as the *cycle basis incidence matrix* of S. As an example, matrix $\mathbf{C}$ for the graph shown in Figure A.13, for the cycle basis,

$$C_1 = (m_1, m_2, m_3)$$

$$C_2 = (m_1, m_4, m_5)$$

$$C_3 = (m_2, m_4, m_6, m_7)$$

is given by

$$\mathbf{C} = \begin{matrix} C_1 \\ C_2 \\ C_3 \end{matrix} \begin{bmatrix} 1 & 1 & 1 & 0 & 0 & 0 & 0 \\ 1 & 0 & 0 & 1 & 1 & 0 & 0 \\ 0 & 1 & 0 & 1 & 0 & 1 & 1 \end{bmatrix}. \tag{A-19}$$

The *cycle adjacency matrix* $\mathbf{D}$ is a $b_1(S) \times b_1(S)$ matrix, each entry d_{ij} of which is 1 if C_i and C_j have at least one member in common and it is 0 otherwise. This matrix is related to the cycle-member incidence matrix by the relationship,

$$\mathbf{CC}^t = \mathbf{D} + \mathbf{W}, \tag{A-20}$$

where $\mathbf{W}$ is diagonal matrix with w_{ii} being the length of the ith cycle and its trace being equal to the total length of the cycles of the basis.

For the above example,

$$\mathbf{CC}^t = \begin{bmatrix} 0 & 1 & 1 \\ 1 & 0 & 1 \\ 1 & 1 & 0 \end{bmatrix} + \begin{bmatrix} 3 & & \\ & 3 & \\ & & 3 \end{bmatrix}. \tag{A-21}$$

An important theorem that is based on the orthogonality property studied in Section A.3.3 can now be stated.

Theorem: Let S have incidence matrix $\mathbf{B}$ and a cycle basis incidence matrix $\mathbf{C}$. Then,

$$\mathbf{CB}^t = \mathbf{0} \text{ (mod 2)}. \tag{A-22}$$

A simple proof of this theorem can be found in Kaveh [113]. Notice that Eq. (A-22) holds, because of the orthogonality property discussed in Section A.3.3. In fact, the above relation holds even if the cutsets or cycles do not form bases or the matrices contains additional cutset and/or cycle vectors.

A.4.3 SPECIAL PATTERNS FOR FUNDAMENTAL CYCLE BASES

For a fundamental cycle basis, with special labelling for its tree members and chords, a matrix $\mathbf{C}$ with a particular 1 pattern.can be obtained. Let S have a tree T whose members are $M(T) = (m_1, m_2, \ldots, m_p)$ and a cotree for which is $M(T^*) = (m_{p+1}, m_{p+2}, \ldots, m_{M(S)})$. Then there is a unique fundamental cycle C_i in $S - M(T^*) + m_i$, $p+1 \leq i \leq M(S)$, and this set of cycles forms a basis for the cycle space of S. For example, for the graph S of Figure A.12(a) whose members are labelled as shown in Figure A.14, the fundamental cycle basis consists of

$$C_1 = (m_1, m_4, m_5, m_8),$$
$$C_2 = (m_2, m_1, m_4, m_5, m_6, m_9), \text{ and}$$
$$C_3 = (m_3, m_2, m_1, m_4, m_5, m_6, m_7, m_{10}),$$

and is given by

$$\mathbf{C} = \begin{matrix} C_1 \\ C_2 \\ C_3 \end{matrix} \begin{bmatrix} 1 & 0 & 0 & 1 & 1 & 0 & 0 & | & 1 & 0 & 0 \\ 1 & 1 & 0 & 1 & 1 & 1 & 0 & | & 0 & 1 & 0 \\ 1 & 1 & 1 & 1 & 1 & 1 & 0 & | & 0 & 0 & 1 \end{bmatrix} = [\mathbf{C}_T \mid \mathbf{I}]. \tag{A-23}$$

$$\phantom{\mathbf{C} = C_2}\ \ \ M(T) \qquad\qquad\ M(T^*)$$

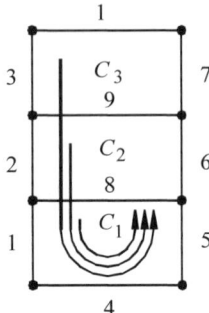

Fig. A.14 A graph with oriented members and cycles.

A.4.4 CUTSET BASES MATRICES

The *cutset-member incidence matrix* $\overline{\mathbf{C}}^*$ for a graph S has a row for each cutset of S and a column for each member. An entry $\overline{c}_{ij}^*$ of $\overline{\mathbf{C}}^*$ is 1 if cutset C_i^* contains member m_j and it is 0 otherwise. This matrix, like $\overline{\mathbf{C}}$, does not determine a graph completely.

Independent rows of $\overline{\mathbf{C}}^*$ for a cutset basis, denoted by $\mathbf{C}^*$, form a matrix known as a *cutset basis incidence matrix*, which is a $\rho(S) \times M$ matrix, $\rho(S)$ being the rank of graph S. As an example, $\mathbf{C}^*$ for the cutset of Figure A.12 with members labelled as in Figure A.15(a), is given below:

$$\mathbf{C}^* = \begin{bmatrix} 0 & 0 & 1 & 0 & 0 & 0 & 0 & 0 & 1 \\ 0 & 0 & 0 & 0 & 0 & 0 & 1 & 0 & 0 & 1 \\ 0 & 1 & 0 & 0 & 0 & 0 & 0 & 1 & 1 \\ 0 & 0 & 0 & 0 & 0 & 1 & 0 & 0 & 1 & 1 \\ 0 & 0 & 0 & 1 & 0 & 0 & 0 & 1 & 1 & 1 \\ 1 & 0 & 0 & 0 & 0 & 0 & 1 & 1 & 1 \\ 0 & 0 & 0 & 0 & 1 & 0 & 0 & 1 & 1 & 1 \end{bmatrix}. \qquad (A\text{-}24)$$

APPENDIX A–BASIC CONCEPTS AND DEFINITIONS   455

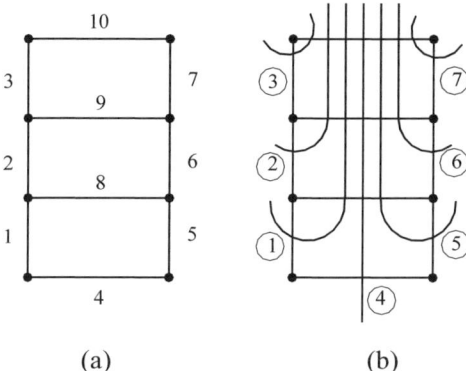

Fig. A.15 A graph with oriented members and cutsets bases.

The *cutset adjacency matrix* **D*** is a $\rho(S) \times \rho(S)$ matrix defined analogously to the cycle adjacency matrix **D**.

A.4.5 SPECIAL PATTERNS FOR FUNDAMENTAL CUTSET BASES

For a fundamental cutset basis with appropriate labelling of the members in T and T^*, as illustrated in Figure A.15(b), if the cutsets are taken in the order of their generators (tree members), the matrix **C*** will have a particular pattern as follows:

$$\mathbf{C}_0^* = \begin{bmatrix} 1 & 0 & 0 & 0 & 0 & 0 & 0 & 1 & 1 & 1 \\ 0 & 1 & 0 & 0 & 0 & 0 & 0 & 0 & 1 & 1 \\ 0 & 0 & 1 & 0 & 0 & 0 & 0 & 0 & 0 & 1 \\ 0 & 0 & 0 & 1 & 0 & 0 & 0 & 1 & 1 & 1 \\ 0 & 0 & 0 & 0 & 1 & 0 & 0 & 1 & 1 & 1 \\ 0 & 0 & 0 & 0 & 0 & 1 & 0 & 0 & 1 & 1 \\ 0 & 0 & 0 & 0 & 0 & 0 & 1 & 0 & 0 & 1 \end{bmatrix} = \begin{bmatrix} \mathbf{I} \mid \mathbf{C}_c^* \end{bmatrix}. \quad \text{(A-25)}$$

From the orthogonality condition, $\mathbf{C}_0 \mathbf{C}_0^{*t} = \mathbf{0}$, that is,

$$\begin{bmatrix} \mathbf{C}_T & \mathbf{I} \end{bmatrix} \begin{bmatrix} \mathbf{I} \\ \mathbf{C}_c^{*t} \end{bmatrix} = \mathbf{0}. \quad \text{(A-26)}$$

Hence $\mathbf{C}_T + \mathbf{C}_c^{*t} = \mathbf{0}(\text{mod } 2)$, and:

$$\mathbf{C}_T = \mathbf{C}_c^{*t} \quad \text{(A-27)}$$

Therefore, for a graph having $\mathbf{C}_0$, one can construct $\mathbf{C}_0^*$ and *vice versa*.

There exists a very simple basis for the cutset space of a graph, which consists of $N - 1$ cocycles of S. As an example, for the graph of Figure A.13, considering n_5 as a datum node, we have

$$\mathbf{C}^* = \begin{bmatrix} 1 & 1 & 0 & 1 & 0 & 0 & 0 \\ 1 & 0 & 1 & 0 & 1 & 0 & 0 \\ 0 & 1 & 1 & 0 & 0 & 1 & 0 \\ 0 & 0 & 0 & 1 & 1 & 0 & 1 \end{bmatrix}, \qquad (A-28)$$

which is the same as the incidence matrix $\mathbf{B}$ of S. The simplicity of the displacement method of structural analysis is due to the existence of such a simple basis.

A.5 DIRECTED GRAPHS AND THEIR MATRICES

An *oriented* or *directed* graph is a graph in which each member is assigned an orientation. A member is oriented from its initial node (*tail*) to its final node (*head*). The initial node is said to be positively incident on the member, and the final node negatively incident, as shown in Figure A.16(a).

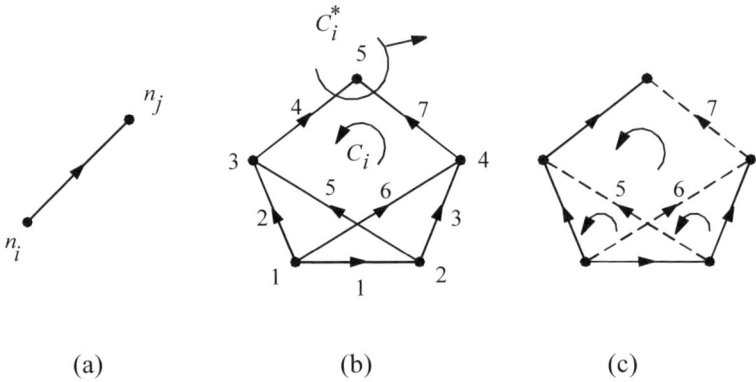

(a)   (b)   (c)

Fig. A.16 An oriented member, a directed graph and a directed tree (with chords shown in dashed lines).

The choice of orientation of members of a graph is arbitrary; however, once it is chosen, it must be retained. Cycles and cutsets can also be oriented as shown in Figure A.16(b).

For example, m_4 is positively oriented in cycle C_i and m_7 is negatively oriented in cutset C_i^*.

APPENDIX A–BASIC CONCEPTS AND DEFINITIONS   457

All the matrices $\bar{\mathbf{B}}$, $\mathbf{B}$, $\mathbf{C}$ and $\mathbf{C}^*$ can be defined as before, with the difference of having +1, −1 and 0 as entries, according to whether the member is positively, negatively and zero incident with a cutset or a cycle.

As an example, for graph S in Figure A.16(b), the matrix $\mathbf{B}$ with n_1 as datum node is formed:

$$\mathbf{B} = \begin{matrix} n_2 \\ n_3 \\ n_4 \\ n_5 \end{matrix} \begin{bmatrix} -1 & 0 & 1 & 0 & 1 & 0 & 0 \\ 0 & -1 & 0 & 1 & -1 & 0 & 0 \\ 0 & 0 & -1 & 0 & 0 & -1 & 1 \\ 0 & 0 & 0 & -1 & 0 & 0 & -1 \end{bmatrix}. \qquad (A\text{-}29)$$

Consider a tree as shown with continuous lines in Figure A.16(c). When the directions of the cycles are taken as those of their corresponding chords (dashed lines), the fundamental cycle basis incidence matrix can be written as

$$\mathbf{C} = \begin{matrix} C_1 \\ C_2 \\ C_3 \end{matrix} \begin{bmatrix} 1 & -1 & 0 & 0 & 1 & 0 & 0 \\ 1 & 0 & 1 & 0 & 0 & 1 & 0 \\ 1 & -1 & 1 & -1 & 0 & 0 & 1 \end{bmatrix}. \qquad (A\text{-}30)$$
$$\underbrace{\hphantom{xxxxxxxxxxx}}_{\mathbf{C}_T} \underbrace{\hphantom{xxxxxxxx}}_{\mathbf{C}_c}$$

It should be noted that the tree members are numbered first, followed by the chords of the cycles in the same sequence as their generation.

Obviously,

$$\mathbf{BC}^t = \mathbf{CB}^t = \mathbf{0}(\mathrm{mod}2), \qquad (A\text{-}31)$$

with a proof similar to that of the non-oriented case.

A cutset basis incidence matrix is similarly obtained:

$$\mathbf{C}^* = \begin{bmatrix} 1 & 0 & 0 & 0 & -1 & 1 & -1 \\ 0 & 1 & 0 & 0 & 1 & 0 & 1 \\ 0 & 0 & 1 & 0 & 0 & 1 & -1 \\ 0 & 0 & 0 & 1 & 0 & 0 & 1 \end{bmatrix}, \qquad (A\text{-}32)$$
$$\underbrace{\hphantom{xxxxxxxxxx}}_{\mathbf{C}_T^*} \underbrace{\hphantom{xxxxxxxx}}_{\mathbf{C}_c^*}$$

where the direction of a cutset is taken as the orientation of its generator (the corresponding tree member).

458   OPTIMAL STRUCTURAL ANALYSIS

It can easily be proved that

$$\mathbf{C}_T = -\mathbf{C}_c^{*t}. \tag{A-33}$$

For a directed graph, Eq. (A-13) becomes

$$\mathbf{BB}^t = \mathbf{A} - \mathbf{V}, \tag{A-34}$$

Similarly, Eq. (A-20) for the directed case becomes

$$\mathbf{CC}^t = \mathbf{D} - \mathbf{W}. \tag{A-35}$$

A.6 GRAPHS ASSOCIATED WITH MATRICES

Matrices associated with graphs are discussed in the previous sections. Sometimes it is useful to consider the reverse of this process and think of the graph associated with an arbitrary matrix **A**. Such a graph has a node associated with each row of the matrix and, if a_{ij} is non-zero, then there is a connecting member from node i to node j. In the case of a symmetric matrix, there is always a connection from i to j whenever there is one from j to i; therefore, one can simply use undirected members. Two simple examples are illustrated in Figure A.17 and Figure A.18. The directed graph associated with a non-symmetric matrix is usually called a *digraph* and the word graph is used for the undirected graph associated with a symmetric matrix:

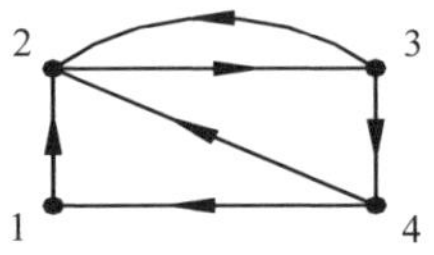

Fig. A.17 A non-symmetric matrix and its associated digraph.

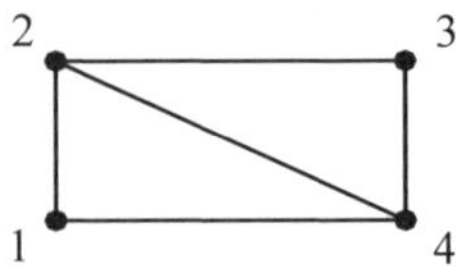

Fig. A.18 A symmetric matrix and its associated graph.

It should be noted that if **A** is required to be taken as an adjacency matrix, then a single loop should be added at each joint, but if it is treated as **A** + **I**, then no addition is needed and the graph can still be considered as a simple one.

APPENDIX A–BASIC CONCEPTS AND DEFINITIONS 459

With a rectangular matrix **E**, a bipartite graph $S = (A, M, B)$ can be associated. With each row of **E**, a node of A is associated and with each column of **E** a node of B is associated. Two nodes of A and B are connected with a member of S if e_{ij} is non-zero. An example of this is shown in Figure A.19.

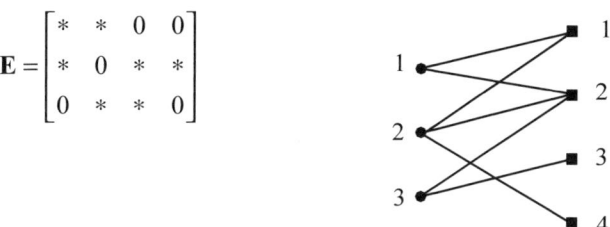

Fig. A.19 A rectangular matrix and its associated bipartite graph.

A.7 PLANAR GRAPHS; EULER´S POLYHEDRON FORMULA

Graph theory and properties of planar graphs were first discovered by Euler in 1736. After 190 years, Kuratowski found a criterion for a graph to be planar. Whitney developed some important properties of embedding graphs in the plane. MacLane expressed the planarity of a graph in terms of its cycle basis. In this section, some of these criteria are studied, and Euler´s polyhedron formula is proved.

A.7.1 PLANAR GRAPHS

A graph S is called *planar* if it can be drawn (embedded) in the plane in such a way that no two members cross each other. For example, a complete graph K_4, shown in Figure A.20, is planar since it can be drawn in the plane as shown.

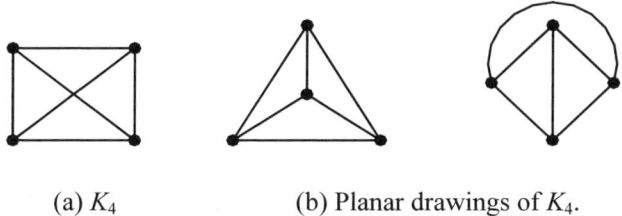

(a) K_4 (b) Planar drawings of K_4.

Fig. A.20 K_4 and two of its drawings.

On the other hand K_5, shown in Figure A.21, is not planar, since every drawing of K_5 contains at least one crossing.

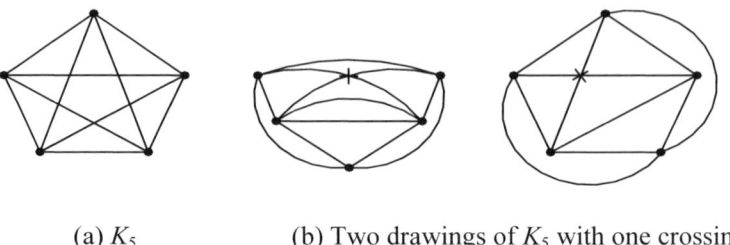

(a) K_5 (b) Two drawings of K_5 with one crossing.

Fig. A.21 K_5 and two of its drawings.

Similarly, $K_{3,3}$ is not planar, as illustrated in Figure A.22.

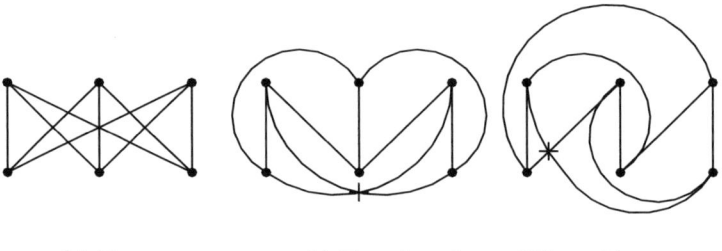

(a) $K_{3,3}$ (b) Two drawings of $K_{3,3}$ with one crossing.

Fig. A.22 $K_{3,3}$ and its drawings.

A planar graph S drawn in the plane divides the plane into regions that are all bounded except one. If S is drawn on a sphere, all the regions will be bounded; however, the number of regions will not change. The cycle bounding a region is called a *regional cycle*. Obviously, the sum of the lengths of regional cycles is twice the number of members of the graph.

There is an outstanding formula that relates the number of regions, members and nodes of a planar graph, in the form

$$R(S) - M(S) + N(S) = 2,$$

where $R(S)$, $M(S)$ and $N(S)$ are the numbers of regions, members and nodes of planar graph S. This formula shows that, for different drawings of S in the plane, $R(S)$ remains constant.

Originally, the above relationship was given for polyhedra, in which $R(S)$, $M(S)$ and $N(S)$ correspond to faces, edges and corners of a polyhedron, respectively. However, the theorem can easily be expressed in graph-theoretical terms as follows:

Theorem (Euler [42]): Let S be a connected planar graph. Then,

$$R(S) - M(S) + N(S) = 2. \tag{A-36}$$

Proof: For a proof, S is re-formed in two stages. In the first stage, a spanning tree T of S is considered in the plane for which $R(T) - M(T) + N(T) = 2$. This is true, since $R(T) = 1$ and $M(T) = N(T) - 1$. In the second stage, chords are added one at a time. Addition of a chord increases the number of members and regions each by unity, leaving the left-hand side of Eq. (A-36) unchanged during the entire process, and the result follows.

A.7.2 THEOREMS FOR PLANARITY

To check the planarity of a graph, different approaches are available that are based on the following theorems. These theorems are only stated and the reader may refer to textbooks on graph theory for proofs.

Theorem (Kuratowski [143]): A graph is planar if and only if it does not contain a subgraph that has K_5 or $K_{3,3}$ as a contraction.

For contracting a member, one of its nodes is brought closer to the other end node until the two ends coincide. Then multiple members are replaced by a single member, as in Figure A.23(a). A *contraction* of a graph is the result of a sequence of member contractions; see Figure A.23(b).

(a) Contraction of a member m_k.

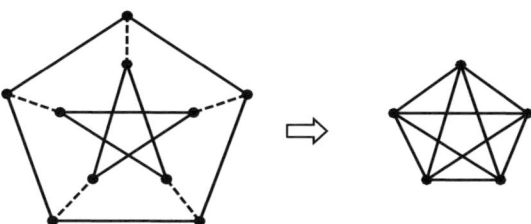

(b) Contraction of the Petersen graph to K_5.

Fig. A.23 Contraction of a member and a graph.

Theorem (MacLane [159]): A connected graph is planar if and only if every block of S with at least three nodes has a cycle basis $C_1, C_2, \ldots, C_{b_1(S)}$ and one additional cycle C_0 such that every member is contained in exactly two of these $b_1(S) + 1$ cycles.

A *block* is a maximal non-separable graph, and a *non-separable* graph is a graph that has no cut-points. A *cut-point* is a node whose removal increases the number of components, and a *bridge* is a member with the same property.

Theorem (Whitney [229]): A graph is planar if and only if it has a combinatorial dual.

For a connected planar graph S, the dual graph S^* is constructed as follows:

To each region r_i of S there is a corresponding node r_i^* of S^* and to each member m_j of S there is a corresponding member m_j^* in S^* such that, if member m_j occurs on the boundary of two regions r_1 and r_2, then the member m_j^* joins the corresponding nodes r_1^* and r_2^* in S^*; see Figure A.24.

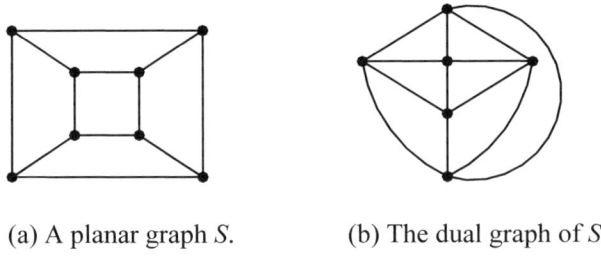

(a) A planar graph S. (b) The dual graph of S.

Fig. A.24 A planar graph and its dual.

A.8 MAXIMAL MATCHING IN BIPARTITE GRAPHS

A.8.1 DEFINITIONS

As defined before, a graph is bipartite if its set of nodes can be partitioned into two sets A and B such that every member of the graph has one end node in A and the other in B. Such a graph is denoted by $S = (A, B)$. A set of members of S is called a *matching* if no two members have a common node. The size of any largest matching in S is called the *matching number* of S, denoted by $\psi(S)$. A subset $N'(S) \subseteq N(S)$ is the *node cover of S* if each member of S has at least one end node in $N'(S)$. The cardinality of any smallest node cover, denoted by $\tau(S)$, is known as the *node covering number* of S.

APPENDIX A–BASIC CONCEPTS AND DEFINITIONS   463

A.8.2 THEOREMS ON MATCHING

The following three theorems are stated, and the proofs may be found in the book by Lovász and Plumner [155]:

Theorem 1 (König [140]): For a bipartite graph S, the matching number $\Psi(S)$ is equal to the node covering number $\tau(S)$.

Theorem 2 (Hall [72]): Let $S = (A, B)$ be a bipartite graph. Then S has a complete matching of A into B if and only if $|\Gamma(X)| \geq |X|$ for all $X \subseteq A$.

$\Gamma(X)$ is the image of X, that is, those elements of B that are connected to the elements of X in S. Figure A.25(a) shows a bipartite graph for which matching exists and Figure A.26(b) illustrates a case in which matching does not exist, because $X = (a_1, a_2)$ are matched to b_1, that is, $|\Gamma(X)| \leq |X|$:

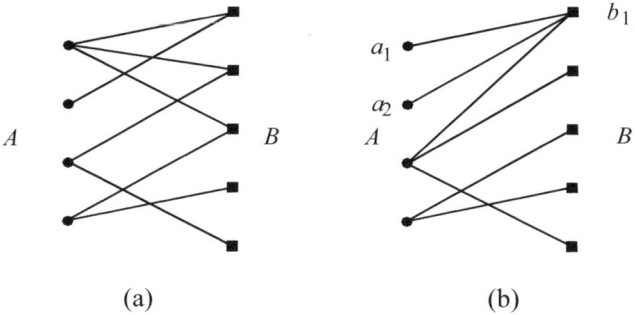

(a)   (b)

Fig. A.25 Matching in bipartite graphs.

A *perfect matching* is a matching that covers all nodes of S.

Theorem 3 (Frobenius [53]): A bipartite graph $S = (A, B)$ has a perfect matching if and only if $|A| = |B|$ and for each $X \subseteq A$, $|X| \leq |\Gamma(X)|$.

Therefore, Frobenius's theorem characterizes those bipartite graphs that have a perfect matching. Hall's theorem characterizes those bipartite graphs that have a matching of A into B. König's theorem gives a formula for the matching number of a bipartite graph.

A.8.3 MAXIMUM MATCHING

Let M be any matching in a bipartite graph $S = (A, B)$. A path P is called an *alternating path with respect to M* or an *M-alternating path* if its members (edges) are alternately chosen from the matching M and outside M. A node is *exposed* (unmatched, not covered) with respect to matching M if no member of M is incident with that node. An *alternating tree* relative to the matching is a tree that satisfies the following two conditions: first, the tree contains exactly one exposed node from A, which is called its *root*, and second, all paths between the root and any other node in the tree are alternating paths.

An M-alternating path joining two exposed nodes is called an *M-augmenting path*. For every such path, the corresponding matching can be made larger by discarding the members of $P \cap M$ and adding those of $P - M$, where P is an M-alternating path $b_2a_1b_1a_3b_3a_4$; see Figure A.26(b). Thus, if S contains any M-alternating path P joining two exposed nodes, then M cannot be a maximum matching, for one can readily obtain a larger matching M' by discarding the members of $P \cap M$ and adding those of $P - M$.

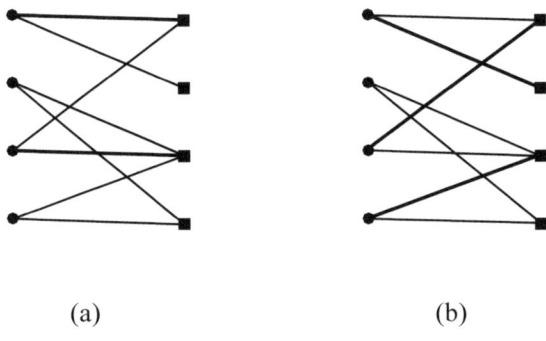

(a)   (b)

Fig. A.26 M-alternating path.

Theorem 4 (Berge [12]): Let M be a matching in a graph S. Then M is a maximum matching if and only if there exists no augmenting path in S relative to M.

The above result provides a method for finding a maximum matching in S. The computational procedure for the construction of a maximum matching begins with considering any feasible matching, possibly the empty matching. Each exposed node of A is made the root of an alternating tree, and nodes and members are added to the trees by means of a labelling technique. Eventually, the following two cases must occur: either an exposed node in B is added to one of the trees or else it is not possible to add more nodes and members to any of the trees. In the former case, the matching is augmented and the formation of trees is repeated with respect to the new matching. In the latter case, the trees are said to be *Hungarian* and the process is terminated.

There are many efficient algorithms for bipartite matching, and the reader may refer Kaveh [94], or to the original paper of Hopcroft and Karp [82], or to the excellent book by Lawler [148].

APPENDIX B

Greedy Algorithm and its Applications

In this appendix, the Greedy Algorithm developed by Edmonds [41] for selecting an optimal base of a matroid is described. This is a powerful method for most combinatorial optimisation problems, and has found many applications in structural optimisation. Some applications of this algorithm in structural mechanics are briefly discussed.

B.1 AXIOM SYSTEMS FOR A MATROID

A matroid may be defined in different interrelated forms, several of which were described in Whitney's original paper [230]. Here the definitions in terms of the concepts of independence, bases and circuits are presented.

DEFINITION IN TERMS OF INDEPENDENCE

A *matroid* M is a set of elements $S = \{s_1, s_2, ..., s_m\}$ and a collection F of subsets of S (called *independent sets*) such that

7(I1) $\emptyset \in F$, where $\emptyset$ is the empty set.
(I2)   If $X \in F$ and $Y \subseteq X$, then $Y \in F$.
(I3)   If $X \in F$ and $Y \in F$ with $|X| = |Y|+1$, then there exists $s \in X - Y$ such that $Y + s \in F$.

Here, $|X|$ and $|Y|$ denote the cardinalities of the sets X and Y, respectively.

For a *matroid* M = (S,F), those subsets of S belonging to F are called *independent* and those that do not belong to F are known as *dependent*. A maximal independent subset of a matroid is known as a *base* of M.

DEFINITION IN TERMS OF BASES

M(S,F) is a matroid if the collection of bases of M, denoted by B, satisfies the following conditions:

(B1) $B \neq \emptyset$.
(B2) $|B_1| = |B_2|$ for every $B_1, B_2 \in B$.
(B3) If $B_1, B_2 \in B$ and $s_1 \in B_1$, then there exists a $s_2 \in B_2$ such that $(B_1 - s_1 + s_2) \in B$.

A *circuit* of a matroid M is a minimal dependent set of S.

DEFINITION IN TERMS OF CIRCUITS

M(S,F) is a *matroid* if the collection of circuits of M, denoted by C, satisfies the following postulates:

(C1) No proper subset of a circuit is a circuit.
(C2) If C_1 and C_2 are distinct circuits of C and $s \in C_1 \cap C_2$, then there exists a circuit C_3 of C such that $C_3 \subseteq (C_1 \cup C_2) - s$.

Corresponding to each subset F_i, a number $r(F_i) \in Z$ is defined, which is known as the *rank* of F_i, as follows:

$$r(F_i) = \text{Max } \{|X|: X \subseteq F_i, X \in F\}. \tag{B-1}$$

Now it is obvious that knowledge of the bases, or circuits, or rank functions is sufficient to uniquely determine the corresponding matroid.

Example 1: Consider the following matrix:

$$\mathbf{A} = \begin{matrix} & 1 & 2 & 3 & 4 & 5 \\ & \begin{bmatrix} 1 & 0 & 0 & 1 & 1 \\ 0 & 1 & 0 & 0 & 1 \end{bmatrix} \end{matrix}$$

over the field $\mathbb{R}$ of real numbers. The column set $\{1,2,3,4,5\}$ of **A** and its independent subsets form a matroid M(**A**). The set of independent subsets of this

APPENDIX B–GREEDY ALGORITHM AND ITS APPLICATIONS   467

matroid is obtained as $I = \{\varnothing, \{1\}, \{2\}, \{4\}, \{5\}, \{1,2\}, \{2,4\}, \{2,5\}, \{4,5\}\}$ and the set of its circuits is $C = \{\{3\}, \{1,4\}, \{1,2,5\}, \{2,4,5\}\}$.

Example 2: Let S be the graph as shown in Figure B.1. Consider a matroid M(S), formed on the members $\{m_1, m_2, m_3, m_4, m_5\}$ of S with circuit set as

$C = \{\{m_3\}, \{m_1, m_4\}, \{m_1, m_2, m_5\}, \{m_2, m_4, m_5\}\}$. This matroid is known as the *cycle matroid* of the graph S, as defined in the next section.

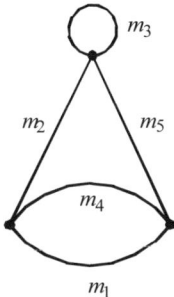

Fig. B.1 A planar graph S

Now compare M(S) with M($\mathbf{A}$) in Example 1. It is seen that, under bijection Ψ from $\{1,2,3,4,5\}$ to $\{m_1, m_2, m_3, m_4, m_5\}$ defined by $\Psi(i) = m_i$, a set X is a circuit in M($\mathbf{A}$) if and only if $\Psi(X)$ is a circuit in M(S). Equivalently, a set Y is independent in M($\mathbf{A}$) if and only if $\Psi(Y)$ is independent in M(S). Thus, matroids M($\mathbf{A}$) and M(S) have the same structure or are *isomorphic*. A matroid that is isomorphic to the cycle matroid of a graph is called *graphic*. Therefore, the matroid M($\mathbf{A}$) in Example 1 is graphic.

B.2 MATROIDS APPLIED TO STRUCTURAL MECHANICS

Matroids have been applied to some problems in structural analysis and the study of the rigidity of skeletal structures. In this section, examples of such matroids are considered, and the properties associated with each one are discussed.

A BASIS FOR A FINITE VECTOR SPACE

A conceptual study of structural analysis using vector spaces has been made by Maunder [162]. One can easily obtain a matroidal version of this study by constructing matroids of the following kind.

Let V be a finite vector space and F be the collection of linearly independent subsets of vectors of V. Then, M = (V,F) forms a matroid. The rank function of this matroid is the dimension of V, and its base forms a basis of the vector space.

Although a finite vector space always constitutes a matroid, not all matroids are realisable as vector spaces.

A BASIS FOR CYCLE SPACE OF A GRAPH

A cycle basis of a graph is defined in Appendix A, and its application for the formation of a statical basis of a structure is described in Chapter 3. In this section, a cycle space and its bases are formulated in terms of matroids.

Let C contain all simple cycles of a graph S, and F be the collection of mod 2 independent cycles of S. Then (C,F) forms a matroid, defined as *cycle space matroid* $M_s(S)$ of S. A base of $M_s(S)$ is a cycle basis of S, and its rank is $M(S) - N(S) + b_0(S)$.

The above matroid can be defined using the member-cycle incidence matrix of a graph. Each row of this matrix corresponds to a member, and each column represents a cycle; see Kaveh [94,96,112].

The columns of a member-cycle incidence matrix are either dependent or independent. Take the columns of the matrix as elements of C, and independent subsets of columns as elements of F. Then, (C,F) forms a cycle space matroid $M_s(S)$ of S.

Example: Consider a graph S as shown in Figure B.2. This graph contains 3 cycles C = $\{C_1, C_2, C_3\}$ and F = $\{(C_1),(C_2),(C_3),(C_1,C_2),(C_2,C_3),(C_1,C_2,C_3)\}$. The rank of $M_s(S) = 8 - 6 + 1 = 3$, and $\{C_1, C_2\}$ is a typical base of this matroid.

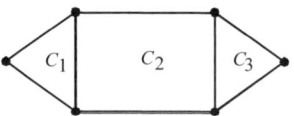

Fig. B.2 A graph S and its cycles.

A BASIS FOR CUTSET SPACE OF A GRAPH

A cutset space of a graph is defined in Chapter 1, and its application for the formation of a kinematical basis of a structure when the displacement method is used is described in Chapter 4. In this section, a cutset space and its bases are defined in terms of matroids.

APPENDIX B–GREEDY ALGORITHM AND ITS APPLICATIONS

Let C* contain all cutsets of a graph S, and F be the collection of mod 2 independent cutsets of S. Then (C*,F) forms a matroid, defined as *cutset space matroid* $M_c(S)$ of S. A base of $M_c(S)$ is a cutset basis of S and its rank is given by $N(S) - b_0(S)$. This matroid can also be defined using the member-cutset incidence matrix of S. The rows and columns of this matrix correspond to members and cutsets, respectively. The columns of this matrix are either dependent or independent. Take the columns of the matrix as elements of C*, and independent subsets of columns as elements of F. Then, (C*,F) forms a cutset space matroid $M_c(S)$ of S.

Example: Consider a graph as shown in Figure B.3. The set C* contains 4 cutsets $\{C_1^*, C_2^*, C_3^*, C_4^*\}$ and F = $\{(C_1^*), (C_2^*), (C_3^*), (C_4^*), (C_1^*, C_2^*), (C_1^*, C_3^*), (C_2^*, C_3^*), (C_1^*, C_4^*), \ldots, (C_2^*, C_3^*, C_4^*)\}$. A typical base of the cutset space matroid, can be taken as $B_1 = \{C_1^*, C_2^*, C_4^*\}$.

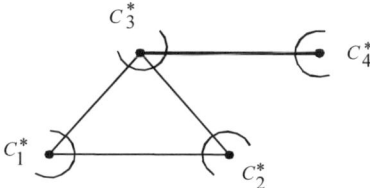

Fig. B.3 A graph and its cutsets.

CYCLE MATROID OF A GRAPH

Spanning trees of a connected graph (spanning forest when S is not connected) have various applications. Some of its applications in structural engineering are described in Chapters 3, 5 and 6. In the following text, a cycle matroid is defined in different interrelated forms, a base of which is a spanning tree of S.

Let S be a graph. Consider S as the set of members of S and let $X \in$ F if and only if X does not contain a cycle of S, that is, it is a cycle-free subgraph (subtree if S is connected and subforest if it is disconnected). Then F is a collection of independent sets of a matroid in S, known as the *cycle matroid* of S, denoted by M(S). This matroid is also known as a *polygon matroid*.

Alternatively, let S be a graph and consider the set of all spanning forests of S as B. It can easily be shown that B is a base set of a matroid M = (S,F) on member set $M(S)$ of S, known as the *cycle matroid* of S.

Similarly, let C denote the set of simple cycles of a graph S, then C is the set of circuits of a matroid M on member set $M(S)$, called *a cycle matroid* of S. The rank of $M(S)$ is $N(S) - b_0(S)$ and for a connected graph it is $N(S) - 1$.

Example: Consider a graph as shown in Figure B.4(a). The sets S and F for the cycle matroid of S are as follows:

$S = \{m_1, m_2, m_3, \ldots, m_7\}$ and $F = \{(m_1), (m_2), \ldots, (m_7), (m_1, m_2), (m_2, m_3), \ldots, (m_1, m_2, m_4), \ldots, (m_5, m_6, m_7), \ldots, (m_1, m_2, m_4, m_5)\}$.

A typical base of $M(S)$ may be considered as $B_1 = \{m_1, m_2, m_4, m_5\}$ and a typical circuit of $M(S)$ can be taken as $\{m_1, m_2, m_3\}$ or $\{m_1, m_2, m_5, m_6\}$.

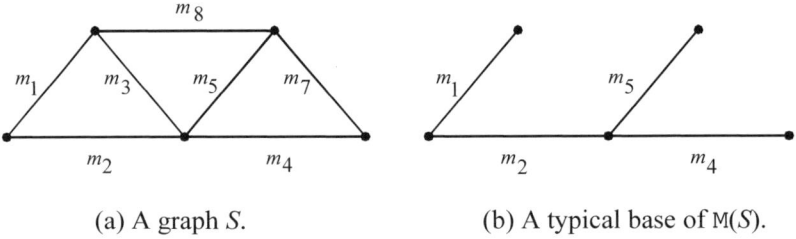

(a) A graph S. (b) A typical base of $M(S)$.

Fig. B.4 A graph S and a typical base of its cycle matroid.

B.3 COCYCLE MATROID OF A GRAPH

Let S be a graph and let C* denote the set of cutsets of S. Then C* is the set of circuits of a matroid on $M(S)$, called a *cocycle* or *cutset matroid* of S, denoted by $M^*(S)$. Obviously, a set X of members of S is a base of the cocycle matroid $M^*(S)$, if and only if $M(S) - X$ is a spanning forest of S. For a connected graph, the members of $M(S) - T$ are known as *cotrees* of S. The rank of $M^*(S)$ is given as $r(M^*(S)) = M(S) - N(S) + b_0(S)$.

Definition: Let M be a matroid on S, whose bases are B_i. The collection of sets $S - B_i$ are bases of another matroid M* on S, known as the *dual* matroid of M. This dual matroid is unique for an M, and the dual of a dual matroid is M itself. Circuits of M are called *cocircuits* or *cutsets* of M*.

By definition, it follows that the cycle matroid $M(S)$ is the dual of the cocycle matroid $M^*(S)$ of a graph S.

Example: Let S be a connected graph as shown in Figure B.5.

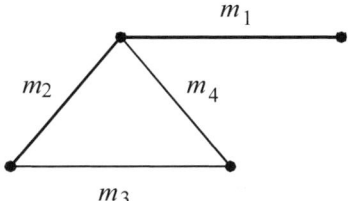

Fig. B.5 A connected graph S.

The circuits of $M(S)$ and $M^*(S)$ are as follows:

$$C\ (S) = \{m_2, m_3, m_4\},$$

and

$$C^*(S) = \{(m_1), (m_2, m_4), (m_3, m_4), (m_2, m_3)\}.$$

B.4 MATROID FOR NULL BASIS OF A MATRIX

Matroids are employed in combinatorial approaches to the force method of structural analysis, as described in Section B.6. Matroids are also used in algebraic force methods [75], a brief description of which is given here.

Let $\mathbf{Ax} = \mathbf{b}$ be the equilibrium equations of a structure, where $\mathbf{x}$ and $\mathbf{b}$ are the vectors of internal forces and applied loads. For a statically indeterminate structure, $\mathbf{A}$ is an $m \times n$ rectangular matrix with $n < m$ and rank $\mathbf{A} = m$.

Take the columns of $\mathbf{A}$ as the elements of S of a matroid $M = (S, F)$ whose independent subsets are linearly independent subsets of the columns of $\mathbf{A}$. A circuit is a minimal dependent subset of columns. Generate all such circuits and consider it as $C = \{C_1, C_2, ..., C_r\}$. Now form another matroid $M_n = \{C, F_n\}$, where F_n consists of subsets of independent circuits of C. A base of M_n is a null basis of $\mathbf{A}$, that is columns of a matrix $\mathbf{B}_1$ such that $\mathbf{AB}_1 = \mathbf{0}$.

For an efficient analysis, special null bases are required, which correspond to sparse flexibility matrices. The formation of such bases becomes feasible using the combinatorial optimisation method of Section B.5.

B.5 COMBINATORIAL OPTIMISATION: THE GREEDY ALGORITHM

In 1926, Boruvka solved the following problem:

> Given a matrix of order n, having distinct positive real coefficients with $A_{ii} = 0$ and $A_{ij} = A_{ji}$, it is possible to find a set of coefficients such that:
>
> 1. There exist two randomly natural numbers k_1, k_2 ($\leq n$) belonging to the set of the form $A_{k1h2}, A_{h2h3}, ..., A_{h-2q-1}, A_{hq-1k2}$.
>
> 2. The sum of the terms of this set is minimal.

In graph-theoretical terms, the above problem can be stated as follows:

For a connected graph with distinct positive real numbers assigned to its members, there is a shortest spanning tree, where the length of the tree is the sum of the numbers assigned to its branches.

After 30 years, Kruskal [142] in 1956 stated the above problem and gave three interrelated efficient algorithms for the selection of a shortest spanning tree of a connected graph. The uniqueness of the existence of such a tree was also proved in his paper. One of these methods is summarised as follows:

Let $\{m_i; i = 1,2, ..., M(S)\}$ be the member set of a graph S. Perform the expansion,

$$m_1 \to m_1 \cup m_2 \to ... \to T, \qquad (B\text{-}2)$$

where m_{i+1} is chosen such that it has the smallest weight and does not form a cycle with $m_1 \cup m_2 \cup ... \cup m_i$. A shortest spanning tree will then be obtained.

This method formed a basis for the Greedy Algorithm for matroids, independently proved by three different authors [41,54,227].

GREEDY ALGORITHM

Let M = (S,F) be a matroid. Assign positive values to each element of S, denoted by $W(s)$, $s \in $ S. For a subset $X \in $ F, define a weight function as

$$W(X) = \Sigma W(s_i), \qquad (B\text{-}3)$$

where summation is taken over all elements s_i of $X \subseteq $ S.

APPENDIX B–GREEDY ALGORITHM AND ITS APPLICATIONS   473

The problem is finding a subset X_{opt} of S such that $X_{opt} \in F$ and $W(X_{opt})$ is minimum (or maximum) over all elements of S. The Greedy Algorithm proceeds as follows:

Select an element s_1 of minimal (maximal) measure (weight) from S, and let $F_1 = \{s_1\}$. Form F_2 from F_1 by adding an element s_2 of minimal (maximal) measure such that F_2 is an independent set from $S - \{s_1\}$, and let $F_2 = F_1+\{s_2\}$. Subsequently choose s_{i+1} of minimal (maximal) measure from $S - \{s_1,s_2,...,s_i\}$ such that F_{i+1} is still an independent set. This process is clearly finite, and the finally selected set is a base of minimal (maximal) measure for M.

An elegant proof of the minimality of the selected base may be found in Welsh [227].

Example: Consider a graph S as shown in Figure B.6(a), with some positive weights assigned to its members. A base of minimal measure for cycle matroid M(S), which is a spanning tree of minimal weight, is selected as depicted in Figure B.6(b).

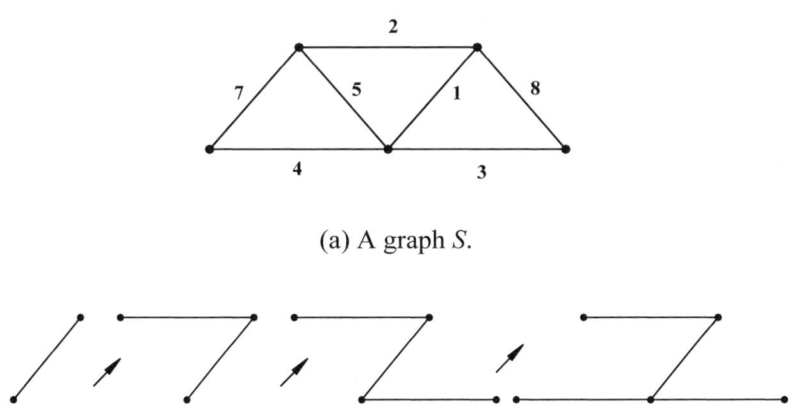

(a) A graph S.

(b) Expansion process of the Greedy Algorithm.

Fig. B.6 A graph and the selected minimal base for its $M(S)$ matroid.

B.6 APPLICATION OF THE GREEDY ALGORITHM

In Chapter 3, it is shown that, for an efficient force method, the sparsity of the flexibility matrix of a structure, which is pattern equivalent to the generalised cycle adjacency matrix of its graph model, should be maximised. This can be achieved by the use of a generalised cycle basis of minimal measure, where the weight of a

γ-cycle is taken as its length (the number of its members). The Greedy Algorithm is a powerful means for this purpose. However, its application engenders certain difficulties, which are discussed in the following text.

Let a γ-cycle of a graph S be defined as a minimal subgraph C_i of S, which is rigid, and $\gamma(C_i) = a$. A maximal set of independent γ-cycles of S is known as a *generalised cycle* basis of S, the dimension of which is equal to $\eta(S) = \gamma(S)/a$. The integer "a" is defined in Table 2.1.

The set of all γ-cycles of S, together with F containing independent subsets of γ-cycles, form a matroid $M_{gc}(S)$, called the *generalised cycle space matroid*. A base of this matroid is a generalised cycle basis of S. Therefore, the Greedy Algorithm selects a minimal generalised cycle basis of S.

Algorithm: Let the weight of a γ-cycle be measured by the number of its members. Select all γ-cycles of S, denoted by C, and proceed as follows:

Step 1. Select the first γ-cycle of minimal length from C.

Step 2. Take the second independent γ-cycle of minimal length from $C - \{C_1\}$.

Step k. Subsequently choose a γ-cycle C_k of the least length from $C - \{C_1, C_2, ..., C_{k-1}\}$, which is independent of the previously selected γ-cycles. Continue the process until $\eta(S)$ of γ-cycles, forming a minimal generalised cycle basis, is generated.

Simple proof for the minimality of selected basis by the above procedure can be found in Kaveh [98,114].

B.7 FORMATION OF SPARSE NULL BASES

The bipartite graph $B(\mathbf{A})$ of a matrix $\mathbf{A}$ can be constructed by associating one row-node with each row i and one column-node with each column j if the corresponding entry a_{ij} of $\mathbf{A}$ is non-zero. For example, the matrix $\mathbf{A}$ and the corresponding bipartite graph $B(\mathbf{A})$, together with a matching in $\mathbf{A}$ and its graph, are shown in Figure B.7.

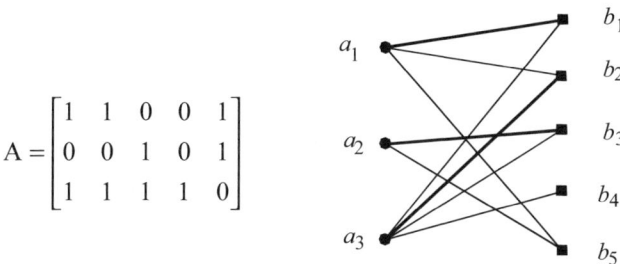

Fig. B.7 A rectangular matrix **A** and its bipartite graph $B(\mathbf{A})$.

It can be proved that a matrix **A** has complete matching if and only if it has the Hall property; that is if subsets of its rows have non-zeros in at least as many columns. It has also been proved that the matching number of a matrix is greater than or equal to its rank. Therefore, a matrix with full rank has a complete matching.

Null Bases Formation: A complete matching of $B(\mathbf{A})$ partitions the columns of **A** into the set of matched columns M and a set of unmatched columns U. For example in Figure B.7, $\{b_1, b_2, b_3\}$ are matched and $\{b_4, b_5\}$ are unmatched columns. It is shown that for a column $u \in U$, a circuit can be constructed using an alternating path algorithm (see Appendix A).

An M-alternating path is a path whose members are alternately from the matching M and outside M. For example, in Figure B.7, $\{b_5, a_1, b_1, a_3, b_2\}$ is an M-alternating path in **A**. We say b_5 is *reachable* from b_1 and b_3, and show it with $b_5 \to b_1$ and $b_5 \to b_2$. An augmenting path is an alternating path that begins and ends with unmatched nodes. The size of corresponding matching can be increased by making the members in the matching unmatched and vice versa.

For a member $u \in U$, a dependent set $n(u)$ containing u can be considered, which is a circuit if **A** has the *Weak Haar Property* (WHP). A matrix has WHP if every set of columns C satisfies rank$(C) = \Psi(C)$, where $\Psi(C)$ is the matching number. This property ensures that $n(u)$ will be a circuit for all numeric values of the columns of **A**. For a particular set of numeric values of the non-zero entries of **A**, numerical cancellation may occur, in which case the set $n(u)$ will contain a circuit.

Therefore, for finding a circuit of a matrix with WHP, a complete matching M should be constructed and an unmatched column u should be selected. A circuit $n(u)$ is formed by following all M-alternating paths from u and adding columns visited to $n(u)$, that is, $n(u) = u + \{v \in M: u \to v\}$. For example, two circuits $n(b_4) = \{b_4, b_2, b_1\}$ and $n(b_5) = \{b_5, b_3, b_2, b_1\}$ can easily be formed. However, if $n(u)$

does not have WHP, then it contains a circuit, which should be identified by numerical factorisation.

Once the formation of a circuit becomes possible, different algorithms can be designed for the formation of a null basis. Two such algorithms are given by Coleman and Pothen [28, 29], and in the following text, an algorithm for the formation of a fundamental null basis is briefly discussed.

Let N be the empty set. Find a complete matching M of **A**, partitioning the columns of **A** as **A** = [**M U**]. Then, for each $u \in U$, construct a circuit $n(u)$. Augment the null basis N with the computed null vector. This process should be repeated for all members of U to obtain a fundamental null basis of **A** with WHP. When **A** does not have WHP, then a fundamental basis can be computed only when M has full rank. Therefore, one should choose M by a matching, but should ensure that M has full rank while factoring it to compute the null vectors. When it is rank deficient, the dependent columns in M should be rejected, and a new maximum matching should be found. This will ensure the formation of a basis and will always succeed when **A** has full row rank.

References

1. Adeli, H. and Kamal, O., *Parallel Processing in Structural Engineering*, Elsevier Applied Science, New York, 1993.

2. Akhras, G. and Dhatt, G., An automatic relabelling scheme for minimizing a matrix or network bandwidth, *Int. J. Numer. Methods Eng.*, 10(1976)787–797.

3. Akin, J.E. and Pardue, R.M., Element resequencing for frontal solutions, in *The Mathematics of Finite Elements and Applications*, Whiteman J.R. ed. Academic Press, London, England, pp. 535–541, 1975.

4. Al-Nasra, M. and Nguyen, D.T., An algorithm for domain decomposition in finite element analysis, *Comput. Struct.*, 39(1991)277–289.

5. Argyris, J.H. and Kelsey, S., *Energy Theorems and Structural Analysis*, Butterworth, London, England, 1960.

6. Argyris et al, Finite element method: The natural approach, *Comput. Methods Appl. Mech. Eng.*, 7(1979)1–106.

7. Armstrong, B.A., Near-minimal matrix profile and wavefronts for testing nodal resequencing algorithms, *Int. J. Numer. Methods Eng.*, 21(1985)1785–1790.

8. Baase, S., *Computer Algorithms, Introduction to Design and Analysis*, 2nd edition, Addison Wesley, USA, 1993.

9. Bathe, K.H. and Wilson, E.L., *Numerical Methods in Finite Element Analysis*, Englewood Clifs, NJ, Prentice Hall, 1976.

10. Behzad, M., A characterization of total graphs, *Proc. Amer. Math. Soc.*, 26(1970)383–389.

11. Benfield, W.A. and Hruda, R.F., Vibration analysis of structure by component mode substitution, *AIAA J.*, 9(1971)1255–1261.

12. Berge, C., *Graphs and Hypergraphs*, North-Holland Publishing Co, Amsterdam, NY, 1973.

13. Biggs, N.L., *Algebraic Graph Theory*, 2nd edition, Cambridge University Press, Cambridge, England, 1993.

14. Bolker, E.D. and Crapo, H., Bracing rectangular framework I, *SIAM J. Appl. Math.*, 36(1979)473–490.

15. Bondy, A. and Murty, S.R. eds, *Graph Theory and Related Topics*, Academic Press, New York, 1979.

16. Brualdi, R.A. and Ryser, H.J., *Combinatorial Matrix Theory*, Cambridge University Press, Canada, 1992.

17. Brusa, L. and Riccio, F., A frontal technique for vector computers, *Int. J. Numer. Methods Eng.*, 28(1989)1635–1644.

18. Burgess, L.W. and Lai, P.K.F., A new mode renumbering algorithm for bandwidth reduction, *Int. J. Numer. Methods Eng.*, 23(1986)1693–1704.

19. Bykut, A., A note on an element ordering scheme, *Int. J. Numer. Methods Eng.*, 11(1977)194–198.

20. Carey, G.F. and Oden, J.T., *Finite Elements: Computational Aspect*, Vol. III, Prentice Hall, Englewood Cliffs, NJ, 1984.

21. Cassell, A.C., Conditioning of structural flexibility matrices, *J. Struct. Div., ASCE*, 101(1975)1620–1621.

22. Cassell, A.C., An alternative method for finite element analysis; A combinatorial approach to the flexibility method, *Proc. R. Soc. London*, A352(1976)73–89.

23. Cassell, A.C., Henderson, J.C. de C. and Kaveh, A., Cycle bases for the flexibility analysis of structures, *Int. J. Numer. Methods Eng.*, 8(1974)521–528.

24. Castigliano, A., *Theorié de l'équilibre des systèmes élastique et ses applications*, Published by A.F. Negro, Turin, Italy, 1879.

25. Chadha, H.S. and Baugh J.W. Jr, Network-distributed finite element analysis, *Adv. Eng. Software*, 25(1996)267–280.

26. Cheng, K.Y., Note on minimizing the bandwidth of sparse symmetric matrices, *Comput. J.*, 11(1973)27–30.

27. Chung, J.J., *Spectral Graph Theory*, CBMS, Number 92, American Mathematical Society, USA, 1997.

28. Coleman, T.F. and Pothen, A., The null space problem I; Complexity, *SIAM J. Alg. Disc. Methods*, 7(1986)527–537.

29. Coleman, T.F. and Pothen, A., The null space problem II; Algorithms, *SIAM J. Alg. Disc. Methods*, 8(1987)544–561.

30. Craig, R.R. Jr. and Chang, C.J., Free interface methods of substructuring coupling for dynamic analysis, *AIAA J.*, 14(1976)1633–1635.

31. Crane, H.L., Gibbs, N.E., Poole, W.G. Jr. and Stockmeyer, P.K., Algorithm 508: Matrix bandwidth and profile reduction, *ACM Trans. Math. Software*, 2(1976)375–377.

32. Crapo, H., Structural rigidity, *Struct. Topol.*, 1(1979)26–45.

33. Cuthill, E. and McKee, J., Reducing the bandwidth of sparse symmetric matrices, *Proceedings of the 24th National Conference ACM*, Bradon System Press, New Jersy, pp. 157–172, 1969.

34. Cvetković, D.M., Dobb, M. and Sachs, H., *Spectra of Graphs*, Academic Press, New York, 1980.

35. Cvetković, D.M., Lučić, R., A new generalization of the p-sum of graphs, *Univ. Beograd Publ. Elektrotech. Fak., Ser. Mat. Fiz*, 302–319(1974)67–71.

36. Denke, P.H., *A General Digital Computer Analysis of Statically Indeterminate Structures*, NASA-TD-D-1666, USA, 1962.

37. Dorr, M.R., Domain decomposition via Lagrange multipliers, Report No. UCRL-98532, Lawrence Livermore National Laboratory, 1988.

38. Duff, I.S., The influence of vector and parallel processors on numerical analysis, *Proceedings of the IMA/SIAM Conference. The State of the Art in Numerical Analysis*, University of Birmingham, Birmingham, England, pp. 359–407, 1986.

39. Duff, I.S. Erisman, A.M. and Reid, J.K., *Direct Methods for Sparse Matrices*, Oxford Science Publishing, Clarendon Press, Oxford, England, 1986.

40. Duff, I.S., Reid, J.K. and Scott, J.A., The use of profile reduction algorithms with a frontal code, *Int. J. Numer. Methods Eng.*, 28(1989)2555–2568.

41. Edmonds, J., Matroids and greedy algorithm, *Math. Programming*, 1(1971)127–136.

42. Euler, L., Solutio problematic ad Geometrian situs pertinentis, *Comm. Acad. Petropolitanae*, 8(1736)128–140, Translated in Speiser klassische Stücke der Mathematik, Zürich (1927)127–138.

43. Everstine, G.C., A comparison of three resequencing algorithms for the reduction of profile and wavefront, *Int. J. Numer. Methods Eng.*, 14(1979)837–853.

44. Farhat, C., A simple and efficient automatic FEM domain decomposer, *Comput. Struct.*, 28(1988)579–602.

45. Farhat, C. and Lesoinne, M., Automatic partitioning of unstructured meshes for the parallel solution of problems in computational mechanics, *Int. J. Numer. Methods Eng.*, 36(1993)745–764.

46. Farhat, C., Maman, N. and Brown, G.W., Mesh partitioning for implicit computations via iterative domain decomposition: impact and optimization of the subdomain aspect ratio, *Int. J. Numer. Methods Eng.*, 38(1995)989–1000.

47. Farhat, C. and Roux, F.X., A method of finite element tearing and interconnecting and its parallel solution algorithm, *Int. J. Numer. Methods Eng.*, 32(1991)1205–1227.

48. Farhat, C. and Wilson, E., A new finite element concurrent computer program architecture, *Int. J. Numer. Methods Eng.*, 24(1987)1771–1792.

49. Felippa, C.A., Solution of linear equations with skyline-stored symmetric matrix, *Comput. Struct.*, 5(1975)13–29.

50. Fenves, S.J. and Law, K.H., A two-step approach to finite element ordering, *Int. J. Numer. Methods Eng.*, 19(1983)891–911.

51. Fiedler, M., Algebraic connectivity of graphs, *Czech. Math. J.*, 23(1973)298–305.

52. Filho, F.V., Orthogonalization of internal forces and strain systems, *Proceedings, Conference on Matrix Methods in Structural Mechanics*, AFFDL-TR-68-150, Wright Patterson Air Base, Ohio, OH, pp.891–911; 1968.

53. Frobenius, G., Über zerlegbare Determinanten, *Sitzungsber. König. Preuss. Akad. Wiss.*, XVIII(1917)274–277, Jbuch. 46.144 (XV, 6).

54. Gale, D., Optimal assignments in an ordered set; An application of matroid theory, *J. Comb. Theory*, 4(1968)176–180.

55. George, A., Computer implementation of the finite elements, Ph.D. thesis, Computer Science Department, Stanford University, San Francisco CA, *Report STAN-CS-71-208*, 1971.

56. George, A., Solution of linear system of equations; direct methods for finite element problems, in *Lecture Notes in Mathematics*, Dold, A., and Eckmann, E., eds. Springer-Verlag, 572(1977)52–101.

57. George, A. and Liu, J.W.H., Algorithms for partitioning and numerical solution of finite element systems, *SIAM J. Numer. Anal.*, 15(1978)297–327.

58. Gibbs, N.E., Algorithm 509: A hybrid profile reduction algorithm, *ACM Trans. Math. Software*, 2(1976) 378–387.

59. Gibbs, N.E., Poole, W.G. and Stockmeyer, P.K., An algorithm for reducing the bandwidth and profile of a sparse matrix, *SIAM J. Numer. Anal.*, 12(1976)236–250.

60. Gilbert, J.R. and Heath, M.T., Computing a sparse basis for the null space, *SIAM J. Alg. Disc. Methods*, 8(1987)446–459.

61. Godsil, C. and Royle, G., *Algebraic Graph Theory*, Springer-Verlag, New York, 2001.

62. Goldberg, M.K. and Gardner, R., On the minimal cut problem, *Progress in Graph Theory*, Academic Press, San Diego, CA, pp. 295–305, 1984.

63. Gondran, M. and Minoux, M., in *Graphs and Algorithms*, Translated by Vajda S. ed. John Wiley and Sons, 1984, Eyrolles, Paris, 1979.

64. Goodspeed, C.H. and Martin, M., Conditioning of structural flexibility matrices, *J. Struct. Div., ASCE*, 100(1974)1677–1684.

65. Gould, P., The geographical interpretation of eigenvalues, *Trans. Inst. Br. Geogr.*, 42(1967)53–58.

66. Grimes, R.G., Pierce, D.J. and Simon, H.D., A new algorithm for finding a pseudo-peripheral node in a graph, *SIAM J. Anal. Appl.*, 11(1990)323–334.

67. Grooms, H.R., Algorithm for matrix bandwidth reduction, *J. Struct. Div., ASCE*, 98(1972)203–214.

68. Grooms, H.R. and Rowe, J., Substructuring and conditioning, *J. Struct. Div., ASCE*, 103(1977)507–514.

69. Gross J. and Yellen, J., *Graph Theory and its Applications*, CRC Press, New York, 1998.

70. Hale, A.L. and Meirovitch, L., A procedure for improving discrete substructuring representation in dynamic synthesis, *AIAA J.*, 20(1882)1128–1136.

71. Hall, M. Jr., An algorithm for distinct representatives, *Amer. Math. Monthly*, 63(1956)716–717.

72. Hall, K., R-dimensional quadratic placement algorithm, *Manage. Sci.*, 17(1970)219–229.

73. Harary, F., *Graph Theory*, Addison Wesley, Reading, MA, 1969.

74. Harary, F. and Wilcox, G.W., Boolean operations on graphs, *Math. Scand.*, 20(1967) 41–51.

75. Heath, M.T., Plemmons, R.J. and Ward, R.C., Sparse orthogonal schemes for structural optimization using the force method, *SIAM J. Sci. Statist. Comput.*, 5(1984)514–532.

76. Henderson, J. C. de C., Topological aspects of structural analysis, *Aircr. Eng.*, 32(1960)137–141.

77. Henderson, J. C. de C. and Bickley, W.G., Statical indeterminacy of a structure, *Aircr. Eng.*, 27(1955)400–402.

78. Henderson, J.C. de C. and Maunder, E.W.A., A problem in applied topology, *J. Inst. Math. Applic.*, 5(1969)254–269.

79. Hendirckson B., Leland, R., A multilevel algorithm for partitioning graphs, Technical Report SAND93-1301, Sandia National Laboratories, 1993.

80. Henneberg, L., *Statik der Starren Systeme*, Darmstadt, Germany, 1896.

81. Hintz, R.M., Analytical methods in component modal synthesis, *AIAA J.*, 13(1975)1007–1016.

82. Hopcroft, J.E. and Karp, R.M., An $n^{5/2}$ algorithm for maximum matching in bipartite graphs, *SIAM J. Comput.*, 2(1973)225–231.

83. Horton, J.D., A polynomial time algorithm to find the shortest cycle basis of a graph, *SIAM J. Comput.*, 16(1987)358–366.

84. Hubicka, E. and Syslø, M.M., Minimal bases of cycles of a graph, in *Recent Advances in Graph Theory*, Fiedler M. ed. Academia Praha, Prague, Czechoslovakia, pp. 283–293, 1975.

85. Hurty, W.C., Dynamic analysis of structural systems using component modes, *AIAA J.*, 3(1965)678–685.

86. Imrich, W. and Klavžar, S., *Product Graphs; Structure and Recognition*, John Wiley, New York, 2000.

87. Irons, B.M., A frontal solution program for finite element analysis, *Int. J. Numer. Methods Eng.*, 2(1970)5–32.

88. Jennings, A., A compact storage scheme for the solution of symmetric linear simultaneous equations, *Comput. J.*, 9(1966)281–285.

89. Jennings, A., *Matrix Computation for Engineers and Scientists*, John Wiley and Sons, Chchister, England, 1976.

90. Jennings, A. and McKeown J.J., *Matrix Computation*, 2nd edition, John Wiley and Sons, New York, 1992.

91. Johan Z. and Mathur K.K., Scalability of finite element applications on distributed-memory parallel computers, *Comput. Methods Appl. Mech. Eng.*, 119(1994)61–72.

92. Kaneko, I., Lawo, M. and Thierauf, G., On computational procedures for the force methods, *Int. J. Numer. Methods Eng.*, 18(1982)1469–1495.

93. Kardestuncer, H., *Elementary Matrix Analysis of Structures*, McGraw-Hill, New York, 1974.

94. Kaveh, A., Application of topology and matroid theory to the analysis of structures, Ph.D. thesis, London University, IC, March 1974.

95. Kaveh, A., On 2-cell embedding of a graph in *Proceedings of the 6th Annual Iranian National Mathematics Conference*, Ahwaz, Iran, pp. 106–113, 1975.

96. Kaveh, A., Improved cycle bases for the flexibility analysis of structures, *Comput. Methods Appl. Mech. Eng.*, 9(1976)267–272.

97. Kaveh, A., Topological study of the bandwidth reduction of structural matrices, *J. Sci. Tech.*, 1(1977)27–36.

98. Kaveh, A., A combinatorial optimization problem; Optimal generalized cycle bases, *Comput. Methods. Appl. Mech. Eng.*, 20(1979)39–52.

99. Kaveh, A., A note on a two-step approach to finite element ordering, *Int. J. Numer. Methods Eng.*, 19(1984)1753–1754.

100. Kaveh, A., An efficient flexibility analysis of structures, *Comput. Struct.*, 22(1986) 973–977.

101. Kaveh, A., Ordering for bandwidth reduction, *Comput. Struct.*, 24(1986)413–420.

102. Kaveh, A., Multiple use of a shortest route tree for ordering, *Commun. Numer. Methods Eng.*, 2(1986)213–215.

103. Kaveh, A., An efficient program for generating subminimal cycle bases for the flexibility analysis of structures, *Commun. Numer. Methods Eng.*, 2(1986)339–344.

104. Kaveh, A., Subminimal cycle bases for the force method of structural analysis, *Commun. Numer. Methods Eng.*, 3(1987)277–280.

105. Kaveh, A., Suboptimal cycle bases of graphs for the flexibility analysis of skeletal structures, *Comput. Methods Appl. Mech. Eng.*, 71(1988)259–271.

106. Kaveh, A., Topological properties of skeletal structures, *Comput. Struct.*, 29(1988)403–411.

107. Kaveh, A., Topology of skeletal structures, *Z. Angew. Math. Mech.*, 68(1988)347–353.

108. Kaveh, A., On optimal cycle bases of graphs for mesh analysis of networks, *Int. J. NETWORKS*, 19(1989)273–279.

109. Kaveh, A., Optimizing the conditioning of structural flexibility matrices, *Comput. Struct.*, 41(1991)489–494.

110. Kaveh, A., A connectivity coordinate system for node and element ordering, *Comput. Struct.*, 41(1991)1217–1223.

111. Kaveh, A., Recent developments in the force method of structural analysis, *Appl. Mech. Rev.*, 45(1992)401–418.

112. Kaveh, A., Matroids applied to the force method of structural analysis, *Z. Angew. Math. Mech.*, 73(1993)T355–T357.

113. Kaveh, A., *Structural Mechanics: Graph and Matrix Methods*, 2nd edition, RSP (John Wiley), Edinburgh, UK, 1995.

114. Kaveh, A., *Optimal Structural Analysis*, 1st edition, RSP (John Wiley), United Kingdom, 1997.

115. Kaveh, A. and Behfar, S.M.R., Finite element nodal ordering algorithms, *Commun. Appl. Numer. Methods.*, 11(1995)995–1003.

116. Kaveh, A. and Davaran, A., A mixed method for subdomain generation for parallel processing, *Advances in Computational Structural Technology*, Civil-Comp Press, Edinburgh, UK, pp. 259–264, 1996.

117. Kaveh A., and Davaran, A., Spectral bisection of adaptive finite element meshes for parallel processing, *Comput. Struct.*, 70, 315–324, 1999.

118. Kaveh, A. and Ghaderi, I., Conditioning of structural stiffness matrices in *Proceedings of the Civil-Comp93*, Edinburgh, UK, pp.39–45, 1993.

119. Kaveh, A. and Mokhtar-zadeh, A., A comparative study of the combinatorial and algebraic force methods in *Proceedings of the Civil-Comp93*, Edinburgh, UK, pp. 21–30, 1993.

120. Kaveh, A. and Rahami, H., Algebraic graph theory for sparse flexibility matrices, *J. Math. Model. Algorithms*, 2(2003)131–142.

121. Kaveh, A. and Rahami, H. A new spectral method for nodal ordering of regular space structures, *Fin. Elem. Anal. Des.*, 2004, 40(2004)1931–1945.

122. Kaveh, A. and Rahami, H., An efficient method for decomposition of regular structures using graph products, *Int. J. Numer. Methods Eng.*, 2004, 61(2004)1797–1808.

123. Kaveh, A., and Rahimi Bondarabady, H.A., Finite element mesh decompositions using complementary Laplacian matrix, *Commun. Numer. Methods Eng,*, 16(2000)379–389.

124. Kaveh, A. and Rahimi Bondarabady, H.A. Spectral trisection of finite element models, *Int. J. Numer. Methods Heat Fluid Flow*, 11(2001)358–370.

125. Kaveh, A. and Rahimi Bondarabady, H.A. A multi-level finite element nodal ordering using algebraic graph theory, *Fin. Elem. Anal. Des.*, 38(2002)245–261.

126. Kaveh, A. and Rahimi Bondarabady, H.A.: A hybrid graph-theoretical method for domain decomposition, *Fin. Elem. Anal. Des.*, 39, No. 13(2003)1237–1247.

127. Kaveh, A., and Rahimi Bondarabady, H.A., Bisection for parallel computing using Ritz and Fiedler vectors, *Acta Mech.*, 167, No 3–4(2004)131–144.

128. Kaveh, A. and Ramachandran, K., Graph theoretical approach for bandwidth and frontwidth reductions, in *Proceedings of the 3rd International Conference Space Structures*, Nooshin H. ed. Surrey University, UK, pp. 244–249, 1984.

129. Kaveh, A. and Roosta, G.R., Revised greedy algorithm for the formation of minimal cycle basis of a graph, *Commun. Numer. Methods Eng.*, 10(1994) 523–530.

130. Kaveh, A. and Roosta, G.R., A graph theoretical method for decomposition in finite element analysis, in *Proceedings of the Civil-Comp94*, Edinburgh, UK, pp. 35–42, 1994.

131. Kaveh, A. and Roosta, G.R., An efficient method for finite element nodal ordering, *Asian J. Struct. Eng.*, 1(1995)229–242.

132. Kaveh, A. and Roosta, G.R., Graph-theoretical methods for substructuring, subdomaining and ordering, *Int. J. Space Struct.*, 10, No. 2(1995)121–131.

133. Kaveh, A. and Roosta, G.R., Comparative study of finite element nodal ordering methods, *Struct. Eng. Rev.*, 20(1998)86–96.

134. Keyes, D.E and Gropp, W.D., A comparison of domain decomposition techniques for elliptic partial differential equations and their parallel implementation, *SIAM J. Scient. Statist. Comput.*, 8(1987)166–202.

135. Khan A.I. and Topping, B.H.V., Parallel adaptive mesh generation, *Comput. Syst. Eng.*, 2, No. 1(1991)75–102.

136. Khan A.I. and Topping, B.H.V., Sub-domain generation for parallel finite element analysis, in *Proceedings of the Civil Comp 93*, Edinburgh, UK, 1993.

137. King, I.P., An automatic reordering scheme for simultaneous equations derived from network systems, *Int. J. Numer. Methods Eng.*, 2(1970)523–533.

138. Kirchhoff, G., Über die Auflösung der Gleichungen auf welche man bei der Untersuchung der Linearen Verteilung Galvanischer Ströme geführt wird, *Ann. der Physik und Chemie*, 72(1847)497–508, English translation, *IRE Trans. Circuit Theory*, CT5(1958)4–7.

139. Kolasinska, E., On a minimum cycle basis of a graph, *Zastos. Math.*, 16(1980)631–639.

140. König, D., *Theory der endlichen und unendlichen Graphen*, Chelsea, 1950, 1st edition, Akademische Verlagsgesellschaft, Leipzig, Germany, 1936.

141. Kron, G., Diakoptics, piecewise solution of large-scale systems, A series of 20 chapters in the *Electrical J.*, (London), 158(1957)1673–1677 to 162(1959)431–436.

142. Kruskal, J.B., On the shortest spanning subtree of a graph and travelling salemen problem, *Proc. Amer. Math. Soc.*, 7(1956)48–50.

143. Kuratowski, K., Sur le problème des courbes gauches en topologie, *Fund. Math.*, 18(1930)271–283.

144. Laman, G., On graphs and rigidity of plane skeletal structures, *J. Eng. Math.*, 4(1970)331–340.

145. Langefors, B., Analysis of elastic structures by matrix transformation with special regard to semimonocoque structures, *J. Aero. Sci.*, 19(1952)451–458.

146. Langefors, B., Algebraic topology and elastic networks, *SAAB TN49*, Linköping, Sweden, 1961.

147. Law, K.H., A parallel finite element solution method, *Comput. Struct.*, 23(1986)845–858.

148. Lawler, E.L., *Combinatorial Optimization; Networks and Matroids*, Holt, Rinehart and Winston, New York, 1976.

149. Lesoinne, M., Farhat, C. and Geradin, M., Parallel/Vector improvements of the frontal method, *Int. J. Numer. Methods Eng.*, 32(1991)1267–1281.

150. Lim, I.L., Johnston, I.W. and Choi, S.K., A comparison of algorithms for profile reduction of sparse matrices, *Comput. Struct.*, 57(1995)297–302.

151. Liu, W.H. and Sherman, A.H., Comparative analysis of the Cuthill-McKee and Reverse Cuthill-McKee Ordering Algorithms for Sparse Matrices, Report 28, Department of Computer Science, Yale University, New Haven, CT, 1975.

152. Livesley, R.K., *Matrix Methods of Structural Analysis*, 2nd edition, Pergamon Press, New York, 1975.

153. Livesley, R.K. and Sabin, M.A., Algorithms for numbering the nodes of finite element meshes, *Comput. Syst. Eng.*, 2(1991)103–114.

154. Löhner, R., Some useful renumbering strategies for unstructured grids, *Int. J. Numer. Methods Eng.*, 36(1993)3259–3270.

155. Lovász, L. and Plummer, M.D., *Matching Theory*, Math. Studies, 121, North-Holland Publishing Co, Amesterdam, NY, 1986.

156. Lovász, L. and Yemini, Y., On generic rigidity in the plane, *SIAM J. Alg. Disc. Methods*, 3(1982)91–98.

157. Maas, C., Transportation in graphs and the admittance spectrum, *Discr. Appl. Math.*, 16(1987)31–49.

158. Mackerle, J.: Parallel finite element and boundary element analysis: Theory and applications-A biography (1997–1999), *Fin. Elem. Anal. Des.*, 35(2000)283–296.

159. MacLane, S., A structural characterization of planar combinatorial graphs, *Duke Math. J.*, 3(1937)460–472.

160. Malone, J.G., Automated mesh decomposition and concurrent finite element analysis for hypercube multiprocessors computers, *Comput. Methods Appl. Mech. Eng.*, 70 (1988)27–58.

161. Mauch, S.P. and Fenves, S.J., Release and constraints in structural networks, *J. Struct. Div., ASCE*, 93(1967)401–417.

162. Maunder, E.W.A., Topological and linear analysis of skeletal structures, Ph.D. thesis, London University, IC, 1971.

163. McGuire, W. and Gallagher, R.H., *Matrix Structural Analysis*, John Wiley and Sons, New York, 1979.

164. Meek, J., *Matrix Structural Analysis*, McGraw-Hill, New York, 1971.

165. Mohar B., Eigenvalues, diameter and mean distance in graphs, *Graphs and Combinatorics*, 7(1991)53–64.

166. Mohar, B., The Laplacian spectrum of graphs, in *Graph Theory, Combinatorics, and Applications (Proceedings of the 6th International Conference on the Theory and Applications of Graphs)*, Vol. 2, Alavi Y., et al, John Wiley and Sons, New York, pp.871–898, 1991.

167. Müller-Breslau, H., *Die Graphische Statik der Baukonstruktionen*, Alfred Kröner Verlag, 1907 und Leipzig, Germany, 1912.

168. Nash-Williams, C. St. J.A., Decomposition of finite graphs into forests, *J. Lond. Math. Soc.*, 39(1964)12.

169. Noor, A.K., Kamel, H.A. and Fulton, R.E., Substructuring technique–status and projections, *Comput. Struct.*, 8 (1978)621–632.

170. Papademetrious, C.H., The NP-completeness of bandwidth minimization problem, *Comput. J.*, 16(1976)177–192.

171. Parter, S.V., The use of linear graphs in Gauss elimination, *SIAM Rev.*, 3(1961)119–130.

172. Patnaik, S.N., Integrated force method versus the standard force method, *Comput. Struct.*, 22(1986)151–164.

173. Patnaik, S.N. and Yadgiri, S., Frequency analysis of structures by integrated force method, *J. Sound. Vib.*, 83(1982)93–109.

174. Paulino G.H., Menezes, I.F.M., Gattass, M. and Mukherjee, S, Node and element resequncing using the Laplacian of a finite element graph: part I-general concepts and algorithms, *Int. J. Numer. Methods Eng.*, 37(1994)1511–1530.

175. Pellegrino, S. and Calladine, C.R., Matrix analysis of statically and kinematically indeterminate frameworks, *Int. J. Solids Struct.*, 22(1986)409–422.

176. Pestel, E.C. and Leckie, F.A., *Matrix Methods in Elastomechanics*, McGraw-Hill, 1963.

177. Pina, H.L.G., An algorithm for frontwidth reduction, *Int. J. Numer. Methods Eng.*, 17(1981)1539–1546.

178. Pissanetskey, S., *Sparse Matrix Technology*, Academic Press, 1984.

179. Plemmons, R.J. and White, R.E., Substructuring methods for computing the null space of equilibrium matrices, *SIAM J. Matrix Anal. Appl.*, 11(1990)1–22.

180. Plesek, J., Near optimum element renumbering scheme for frontal solver, *Int. J. Numer. Methods Eng.*, 1996.

181. Pothen, A., Sparse null basis computation in structural optimization, *Numer. Math.*, 55(1989)501–519.

182. Pothen, A., Simon H. and Liou K. P., Partitioning sparse matrices with eigenvectors of graphs, *SIAM J. Matrix Anal. Appl.*, 11(1990)430–452.

183. Prezemieniecki, J.S., Matrix structural analysis of substructures, *AIAA J.*, 1(1963)138–147.

184. Prezemieniecki, J.S., *Theory of Matrix Structural Analysis*, McGraw-Hill, New York, 1968.

185. Razzaque, A., Automatic reduction of frontwidth for finite element analysis, *Int. J. Numer. Methods Eng.*, 15(1980)1315–1324.

186. Robinson, J., *Integrated Theory of Finite Element Methods*, John Wiley and Sons, New York, 1973.

187. Robinson, J. and Haggenmacher, G.W., Some new developments in matrix force analysis, in *Recent Advances in Matrix Methods of Structural Analysis and Design*, University of Alabama, USA, pp.183–228, 1971.

188. Rosanoff, R.A. and Ginsburg, T.A., Matrix error analysis for engineers, *Proceedings of the Conference Matrix Methods in Structural Mechanics, AFFDL-TR-68-150*, Wright Patterson Air Base, Ohio, OH, pp.1029–1060, 1968.

189. Rosen, R., Matrix bandwidth minimization, *Proceedings of the 23rd National Conference of the ACM*, Brandon System Press, 1968, pp. 585–595.

190. Roth, R. and Whiteley, W., Tensegrity frameworks, *Trans Amer. Math. Soc.*, 265(1981)419–446.

191. Rubin, S., Improved component mode representation, *AIAA J.*, 13(1975) 995–1006.

192. Russopoulos, A.I., *Theory of Elastic Complexes*, Elsevier Publishing, Amsterdam, 1965.

193. Sabidussi, G., Graph multiplication, *Math. Z.*, 72(1960)446–457.

194. Schwenk, A.J. and R.J. Wilson, On the eigenvalues of a graph, in *Selected Topics in Graph Theory*, Beineke L.W. and Wilson R.J. eds, Academic Press, London, pp. 271–305, 1978.

195. Seale C. and Topping B.H.V., Parallel implementation of recursive spectral bisection, *Proceedings of the Civil-Comp95*, Edinburgh, UK, pp. 459–473, 1995.

196. Seidel, J.J., *Graphs and Their Spectra, Combinatorics and Graph Theory*, PWN Polish Scientific Publishing, Warsaw, Poland, pp. 147–162, 1989.

197. Shah, J.M., Ill-conditioned stiffness matrices, *J. Struct. Div., ASCE*, 92(1966)443–457.

198. Shang-Hou, N., Review of modal synthesis techniques and a new approach, *Shock. Vib. Bull.*, 40(1969)25–30.

199. Shang-Hsien Hsieh, Glaucio Paulino, H. and John F. Abel, Recursive spectral algorithms for automatic domain partitioning in parallel finite element analysis, *Comput. Methods Appl. Mech. Eng.*, 121(1995)137–162.

200. Shee, S.C., A not on the C-poduct of graphs, *Nanta Math.*, 7(1974)105–108.

201. Silvester P.P., Auda, H.A. and Stone, G.D., A memory-economic frontwidth reduction algorithm, *Int. J. Numer. Methods Eng.*, 20(1984)733–743.

202. Simon, H.D., Partitioning of unstructured problems for parallel processing, *Comput. Syst. Eng.*, 2(1991)135–148.

203. Sloan, S.W., An algorithm for profile and wavefront reduction of sparse matrices, *Int. J. Numer. Methods Eng.*, 23(1986)1693–1704.

204. Sloan, S.W., A Fortran program for profile and wavefront reduction, *Int. J. Numer. Methods Eng.*, 28(1989)2651–2679.

205. Sloan, S.W. and Ng, W.S., A direct comparison of three algorithms for reducing profile and wavefront, *Comput. Struct.*, 33(1989)411–419.

206. Sloan, S.W. and Randolph, M.F., Automatic element reordering for finite element analysis with frontal solution schemes, *Int. J. Numer. Methods Eng.*, 19(1983) 1153–1181.

207. Smith, J.D., *Design and Analysis of Algorithms*, PWS-KENT Publishing Company, Boston, MA, 1989.

208. Souza, C.C. de Keunings, R., Wolsey, L.A. and Zone, O., A new approach to minimising the frontwidth in finite element calculations, *Comput. Methods Appl. Mech. Eng.*, 111(1994)323–334.

209. Souza, L.T. and Murray, D.W., An alternative pseudoperipheral node finder for resequencing schemes, *Int. J. Numer. Methods Eng.*, 36(1993)3351–3379.

210. Souza, L.T., and Murray D.W., A unified set of resequencing algorithms, *Int. J. Numer. Methods Eng.*, 38(1995), 565–581.

211. Soyer, E. and Topçu, A., Sparse self-stress matrices for the finite element force method, *Int. J. Numer. Methods Eng.*, 50(2001)2175–2194.

212. Stepanec G.F., Basis systems of vector cycles with extremal properties in graphs, *Uspekhi. Mat. Nauk.*, 19(1964)171–175 (in Russian).

213. Straffing, P.D., Linear algebra in geography; Eigenvectors of networks, *Math. Magazine*, 53(1980)269–276.

214. Sugihara, K., On some problems in the design of plane skeletal structures, *SIAM J. Alg. Disc. Methods*, 4(1983)355–362.

215. Sziveri J. and Topping B.H.V., Parallel subdomain generation method, in *Proceedings of the Civil-Comp 1995*, Edinburgh, UK, pp.449–457.

216. Tay, T.S., Rigidity of multi-graphs II, in *Lecture Notes in Mathematics*, Graph Theory in Singapore 1983, Koh K.M. and Yap H.P., Springer-Verlag, pp. 129–134, 1984.

217. Timoshenko, S. and Young, D.H., *Theory of Structures*, McGraw-Hill, New York, 1945.

218. Tinney, W.F., in Comments on using sparsity technique for power systems problem, in *Sparse Matrix Proceedings*, Willoughby R. ed. IBM Watson Research Center, RAI1707, pp. 25–34, 1969.

219. Topçu, A., *A Contribution to the Systematic Analysis of Finite Element Structures Using the Force Method (in German)*, Doctoral dissert., Essen University, Essen, 1979.

220. Topping, B.H.V. and Khan, A.I, *Parallel Finite Element Computations*, Saxe-Coburg Publications, Edinburgh, UK, 1995.

221. Topping, B.H.V. and Sziveri J, Parallel sub-domain generation method, in *Proceedings of the Civil Comp 95*, Edinburgh, UK, 1995.

222. Vanderbilt, M.D., *Matrix Structural Analysis*, Quantum Publishers Inc, New York, 1974.

223. Vanderstraeten, D. and R. Keunings, Optimized partitioning of unstructured finite element meshes, *Int. J. Numer. Methods Eng.*, 38(1995)433–450.

224. Vanderstraeten D., Zone O. and Keunings R., Non-deterministic heuristic for automatic domain decomposition in direct parallel finite element calculation, *Proceedings of the 6th SIAM Conference on Parallel Processing*, SIAM, 1993, pp. 929–932.

225. Webb, J.P. and Froncioni, A., A time-memory trade-off frontwidth reduction algorithm for finite element analysis, *Int. J. Numer. Methods Eng.*, 23(1986)1905–1914.

226. Weichsel, P.M., The Kronecker product of graphs, *Proc. Amer. Math. Soc.*, 13(1962)47–52.

227. Welsh, D. J.A., Kruskal's theorem for matroids, *Proc. Camb. Phil. Soc.*, 64 (1968)3–4.

228. West, D.B., *Introduction to Graph Theory*, Prentice Hall, New Jersy, 1996.

229. Whitney, H., Non-separable and planar graphs, *Trans Amer. Math. Soc.*, 34 (1932)339–362.

230. Whitney, H., On the abstract properties of linear dependence, *Amer. J. Math.*, 57(1935)509–533.

231. Wilson, R.J. and Beineke, L.W. eds, *Applications of Graph Theory*, Academic Press, London, England, 1979.

232. Wilson, E. and Dovey, H., Solution or reduction of equilibrium equations for large complex structural systems, *Adv. Eng. Software*, 1(1978)19–25.

233. Yee, E.K.L. and Tsuei, Y.G., Direct component modal synthesis technique for system dynamic analysis, *AIAA J.*, 27(1989)1083–1088.

234. Ziegler, F., *Mechanics of Solids and Fluids*, 2nd edition, Springer-Verlag, Vienna and New York, 1995.

235. Zienkiewicz, O.C., *The Finite Element Method in Engineering*, 3rd edition, Maidenhead Berkshire, England, McGraw-Hill, 1977.

236. Zone, O. and Keunings, R., Direct solution of two-dimensional finite element equations on distributed memory parallel computers, in *High Performance Computing II*, Durand M. and Dabaghi F. EL eds. North-Holland, Amsterdam, NY, pp. 333–344, 1991.

237. Zykov, A.A., *Theory of Finite Graphs*, Nuaka, Novosibirsk, 1969 (in Russian).

Index

A

Abstract graph, 456
Accessibility index, 92, 276
Active node, 215
Active status, 228
Acyclic, 460
Adaptive, 371–78
Adjacency, 456
Adjacency list, 350, 435
Adjacency matrix, 273–79
Admissible γ-cycle, 124
Admissible drawing, 35–37
Algebraic connectivity, 281
Algebraic force method, 131–38
Algebraic graph theory, 350–63
Alternating path, 481
Alternating tree, 482
Analysis of algorithms, 207, 213, 220, 249
Analysis, structure, 4
Aspect ratio, 264
Assembling, 158, 160
Associate graph, 119–22, 257
Augmenting path, 482
Axioms, bases, 483–85
circuits, 484–85
independence, 483

B

Banded, 194
Banded form, 193
Banded rectangular matrix, 260
Bandwidth, 193, 194
Bandwidth reduction, 197–203
Bar element, 148, 161–62
Base, 483, 491
Basic structure, 54
Basis, 383
fundamental cutset, 465–66
fundamental cycle, 464–65
generalized cycle, 115–19
minimal cycle, 71–79
null, 489
optimal cycle, 71–72
statical, 55
Beam element, 163–65
Berge's theorem, 482
Betti number, first, 70, 74
Betti number, zero, 464
Betti's law, 19
Bipartite graph, 45, 46, 48, 462
matching, 45
Bisection algorithm width, 353
Block, 480
Boolean operation, 404
Boundary conditions, 168–69
Branch, 460
Breadth-first-algorithm, 460
Breadth-first-search algorithm, 198
Bridge, 480

C

Cartesian product, 404–7, 411–14
    examples for, 426–30
Castigliano's theorem, 14, 150–52
Characteristic polynomial, 274, 467
Chord, 460
Circuit, 484–85
Closed trail, 459
    walk, 459
Cocircuit, 488
Cocycle, 459
Cocycle basis, 176
Cocycle matroid, 488
Combinatorial optimisation, 490–91
Communication number, 379
Compact adjacency list, 435
Compatibility, 5, 60–64, 157
    matrix, 157
Complementary Laplacian, 284
    solution, 55
    solution space, 57
    strain energy, 13
    work, 12
Complete bipartite
    graph, 462
    matching, 45–47
Complexity analysis, 206, 214, 220, 252, 310–12, 323–25
Component, 459
Computational procedure, 176–80
Condition number, 98–101
Conditioning, 97–115
Connected, 458
Connectivity, 456
    member, 50
    node, 50
Constant matrix, 42
Constant stress triangular element, 125
Contour, 196
Contraction, 479
Contragradient principle, 18–19
Coordinate system, local, 142, 143
    global, 142, 143

Cotree, 460
Covering subgraph, 44
Crossing, 36
Crossing number, 36
Cut-point, 480
Cutset, 459
Cutset adjacency matrix, 472
    basis, 465–66
    basis incidence matrix, 472
    matroid, 488
    rank, 465
    space, 464
    space matroid, 486–88
Cutset space of S, 463
Cycle, 459
Cycle adjacency matrix, 470
    admissible, 75
    basis, 91–93
    basis incidence matrix, 470
    exchange, 107
    graph, 91
    length number, 82
    matroid, 485
    regional, 102
    set vector, 463
    space, 463
    space matroid, 487–88
Cycle basis, 464
Cycle length number (CLN), 82
Cycle set vector, 463
Cycle space, 463

D

Datum node, 468
Decomposition, 134, 136
    Method, 47–48
Degree, 456
Degree of kinematical indeterminacy (DKI), 24, 27, 142
    matrix, 5, 279
    of statical indeterminacy, 27–33
Degrees of freedom, 24, 42, 142
Design, structure, 4

INDEX 497

Diagonal storage, 194
Diameter, 198
Digraph, 476
Dimension, 117, 120
Direct product, 409–10, 417–19
    examples for, 431–33
Directed graph, 474–76
Displacement function, 174
    method, 5, 166–73
Distance, 458
Distance number, 257–58
DKI. *See* Degree of kinematical indeterminacy (DKI)
Domain decomposition, 305–30, 393, 404
Drawing, 35
Drawing, admissible, 35–37
    optimal, 36
DSI, 24, 27
Dual graph, 480
Dummy load method, 15–17
    displacement method, 17
    displacement theorem, 17
Dynamic analysis, 336–47
    equation, 338–42

E

Eccentricity, 198
Edge connectivity, 281
Eigenvalue, 98, 273, 274, 372, 410, 419–20
    cycle, 421–26
    path, 421–26
Eigenvector, 410
Element adjacency list, 243
    clique graph, 204–5, 230
    ordering, 241–56
    star graph, 208–9
    wheel graph, 461
Elementary subgraph, 115
    adjacency list, 316
Elimination graph, 267
Energy, 11–13, 150

Engineering based method, 316–17
Equilibrium, 4, 54–57, 142, 156
Equilibrium approach, 5
Equilibrium matrix, 156
Equivalent loads, 171
Euler's polyhedra formula, 477–80
Euler's theorem, 33, 479
Exchange, 107
Expansion process, 28–29
Exposed node, 481
Extreme eigenvalues, 98

F

Factorisation, 133, 136
F-admissible, 101
Fiedler vector, 282, 389–401, 403
Fill-ins, 202, 249
Finite element models, 264
Finite element ordering, 203–41
First Betti number, 70, 74, 464
Flexibility matrices, 256
Flexibility, method, 5
    member, 57–60
    unassembled, 131
Force method, 5, 53, 69, 70, 119, 125–38
Forest, 109, 459, 460
Frequency, 337
Frobenius, 481
Front matrix, 249
Frontal method, 249
Frontwidth, 226
Frontwidth reduction, 241–56
Fully triangulated, 120, 122
Function, unifying, 28
Fundamental cutset, 465
Fundamental cutset basis, 465–66
Fundamental γ-cycle, 117
Fundamental cycle basis, 464
Fundamental generalised cycle basis, 117

G

Gaussian elimination, 192, 193, 195, 266–68, 267
Gauss-Jordan, 132
General loading, 67, 169–73
Generalised cycle basis (GCB), 117
Generalized cycle adjacency matrix, 118
    cycle basis, 118
    cycle basis incidence matrix, 119
Generators, 73, 404, 405
Generic independence, 44, 45
    rigidity matroid, 485
Generically independent, 44
Genre structure, 316
Global coordinate system, 142, 143
Graph based method, 307–9
Graph matrices
    eigen values of, 410–20
Graph products
    definitions, 404–10
    second eigen values for, 419–20
Graph, abstract, 456
    associate, 119–22
    bipartite, 462
    bsection, 364–68
    coarsening, 392–93
    complete, 461
    complete bipartite, 462
    cycle, 461
    directed, 474–76
    dual, 480
    element clique, 204–5
    element star, 208–9
    element wheel, 209–10
    incidence, 217–18
    line, 462
    natural associate, 214–17
    null, 461
    partially triangulated, 211–12
    path, 461
    planar, 477–80
    regular, 462
    representative, 218–20
    simple, 456
    skeleton, 206–7
    star, 461
    triangulated, 212–14
    trisection, 381–86
    wheel, 461
Graphic, 485
Greedy Algorithm, 490–91
Grid, 2
Grid-form truss, 48–49
Ground node, 25, 110
    tree, 25

H

Haar property, 493
Hall's theorem, 481
Heeled SRT ($HSRT_c^{n_0}$), 242
Height, 197
Hinged cycle, 459
Homeomorphic, 457
Hooke's law, 8
Hungarian, 482
Hybrid method, 284–90, 300

I

Ill-conditioned, 97
Image, 35
Improved spectral bisection, 371
IN, 82
Inactive status, 228
Incidence, 456
    graph, 217–18
    matrix, 468
    number, 82
Independence, 483
    control, 74
Index vector, 316
Infinitesimal displacement, 42
Influence matrix. *See* Overall flexibility matrix

Integrated force method, 54
Interior boundary, 305
Internal node, 33
Intersection, 457
Intersection theorem, 29–30
Isomorphic, 485
Isomorphism, 470
Isoperimetric number, 353

K

Kaveh's theorem, 29
King's algorithm, 227
König's theorem, 481
Kronecker product, 409, 410–11
K-total graph, 260, 261, 264
Kuratowski's theorem, 479

L

Labelled graph, 404
Laman's theorem, 44
Lanczos, 354–59
Laplacian matrix, 279–84, 350, 414
Laplacian of regular models
    computing λ_2 for, 424–25
    algorithm, 425
Length, 457
Length, cycle, 116
    shortest route tree, 198
Level structure, 194, 243, 316
    tree, 316
    tree separator, 294
Linear stress element, 129
Link, 459
Local coordinate system, 142, 143
Longest SRT, 197
Loop, 456
Lovazs and Yemini's theorem, 47, 50
LU-factorisation, 134, 135, 136

M

M-alternating path, 482
M-augmenting path, 482
MacLane's theorem, 480
Mass matrix, 336
Matching, 480–82
    complete, 481
    maximal, 480–82
    number, 480, 481
    perfect, 481
Mathematical model, 25
Matrix, constant, 42
    cutset adjacency, 472
    cutset basis incidence, 475
    cycle adjacency, 470
    cycle basis incidence, 470
    degree, 279
    equilibrium, 220
    GCB incidence, 119
    Laplacian, 279, 382
    Mass, 336
    node adjacency, 279
    node-member incidence, 467–68
    null basis, 131
    overall flexibility, 64, 131
    self-stress, 131
    transfer, 338–39
Matroid, 79, 122, 483
Matroid, cocycle, 488
    cutset, 488
    cutset space, 486–87
    cycle, 487
    cycle space, 487–88
    dual, 488
    Gen. cycle space, 491
    polygon, 487
Member connectivity, 50
    displacements, 6, 146
    flexibility matrix, 7, 8–11
    forces, 6, 146
    list, 468–69
    stiffness matrix, 7, 8–11
Members, 456

500   INDEX

Method, displacement, 5, 166–73
    force, 5, 125–38
Minimal base, 491
Minimal cycle, 72, 73
Minimal cycle basis, 71–79
    cycle, 71
    GCB, 118–19, 122
Mixed method, 39
Mode shapes, 337
Multiple member, 456
Multiple root SRT, 242, 251

N

Narrowest SRT, 197
Nash-William's theorem, 47
Natural associate graph, 214–17
    forces, 129
    frequency, 337
Neighbouring nodes, 242
NEPS (non-compact extended p-sum), 435
Nested dissection, 294
Nodal decomposition, 201
    ordering, 284–90
Node adjacency matrix, 466–67
    connectivity, 281
    cover, 351
    covering number, 480
    datum, 468
    element list, 316
    exposed, 481
    ground, 110
Node connectivity, 281
Node-element list, 317
Node-member incidence matrix, 467
Nodes, 456
Non-compact extended p-sum, 435–36
Non-separable, 480
NP-hard, 354
Null basis, 131, 489, 492–94
    basis matrix, 131
    graph, 461
    vector, 131

Nullity, 464
Number, Betti, 24, 464
    condition, 98–101
    crossing, 36
    cycle length, 82
    distance, 257–58
    incidence, 82
    width, 197

O

Optimal cycle basis, 71, 81–84
    Displacement method, 141–86
    drawing, 36
    force method, 69
    transversal, 201
Optimally conditioned cutset basis, 180–86, 181
Optimally conditioned cycle basis, 101–3
Ordering, 191–268, 202, 203, 224, 278
Orientation coefficient, 110, 112
Oriented graph, 350
Orthogonality, 464
Overall flexibility matrix, 132
    stiffness matrix, 159

P

Partially triangulated graph, 211–12
Particular solution, 55
Particular solution space, 55
Partitioning, 352
Passive node, 215
Path, 457–59, 457
Path, alternating, 481
    augmenting, 482
    M-alternating, 482
    shortest, 458
Pattern equivalence, 70, 192, 262
Perfect matching, 481
Perron-Frobenious theorem theorem, 92
Planar drawing
    graph, 477–79
    truss, 38–39

Planting, 158, 160
Polygon matroid, 487
Positive definite, 154
Post-active status, 228
Potential energy, 131
Pre-active status, 228
Primary structure, 54, 68
Prime cutest, 459
Principle of virtual work, 15–17
Priority, 215, 218, 230
    queue, 227, 228, 229
Profile optimisation, 224–41, 433–34
Profile reduction, 226
    spectral method for, 433–34
Pseudo-peripheral node, 194, 199, 228, 246, 274
    determination of, 391
    diameter, 199
    peripheral node, 199, 274

Q

QR factorisation, 136, 137

R

Rank, 465
Rayleigh quotient, 420
Rayleigh's Quotient, 100, 419
Reachable, 493
Reciprocal theorem, 19–20
Rectangular element, 129
    matrices, 260–66
Recursive spectral, 359–63
    sequential-cut partitioning, 362
    two-way partitioning, 362
Reduced stiffness matrix, 154, 168
Redundancy, 27
Redundants, 5, 54, 55, 57
Regional cycle, 102, 478
Regular structures
    decomposition and nodal ordering using graph products, 403–36
Released structure, 54

Removable subgraph, 116
Representative graph, 218–20
Reverse Cuthill-McKee, 224
Rigid, 42, 43
Rigidity, 41–49
Ring sum, 457
Ritz vector, 284, 290, 389–401
Root selection, 246–49
Root-mean-square, 226, 244, 250
Rotation, 143, 144

S

S-admissible, 101
Self-equilibrating stress systems (SESs), 55
Self-stress matrix, 131
Set of modal force equations, 341
Shortest path, 458
    route tree, 194, 196–97, 459–61
Shortest root tree (SRT_{n_0}), 242
Simple graph, 456
Skeletal structures, 2
    mathematical model of, 25
Skeleton graph, 206–7
Skyline scheme, 224
Skyline storage, 226
Space structures, 35–41
Space
    frame, 37
    truss, 38
Space, cutset
    cycle, 464
    vector, 464
Spanning tree, 193, 242, 459–61, 459
    Forest, 47, 109
Sparse, 71
Sparsity coefficient, 71
Spectral bisection, 371–78
    trisection, 378–89
Spectral radius, 275
Spectrum, 274, 467
SRT. *See* Shortest path, route tree
Start column, 136

Starting node, 196, 198, 277
Statical basis, 55, 70
Statical indeterminacy, 70
Stiff, 43, 44
Stiffness, matrix, 146, 153–58, 153, 158, 159, 160
    matrix, 153–58
    method, 5
    reduced, 154, 168
    of the structure, 164
    unassembled, 153
Strain energy, 13, 152
Strong Cartesian product, 407–9, 414–16
Structural analysis, 4
    design, 4
    examples for, 430–31
    steps of, 4
Structural design, 4
Structure
    continua, 3
    definition, 1, 54
    flexibility matrix of, 64
    skeletal, 2
Sturm, 355
Subcontour, 242
Subdomaining, 350–71
    engineering based, 316–17
    graph based method, 307–9
    mixed method, 364–68
Subgraph, 457
    elementary, 115
Subminimal cycle basis, 71
    GCB, 118, 123–25
Suboptimal, cycle basis, 72–79, 72–79, 81–84
    GCB, 119–22
Suboptimally conditioned cutest basis, 182–83
    conditioned cycle basis, 104–7
Substructuring direct method, 298–99
    displacement method, 296–98
    dynamic analysis, 336–47
    force method, 330–36
    hybrid method, 300
    iterative method, 299–300

Sugihara, 46
Symmetry, 151

T

Theorem, Berge, 482
    Castigliano, 14
    Euler, 479
    Frobinuis, 481
    Hall, 481
    Intersection, 29–30
    Kaveh, 29
    König, 481
    Kuratowski, 479
    Laman, 25
    Lovazs and Yemini, 47, 50
    MacLane, 480
    Nash-Williams, 47
    Reciprocal, 19–20
    Sugihara, 46
    Whitney, 477
Topological force method, 54, 71
    properties, 23, 24, 35
Total graph, 260, 261
    degrees of freedom, 42
Trail, 457–59
Transfer matrix, 338–39
Transformation, 142–46
Transversal, 201, 277–78
    minimal, 201
    optimal, 201
Tree, shortest route, 194, 196–97, 460
    length, 198
    spanning, 459–61
    width, 197
Triangular element, 173–76
Triangulated graph, 212–14
Truss, 2, 5
Turn-back, 87, 136, 137

U

Unassembled flexibility matrix, 130
  stiffness matrix, 153
Unifying function, 28
Union, 457
Union-intersection method, 30
Unit displacement method, 17
Unit load method, 15, 16

V

Valency, 456
Variable banded form, 224
Vector space, 463–66
Vertices, 203
Virtual work, principle, 15–17

W

Walk, 457–59
Wavefront, 226
Weak Haar, 493
Weight, 82, 83, 101, 102, 103, 305
Weighted adjacency matrix, 91
  associate graph, 375
  graph, 101, 371–78
  incidence graph, 380–81, 390
  Laplacian, 390
Weighted associate graph, 257
Weighted incidence graph, 380
Well-conditioned, 69
Whitney's theorem, 480
Width, 197
Width number, 197
Work, 11–13, 157

Z

Zero Betti number, 464

γ-chord, 116
γ-cotree, 116
γ-cycle, 116, 117
γ-path, 116
γ-tree, 115

Index of Symbols

The important symbols, their definitions and the page upon which each symbol occurs are listed in the following.

A	node adjacency matrix	466
A*	weighted adjacency matrix	91
a, b, c	integer coefficients	28
A_i	$= S^{i-1} \cap S_i$	29, 31
AL	adjacency list	469
$\overline{\mathbf{B}}$	node-member incidence matrix	467
$b(\mathbf{A})$	bandwidth of matrix **A**	196
$B(S)$	bipartite graph of S	48
$B(S) = (A, E, B)$	bipartite graph with node sets A and B and member set E	45
B	collection of bases of a matroid M	484
B	incidence matrix equilibrium matrix	137, 156, 468, 471, 474
$b_0(S)$	zero Betti number of S	464
$\mathbf{B}_0\mathbf{p}$	particular solution	55
$b_1(S)$	first Betti number of S	464

Optimal Structural Analysis   A. Kaveh
© 2006 Research Studies Press Limited

INDEX OF SYMBOLS

$\mathbf{B}_1\mathbf{q}$	complementary solution	55
$c(S^p)$	crossing number of a planar drawing S^p of S	36, 37
C^*	collection of cocycles of a matroid	488
$\mathbf{C}^*$	cutset basis incidence matrix	472
C	collection of circuits of a matroid	484
$\mathbf{C}$	cycle basis incidence matrix	470, 471
$\overline{\mathbf{C}}^*$	cutset-member incidence matrix	472
$\overline{c}_{ij}$	entry of $\overline{\mathbf{C}}$	469
$\overline{\mathbf{C}}$	cycle-member incidence matrix	469
C_i	cycle	464
C_k	cycle graph with k nodes	461
$d(n_i, n_j)$	distance between two nodes n_i and n_j	198
$\mathbf{D}(S)$	degree matrix of a graph S	279
$\mathbf{D}^*$	cutset adjacency matrix	473
$\mathbf{D}$	cycle adjacency matrix	470
$\deg(n_i)$	degree of a node n_i	279, 456
$e(n_i)$	eccentricity of n_i	198
$e(S)$	member connectivity	50
E	elastic modulus	163
$f(S)$	degree of freedom	42
F	collection of independent subsets	487
FEA	fixed end action	67
$\mathbf{f}_m$	member flexibility	7

INDEX OF SYMBOLS 507

$\mathbf{F}_m, \mathbf{f}$	unassembled flexibility matrix of a structure	60
$F_{\max}, W_{\max}$	maximum frontwidth	226, 244
$\mathbf{G}$	overall flexibility matrix of a structure	69, 70, 131
G	shear modulus	60
$\mathbf{H}$	constant matrix	42
$\mathbf{I}$	unit matrix	7
I_{jk}^c, CLN	cycle length number of the member jk	82
I_{jk}, IN	incidence number of the member jk	82
J	Saint-Venant torsion constant	60
$\overline{k}$	stiffness matrix of an element in its local coordinate system	152
$\mathbf{K}$	stiffness matrix of the structure	153
$\mathbf{k}$	unassembled stiffness matrix	153
$\mathbf{K}_{ff}$	reduced stiffness matrix	168
$\mathbf{k}_i$	stiffness matrix of element i	249
k_{ij}	stiffness coefficient	151, 153
$\mathbf{k}_m$	member stiffness	7
K_N	complete graph with N nodes	461
$K_{r,s}$	bipartite graph	462
$K\text{-}T(S)$	K-total graph	261
$L(C)$	total length of the cycles (generalized cycles) of a basis	71, 79, 80, 118
$L(P_k)$	length of the path P_k	458
$\mathbf{L}$	Laplacian matrix	279, 280

INDEX OF SYMBOLS

$\mathbf{L}_c$	complementary Laplacian matrix	284
M	matroid	483
M(S)	cycle matroid of S	487
M(S),M	number of members of S	27, 28, 41
M(S,F)	matroid	484
$M(S_i)$	number of members of S_i	457
M*(S)	dual matroid of M(S)	488
$\mathrm{M}_c(\mathrm{S})$	cutset space matroid of S	487
$M_c(S)$	number of members for triangulation of S	33
$M_c(S^P)$	number of members for full triangulation	38, 39
ML	member list	468
$\mathrm{M}_s(S)$	cycle space matroid of S	486
N(S),N	number of nodes of S	27, 28, 41
$N(S_i)$	number of nodes of S_i	457
$N_i(S)$	number of internal nodes of S	33
n_i	node	456
N_k	null graph with k nodes	461
NP-complete		196
P,Q	permutation matrices, transformation matrix	192
p	joint load vector	55
P	vector of pointers, external forces	15, 150, 469
P,P_i	path	458
P_k	path graph	461

INDEX OF SYMBOLS   509

q	redundants vector	54, 131
$R(S)$	number of regions of embedded graph S	478
R	member forces	15, 18
$\bar{\mathbf{r}}$	element force vector in local coordinate system	145
$\mathbf{r}, \mathbf{r}_m$	member (element) force vector in global coordinate system	58, 59, 128
$R_i(S)$	number of internal regions	35
S	graph model of a structure	24, 25
$\bar{S}$	covering subgraph	44
S	set of elements of a matroid	483
S_i	subgraph of S	457
$S^k = \bigcup\limits_{i=1}^{k} S_i$	that is, union of k subgraphs	457
S^*	elimination graph	267
S^p	planar drawing of S	37
SRT	shortest route tree	460
T^*	cotree	460, 465
t	trail	457
T	transformation matrix	145
T	tree	459, 460
T_{ij}	coefficient of the transformation matrix **T**	162
$\bar{\mathbf{u}}$	element distortion vector in local coordinate system	163
U^*	complementary strain energy	13
u, **u**$_m$	member distortion vector in global coordinate system	7, 15, 18

INDEX OF SYMBOLS

U	strain energy	13
V	degree (valency) matrix	468
$\mathbb{V}$	finite vector space	486
v	joint displacement vector	153
v$_2$	Fiedler vector	285
v$_c$	relative displacements of cuts	61
$W(m_i)$	weight of member m_i	101
W^*	complementary work	12, 13
w	a walk in S	457
W	work done by external loads	12
W_k	wheel graph	461
$|X|$	cardinality of X	44
x, y, z	global coordinate system	143
$\overline{xyz}$	local coordinate system	143
Z_2	integers modulo 2, denoted by GF(2)	70

Greek Symbols

α	integer equal to 3 or 6	27
$\gamma(S)$, DSI	degree of statical indeterminacy	27
$\gamma_0(S)$	number of components	28
$\overline{\gamma}(A_i)$	$= aM(A_i) + bN(A_i) + c$	29, 30
$\mu_S(S)$	$-\gamma(S)$ for planar trusses	44
$\rho_S(X)$	rank of a submatrix of **H** corresponding to the members of X	44

INDEX OF SYMBOLS 511

χ	sparsity coefficient	71
$\sigma_i(\mathbf{C})$	intersection coefficient	72, 118
$\gamma(S)$	unifying linear function, node connectivity	28
λ_{max}, λ_{min}	largest and smallest eigenvalues of a matrix	98
$\eta(S)$, DKI	degree of kinematical indeterminacy	142
β	degrees of freedom of a node	192
$\delta(S)$	diameter of S	198
$\lambda_2 = \alpha(S)$	algebraic connectivity, second eigenvalue	281
$\bar{\phi}$	an approximation to the Fiedler vector	284
$\rho(S)$	rank of the graph S	464, 465, 472
$\Gamma(X)$	image of X	481
$\tau(S)$	node covering number	480
$\Psi(S)$	matching number	481

Logical symbols

{ }	a vector or a set	6
[ ]^t	Transposition of a matrix	6
|·|	cardinality	44
$\subseteq$	containment symbol: a subset of a set	43
$\rightarrow$	expansion	29
$\times$	Cartesian product	405
$\boxtimes$	strong Cartesian product	407
$*$	direct product	409

INDEX OF SYMBOLS

$\otimes$	Kronecker product	410
$\oplus$	ring sum	457
$\cup$	union	29, 457
$\cap$	intersection	29, 457